Reactive Oxygen Species and Oxidative Damage in Plants Under Stress

Dharmendra K. Gupta • José M. Palma •
Francisco J. Corpas
Editors

Reactive Oxygen Species and Oxidative Damage in Plants Under Stress

Springer

Editors
Dharmendra K. Gupta
Gottfried Wilhelm Leibniz Universität Hannover
Institut für Radioökologie und Strahlenschutz (IRS), Gebäude 4113
Hannover, Germany

José M. Palma
Department of Biochemistry, Cell and Molecular Biology of Plants
Estación Experimental del Zaidín (EEZ), Spanish National Research Council (CSIC)
Granada, Spain

Francisco J. Corpas
Department of Biochemistry, Cell and Molecular Biology of Plants
Estación Experimental del Zaidín (EEZ), Spanish National Research Council (CSIC)
Granada, Spain

ISBN 978-3-319-20420-8 ISBN 978-3-319-20421-5 (eBook)
DOI 10.1007/978-3-319-20421-5

Library of Congress Control Number: 2015950068

Springer Cham Heidelberg New York Dordrecht London

Printed on acid-free paper

Springer International Publishing AG Switzerland is part of Springer Science+Business Media (www.springer.com)

This book is dedicated in the memory of ***Prof. Emeritus Kozi Asada****, Kyoto University, Japan, for his large contribution in the field of plant ROS, who passed away on 15th December, 2013, and also to* ***Prof. Paul Bolwell****, Royal Holloway, University of London, UK, for his long contribution in plant oxidative burst in response to pathogens who also passed away on 13th of April, 2012.*

Preface

In plants as well as in all aerobic organisms, reactive oxygen species (ROS) are produced commonly as a by-product of aerobic metabolism. It depends on the formation and nature of ROS; some are toxic and easily destroyed/detoxified by several enzymatic and nonenzymatic mechanisms in the plant cells. However, lately, the role of ROS as second messengers participating in signaling processes under normal and certain stress conditions was postulated (Foyer and Noctor 2003). Environmental stresses such as heat, cold, drought, salinity, heavy metal toxicity, ozone, and ultraviolet radiation as well as pathogens/contagion attack lead to enhanced generation of ROS in plants due to disruption of cellular homeostasis. When the increment of ROS in plant cells rapidly increased and the scavenging systems of ROS do not operate properly, a situation of oxidative stress and oxidative injury occurs. The toxicity caused by heavy metals leads to intervention with metabolism and other biological activities through the generation of ROS such as superoxide radicals ($O_2^{\bullet -}$), hydroxyl radicals ($^{\bullet}OH$), and hydrogen peroxide molecule (H_2O_2). Under certain conditions which involve the presence of transition metal ions, basically Cu^{2+} and Fe^{3+}, H_2O_2 may be reduced to $^{\bullet}OH$ radicals by superoxide and generates oxidative damage to the plants. One of the major consequences of heavy metals action in the cell is the enhanced generation of ROS which usually damage the cellular components such as membranes, nucleic acids, chloroplast pigments, and alteration in enzymatic and non-enzymatic antioxidants (Gupta et al. 2013a). Stress-induced increases in ROS level can cause different degree of oxidation of cell components and a gross change in redox status. Thus, an oxidative outburst as a consequence of stress is reflected in the levels of ROS molecules ($O_2^{\bullet -}$, H_2O_2 and $^{\bullet}OH$), which are biochemically connected through metabolic reactions (Halliwell and Gutteridge 2007).

ROS generation is evident in chloroplast, mitochondria, peroxisomes, plasma membrane, and apoplast adjacent to membrane. In green plants, chloroplast is the most important among the organelles in respect of ROS generation as O_2 is continuously provided through the water autolysis and readily available inside the organelle. Several reports showed that ROS induction can take place in response to cadmium stress in pepper (*Capsicum annum* L.) (León et al. 2002) and *Arabidopsis thaliana* (Remans et al. 2010), Pb and As stress in *Zea mays* and in *A. thaliana* (Gupta et al. 2009, 2013b), Cd and Cu stress in pea (*Pisum sativum*) and *A. thaliana* (Palma et al. 1987; Remans et al. 2010), Ni stress in wheat (*Triticum aestivum*) (Hao et al. 2006), and Zn stress in Brassica (Feigl et al. 2014). Since now there is no evidence that cytoplasmic phytochelatins (PCs) have a role in prevention of ROS induction at plasma membrane or associated ROS formation at apoplast. However, it is acceptable that cell-wall-associated peroxidase catalyzes formation of membrane-permeable H_2O_2 in apoplast and then makes it possible to interact with cytosolic PCs and other thiol peptides.

Since last three decades, it's indeed a big boom in the field of ROS and its role/function in plants. The main purpose of the book is to provide detailed and comprehensive knowledge to the academicians and researchers who are interested in the field of oxidative damage caused by stresses in plants with special reference to the metabolism of ROS and site of production of ROS in plant systems. Other key features of this book are ROS signaling, ROS and disease resistance, redox regulation, and antioxidant defense during stresses, heavy metal-induced oxidative stresses, and heavy metal toxicity and detoxification mechanism. Some chapters are also focusing on hormones/polyphenols as antioxidants and the future of transgenic plants in antioxidative defense. The functional interaction between ROS and the reactive nitrogen species (RNS) is also addressed in this volume. In the nutshell, the information compiled in this book will bring very deep knowledge and advancement in the field of ROS and oxidative damages caused by stresses in current years in plant sciences.

Dr. Dharmendra K. Gupta, Prof. José M. Palma, and Dr. Francisco J. Corpas personally thank the authors for contributing their valuable time, knowledge, and enthusiasm to bring this book into the present shape.

Hannover, Germany
Granada, Spain

Dr. Dharmendra K. Gupta
Prof. José M. Palma
Dr. Francisco J. Corpas

References

Feigl G, Lehotai N, Molnár Á, Ördög A, Rodríguez-Ruiz M, Palma JM, Corpas FJ, Erdei L, Kolbert Z (2014) Zinc induces distinct changes in the metabolism of reactive oxygen and nitrogen species (ROS and RNS) in the roots of two Brassica species with different sensitivity to zinc stress. Ann Bot pii:mcu246

Foyer CH, Noctor G (2003) Redox sensing and signalling associated with reactive oxygen in chloroplasts, peroxisomes and mitochondria. Physiol Planta 119:355–364

Gupta DK, Nicoloso FT, Schetinger MRC, Rossato LV, Pereira LB, Castro GY, Srivastava S, Tripathi RD (2009) Antioxidant defence mechanism in hydroponically grown *Zea mays* seedlings under moderate lead stress. J Hazard Mat 172:479–484

Gupta DK, Corpas FJ, Palma JM (2013a) Heavy metal stress in plants. Springer, Germany

Gupta DK, Inouhe M, Rodríguez-Serrano M, Romero-Puerta MC, Sandalio LM (2013b) Oxidative stress and arsenic toxicity: role of NADPH oxidases. Chemosphere 90:1987–1996

Halliwell B, Gutteridge JMC (2007) Free radicals in biology and medicine. Oxford University Press, Oxford, UK

Hao F, Wang X, Chen J (2006) Involvement of plasma membrane NADPH oxidase in nickel-induced oxidative stress in roots of wheat seedling. Plant Sci 170:151–158

León AM, Palma JM, Corpas FJ, Gómez M, Romero-Puertas MC, Chatterjee D, Mateos RM, del Río LA, Sandalio LM (2002) Antioxidative enzymes in cultivars of pepper plants with different sensitivity to cadmium. Plant Physiol Biochem 40:813–820

Palma JM, Gómez M, Yáñez J, del Río LA (1987) Increased levels of peroxisomal active oxygen-related enzymes in copper-tolerant pea plants. Plant Physiol 85:570–574

Remans T, Opdenakker K, Smeets K, Mathijsen D, Vangronsveld J, Cuypers A (2010) Metal-specific and NADPH oxidase dependent changes in lipoxygenase and NADPH oxidase gene expression in *Arabidopsis thaliana* exposed to cadmium or excess copper. Funct Plant Biol 37:532–544

Contents

Production Sites of Reactive Oxygen Species (ROS) in Organelles from Plant Cells

Francisco J. Corpas, Dharmendra K. Gupta, and José M. Palma

Contents

Abstract Reactive oxygen species (ROS) have been considered for a long time as undesirable by-product of the cellular metabolism, but recently the role of ROS in molecular signaling processes has been reported. Consequently, the cell must keep a fragile equilibrium between ROS production and the antioxidant defenses that protect cells in vivo against potential damages (oxidative stress) and, alternatively, allow the inter- and intra-cell communications. This equilibrium may become disturbed under different array of adverse conditions by an excessive generation of ROS or by an impaired antioxidant defenses. Plant cells have a compartmental-ization of ROS production in the different organelles including chloroplasts,

F.J. Corpas (✉) • J.M. Palma (✉)
Group of Antioxidants, Free Radicals and Nitric Oxide in Biotechnology, Food and Agriculture, Department of Biochemistry, Cell and Molecular Biology of Plants, Estación Experimental del Zaidín, CSIC, Apartado 419, E-18008 Granada, Spain
e-mail: javier.corpas@eez.csic.es; josemanuel.palma@eez.csic.es

D.K. Gupta
Institut für Radioökologie und Strahlenschutz (IRS), Gottfried Wilhelm Leibniz Universität Hannover, Herrenhäuser Str. 2, Gebaüde 4113, 30419 Hannover, Germany

D.K. Gupta et al. (eds.), *Reactive Oxygen Species and Oxidative Damage in Plants Under Stress*, DOI 10.1007/978-3-319-20421-5_1

mitochondria, or peroxisomes, and they also have a complex battery of antioxidant enzymes usually close to the site of ROS production. Cell compartmentalization has been demonstrated to be an additional mechanism of cellular ROS modulation for signaling purposes. This chapter will provide a general overview of the main system of ROS production/regulation in plant cells.

Keywords Reactive oxygen species • Chloroplasts • Mitochondria • Peroxisomes

1 Introduction

Reactive oxygen species (ROS) is a term which includes radical and non-radical oxygen species formed by the partial reduction of oxygen. The main ROS mostly investigated are superoxide radical ($O_2^{\bullet -}$), hydroxyl radical ($^{\bullet}OH$), alkoxyl ($RO^{\bullet}$) and peroxyl ($ROO^{\bullet}$) as radicals molecules, and hydrogen peroxide (H_2O_2), singlet oxygen ($^1O_2$), ozone (O_3), and hypochlorous acid (HClO) as non-radical. Under normal conditions, these molecules are produced in many metabolic pathways as normal by-product, being the respective electron transport chains present in chloroplasts and mitochondria the main sources of these ROS (Halliwell 2006; del Río 2015). However, the presence of free metals, such as iron, copper, and manganese, released from metalloprotein complexes can also contribute to ROS production. Plant cells enclose a wide range of enzymatic and nonenzymatic antioxidant systems which usually are nearby the ROS production site being an excellent mechanism to avoid the undesirable potential negative effects of ROS (oxidative stress) but also to modulate their signaling role.

In parallel, plant cells contain a series of ROS-scavenging nonenzymatic antioxidants such as ascorbic acid, glutathione (GSH), carotenoids, and others, as well as a wide battery of enzymes such as superoxide dismutase (SOD), catalase, glutathione peroxidase (GPX), peroxiredoxin (Prx), and the ascorbate–glutathione cycle. All these latter elements have multiple isozymes located in all cell compartments which provide a highly efficient system for detoxifying ROS. The main goal of this chapter is to offer a general overview of the main system of ROS production/scavenging in the principal plant organelles.

2 Chloroplasts

Due to their abundance and diversity of pigments, chloroplasts are the cell organelles more susceptible to be attacked by ROS. These photosynthetic compartments are also great sources of ROS production, including basically $O_2^{\bullet -}$ and singlet oxygen ($^1O_2$). Chloroplasts harbor in thylakoids the key elements to fully carry out the photosynthesis, with the structures involved in the light-dependent phase being mainly responsible for the ROS generation (Tripathy and Oelmüller 2012). Complementarily, these organelles contain powerful antioxidant systems to counterbalance the ROS production under normal conditions.

2.1 Production of Reactive Oxygen Species

The major site of superoxide radical's production is linked to the photosystem I (PSI). Under illumination conditions, O_2 is continuously provided by the water autolysis performed in the PSII as indicated in reaction [1], so light would favor the superoxide radical formation reaction [2] at the PSI location. There, under excessive reduced ferredoxin and low NADP availability, the autoxidation of this iron–sulfur protein occurs with the formation of $O_2^{\bullet -}$, as depicted in reaction [3].

Reaction [1] $2H_2O \rightarrow 4e^- + O_2 + 4H^+$
Reaction [2] $2O_2 + 2e^- \rightarrow 2O_2^{\bullet -}$
Reaction [3] $\text{Fdred} + O_2 \rightarrow \text{Fdox} + O_2^{\bullet -}$

If the conditions persist, the reduced ferredoxin is able to react with superoxide radicals to form hydrogen peroxide, and this is what Mehler (1951) found when he performed his experiments with illuminated chloroplasts (reaction 4).

Reaction [4] $\text{Fdred} + O_2^{\bullet -} + 2H^+ \rightarrow \text{Fdox} + H_2O_2$

Asada and colleagues (1974) corroborated later that all the H_2O_2 formation attributable to chloroplasts was a consequence of the disproportionation of superoxide radicals previously formed. It has been found that the H_2O_2 photo produced via $O_2^{\bullet -}$ accumulates in thylakoids, whereas in intact chloroplasts this ROS does not accumulate (Asada 2006). The steady-state level of H_2O_2 in chloroplasts was determined to be about 0.5 μM, with increases under stress conditions up to 1–15 μM.

The direct production of $O_2^{\bullet -}$ to a lower extent at the level of the PSI was also reported, and it was postulated that, when the NADP availability lowers and the Calvin–Benson cycle does not operate properly, the ferredoxin autoxidation takes place initially and afterwards the direct formation of superoxide radicals from the PSI (Halliwell and Gutteridge 2007). Simultaneously, another source of superoxide radicals is also associated to PSII, for instance, through the autoxidation of PSII electronic acceptors and mostly at the level of the plastoquinone (Gupta and Igamberdiev 2015). The superoxide radical's production in chloroplasts is promoted above the normal conditions under certain circumstances, basically stress situations which proceed with stomata closure. Then, the CO_2 availability decreases and the photosynthetic carbon reductive pathway (Calvin–Benson cycle) is somehow impaired, with the concomitant lower provision of NADP for the thylakoid-linked ferredoxin-NADP reductase. Accordingly, reduced ferredoxin accumulates and develops the scenario described above. Overall, the rate of $O_2^{\bullet -}$ production in isolated chloroplasts was initially reported to be about 30 μmol mg^{-1} Chl h^{-1} (Asada 1992). Later, it was probed to that the superoxide radical's generation increased from 240 to 720 μM s^{-1} under stress conditions (Polle 2001).

Singlet oxygen is produced at the PSII (P680) by excitation of oxygen of the ground (triplet) state $^3O_2$ till singlet state ($^1O_2$), as indicated in reaction [5].

Reaction [5] $^3O_2 + {}^3P680^*$ (excited P680) $\rightarrow {}^1O_2 + {}^3P680$

Under intense illumination conditions and/or low CO_2 assimilation rate undergone due to environmental stresses or certain physiological conditions, electrons from chlorophyll are excited to a higher energy layer, and this energy excess is transferred to oxygen, thus generating singlet oxygen responsible for photodynamic damages such as bleaching of leaves (Telfer et al. 1994; Hideg et al. 1998; Asada 2006). Additionally, it has been also found that biosynthetic and catabolic intermediates of chlorophyll are photosensitizers which generate singlet oxygen (Wagner et al. 2004; Pruzinska et al. 2005). Although $^1O_2$ is rapidly quenched by water, its lifetime and diffusion distance from the generation site are very short. So, the distance among the generation and the target sites of $^1O_2$ is a critical factor to evaluate the biological effect of this ROS (Asada 2006).

Many herbicides, including methyl viologen (paraquat), diquat, DCMU [3-(3,4-dichlorophenyl)-1,1-dimethylurea], atrazine, and others base their mechanism of action by promoting the generation of ROS. Thus, cationic herbicides such as methyl viologen trigger the formation of superoxide radicals at the level of PSI; other polar compounds like DCMU uncouple the electron fluxes at the PSII level with excitation of chlorophyll and the energy excess of excited chlorophyll being transferred toward the formation of $^1O_2$. It has been demonstrated that many plants (tobacco, tomato, potato, and alfalfa, among others) transfected with additional *SOD* genes showed reduced damage symptoms after being subjected to diverse herbicides.

2.2 ROS Scavenging Systems

Chloroplasts contain a battery of scavengers that not only protect chloroplasts from the direct effects of ROS but also relax the electron excess stress. Thus, a series of antioxidant enzymes and small molecules regulate the endogenous ROS levels, thus allowing a coordinated response under stress conditions (Foyer et al. 1991, 1994; Gill and Tuteia 2010). Chloroplastic membranes are rich of carotenoids (provitamin A) and α-tocopherol (vitamin E), two powerful $^1O_2$ scavengers, so this ROS with high ability to diffuse in hydrophobic environments can be promptly removed by these antioxidants, although ascorbate can also be an active scavenger of this species.

Carotenoids, mainly β-carotene, besides working as complementary light-absorbing pigments, can dissipate the photodynamic effect directly and indirectly. Hence, the energy excess accumulated in the triplet state of chlorophyll as consequence of intense illumination can be transferred to carotenoids which move up to their triplet state. These excited carotenoids go back to their ground state by dissipating their excess energy as heat. On the other hand carotenoids are able to counterbalance the production of $^1O_2$ promoted by the triplet-state chlorophyll. Again, excited carotenoids, as consequence of their interaction with $^1O_2$, dissipate their higher energy as heat rendering the ground-state pigments. Up to 11 molecule of β-carotene have been assigned to the PSII reaction center and antenna subunit complex (Asada 2006). Xanthophylls, a series of molecules framed within the carotenoids group, are also involved in the antioxidant metabolism in a stroma–lumen interaction. This mechanism implies to violaxanthin, antheraxanthin, and zeaxanthin which are interconverted one in another by epoxidation/de-epoxidation reactions, thus giving rise to the so-called xanthophylls cycle (Adams and Demmig-Adams 1992; Demmig-Adams and Adams 2006). The epoxidation pathway (zeaxanthin–antheraxanthin–violaxanthin), carried out at neutral pH under low light in the stroma, depends on the provision of NADPH, whereas the de-epoxidation is achieved in the lumen at acid pH (around 5, high light) with the participation of ascorbate which is converted into dehydroascorbate (Adams and Demmig-Adams 1992; Demmig-Adams and Adams 2006).

Alpha-tocopherol is another molecule which can quench $^1O_2$, although its effectiveness regarding β-carotene is much lower, about two orders of magnitude. After the reaction of α-tocopherol with $^1O_2$, α-tocopherylquinone is formed (Halliwell and Gutteridge 2007), and this can regenerate again α-tocopherol by the reaction with ascorbate. As a result of this reaction chain, monodehydroascorbate is formed, and this is integrated within the enzymatic pathways displayed below (Fig. 1). Tocopherols are also involved in suppressing the lipid peroxidation of thylakoids by trapping lipid radicals (Muller et al. 2006).

From all antioxidant molecules, ascorbate seems to be the most versatile since this compound not only scavenges all types of ROS by itself but also participates in the ascorbate–glutathione cycle (see below) and in the regeneration of other antioxidants as reported above for α-tocopherol. Thus, a very significant role in the chloroplast redox homeostasis is attributed to ascorbate. In fact, chloroplasts are the main cellular pool of ascorbate in spite that this antioxidant is synthesized in mitochondria (Foyer et al. 1991; Smirnoff 2001).

The presence of several superoxide dismutases (SOD; EC 1.15.1.1) has been reported in chloroplasts (Hayakawa et al. 1984; Grace 1990). SODs are a class of metalloenzymes with different nature depending on the heavy metal located in the active site of the protein which catalyze the reaction [6]:

Reaction [6]: $2O_2^{\bullet -} + 2H^+ \rightarrow H_2O_2 + O_2$

Three main SOD types have been described in plants: copper–zinc-, iron-, and manganese-containing superoxide dismutases (CuZn–SODs, Fe–SODs, and Mn–SODs, respectively; Rodríguez-Serrano et al. 2007). Chloroplasts commonly house

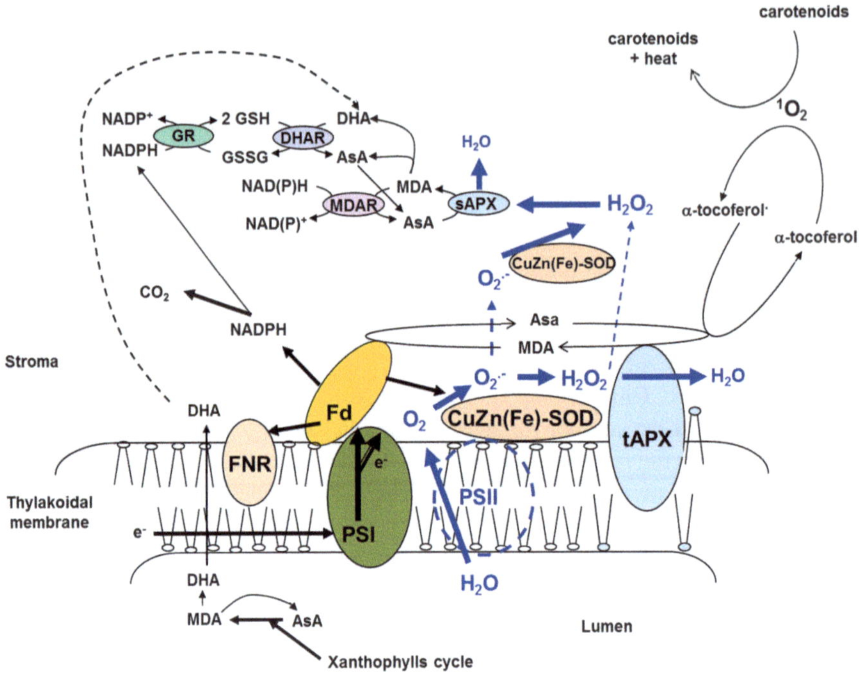

Fig. 1 Integrated model of production and scavenging of reactive oxygen species in chloroplasts. Electrons in PSI are usually "sailing" toward the PSI-linked ferredoxin (Fd) and by action of the NADP–ferredoxin reductase (FNR), NADPH is formed which can be used in the photosynthetic carbon fixation. Subsidiary, superoxide radicals ($O_2^{\bullet -}$) can be generated continuously in the presence of O_2 provided by PSII after H_2O photolysis. $O_2^{\bullet -}$ is then dismutated either by the thylakoid-linked superoxide dismutase (both CuZn–SOD and Fe–SOD) or the soluble forms of these isozymes. The H_2O_2 generated by the action of SOD is decomposed by either the ascorbate peroxidase bound to thylakoid membranes (tAPX) or the soluble isozyme (sAPX), using ascorbate (AsA) as reducing source. sAPX is integrated within the chloroplastic ascorbate–glutathione cycle (AGC) which implies the participation of the monodehydroascorbate reductase (MDAR), the dehydroascorbate reductase (DAR), and glutathione reductase (GR). This redox pathway is involved in the removal of H_2O_2 with expenses of NADPH. AsA could also be used to regenerate α-tocopherol from α-tocopherylquinone (α-tocopheryl radical), after this lipophilic antioxidant has been used as a singlet oxygen ($^1O_2$) quencher. $^1O_2$ can be also scavenged by carotenoids with excess energy being dissipated as heat. Throughout these processes, monodehydroascorbate (MDA) is formed, and this radical can be used to regenerate ascorbate in the stroma by either direct action of reduced Fd or through the AGC. MDA is also produced at the chloroplastic lumen in the xanthophylls cycle. MDA dismutates into ascorbate and dehydroascorbate which can migrate through the thylakoid membrane and be coupled to the stroma AGC. As depicted as blue arrows, a water–water cycle occurs, with consume of water in the lumen and production in the stroma side

CuZn–SOD and Fe–SOD isozymes, although the presence of one Mn–SOD has been reported in chromoplasts from pepper fruits (Martí et al. 2009). Both SOD isoenzyme types have been reported to be attached to the thylakoids near the PSI where $O_2^{\bullet -}$ is produced but also soluble in the stroma (Asada 2006; Mittova et al. 2015) (Fig. 1).

H_2O_2 is mainly removed by the action of the ascorbate peroxidase (reaction [7]; APX; EC 1.11.1.11) which, like SODs, is located either attached to the thylakoid membrane (tAPX) or soluble in the stroma (sAPX) (Yoshimura et al. 1999; Shigeoka et al. 2002; Maruta et al. 2010). In thylakoids, APX is in the vicinity of PSI so the flux of electrons through PSI, SODs, and tAPX forms a thylakoidal scavenging system which functions as the first defense against ROS, with the participation of reduced ferredoxin which directly provides electrons to monodehydroascorbate to regenerate ascorbate (reaction [8]; Fig. 1).

Reaction [7] $H_2O_2 + AsA \rightarrow 2H_2O + 2MDA$
Reaction [8] $2MDA + 2Fdred \rightarrow 2Asa + 2Fdox$

The sAPX is integrated within the ascorbate–glutathione cycle, also called Foyer–Halliwell–Asada cycle, where the enzymes monodehydroascorbate reductase (MDAR; EC 1.6.5.4), dehydroascorbate reductase (DAR; EC 1.8.5.1), and glutathione reductase (GR; EC 1.6.4.2) are involved in the H_2O_2 scavenging associated to the NADPH expense (Corpas and Barroso) (Fig. 1).

Overall, as the result of the series of reactions which involved the formation (reactions 1 and 2) and scavenging (reactions 6, 7 and 8) of ROS in chloroplasts renders the final stoichiometry given in reaction [9]:

Reaction [9] $2H_2O + O_2 \rightarrow O_2 + 2H_2O$

which allows introducing the concept water–water cycle proposed by Professor Kozi Asada (1999) as a unique pathway located in chloroplasts involving the dynamics of oxygen in these organelles and integrating a network of molecules which goes beyond the simple ROS-antioxidant pair.

Peroxiredoxins and thioredoxins are also systems involved in the detoxification of hydrogen peroxide in chloroplasts. Peroxiredoxins are thiol-based peroxidases which may utilize the reducing power provided through thioredoxins to scavenge H_2O_2(Puerto-Galán 2013). Thioredoxins are crucial for the chloroplast redox network, mediating environmental signals to the organelle proteins. Thus, chloroplast thioredoxins have been found to be very versatile and to control the structure and function of proteins by reducing disulfide bridges in the redox active site of a protein (Schürman and Jacquot 2000; Nikkanen, and Rintamaki 2014). A thioredoxin system which gains electrons from the PSI-linked ferredoxin and involves a ferredoxin–thioredoxin reductase has been found. Besides, a thioredoxin that uses NADPH as the reducing source through a NADPH–thioredoxin reductase has been reported (Nikkanen and Rintamaki 2014). Finally, a more complex system where the reducing power from NADPH is successively transferred following the sequence thioredoxin reductase, thioredoxin, and peroxiredoxin to reduce H_2O_2 up to water has been displayed (Dietz 2003). The possibility that this latter system may function as a water–water cycle under certain conditions was already proposed by Asada (2006).

3 Mitochondria

In mammalian cells, mitochondria are the major cell loci for ROS production. In plants, mitochondria constitute one of the main ROS production sites due to unavoidable impairments of the electron transport chain (ETC) responsible of the aerobic respiration which is located at the inner mitochondrial membrane. A short review of the ROS metabolism, both generation and scavenging involved systems, will be given in this chapter, although a wider view of this subject will be displayed in chapter "What Do the Mitochondrial Antioxidant and Redox Systems Have to Say Under Salinity, Drought and Extreme Temperature?" (F. Sevilla and colleagues).

Similarly to what happened in chloroplasts, the first reports on ROS in mitochondria in the mid-1960s revealed that these organelles were able to produce H_2O_2 (Hinkle et al. 1967). Years later, the demonstration of $O_2^{\bullet-}$ generation by submitochondrial particles bearing diverse ETC complexes (Loschen and Azzi 1975), along with the discovery of the presence of SOD activity in the organelle, led to conclude that the original ROS formed in mitochondria were superoxide radicals. About 2–5 % of the consumed O_2 in mitochondria is derived toward the formation of this species. By further research and thanks to the use of inhibitors of the ETC, namely, rotenone and antimycin, it was found that the $O_2^{\bullet-}$ production sites reside in complex I and complex III (Fig. 2a) (Møller 2001; Sweetlove and Foyer 2004;

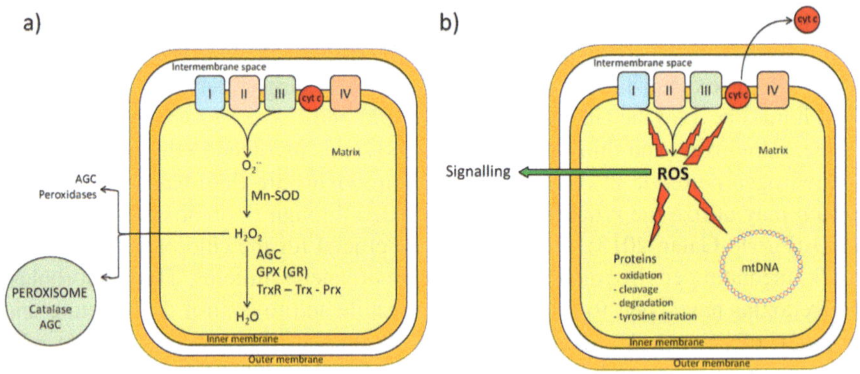

Fig. 2 Production and effects of ROS in mitochondria. (**a**) ROS production in mitochondria. Superoxide radicals ($O_2^{\bullet-}$) are generated at the complexes I and III from the electron transport chain located in the inner membrane. Mn–SOD disproportionates $O_2^{\bullet-}$ into H_2O_2 which, in turn, is removed by the ascorbate–glutathione (AGC) cycle enzymes in plants and in animal cells by a glutathione peroxidase (GPX) and a system involving thioredoxin (Trx), peroxiredoxin (Prx), and a thioredoxin reductase (TrxR). H_2O_2 can also come out of the mitochondria and be either scavenged in the cytosol by soluble peroxidases and the cytosolic AGC or driven to peroxisomes where catalase and AGC decompose it. (**b**) Effects of ROS on mitochondrial macromolecules. Under controlled conditions, ROS produced in mitochondria participates in signaling processes. However, when ROS generation exceeds the scavenging systems, ROS may attack mitochondrial DNA and trigger mutations, promote oxidation, cleavage and degradation or nitration of proteins, and favor the release of cytochrome *c* from the organelle membranes toward the cytosol, as it occurs in apoptosis

Gupta and Igamberdiev 2015). Rotenone inhibits the electron transfer from complex I (NADH–ubiquinone oxidoreductase) to ubiquinone, whereas antimycin binds to complex III (ubiquinol–cytochrome *c* oxidoreductase), thus avoiding this complex capturing electrons from the previous ETC components. A more precise study of the mitochondrial localization of $O_2^{\bullet -}$ production reported that this event develops in two ubiquinone pools: one associated to complex I and the other one linked to complex III (Raha and Robinson 2000; Popov 2015).

According to the mechanism of action of complexes I and III and the position of the respective ubiquinone pools in mammalian cells, it was postulated that $O_2^{\bullet -}$ generated in complex I was disposed of at the matrix of the organelle, whereas complex III dropped this ROS to the intermembrane space (Raha and Robinson 2000; Murphy 2009). In the matrix, $O_2^{\bullet -}$ dismutates by the action of a Mn–SOD (Fig. 2a), characteristic of mitochondria (del Río et al. 2002; Rodríguez-Serrano et al. 2007; Palma et al. 2013), and, in animal cells, the resulting H_2O_2 is detoxified by a selenium-dependent glutathione peroxidase (SeGPX) which, in turn, is coupled to a GR for the continuous provision of reduced glutathione (GSH). However, very few references have reported the presence of a CuZn–SOD in the intermembrane space, and this eventuality is far to be still consensed by the scientific community. H_2O_2 from the matrix can be pumped off to the cytosol through the mitochondrial membranes and then scavenged by diverse detoxifying systems such as peroxidases and the ascorbate–glutathione cycle or enters the peroxisomes, where catalase/ascorbate–glutathione cycle would decompose it. A thioredoxin–peroxiredoxin system located in the matrix could also remove H_2O_2 with the participation of a thioredoxin reductase which would utilize NADPH, provided by a NADP-dependent isocitrate dehydrogenase as electron donor (Murphy 2009). In plants, the presence of all enzyme components of the AGC in mitochondria has been demonstrated (Jiménez et al. 1997), and the participation of this pathway to remove H_2O_2 in this compartment is the most accepted issue for plant biologists (Fig. 2a) (Mittova et al. 2015). The necessary NADPH for the action of the GR is a common metabolite in plant mitochondria (Møller 2001). Alternative oxidase (AOX) has been reported to be activated when the reduction level of ubiquinone increases, so this is a dissipating mechanism which is also useful to prevent the overproduction of superoxide radicals (Maxwell et al. 1999; Rhoads et al. 2006; Gupta and Igamberdiev 2015).

Under certain stress conditions where H_2O_2 production overtakes the scavenging barriers and in the presence of transition metals, basically Fe^{3+} and Cu^{2+}, $^{\bullet}OH$ radicals can be formed in a Fenton-type reaction. Hydroxyl radicals could then be able to attack the mitochondrial genome provoking mutations in many of the ETC components which are encoded by the mitochondrial DNA (Fig. 2b) (Raha and Robinson 2000; Murphy 2009). ROS also damage proteins by diverse mechanisms which include oxidation, cleavage, and degradation of backbones and tyrosine nitration (Gupta and Igamberdiev 2015). Overall, ROS are important molecules to promote redox signaling events in mitochondria (Møller and Sweetlove 2010; Hebelstrup and Møller 2015), but under mitochondrial dysfunction, the overproduction of ROS under stress conditions and senescence ROS may lead to apoptosis (programmed cell death, PCD) and necrosis. PCD is characterized by the release of cytochrome *c* from the inner mitochondrial membrane to the cytosol as a consequence of the damage (lipid peroxidation) undergone in membranes by ROS attack (Fig. 2b) (Murphy 2009).

3.1 Ascorbate Biosynthesis

A very important event in the antioxidant balance is the synthesis of ascorbate. This antioxidant molecule is synthesized by the great majority of phyla, excepting primates, rodents, and some others. Human cells lack the last enzyme of the ascorbate synthesis, the L-gulono-lactone oxidase, that makes human beings strictly dependent on an external ascorbate source, mainly fruits and vegetables. In plants, although several alternative pathways have been described, the main last step of the ascorbate biosynthesis is catalyzed by the L-galactono-lactone dehydrogenase (GalLDH), an enzyme which oxidizes L-galactono-lactone to ascorbic acid without the participation of any redox cofactor (Smirnoff 2001; Valpuesta and Botella 2004). GalLDH has been reported to be located in the inner mitochondrial membrane, neighbor to the ETC, and providing the electrons from the L-galactono-lactone to the terminal oxidase of complex IV (Bartoli et al. 2000). Thus, an interesting issue as a source of the investigation in plant antioxidant arises: ascorbate is synthesized in mitochondria but the major pool of this antioxidant is found in chloroplasts. The presence of ascorbate in other organelles suggest a very complex mechanism by which the ascorbate biosynthesis is triggered under certain stress conditions and how this important molecule is addressed to the diversity of organelles, mainly chloroplasts.

4 Plasma Membrane

Plant membrane-bound NADPH oxidase (NOX), also called respiratory burst oxidase homologue (RBOH), has the capacity to transfer electrons from intracellular NADPH across the plasma membrane to molecular oxygen in the apoplast site and generate $O_2^{\bullet -}$ which can then dismutate through different mechanisms to H_2O_2. *RBOH* genes belong to a multigenic family with 10 members in *Arabidopsis thaliana* (*RBOHA-RBOHJ*) and 9 in rice (*Oryza sativa*) but also with five groups of orthologous sequences (Torres et al. 2002; Sagi and Fluhr 2006; O'Brien et al. 2012; Skelly and Loake 2013).

The plant Rboh protein has two main components: (i) membrane-bound respiratory burst oxidase homologue (Rboh) with a molecular weight between 105 and 112 kDa (being homologue of $gp91^{phox}$ from mammalian phagocyte NAPDH oxidase) and (ii) its cytosolic regulator Rop (Rho-like protein) which is a Rac homologue of plants. Thus, the integral plasma membrane protein is composed of

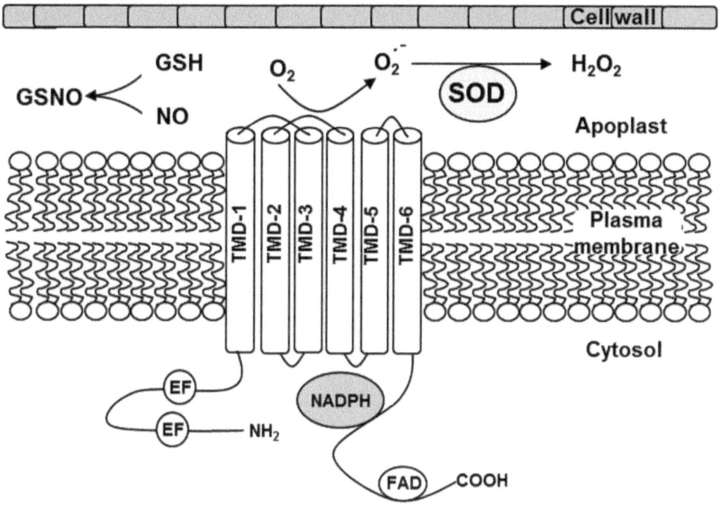

Fig. 3 Simple model of the structure and localization of the components of the plant membrane-bound respiratory burst oxidase homologues (RBOH) protein and other antioxidant elements. EF hand domains, FAD flavin adenine dinucleotide, GSH glutathione, GSNO S-nitrosoglutathione, NADPH reduced form of the nicotinamide adenine dinucleotide phosphate, NO nitric oxide, SOD superoxide dismutase, TMD-1 to TMD-6 transmembrane domains

six transmembrane domains (TMD-1 to TMD-6) connected by five loops (loops A–E) where TMD-3 and TMD-5 contain pairs of His residues required to bind two heme groups, C-terminal FAD and NADPH hydrophilic domains, and two N-terminal calcium-binding (EF-hand) motifs and some phosphorylation target sites (Yoshie et al. 2005; Marino et al. 2012) (Fig. 3). Besides this complex structure, there are also regulatory components involving phosphorylation and Ca^{2+} (Ogasawara et al. 2008) such as calcium-dependent protein kinases (CDPKs are Ser/Thr protein kinases that include a Ca^{2+}-binding calmodulin-like domain) (Kobayashi et al. 2007), Ca^{2+}/CaM-dependent protein kinase (CCaMK) (Shi et al. 2012), and Rop (Wong et al. 2007). Moreover, new mechanisms of regulation have been reported including phosphatidic acid binding (Zhang et al. 2009) and *S*-nitrosylation, which are posttranslational protein modifications mediated by nitric oxide-derived molecules (Corpas et al. 2015). Thus, in the *Arabidopsis* Rboh isoform D (AtRBOHD), the *S*-nitrosylation of Cys 890, thus abolishing the ability to generate $O_2^{\bullet-}$ (Yun et al. 2011), provides a clear interrelationship between reactive oxygen and nitrogen species.

Rboh is involved in many plant processes including cell growth (Foreman et al. 2003), plant development, stomatal closure (Shi et al. 2012), pollen tube growth (Kaya et al. 2014), symbiotic interactions (Marino et al. 2012; Kaur et al. 2014), abiotic stress, and pathogen response (Wojtaszek 1997; Torres et al. 2002; Daudi et al. 2012; Siddique et al. 2014). However, the number of Rboh isozymes which are differentially expressed suggests a certain grade of specialization for each one. For example, in *Arabidopsis thaliana* which has

10 genes, the focus has been pointed toward *AtRbohB*, *AtRbohC*, *AtRbohD*, and *AtRbohF*, especially *AtRbohD*, because it is constitutively and ubiquitously expressed (Kadota et al. 2014); however, the information about the other six *Rboh* genes is very scarce.

On the other hand, the apoplast space seems to be more complex than we could expect because it contains other elements such as SOD (Streller et al. 1997; Vanacker et al. 1998; Kukavica et al. 2005), the antioxidant glutathione (GSH) (Vanacker et al. 1999; Pignocchi and Foyer 2003), and nitric oxide (Stöhr and Ullrich 2002; Bethke et al. 2004). Thus, the SOD must regulate the H_2O_2 production during the dismutation of $O_2^{\bullet-}$ generated by Rboh being a mechanism of regulation of signaling between cells mediated by H_2O_2. Moreover, GSH and NO can interact to form *S*-nitrosoglutathione (GSNO), which is also recognized as a signaling molecule (Corpas et al. 2013), and can mediate the posttranslational modifications of proteins affecting their activities such as it occurs to ascorbate peroxidase (Begara-Morales et al. 2014).

Besides the mechanism of the local production of $O_2^{\bullet-}$ by Rboh, it has been proposed that after some stimuli (i.e., pathogens) and the generation of a local burst of ROS mediated by Rboh in an specific cells, there is a cascade of cell-to-cell communication events that carries a systemic signal over long distances throughout different tissues of the plants (see chapter "ROS as Key Players of Abiotic Stress Responses in Plants" of this book by Suzuki for deeper discussion) which opens a new perspective of the Rboh functions (Marino et al. 2012; Kaur et al. 2014).

5 Peroxisomes

Unlike other subcellular compartments, peroxisome is a single membrane-bounded compartment with a diverse range of specific metabolic functions depending on the tissue localization, the plant developmental step, and the environmental conditions (del Río et al. 2002; Mano and Nishimura 2005; Palma et al. 2009; Hu et al. 2012; Baker and Paudyal 2014). Among the principal functions of peroxisomes in plant cells, the fatty acid β-oxidation, the glyoxylate cycle, the photorespiration cycle, the metabolism of ureides, and the metabolism of reactive oxygen and nitrogen species (ROS and RNS) can be included, being the peroxisomal characteristic enzymes catalase and H_2O_2-generating flavin oxidases, which reflects a prominent oxidative metabolism. Table 1 summarizes the main peroxisomal ROS-producing systems and the involved enzymes.

5.1 H_2O_2-Producing System

Peroxisomal H_2O_2 generation is considered a side product of diverse pathways where peroxisomes are involved; however, the capacity to go through membranes

Table 1 Summary of the main ROS-producing systems and involved enzymes identified in peroxisomes from higher plants

Pathway	Peroxisomal enzyme	Reaction
H_2O_2-producing system		
β-oxidation	Acyl CoA oxidase (EC:1.3.3.6)	Acyl-CoA → *trans*-2-enoyl-CoA + H_2O_2
Photorespiration	Glycolate oxidase (EC 1.1.3.15)	Glycolate + O_2 → glyoxylate + H_2O_2
Sulphite detoxification	Sulfite oxidase (EC 1.8.3.1)	Sulfite + O_2+ H_2O → sulfate + H_2O_2
ROS metabolism	Superoxide dismutase (EC 1.15.11)	$O_2^{\bullet -}$ + $O_2^{\bullet -}$ + H+ → H_2O_2 + O_2
Purine metabolism	Urate oxidase (EC 1.7.3.3)	Uric acid + O_2 + H_2O → 5-hydroxyisourate+ H_2O_2 → allantoin + CO_2
Sarcosine metabolism	Sarcosine oxidase (EC 1.5.3.1)	Sarcosine + O_2 + H_2O → glycine + formaldehyde + H_2O_2 and L-pipecolate → Δ^1-piperideine-6-carboxylate + H_2O_2
Polyamine catabolism	Polyamine oxidase (EC 1.5.3.3)	Spermine + O_2 + H_2O → spermidine + 3-aminopropanal + H_2O_2
Superoxide-generating system		
Purine metabolism	Xanthine oxidase (EC 1.1.3.22)	Xanthine + O_2→ uric acid + $O_2^{\bullet -}$
Peroxisomal membrane polypeptides	PMP32 (membrane monodehydroascorbate reductase)	NADH + PMP32 → $O_2^{\bullet -}$

involves the capacity of this molecule to be used as a signal. Thus, peroxisomal fatty acid β-oxidation allows the breakdown of these molecules to acetyl-CoA and the subsequent conversion of acetyl-CoA to succinate via the glyoxylate cycle. In the β-oxidation pathway, the enzyme acyl-CoA oxidase catalyzes the conversion of acyl-CoA into trans-2-enyl-CoA with the concomitant generation of H_2O_2 (Arent et al. 2008). This pathway has a relevant physiological function because it allows the conversion of triacylglyceride pools in seedlings, the turnover of membrane lipids during senescence or starvation situation, as well the synthesis of fatty acid-derived hormones such as indole acetic acid (IAA), jasmonic acid (JA), and salicylic acid (SA) which consequently are involved in stress response and growth regulation (Poirier et al. 2006; Delker et al. 2007; Baker and Paudyal 2014). Photorespiration involves the light-dependent uptake of O_2 and release of CO_2 during the metabolism of phosphoglycolate, the two-carbon by-product by the oxygenase activity of Rubisco. This pathway involves several organelles (chloroplasts, mitochondria, and peroxisomes) with the peroxisomal glycolate oxidase generating H_2O_2.

There are other peroxisomal H_2O_2-producing enzymes but the available information on their function is still scarce. Thus, sulfite oxidase (SO) catalyzes the conversion of sulfite to sulfate with the concomitant generation of H_2O_2(Hänsch et al. 2006). It has been reported that low concentrations of sulfite inhibit catalase activity (Veljovic-Jovanovic et al. 1998), which could therefore be a means of regulating both enzymes. Sarcosine, also known as *N*-methylglycine, is an intermediate and by-product of glycine synthesis and degradation which also generates H_2O_2. The enzyme responsible is the sarcosine oxidase (SOX) which is a 46-kDa monomer that covalently attaches FAD molecule. Moreover, the SOX activity also catalyzes the conversion of L-pipecolate to Δ^1-piperideine-6-carboxylate plus H_2O_2 being a side branch of lysine catabolism (Goyer et al. 2004). In Arabidopsis, among the family of polyamine oxidases (PAO), it has been identified a peroxisomal isoform (AtPAO4) which is involved in polyamine catabolism especially in roots (Kamada-Nobusada et al. 2008; Planas-Portell et al. 2013).

5.2 *Superoxide-Generating System*

Xanthine oxidoreductase (XOR) is an FAD-, molybdenum-, iron-, and sulfur-containing hydroxylase enzyme that catalyzes the conversion of the purines hypoxanthine and xanthine into uric acid with the concomitant formation of either NADH or $O_2^{\bullet-}$ and plays an important role in nucleic acid degradation in all organisms (Harrison 2002). The enzyme is a homodimer, and each subunit contains one molybdenum atom, one FAD group, and two Fe_2S_2 centers. The molybdenum cofactor (Moco) present in XOR is also shared by other key enzymes that catalyze basic reactions in carbon, nitrogen, and sulfur metabolism, such as aldehyde oxidase, nitrate reductase, and sulfite oxidase (Schwarz and Mendel 2006). XOR exists in two interconvertible forms: an NAD-dependent dehydrogenase or xanthine dehydrogenase (XDH; EC 1.1.1.204), which can be converted into an oxygen-dependent oxidase or xanthine oxidase (XOD; EC 1.1.3.22). The presence of XOD activity in peroxisomes has been reported in different plant species (Sandalio et al. 1988; del Río et al. 1989; Mateos et al. 2003). More recently, additional biochemical and immunological results demonstrate the presence of XOR in leaf peroxisomes, showing that the XOD form, which generates superoxide radicals, is the predominant form in these oxidative organelles being differentially modulated under cadmium-induced oxidative stress (Corpas et al. 2008).

On the other hand, the peroxisomal membrane is another potential source of ROS, specifically $O_2^{\bullet-}$, through the existence of a small electron transport chain using NADH as electron donor. This is composed of a flavoprotein NADH:ferricyanide reductase of about 32 kDa and a cytochrome b (López-Huertas et al. 1999). The identity of the membrane protein of 32 kDa seems to be the enzyme monodehydroascorbate reductase (MDAR) since this enzyme has been described to be present in both matrix and membrane polypeptide of peroxisomes (Leterrier et al. 2005; Lisenbee et al. 2005). Additionally, using NADPH it was found a peroxisomal membrane of 29 kDa that had the capacity to generate $O_2^{\bullet-}$ and to reduce cytochrome *c* (López-Huertas et al. 1999). The identity of this protein is not clear, but it could be related to the family of NADPH:cytochrome P450 reductase (López-Huertas et al. 1999).

5.3 Peroxisomal Antioxidant Systems

Besides the presence of catalase, a well-characterized peroxisomal antioxidant enzyme which keeps the H_2O_2 under control (Palma et al. 2013), there is another complementary antioxidant systems to regulate the content of $O_2^{\bullet-}$ and H_2O_2 in these organelles (Corpas et al. 2001).

In the case of the H_2O_2, plant peroxisomes enclose a particular ascorbate–glutathione cycle (Jiménez et al. 1997; Reumann and Corpas 2010) since its components have a special distribution with some membrane-bound enzymes such as the APX (Bunkelmann and Trelease 1996; Corpas and Trelease 1998) and the MDAR (Leterrier et al. 2005; Lisenbee et al. 2005) and others located in the matrix, such as the GR (Romero-Puertas et al. 2006) and also the DAR (Fig. 4). This peroxisomal system has been described to participate in the mechanism of response to different processes including growth (Narendra et al. 2006), leaf

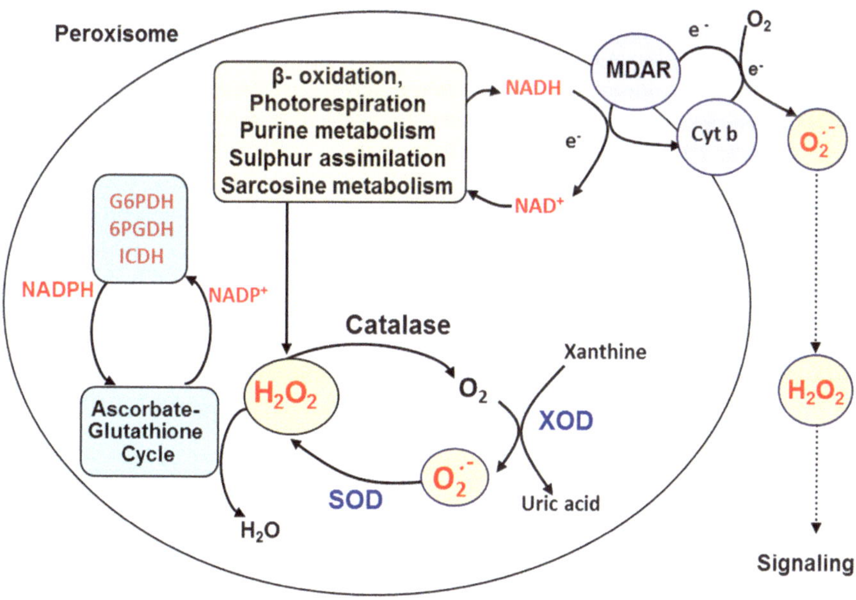

Fig. 4 Model of ROS production in plant peroxisomes. APX ascorbate peroxidase, G6PDH glucose-6-phosphate dehydrogenase, 6PGDH 6-phosphogluconate dehydrogenase, ICDH NADP–isocitrate dehydrogenase, MDAR monodehydroascorbate reductase, XOD xanthine oxidase

senescence (Jiménez et al. 1998; Palma et al. 2006), fruit ripening (Mateos et al. 2003), or heavy metal stress (Leterrier et al. 2005).

In animal cells, peroxisomes have been reported to contain exclusively a CuZn–SOD; however, in plant peroxisomes, it can be found, depending on the tissue and/or plant species and the three types of SOD isozymes, located in the matrix and/or in the membrane. Although the presence of either a CuZn–SOD or a Mn–SOD is the most common issue (del Río et al. 1983; Corpas and Trelease 1998; del Río et al. 2002), there are other cases where the presence of a Mn–SOD plus a CuZn–SOD (del Río et al. 2002) or a Fe–SOD has been demonstrated (Droillard and Paulin 1990).

Additionally, during the last decade, new components related with the peroxisomal metabolism of ROS have been discovered such as a closer family of molecules designated as reactive nitrogen species (RNS) (Corpas et al. 2013). All this indicates that peroxisomes enclose and complex nitro-oxidative apparatus characterized by a relevant flexibility which can adapt to fluctuating conditions.

6 Conclusions

In comparison to animal cells, higher plants have a most complex and active ROS metabolism under optimal environmental conditions which is in part consequence of the photosynthesis and photorespiration processes. ROS are obligated site products of many physiological pathways which are present in all cell compartments, including chloroplasts, mitochondria, plasma membrane, and peroxisomes. Although ROS have been considered as toxic molecules, this concept has changed because under a controlled production ROS are part of the mechanism of signaling or defense. This control is achieved by cellular complex of antioxidative systems which usually are close to the different sites of ROS production at subcellular level. However, under adverse environmental and/or certain physiological conditions, the cellular equilibrium between ROS production and scavenging could be broken and overcome the defense battery, which can provoke oxidative damage with fatal consequences for the normal cell functions. Future research is needed to get deeper knowledge and to decipher new mechanisms of regulation to keep under control the ROS production and their signaling implications in combination with RNS.

Acknowledgments Work in our laboratories is supported by ERDF grants co-financed by the Ministry of Economy and Competitiveness (projects AGL2011-26044, BIO2012-33904) and the Junta de Andalucía (group BIO192) in Spain.

References

Adams WW, Demmig-Adams B (1992) Operation of the xanthophyll cycle in higher plants in response to diurnal changes in incident sunlight. Planta 186:390–398

Arent S, Pye VE, Henriksen A (2008) Structure and function of plant acyl-CoA oxidases. Plant Physiol Biochem 46:292–301

Asada K, Kiso K, Yoshikawa K (1974) Univalent reduction of molecular oxygen by spinach chloroplasts on illumination. J Biol Chem 249:2175–2181

Asada K (1992) Production and scavenging of active oxygen in chloroplasts. In: Scandalios JG (ed) Molecular biology of free radical scavenging system. Cold Spring Harbor Laboratory Press, Plainview, NY, pp 173–192

Asada K (2006) Production and scavenging of reactive oxygen species in chloroplasts and their functions. Plant Physiol 141:391–396

Baker A, Paudyal R (2014) The life of the peroxisome: from birth to death. Curr Opin Plant Biol 22:39–47

Begara-Morales JC, Sánchez-Calvo B, Chaki M, Valderrama R, Mata-Pérez C, López-Jaramillo J, Padilla MN, Carreras A, Corpas FJ, Barroso JB (2014) Dual regulation of cytosolic ascorbate peroxidase (APX) by tyrosine nitration and S-nitrosylation. J Exp Bot 65:527–538

Bethke PC, Badger MR, Jones RL (2004) Apoplastic synthesis of nitric oxide by plant tissues. Plant Cell 16:332–341

Bunkelmann JR, Trelease RN (1996) Ascorbate peroxidase. A prominent membrane protein in oilseed glyoxysomes. Plant Physiol 110:589–598

Corpas FJ, Barroso JB (2014) NADPH-generating dehydrogenases: their role in the mechanism of protection against nitro-oxidative stress induced by adverse environmental conditions. Front Environ Sci 2:55

Corpas FJ, Trelease RN (1998) Differential expression of ascorbate peroxidase and a putative molecular chaperone in the boundary membrane of differentiating cucumber seedling peroxisomes. J Plant Physiol 153:332–338

Corpas FJ, Barroso JB, del Río LA (2001) Peroxisomes as a source of reactive oxygen species and nitric oxide signal molecules in plant cells. Trends Plant Sci 6:145–50

Corpas FJ, Palma JM, Sandalio LM, Valderrama R, Barroso JB, del Río LA (2008) peroxisomal xanthine oxidoreductase: characterization of the enzyme from pea (*Pisum sativum* L.) leaves. J Plant Physiol 165:1319–1330

Corpas FJ, Alché JD, Barroso JB (2013) Current overview of S-nitrosoglutathione (GSNO) in higher plants. Front Plant Sci 4:126

Corpas FJ, Begara-Morales JC, Sánchez-Calvo B, Chaki M, Barroso JB (2015) Nitration and S-nitrosylation: two post-translational modifications (PTMs) mediated by reactive nitrogen species (RNS) which participate in signaling processes of plant cells. In: Gupta KJ, Igamberdiev AU (eds) Reactive oxygen and nitrogen species signalling and communication in plants. Springer, Berlin

Daudi A, Cheng Z, O'Brien JA, Mammarella N, Khan S, Ausubel FM, Bolwell GP (2012) The apoplastic oxidative burst peroxidase in Arabidopsis is a major component of pattern-triggered immunity. Plant Cell 24:275–287

Demmig-Adams B, Adams W (2006) Photoprotection in an ecological context: the remarkable complexity of thermal energy dissipation. New Phytol 172:11–21

del Río LA (2011) Peroxisomes as a cellular source of reactive nitrogen species signal molecules. Arch Biochem Biophys 506:1–11

del Río LA (2015) ROS and RNS in plant physiology: an overview. J Exp Bot 66:2827–2837

del Río LA, Lyon DS, Olah I, Glick B, Salin ML (1983) Immunocytochemical evidence for a peroxisomal localization of manganese superoxide dismutase in leaf protoplasts from a higher plant. Planta 158:216–224

del Río LA, Fernández VM, Rupérez FL, Sandalio LM, Palma JM (1989) NADH induces the generation of superoxide radicals in leaf peroxisomes. Plant Physiol 89:728–31 del Río LA, Corpas FJ, Sandalio LM, Palma JM, Gómez M, Barroso JB (2002) Reactive oxygen species, antioxidant systems and nitric oxide in peroxisomes. J Exp Bot 53:1255–1272

Delker C, Zolman BK, Miersch O, Wasternack C (2007) Jasmonate biosynthesis in Arabidopsis thaliana requires peroxisomal β-oxidation enzymes-additional proof by properties of pex6 and aim1. Phytochemistry 68:1642–1650

Dietz KJ (2003) Plant peroxiredoxins. Annu Rev Plant Physiol Plant Mol Biol 54:93–107

Droillard MJ, Paulin A (1990) Isozymes of superoxide dismutase in mitochondria and peroxisomes isolated from petals of carnation (*Dianthus caryophyllus*) during senescence. Plant Physiol 94:1187–1192

Foreman J, Demidchik V, Bothwell JH, Mylona P, Miedema H, Torres MA, Linstead P, Costa S, Brownlee C, Jones JD, Davies JM, Dolan L (2003) Reactive oxygen species produced by NADPH oxidase regulate plant cell growth. Nature 422:442–446

Foyer CH, Lelandais M, Edwards EA, Mullineaux PM (1991) The role of ascorbate in plants, interaction with photosynthesis, and regulatory significance. In: Pell E, Steffen K (eds) Active oxygen/oxidative stress and plant metabolism. American Society of Plant Physiologists, Rockville, pp 131–143

Foyer CH, Lescure JC, Lefebvre C, Morot-Gaudry JF, Vincentz M, Vaucheret H (1994) Adaptations of photosynthetic electron transport, carbon assimilation, and carbon partitioning in transgenic *Nicotiana plumbaginifolia* plants to changes in nitrate reductase activity. Plant Physiol 104:171–178

Gill SS, Tuteja N (2010) Reactive oxygen species and antioxidant machinery in abiotic stress tolerance in crop plants. Plant Physiol Biochem 48:909–930

Goyer A, Johnson TL, Olsen LJ, Collakova E, Shachar-Hill Y, Rhodes D, Hanson AD (2004) Characterization and metabolic function of a peroxisomal sarcosine and pipecolate oxidase from Arabidopsis. J Biol Chem 279:16947

Grace SC (1990) Phylogenetic distribution of superoxide dismutase supports an endosymbiotic origin for chloroplasts and mitochondria. Life Sci 47:1875–86

Gupta KJ, Igamberdiev AU (2015) Compartmentalization of reactive oxygen species and nitric oxide production in plant cells: an overview. In: Gupta KJ, Igamberdiev AU (eds) Reactive oxygen and nitrogen species signaling and communications in plants. Springer International Publishing, Switzerland, pp 1–14

Hänsch R, Lang C, Riebeseel E, Lindigkeit R, Gessler A, Rennenberg H, Mendel RR (2006) Plant sulfite oxidase as novel producer of H_2O_2: combination of enzyme catalysis with a subsequent non-enzymatic reaction step. J Biol Chem 281:6884–6888

Halliwell B (2006) Reactive species and antioxidants. Redox biology is a fundamental theme of aerobic life. Plant Physiol 141:312–322

Halliwell B, Gutteridge JMC (2007) Free radicals in biology and medicine. Oxford University Press, Oxford, UK

Harrison R (2002) Structure and function of xanthine oxidoreductase: where are we now? Free Radic Biol Med 33:774–797

Hayakawa T, Kanematsu S, Asada K (1984) Occurrence of CuZn-superoxide dismutase in the intrathylakoid space of spinach chloroplasts. Plant Cell Physiol 25:883–889

Hebelstrup KH, Møller I (2015) Mitochondrial signaling in plants under hypoxia: use of reactive oxygen species (ROS) and reactive nitrogen species (RNS). In: Gupta KJ, Igamberdiev AU (eds) Reactive oxygen and nitrogen species signaling and communications in plants. Springer International Publishing, Switzerland, pp 63–77

Hideg E, Kalai T, Hideg K, Vass I (1998) Photoinhibition of photosynthesis in vivo results in singlet oxygen production detection via nitroxide-induced fluorescence quenching in broad bean leaves. Biochemistry 37:11405–11411

Hinkle PC, Butow RA, Rackers E (1967) Partial resolution of the enzymes catalyzing oxidative phosphorylation. XV Reverse electron transfer in the flavin-cytochrome b region of the respiratory chain of beef heart submitochondrial particles. J Biol Chem 242:5169–5173

Hu J, Baker A, Bartel B, Linka N, Mullen RT, Reumann S, Zolman BK (2012) Plant peroxisomes: biogenesis and function. Plant Cell 24:2279–2303
Jiménez A, Hernández JA, del Río LA, Sevilla F (1997) Evidence for the presence of the ascorbate-glutathione cycle in mitochondria and peroxisomes of pea leaves. Plant Physiol 114:275–284
Jiménez A, Hernández JA, Pastori G, del Rio LA, Sevilla F (1998) Role of the ascorbate-glutathione cycle of mitochondria and peroxisomes in the senescence of pea leaves. Plant Physiol 118:1327–1335
Kadota Y, Sklenar J, Derbyshire P, Stransfeld L, Asai S, Ntoukakis V, Jones JD, Shirasu K, Menke F, Jones A, Zipfel C (2014) Direct regulation of the NADPH oxidase RBOHD by the PRR-associated kinase BIK1 during plant immunity. Mol Cell 54:43–55
Kamada-Nobusada T, Hayashi M, Fukazawa M, Sakakibara H, Nishimura M (2008) A putative peroxisomal polyamine oxidase, AtPAO4, is involved in polyamine catabolism in *Arabidopsis thaliana*. Plant Cell Physiol 49:1272–1282
Kaur G, Sharma A, Guruprasad K, Pati PK (2014) Versatile roles of plant NADPH oxidases and emerging concepts. Biotechnol Adv 32:551–563
Kaya H, Nakajima R, Iwano M, Kanaoka MM, Kimura S, Takeda S, Kawarazaki T, Senzaki E, Hamamura Y, Higashiyama T, Takayama S, Abe M, Kuchitsu K (2014) Ca2+-activated reactive oxygen species production by Arabidopsis RbohH and RbohJ is essential for proper pollen tube tip growth. Plant Cell 26:1069–1080
Kobayashi M, Ohura I, Kawakita K, Yokota N, Fujiwara M, Shimamoto K, Doke N, Yoshioka H (2007) Calcium-dependent protein kinases regulate the production of reactive oxygen species by potato NADPH oxidase. Plant Cell 19:1065–1080
Kukavica B, Vucinić Z, Vuletić M (2005) Superoxide dismutase, peroxidase, and germin-like protein activity in plasma membranes and apoplast of maize roots. Protoplasma 226:191–197
Leterrier M, Corpas FJ, Barroso JB, Sandalio LM, del Río LA (2005) Peroxisomal monodehydroascorbate reductase. Genomic clone characterization and functional analysis under environmental stress conditions. Plant Physiol 138:2111–2123
Lisenbee CS, Lingard MJ, Trelease RN (2005) Arabidopsis peroxisomes possess functionally redundant membrane and matrix isoforms of monodehydroascorbate reductase. Plant J 43:900–914
López-Huertas E, Corpas FJ, Sandalio LM, del Río LA (1999) Characterization of membrane polypeptides from pea leaf peroxisomes involved in superoxide radical generation. Biochem J 337:531–536
Loschen G, Azzi A (1975) On the formation of hydrogen peroxide and oxygen radicals in heart mitochondria. Recent Adv Stud Cardiac Struct Metab 7:3–12
Mano S, Nishimura M (2005) Plant peroxisomes. Vitam Horm 72:111–154
Marino D, Dunand C, Puppo A, Pauly N (2012) A burst of plant NADPH oxidases. Trends Plant Sci 17:9–15
Martí MC, Camejo D, Olmos E, Sandalio LM, Fernández-García N, Jiménez A, Sevilla F (2009) Characterisation and changes in the antioxidant system of chloroplasts and chromoplasts isolated from green and mature pepper fruits. Plant Biol 11:613–624
Maruta T, Tanouchi A, Tamoi M, Yabuta Y, Yoshimura K, Ishikawa T, Shigeoka S (2010) Arabidopsis chloroplastic ascorbate peroxidase isoenzymes play a dual role in photoprotection and gene regulation under photooxidative stress. Plant Cell Physiol 51:190–200
Mateos RM, León AM, Sandalio LM, Gómez M, del Río LA, Palma JM (2003) Peroxisomes from pepper fruits (*Capsicum annuum* L.): purification, characterisation and antioxidant activity. J Plant Physiol 160:1507–1516
Maxwell DP, Wang Y, McIntosh L (1999) The alternative oxidase lowers mitochondrial reactive oxygen production in plant cells. Proc Natl Acad Sci U S A 96:8271–8276
Mehler AH (1951) Studies on reactions of illuminated chloroplasts. II. Stimulation and inhibition of the reaction with molecular oxygen. Arch Biochem Biophys 34(2):339–351

Mittova V, Volokita M, Guy M (2015) Antioxidative systems and stress tolerance: insights from wild and cultivated tomato species. In: Gupta KJ, Igamberdiev AU (eds) Reactive oxygen and nitrogen species signaling and communications in plants. Springer International Publishing, Switzerland, pp 89–131

Møller IM (2001) Plant mitochondria and oxidative stress: electron transport, NADPH turnover and metabolism of reactive oxygen species. Annu Rev Plant Physiol Plant Mol Biol 52:561–591

Muller M, Hernández I, Alegre L, Munné-Bosch S (2006) Enhanced alpha-tocopherol quinone levels and xanthophyll cycle de-epoxidation in rosemary plants exposed to water deficit during a Mediterranean winter. J Plant Physiol 163:601–606

Narendra S, Venkataramani S, Shen G, Wang J, Pasapula V, Lin Y, Kornyeyev D, Holaday AS, Zhang H (2006) The Arabidopsis ascorbate peroxidase 3 is a peroxisomal membrane-bound antioxidant enzyme and is dispensable for Arabidopsis growth and development. J Exp Bot 57:3033–3042

Nikkanen L, Rintamäki E (2014) Thioredoxin-dependent regulatory networks in chloroplasts under fluctuating light conditions. Phil Trans R Soc B 369:20130224

O'Brien JA, Daudi A, Butt VS, Bolwell GP (2012) Reactive oxygen species and their role in plant defence and cell wall metabolism. Planta 236:765–779

Palma JM, Jiménez A, Sandalio LM, Corpas FJ, Lundqvist M, Gómez M, Sevilla F, del Río LA (2006) Antioxidative enzymes from chloroplasts, mitochondria, and peroxisomes during leaf senescence of nodulated pea plants. J Exp Bot 57:1747–58

Palma JM, Corpas FJ, del Río LA (2009) Proteome of plant peroxisomes: new perspectives on the role of these organelles in cell biology. Proteomics 9:2301–2312

Palma JM, Gupta DK, Corpas FJ (2013) Metalloenzymes involved in the metabolism of reactive oxygen species and heavy metal stress. In: Gupta DK, Corpas FJ, Palma JM (eds) Heavy metal stress in plants. Springer, Berlin

Bartoli CG, Pastori GM, Foyer CH (2000) Ascorbate biosynthesis in mitochondria is linked to the electron transport chain between complexes III and IV. Plant Physiol 123:335–344

Pignocchi C, Foyer CH (2003) Apoplastic ascorbate metabolism and its role in the regulation of cell signalling. Curr Opin Plant Biol 6:379–389

Planas-Portell J, Gallart M, Tiburcio AF, Altabella T (2013) Copper-containing amine oxidases contribute to terminal polyamine oxidation in peroxisomes and apoplast of *Arabidopsis thaliana*. BMC Plant Biol 13:109

Poirier Y, Antonenkov VD, Glumoff T, Hiltunen JK (2006) Peroxisomal β-oxidation-a metabolic pathway with multiple functions. Biochim Biophys Acta 1763:1413–1426

Polle A (2001) Dissecting the superoxide dismutase-ascorbate-glutathione-pathway in chloroplasts by metabolic modeling. Computer simulations as a step towards flux analysis. Plant Physiol 126:445–462

Popov VN (2015) Feedback loop of non-coupled respiration and reactive oxygen species production in plant mitochondria. In: Gupta KJ, Igamberdiev AU (eds) Reactive oxygen and nitrogen species signaling and communications in plants. Springer International Publishing, Switzerland, pp 79–88

Pruzinska A, Tanner G, Aubry S, Anders I, Moser S, Muller T, Ongania K-H, Kräutler B, Youn J-Y, Liljegren SL et al (2005) Chlorophyll breakdown in senescent Arabidopsis leaves: characterization of chlorophyll catabolites and of chlorophyll catabolic enzymes involved in the degreening reaction. Plant Physiol 139:52–63

Puerto-Galán L, Pérez-Ruiz JM, Ferrández J, Cano B, Naranjo B, Nájera VA, González M, Lindahl AM, Cejudo FJ (2013) Overoxidation of chloroplast 2-Cys peroxiredoxins: balancing toxic and signaling activities of hydrogen peroxide. Front Plant Sci 4:310

Raha S, Robinson BH (2000) Mitochondria, oxygen free radicals, disease and ageing. Trends Biochem Sci 25:502–508

Reumann S, Corpas FJ (2010) The peroxisomal ascorbate-glutathione pathway: molecular identification and insights into its essential role under environmental stress conditions. In: Anjum

NA, Umar S, Chan MT (eds) Ascorbate-glutathione pathway and stress tolerance in plants. Springer, Berlin

Rhoads DM, Umbach AL, Subbaiah CC, Siedow JN (2006) Mitochondrial reactive oxygen species. Contribution to oxidative stress and interorganellar signaling. Plant Physiol 141:357–366

Rodríguez-Serrano M, Romero-Puertas MC, Pastori GM, Corpas FJ, Sandalio LM, del Río LA, Palma JM (2007) Peroxisomal membrane manganese superoxide dismutase: characterization of the isozyme from watermelon cotyledons. J Exp Bot 58:2417–2427

Romero-Puertas MC, Corpas FJ, Sandalio LM, Leterrier M, Rodríguez-Serrano M, del Río LA, Palma JM (2006) Glutathione reductase from pea leaves: response to abiotic stress and characterization of the peroxisomal isozyme. New Phytol 170:43–52

Sagi M, Fluhr R (2006) Production of reactive oxygen species by plant NADPH oxidases. Plant Physiol 141:336–340

Sandalio LM, Fernández VM, Rupérez FL, del Río LA (1988) Superoxide free radicals are produced in glyoxysomes. Plant Physiol 87:1–4

Schürman P, Jacquot JP (2000) Plant thioredoxin systems revisited. Annu Rev Plant Physiol Plant Mol Biol 51:371–400

Schwarz G, Mendel RR (2006) Molybdenum cofactor biosynthesis and molybdenum enzymes. Annu Rev Plant Biol 57:623–647

Shi YC, Fu YP, Liu WQ (2012) NADPH oxidase in plasma membrane is involved in stomatal closure induced by dehydroascorbate. Plant Physiol Biochem 51:26–30

Shigeoka S, Ishikawa T, Tamoi M, Miyagawa Y, Takeda T, Yabuta Y, Yoshimura K (2002) Regulation and function of ascorbate peroxidase isoenzymes. J Exp Bot 53:1305–1319

Siddique S, Matera C, Radakovic ZS, Hasan MS, Gutbrod P, Rozanska E, Sobczak M, Torres MA, Grundler FM (2014) Parasitic worms stimulate host NADPH oxidases to produce reactive oxygen species that limit plant cell death and promote infection. Sci Signal 7(320):ra33

Skelly MJ, Loake GJ (2013) Synthesis of redox-active molecules and their signaling functions during the expression of plant disease resistance. Antioxid Redox Signal 19:990–997

Smirnoff N (2001) L-ascorbic acid biosynthesis. Vitam Horm 61:241–266

Stöhr C, Ullrich WR (2002) Generation and possible roles of NO in plant roots and their apoplastic space. J Exp Bot 53:2293–2303

Streller S, Schinkel H, Wingsle G (1997) Apoplasmic CuZn-superoxide dismutase in *Pinus sylvestris*. Phyton Ann Rei Bot 37:271–276

Sweetlove LJ, Foyer CH (2004) Roles for reactive oxygen species and antioxidants in plant mitochondria. In: Day DA, Millar AH, Whelan J (eds) Plant mitochondria: from genome to function, vol 1, Advances in photosynthesis and respiration. Kluwer Academic, Dordrecht

Telfer A, Dhami S, Bishop SM, Phillips D, Barber J (1994) Beta-carotene quenches singlet oxygen formed by isolated Photosystem-II reaction centers. Biochemistry 33:14469–14474

Torres MA, Dangl JL, Jones JDG (2002) Arabidopsis gp91phox homologues AtrbohD and AtrbohF are required for accumulation of reactive oxygen intermediates in the plant defense response. Proc Natl Acad Sci USA 99:517–522

Tripathy BC, Oelmüller R (2012) Reactive oxygen species generation and signaling in plants. Plant Signal Behav 7:1621–1633

Valpuesta V, Botella MA (2004) Biosynthesis of L-ascorbic acid in plants: new pathways for an old antioxidant. Trends Plant Sci 9:573–577

Vanacker H, Carver TL, Foyer CH (1998) Pathogen-induced changes in the antioxidant status of the apoplast in barley leaves. Plant Physiol 117:1103–1114

Vanacker H, Foyer CH, Carver TLW (1999) Changes in apoplastic antioxidants induced by powdery mildew attack in oat genotypes with race non-specific resistance. Planta 208:444–452

Veljovic-Jovanovic S, Oniki T, Takaham U (1998) Detection of Monodehydroascorbic acid radical in sulfite-treated leaves and mechanism of its formation. Plant Cell Physiol 39:1203–1208

Wagner DE, Przybyla D, op den Camp RG, Kim C, Landgraf F, Lee KP, Wursch M, Laloi C, Nater M, Hideg E, Apel K (2004) The genetic basis of singlet oxygen-induced stress responses of *Arabidopsis thaliana*. Science 306:1183–1185

Wojtaszek P (1997) Oxidative burst: an early plant response to pathogen infection. Biochem J 322:681–692

Wong HL, Pinontoan R, Hayashi K, Tabata R, Yaeno T, Hasegawa K, Kojima C, Yoshioka H, Iba K, Kawasaki T, Shimamoto K (2007) Regulation of rice NADPH oxidase by binding of Rac GTPase to its N-terminal extension. Plant Cell 19:4022–4034

Yoshie Y, Goto K, Takai R, Iwano M, Takayama S, Isogai A, Che FS (2005) Function of the rice gp91phox homologs OsrbohA and OsrbohE genes in ROS-dependent plant immune responses. Plant Biotechnol 22:127–135

Yoshimura K, Yabuta Y, Tamoi M, Ishikawa T, Shigeoka S (1999) Alternatively spliced mRNA variants of chloroplast ascorbate peroxidase isoenzymes in spinach leaves. Biochem J 338(Pt 1):41–48

Yun BW, Feechan A, Yin M, Saidi NB, Le Bihan T, Yu M, Moore JW, Kang JG, Kwon E, Spoel SH, Pallas JA, Loake GJ (2011) S-nitrosylation of NADPH oxidase regulates cell death in plant immunity. Nature 478:264–268

Zhang Y, Zhu H, Zhang Q, Li M, Yan M, Wang R, Wang L, Welti R, Zhang W, Wang X (2009) Phospholipase dalpha1 and phosphatidic acid regulate NADPH oxidase activity and production of reactive oxygen species in ABA-mediated stomatal closure in Arabidopsis. Plant Cell 21:2357–2377

What Do the Plant Mitochondrial Antioxidant and Redox Systems Have to Say Under Salinity, Drought, and Extreme Temperature?

F. Sevilla, A. Jiménez, and J.J. Lázaro

Contents

Abstract Mitochondria are ubiquitous organelles with a notable oxidative metabolism. They are a significant site of reactive oxygen species (ROS) production in plant cells, including superoxide ($O_2^{\bullet-}$) and H_2O_2. In addition to ROS, there are compelling indications that nitric oxide (NO) can be generated in this organelle by both reductive and oxidative pathways. ROS and reactive nitrogen species (RNS) play a key role in signaling but they can also be deleterious via oxidation of cell components when overproduced as a consequence of adverse conditions. The high production of ROS obligates mitochondria to be provided with a set of

F. Sevilla (✉) • A. Jiménez
Department of Stress Biology and Plant Pathology, CEBAS-CSIC, 30100 Murcia, Spain
e-mail: fsevilla@cebas.csic.es

J.J. Lázaro
Department of Biochemistry, Cellular and Molecular Biology of Plants, EEZ-CSIC, 18008 Granada, Spain

D.K. Gupta et al. (eds.), *Reactive Oxygen Species and Oxidative Damage in Plants Under Stress*, DOI 10.1007/978-3-319-20421-5_2

ROS-scavenging mechanisms. The first line of plant mitochondrial antioxidants is composed of superoxide dismutase and the enzymes of the ascorbate–glutathione cycle, which are not only able to scavenge ROS but also to repair cell damage and possibly serve as redox sensors. Besides direct control by antioxidants, mitochondrial ROS production is tightly controlled by multiple redundant systems affecting inner membrane potential such as NADPH-dependent dehydrogenases, alternative oxidase (AOX), and uncoupling proteins. In addition, the presence of specific protein families responsible for dithiol/disulfide exchange reactions such as the thioredoxin (Trx), peroxiredoxin (Prx), and sulfiredoxin (Srx) families in the mitochondria has been recently reported. These proteins are critically important under some abiotic stress conditions by controlling the cellular redox status of thiol groups of cysteinyl residues as well as acting as antioxidant defense mechanisms. Here, we summarize the insights of the involvement of this Trx–Prx–Srx system and the ASC–GSH cycle components in plant tolerance to abiotic stress, more specifically in salinity, drought, and extreme temperatures, as some of the most important unfavorable environmental conditions for plant yield and growth. In the plant response to stress, it seems that not only the antioxidant but also the redox systems are emerging as key components functioning in both redox sensing and signal transduction pathways.

Keywords Abiotic stress • Antioxidants • Drought • Extreme temperatures • Mitochondria • Redox proteins • RNS • ROS • Salinity • Signaling

1 Introduction

Abiotic stress is defined as an environmental condition that reduces growth and yield below optimum levels, and in fact, environmental factors limit crop production by about 60 % (Boyer 1982). Plant responses to abiotic stresses are dynamic and complex and are dependent on the species, tissue, or organ. In addition, the application time, the level, and the duration of stress can have a significant effect on the complexity of the response. To survive under abiotic stress conditions, plants have developed a complex signaling network involving interactions and cross talk with many molecular pathways in which intracellular organelle signals and their interactions during stressful conditions represent one of the primary defense responses (Agrawal et al. 2011; del Río 2011). One of the most common molecular responses and one of the earliest signals in many abiotic stresses involve reactive oxygen species (ROS) and reactive nitrogen species (RNS), which modify enzyme activity and are involved in gene regulation. For instance, heavy metals, extreme temperatures (both heat and cold) or drought, and salinity are known to induce the production of ROS/RNS that play important roles in plant growth and development as well as in sensing environmental cues (Sandalio et al. 2001; Camejo et al. 2013; Foyer and Noctor 2013). However, ROS can also cause loss of redox homeostasis in tissues, resulting in oxidative injuries in plants (Rodríguez-Serrano et al. 2009; Lázaro et al. 2013). In green tissues, although the main sources of ROS are chloroplasts and peroxisomes, mitochondria account for the total production generating ROS as a product of respiration (Moller 2001). The exposure to abiotic stress conditions causes a deregulation, an overflow, or even a disruption of electron transport chains (ETC) in mitochondria and chloroplasts, giving rise to ROS accumulation. Singlet oxygen ($^1O_2$), the hydroxyl radical ($^\bullet OH$), the superoxide radical ($O_2^{\bullet -}$), and hydrogen peroxide (H_2O_2) are all strongly oxidizing compounds and therefore potentially harmful for cell integrity. Among them, H_2O_2 is the most stable ROS formed in the reaction of $^1O_2$ with $O_2^{\bullet -}$ and as a product of spontaneous dismutation of $O_2^{\bullet -}$ (Foyer and Noctor 2009; Lázaro et al. 2013).

Mitochondria are central organelles in setting the cellular redox balance and homeostasis (Noctor et al. 2007). Changes in ROS levels due to the perturbation of the respiratory complex I have been proposed to trigger a mitochondrial retrograde signal (Rhoads and Subbaiah 2007) in which alternative oxidase (AOX) is one of the main components (Maxwell et al. 1999). Redox signals are probably transduced by oxidation of proteins such as ROS-activated transcription factors and kinases (Mittler et al. 2011). Moreover, changes in the thiol status of proteins are also probably involved in redox signaling under abiotic stress (Sevilla et al. 2015). Other molecules, including the antioxidants ascorbate (ASC) and glutathione (GSH), lipids, and organic acids, are modified by ROS, with implications for their signaling functions (Bartoli et al. 2004; Farmer and Mueller 2013). Similar to ROS, nitric oxide (NO) and NO-related molecules such as *S*-nitrosoglutathione (GSNO) also act as possible signals. These molecules regulate proteins, genes, and phytohormone signaling during environmental stress (del Río et al. 2002; Yu et al. 2014). The ROS detoxification system is essential for protecting organelles from the oxidative damage that can occur under abiotic stress conditions and transduction of the signal is considered as a primary response of defense (Camejo et al. 2013; Romero-Puertas et al. 2013).

In this context, the analysis of changes in gene expression and proteins or metabolites of a given cell compartment is a useful tool for understanding the effect of a specific stress situation and the response exerted by the different defense systems. Plant mitochondria contain a battery of nonenzymatic and enzymatic compounds able to modulate the ROS/RNS content produced during the normal metabolism and increased under stress conditions (Gómez et al. 1999; Mittler et al. 2004; Camejo et al. 2013). The first line of defense is Mn-SOD, which catalyzes the dismutation of $O_2^{\bullet -}$ to molecular oxygen and H_2O_2, and the ascorbate–glutathione (ASC–GSH) cycle. This pathway is in charge of the detoxification of H_2O_2, using NADPH as a reductant as well as the antioxidants ascorbate and glutathione (Jiménez et al. 1997, 1998a) and specific enzymes involved in the regeneration of the reduced forms of these antioxidants. Other enzymes involved in H_2O_2 and other hydroperoxides' detoxification are peroxiredoxins (Prxs), which, together with thioredoxin (Trx), thioredoxin reductase (TR), sulfiredoxin (Srx), and NADPH, constitute a system that ensures not only redox homeostasis in the mitochondria but also signaling during normal and stress conditions (Pulido et al. 2009; Lázaro et al. 2013).

In this chapter, we review current knowledge on the involvement of plant mitochondria in abiotic stress, more specifically, in the response to salt, drought, and extreme temperature conditions. Special attention is paid to the ascorbate–glutathione cycle, AOX, and the Trxs/Prx/Srx response under these environmental conditions.

2 Mitochondria as Central Organelles in Stress

Mitochondria are highly dynamic, metabolically active cell organelles that are characterized by their involvement in processes like the tricarboxylic acid cycle and the production of ATP during oxidative phosphorylation. Mitochondria also have an important role in cellular proliferation, development, growth, and programmed cell death (Rhoads and Subbaiah 2007; Mittler et al. 2011). From a functional point of view, plant mitochondria differ from their animal counterparts by playing additional roles, as they house enzymes involved in vitamin biosynthesis and some enzymes of the photorespiratory pathway in photosynthetic tissues, as well as some additional alternative complexes which play an important role when the cells are exposed to metabolic dysfunction and/or stress (Rasmusson and Moller 1991; Ribas-Carbó et al. 2005a).

In the past two decades, it has become clear that mitochondria have developed mechanisms to communicate their biosynthetic and bioenergetic fitness to the rest of the cells, meaning that they also have signaling functions. Such mitochondria-to-nucleus communication named mitochondrial retrograde regulation (MRR) is necessary as the nucleus encodes most organelle proteins. Upon disturbance of the mitochondrial function caused by abiotic stress, mitochondria signal the nucleus to trigger the expression of responsive genes. In this process, organellar redox state and Ca^{2+} and ROS metabolism have been proposed as sources for retrograde signals, which provide the metabolic flexibility that, during stress conditions, plays an important role in the acclimation of plants (Rhoads and Subbaiah 2007; Woodson and Chory 2008). A common effect of all abiotic stresses is increased ROS and RNS production, which are particularly important in mitochondria, acting as a part of the signal transduction induced by stress conditions (Cvetkovska and Vanlerberghe 2012).

The mitochondrial electron transport chain is composed of four large protein complexes, some of them organized into supracomplexes: a proton-pumping NADH dehydrogenase (complex I), a succinate dehydrogenase (complex II), a Cyt bc1 complex (complex III), and a Cyt c oxidase (complex IV). Electron flow through complexes I, III, and IV results in the formation of an electrochemical proton gradient across the inner mitochondrial membrane. This drives the ATP synthase (complex V) to convert ADP to ATP. In addition to this classic ETC, plant mitochondria contain two additional non-phosphorylating pathways including alternative NADPH dehydrogenases (type II NDH) and alternative oxidase (AOX) (Maxwell et al. 1999). Type II NDH bypass complex I and supply electrons to the ubiquinone pool and thus do not contribute to the generation of the proton motive force needed for ATP synthesis The cyanide-insensitive AOX oxidizes ubiquinol and reduces oxygen to water without translocating protons across the inner membrane. This alternative pathway bypasses the main proton-translocating complexes III and IV and oxidizes NADPH with varying levels of energy conservation and electron transport to help balance the energy status of the cell (Ribas-Carbó et al. 1997; Millar et al. 2011; Moore et al. 2013).

3 Mitochondrial ROS and RNS Production

3.1 ROS Production

An inevitable feature of mitochondrial biochemistry is the generation of ROS, such as $O_2^{\bullet-}$ and H_2O_2, associated with the electron transport chain. In plant cells, the first report that the respiratory chain produced ROS came in 1978 (Rich and Bonner 1978), 12 years after the pioneering work of Jensen (1966), who demonstrated $O_2^{\bullet-}$ generation in mammalian mitochondria. Within mitochondria, $O_2^{\bullet-}$ is produced by the one-electron reduction of O_2, and although it is a reaction thermodynamically favored, only a small proportion of mitochondrial electron carriers do so (Murphy 2009). These carriers include the redox-active prosthetic groups within proteins and complex I and complex III are probably the primary sites of $O_2^{\bullet-}$ generation (Noctor et al. 2007). Under specific conditions, complex II may also be involved in ROS generation, in the course of reverse electron transport (Turrens 2003). This superoxide can, in turn, act as substrate for the generation of H_2O_2 and hydroxyl radicals ($^{\bullet}OH$). An important generalization is that ROS formation increases as the ETC becomes more highly reduced (Moller 2001). Mitochondrial ROS generation is therefore higher in ADP-limiting conditions that increase the mitochondrial transmembrane potential and decrease when ADP is being actively phosphorylated by artificial uncoupler proteins (UCPs) that dissipate the transmembrane potential (Collins et al. 2012). Accepting electrons from ubiquinone, AOX may prevent the over-reduction of the ETC at complex I and/or III. Hence, this route of electron transport could be important for dampening ROS formation under conditions in which Cyt pathway (CP) components have suffered stress-induced damage or, since AOX respiration is less tightly coupled to ATP production, under conditions in which ADP availability is limiting (Maxwell et al. 1999; Millar et al. 2011). Thus, dissipation of the membrane potential directly by uncoupler components may be important in tissues with low AOX expression and/or activities (Trono et al. 2004). ROS accumulation in mitochondria can also be influenced by type II NDH, which could function as "safety valves," to limit ROS production by keeping the ETC relatively oxidized (Rasmusson and Wallström 2010; Millar et al. 2011), by the activity and modification of antioxidant systems and oxygen concentration (Jiménez et al. 1998b), and by posttranslational modifications of respiratory complexes (Beer et al. 2004).

3.2 NO Production

In recent years, nitric oxide (NO) has emerged as an intracellular and intercellular signal molecule and an important metabolite in plants, where two major enzymatic pathways have been proposed as participating in its formation: the oxidation of L-arginine to L-citrulline by a nitric oxide synthase (NOS)-like enzyme and the reduction of nitrite to NO by a nitrate reductase (NR) (Fröhlich and Durner 2011; Gupta et al. 2011).

In the past decade, the presence of NOS-like activity has been demonstrated in plant peroxisomes. However, the characterization of such an enzyme is still unresolved (del Río et al. 2002). The reduction of nitrite to NO by mitochondrial ETC has been reported to produce small amounts of ATP under hypoxic conditions (Stoimenova et al. 2007). NO production by a mitochondrial nitrite-reducing activity (NR) has been detected in different photosynthetic sources and mitochondria isolated from roots of diverse plant species. Nitrite reduction occurs at the site of cytochrome c oxidase (COX), but it can also occur at the sites of complex III and AOX although a clear mechanism has only been established for COX under hypoxia. However, this may become increasingly important as partial pressures of oxygen are reduced from the ambient level (Igamberdiev et al. 2014; Yu et al. 2014). In mitochondria, the formation of RNS is due to NO's short half-life and reactivity. NO can gain or lose an electron, to form the nitrosyl anion (NO^-) or the nitrosonium cation (NO^+) and to form higher oxides including NO_2 and N_2O_3, or can react immediately with $O_2^{\bullet-}$ originated from ETC, to form peroxynitrite ($ONOO^-$). Through this reaction, superoxide probably plays a role in regulating free NO levels (Leitner et al. 2009). Moreover, NO can react with mitochondrial glutathione (GSH) to form GSNO or with thiols or the catalytic metal centers of proteins. This can result in covalent modification of cysteine residues in a process called S-nitrosylation and, in the NO binding to transition metals, through metal nitrosylation (Yu et al. 2014). The subcellular localization of GSNO in pea mitochondria has been reported (Camejo et al. 2013; Corpas et al. 2013) and it is considered as the most abundant low-molecular-mass *S*-nitrosothiol (SNO) and also a vehicle of NO in the cell, which enables NO biological activity to expand.

Prime targets of NO and its derivates in mitochondria are the electron transport components and enzymes, producing an inhibition of CP, whereas the alternative pathway is only partially inhibited (Day et al. 1996; Martí et al. 2012). These inhibitions may potentially be involved in the regulation of energy metabolism, generation of ROS, cell death, and response to stress. Recent evidences showed that pea Mn-SOD was not inactivated by NO upon mitochondrial treatment with the NO donor DETA-NONOate (Martí et al. 2012). The insensitivity of AOX to NO and the potential capacity of Mn-SOD to bind and stimulate NO decay under anaerobic and aerobic conditions (Filipovic et al. 2007) represent mechanisms to prevent its deleterious effects on the respiratory activity.

4 Antioxidant and Redox Systems in Plant Mitochondria

As previously noted, mitochondria are important cellular organelles for orchestrating the plant stress response. The mitochondrial antioxidant system, through ROS and RNS detoxification and/or regulation, has a pivotal role affecting redox signaling.

4.1 AOX

As described above, this enzyme is a quinol oxidase that catalyzes the reduction of oxygen to water by directly accepting electrons from the ubiquinone pool, dissipating the energy as heat, and lowering the ADP/O ratio. The AOX is a nuclear-encoded protein linked to the mitochondrial inner membrane present in all kingdoms of life except in archaebacteria (McDonald and Vanlerberghe 2006). In angiosperms, a multigene family has been shown to encode the AOX protein (McDonald et al. 2009), and the genes are both tissue and development specific. In higher plants, AOX exists as a dimer and can be redox regulated by the formation of a disulfide bridge between conserved cysteine residues of each monomer (Vanlerberghe 2013). The reduction level and amount of UQ pool and, in some situations, the concentration of AOX protein are thought to be important in determining the extent of AOX activity (Ribas-Carbó et al. 1995, 1997). Nevertheless, the AOX protein can be activated by posttranslational modifications. Biochemical characterizations have reported that AOX can be both reduced and activated by mitochondrial thioredoxin (PsTrx*o*1) by using its effector pyruvate (Gelhaye et al. 2004). Similarly, pea mitochondrial AOX homodimers are specifically reduced by PsTrx*o*1. This thioredoxin was also able to produce the activation of oxygen consumption by the AOX pathway, using NADPH/NTR system (Martí et al. 2009). While the redox regulation of AOX has been well documented in vitro, its impact on the in vivo activity of alternative oxidase pathway remains controversial (Millenaar and Lambers 2003). A better understanding of the regulation of the AOX protein has been provided by recent advances made in its molecular characterization (McDonald et al. 2009; Kido et al. 2010) that opens new exciting opportunities for the AOX research on the relationship between its structure, regulation, and function.

4.2 Mn-SOD

Manganese superoxide dismutase (Mn-SOD) is a tetrameric isoenzyme initially purified and characterized in pea leaves, as a result of a pioneering work led by Prof. LA del Río (Sevilla et al. 1982). The enzyme is important in providing protection against oxidative stress controlling H_2O_2 production, thus avoiding the formation of $^{\bullet}OH$ radicals (Marques et al. 2014). Mn-SOD suppression in *Arabidopsis* causes decreased growth, although the respiration rate was not affected, whereas the mitochondrial redox balance and some of the tricarboxylic acid cycle enzymes were altered. Unexpectedly, Mn-SOD mutant plants displayed an increased antioxidant capacity, suggesting the existence of a retrograde pathway trying to compensate the lack of this antioxidant enzyme (Morgan et al. 2008). As a result of Mn-SOD, the newly formed H_2O_2 can be decomposed by the mitochondrial peroxidase activities that depend on the ascorbate and glutathione through the ASC–GSH cycle or/on the Trx–Prx–Srx system.

4.3 ASC–GSH Cycle

The first publications describing the presence of the some components of the so-called ascorbate–glutathione cycle or Foyer–Halliwell–Asada pathway in mitochondria (Fig. 1) appeared in 1981 and 1990 with MDHAR and GR of potato and pea mitochondria, respectively (Arrigoni et al. 1981; Edwards et al. 1990). This pathway is considered to be a crucial mechanism for H_2O_2 metabolism in which the heme-containing ascorbate peroxidase (APX) scavenges H_2O_2 using reduced ascorbate that is oxidized to monodehydroascorbate (MDA) or dehydroascorbate (DHA). These oxidized forms are reduced by FAD-containing monodehydroascorbate reductase (MDHAR), using NADPH as reductant, and dehydroascorbate reductase (DHAR), using GSH as electron donor, respectively. The flavoprotein glutathione reductase (GR) is the enzyme in charge of the reduction of oxidized GSSG in a NADPH-dependent manner (Foyer and Halliwell 1976; Asada 1999). In mitochondria, the final proof of the existence of a complete cycle, similar to that in chloroplasts, was described in pea leaves (Jiménez et al. 1997) and further in *Arabidopsis* (Chew et al. 2003). Biochemical and enzymatic latency assays indicated that the mitochondrial APX activity resulted in at least two isozymes in pea mitochondria with different degrees of substrate specificity and sensitivity to inhibitors, when compared to that found in peroxisomes and chloroplasts (Jiménez et al. 1998a). These APX isozymes were linked to both the inner and the outer face of the external membrane in concert with membrane-associated MDHAR, while DHAR and GR were located in the matrix, together with the antioxidants ascorbate and glutathione (Fig. 1). The possible presence of the APX isozymes linked to the inner face of the external membrane was reported in *Arabidopsis* mitochondria by Chew et al. (2003). It was reported that the differential sub-compartmental

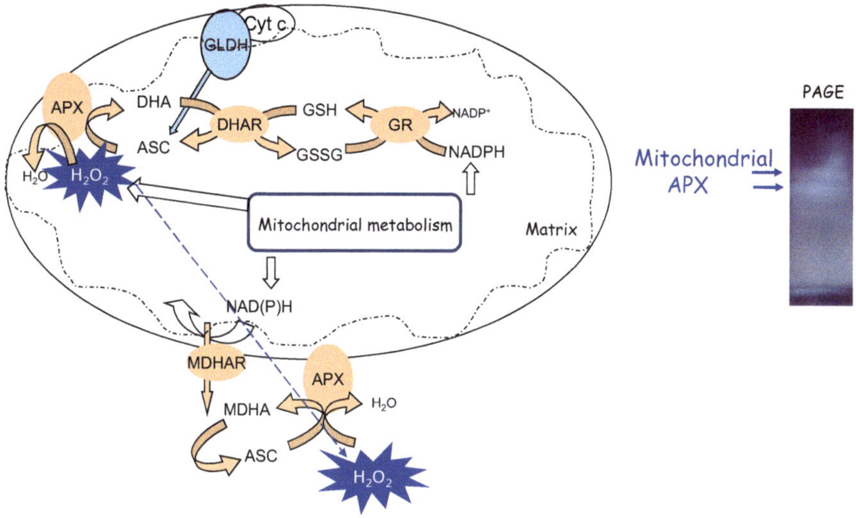

Fig. 1 Mitochondrial ascorbate–glutathione cycle. The hydrogen peroxide in the mitochondria produced by the organelle metabolism is reduced by APX at the expense of ASC to produce DHA that is reduced to ASC by DHAR using GSH as electron donor. The resulting GSSG is regenerated by GR and NADPH. MDHA is reduced back to ASC by MDHAR using NADPH from the mitochondrial metabolism. In the gel, APX isoenzymes detected in isolated pea mitochondria are revealed after PAGE and NBT staining (Jiménez et al. 1998a)

localization of the two APX isoenzymes may indicate slightly divergent roles in ASC recycling: (1) protection against H_2O_2 as it leaks out of mitochondria, thereby preventing oxidative damage to other cell compartments, acting together with MDHAR, and (2) clearance of intramitochondrial H_2O_2 in concert with DHAR and GR (Jiménez et al. 1998a; del Río et al. 1998). The presence of one isozyme of APX on the intermembrane space side of the inner membrane may be also useful for using the ASC generated at this site. Reports including those from Christine Foyer's group described that ascorbate is produced in mitochondria by the terminal enzyme L-galactono-1,4-lactone dehydrogenase (GalLDH) and showed its coupling to the CP has been demonstrated (Bartoli et al. 2000). This enzyme is also attached to the inner mitochondrial membrane, localized on a subset of respiratory chain complex I, and its activity is highly dependent on the availability of oxidized Cyt c (Millar et al. 2003). In addition to the reductive GSH-dependent DHA reduction, ASC regeneration may also be attributed to the respiratory electron transport chain (Szarka et al. 2007) or linked to other redox compounds such as the glutaredoxin (Grx) and Trx systems (Meyer et al. 2012).

The antioxidant role of ascorbate in plants is evident from the relatively high concentrations found in many tissues and in subcellular organelles. On average, the ASC concentration in *Arabidopsis* cells is 5.6 mM (Zechmann et al. 2011). The cellular ASC abundance and homeostasis has profound effects on gene transcription, mainly those regulating plant growth and defense. This is mediated, at least in part, through changes in the plant hormone, abscisic acid (ABA) (Kerchev et al. 2011). Together with the mitochondria, ASC is found in many subcellular organelles, including peroxisomes, vacuoles, cytosol, the cell wall, and chloroplasts, where concentrations can reach up to 50 mM (Rautenkranz et al. 1994; Jiménez et al. 1997; Takahama 2004). The mitochondrial metabolism has been described as being involved in the modulation of photosynthesis through the suggested complementation of AOX and ASC metabolism in this way protecting against photoinhibition. Changes in mitochondrial ASC synthesis dependent on respiration could regulate retrograde signaling as a common signal from both mitochondria and chloroplasts (Talla et al. 2011). A good example of such an inter-organelle communication is the ASC produced in the mitochondria and then transported into the apoplast. Interestingly, ASC appears to exert its greatest influence by setting thresholds for apoplastic and cytoplasmic signaling (Munné-Bosch et al. 2013). When the ASC and GSH contents are depleted, balanced mechanisms exist to maintain the processes dependent on these antioxidants as well as related signaling responses in specific compartments (Foyer and Noctor 2011).

Glutathione is a key component in the regulation of cellular thiol redox homeostasis. It also plays key roles in the ROS detoxification, redox signaling, modulation of defense gene expression, and the regulation of enzymatic activities (Schnaubelt et al. 2013). In this way glutathione may influence plant development not only under optimal conditions but also in stress situations (extensively reviewed by Noctor et al. 2012; Kocsy et al. 2013; Schmitt et al. 2014). Additionally, glutathione through the posttranslational modification known as glutathionylation is involved in protecting proteins from oxidation (Zaffagnini et al. 2012). Although cellular compartments and changing growth conditions may influence its levels, the biosynthesis of GSH is mainly localized in cytosol and plastids: glutamate cysteine ligase is stromatic, whereas GSH synthetase is targeted to the stroma and cytosol (Preuss et al. 2014). Once synthesized, GSH is transported to mitochondria, although the nature and regulation of these transporters are still unclear (Bachhawat et al. 2013). The presence of glutathione was reported in pea and *Arabidopsis* mitochondria, and in fact, immunolabeling studies have localized this antioxidant in *Arabidopsis* in both mitochondria and chloroplast, containing about 15–25 % and 62–75 %, respectively, of the total pool of GSH (Jiménez et al. 1997; Fernández-García et al. 2009). The role of glutathione in mitochondria is quite an important issue. It seems that the depletion and oxidation of glutathione in mitochondria favors the accumulation of ROS found in senescent leaves (Jiménez et al. 1998b) and could be one reason for the induction of the programmed cell death events after pathogen infection in *Nicotiana tabacum* plants (Zechmann et al. 2014). The involvement of mitochondrial GSH in cell development has been reported in studies using both *rml1* and *pad2-1* mutant (see Zechmann and Müller 2010) pointing to the mitochondrial glutathione level as a key component for the proper plant development in situations when this antioxidant is depleted. Another example of the importance of glutathione in mitochondria is its role in NO scavenging forming GSNO, which can be further metabolized to oxidized glutathione (GSSG) and NH_3 (Wilson et al. 2008).

As in the case of ASC, a link between complex I (CI) activity and GSH has also been shown in CI *Arabidopsis* mutants. These mutants were insensitive to a GSH biosynthesis inhibitor and showed higher levels of GSH, implying an as yet unexplained effect of mitochondrial respiration on GSH homeostasis (Koprivova et al. 2010).

4.4 Peroxiredoxin System

Peroxiredoxin, thioredoxin, thioredoxin reductase, sulfiredoxin, and NADPH all together constitute a system that ensures not only redox homeostasis in the mitochondria but also signaling. Prxs are thiol peroxidases involved in peroxide detoxification and signaling. They are ubiquitous, are located in different cellular compartments, and are expressed as different isoforms. Plant mitochondria, unlike its mammalian counterpart, have only one type of Prx, PrxIIF, an atypical Prx highly conserved between different species that contains two residues of cysteine (Barranco-Medina et al. 2007; Dietz 2011). Both residues are essential for efficient catalysis, and after hydroperoxide reduction, they form an intramolecular disulfide bridge that is reduced by mitochondrial Trx*o*. Mitochondrial PrxIIF crystallizes as hexamers (Barranco-Medina et al. 2006) which are favored in oxidant conditions but dissociate to dimers upon reduction. The presence of peroxidatic cysteine is critical for hexamer formation, whereas the resolving cysteine is not necessary (Barranco-Medina et al. 2007, 2008b). Posttranslational modification by S-nitrosylation of mitochondrial PrxIIF induces a conformational change in the protein and provokes a reduction in its peroxidase activity while acquiring a novel function as transnitrosylase that is fundamental for the signal transduction role of NO in plants (Camejo et al. 2015).

The involvement of Trx*o* in the detoxification of ROS via PrxIIF has been demonstrated in vivo by interaction between both mitochondrial proteins and in vitro with recombinant proteins (Barranco-Medina et al. 2008a; Martí et al. 2009). Trx harbors a dithiol active site and regulates target enzymes by thiol/disulfide exchange. The presence of Trx in plant mitochondria was shown by Laloi et al. (2001) in *Arabidopsis* and was classified as Trx*o* type. More recently, a pea Trxo was described in both mitochondria and nucleus under normal conditions, and similarly to PrxIIF, pea Trx*o* reduced mitochondrial alternative oxidase (AOX) homodimers (Martí et al. 2009). As we have described previously, Trx*o* through activation of PrxIIF and AOX could also play a role in linking ROS and redox signaling in mitochondria. Mitochondrial Trx*o* is reduced by a homodimeric TR firstly reported by Laloi et al. (2001), which in turn is reduced by NADPH

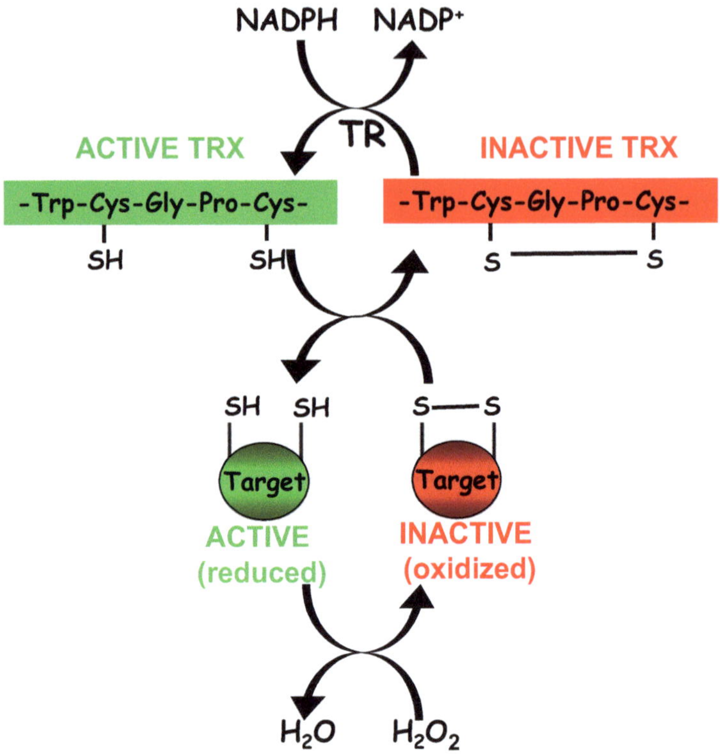

Fig. 2 Thioredoxin (Trx) system. Trx is reduced by NADPH-dependent Trx reductase (TR). Reduced Trx can reduce in turn target proteins from their oxidized forms

(Gelhaye et al. 2005) (Fig. 2). In *Arabidopsis*, there are three TRs, namely, TRA, TRB, and TRC, localized to different cellular compartments, cytosol, mitochondria, and chloroplast, respectively (Serrato et al. 2004).

Under severe oxidative stress, PrxIIF overoxidizes to the inactive sulfinic form, which could serve as a sensor of the hyperoxidative conditions inside the mitochondria. An enzyme called Srx, a special type of ATP-dependent reductase, located in chloroplasts and mitochondria is able to retroreduce the overoxidized form of chloroplastic 2-Cys Prx and mitochondrial PrxIIF (Biteau et al. 2003; Iglesias-Baena et al. 2010, 2011). The interaction between Srx and its mitochondrial targets PrxIIF and Trxo has been proved, and the mechanism of reduction of the sulfinic PrxIIF is similar to the sulfinic 2-Cys Prx, via formation of a phosphoryl intermediate on the sulfinyl moiety, a thiosulfinate between Srx and PrxIIF, and finally, a heterocomplex Srx–Trx. Both complexes strengthen the mechanism for sulfinic PrxIIF reduction by Srx into the plant mitochondria (Iglesias-Baena et al. 2011) (Fig. 3).

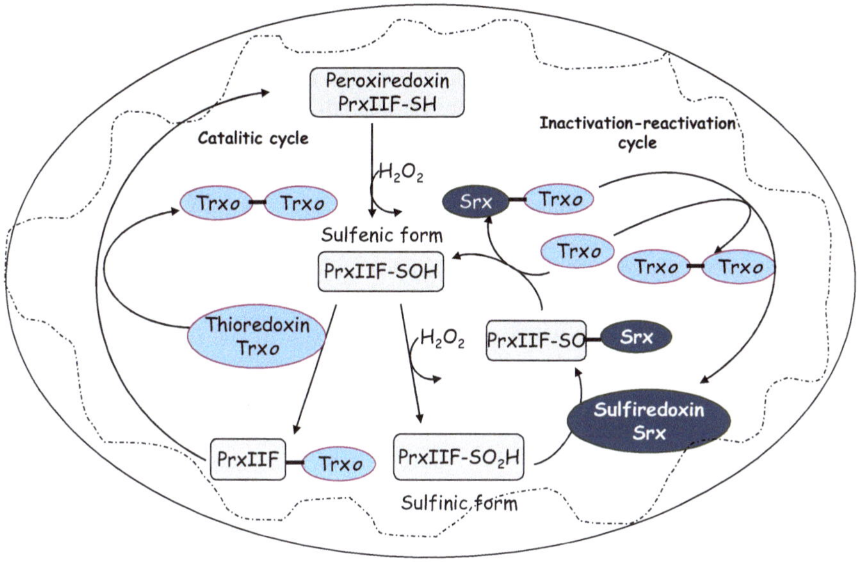

Fig. 3 Catalytic cycle of mitochondrial PrxIIF overoxidation. Under physiological conditions, mitochondrial PrxIIF is oxidized to its sulfenic form acting as peroxidase. At high concentration of H_2O_2, PrxIIF may be overoxidized to the inactive sulfinic form (PrxIIF-SO_2H) that is converted into sulfinate (PrxIIF-SO-S-Srx) with Srx. Mitochondrial Trx*o* reduces the heterocomplex to release PrxIIF-SOH and Srx-Trx*o* which is reduced to Srx-SH by Trx*o*. The sulfenic form of PrxIIF is reduced by Trx*o* that forms PrxIIF-Trx*o* and the active PrxIIF-SH is released by another Trx*o* that forms the dimer Trx*o*–Trx*o*

5 Mitochondrial Antioxidant and Redox System Are Involved in Abiotic Stress Response

Environmental factors have important consequences for plant growth and survival and seriously limit crop production. FAO reports have informed that almost 70 % of the global land area is affected by some environmental constraint (see review by Cramer et al. 2011). Thus, abiotic stresses including salinity, temperature extremes, and drought are a subject of increasing interest due to the significant impact on food yields in a scenario of continued reduction of arable land and water resources and climate change (Lobell et al. 2011).

5.1 Mitochondrial Response Under Salinity

Soil salinization is regarded as a problem for agriculture in the world. In these conditions, the ability of plants to take up enough water decreased, while the Na^+ and Cl^- taken up by roots impair growth by causing an osmotic imbalance, ion toxicity, the alteration of membranes, redox imbalance, cellular energy depletion, and/or the disruption of photosynthesis (Abogadallah 2010; Maathuis et al. 2014). The response of plants to salinity is complicated and involves metabolic and genetic adjustments. This response is dependent on the salt tolerance of the plant and the severity and duration of the stress. Among the different mechanisms involved in the response, ROS are considered as essential components of salt tolerance, and several examples are known of enhanced capacities to scavenge ROS in mutants or transgenic plants showing higher salt tolerances (Hasegawa et al. 2000). ROS generation generally increases in mitochondria under salinity due to a dysfunction in the respiratory electron transport chain, increasing the rate, with consequent electron leakage to O_2 (Hernández et al. 1993, 2001). In pea plants grown under salinity, the generation of $O_2^{\bullet-}$ and H_2O_2 has been demonstrated in mitochondria of two pea cultivars differing in NaCl sensitivity (Hernández et al. 1993). In tobacco, it was shown that short-term salt stress resulted in a strong increase in leaf $O_2^{\bullet-}$, with only a moderate increase of H_2O_2 (Andronis and Roubelakis-Angelakis 2010). Most reports on mitochondria isolated from plants growing in saline conditions demonstrate an inhibition of the respiratory rate with a decreased enzymatic capacity (see review by Jacoby et al. 2010), although this does not mean a decrease in in vivo respiratory activity. As an example, in *Arabidopsis* after 6 h of saline stress, cytochrome c oxidase decreased transcript and protein abundance, while NDH increased its transcript level but decreased the protein level (Jacoby et al. 2010). The presence of NDH and AOX confers a higher flexibility and complexity to plant ETC than its mammalian counterpart. AOX could represent a key part of a larger stress response at mitochondrial level (Clifton et al. 2006; Elhafez et al. 2005). Oxygen isotope fractionation is presented as the most reliable technique for studying the regulation of the electron partitioning between the two respiratory pathways (Lambers et al. 2005). When the effects of salinity on respiration and partitioning were investigated, the alternative pathway (AP) was found to increase its contribution to respiration during long-term (14 days) NaCl stress conditions in pea plants (Martí et al. 2011) similar to that found in severe drought conditions (Ribas-Carbó et al. 2005b). However, an observed decrease in total respiration was mainly due to a decrease in the CP activity, whereas AP activity remained constant (Florez-Sarasa et al. 2007). This decrease in CP was thought to be due to a reduced energy demand for growth and to the control exerted by an increase in the ATP/ADP ratio (Munns 2002), as was observed from the significant reduction in leaf growth under saline conditions. The maintained AP activity may diminish ROS generation in the mitochondria under saline conditions and reflect the presence of the sustainable active form of AOX, in which the increased PsTrxo1 activity observed could play a role in the response of AOX to salinity which is also reflected at transcript and protein content levels (Martí et al. 2009). Moreover, the analysis of co-expression of AOX genes and other components of the ETC, as external and internal NDH, has revealed that these respiratory components could work together to oxidize external NADPH under a number of conditions including salinity. Another interesting point is that the lack of AOX in antisense *AOX1a Arabidopsis* plants under salinity and drought altered transcript abundance of other mitochondrial proteins such as glutaredoxin (Grx) and APX6 as well as several antioxidative chloroplast and cytosol proteins (Umbach et al. 2005; Giraud et al. 2008). These changes pointed AOX as a key component in adaptation through the coordination to other antioxidative defenses in the different cell compartments (Clifton et al. 2006). However, the correlation between AOX expression and activity is not always positive and both were induced by salt in shoots, roots, and cell cultures of different plants (see reviews by Jacoby et al. 2010; Lázaro et al. 2013) while discrepancies have been reported and discussed (Guy and Vanlerberghe 2005; Ribas-Carbó et al. 2005b; Vidal et al. 2007; Rasmusson et al. 2009; Martí et al. 2011), revealing the importance of the posttranscriptional and posttranslational regulation of AOX under salinity stress conditions. Another interesting aspect of the response of mitochondria to salinity is the increased abundance of proteins involved in mitochondrial ROS defense, pointing to a huge impact of salinity on the organelle function in vivo. Under salt stress, this enhancement of antioxidants is necessary to defend the mitochondrial components, including respiration function, against increased ROS generation, thus preventing higher damages (Hernández et al. 1993; Sweetlove et al. 2002; Tan et al. 2010). Again, the response is variable. Thus, different antioxidative enzyme isoforms in the cell compartments can present specific profile activity in cultivars differing in salt tolerance (Hernández et al. 1993, 2000; Olmos et al. 1994; Gueta-Dahan et al 1997; Ashraf 2009). Among them, Mn-SOD is a key stress-responsive enzyme in mitochondria. Under salt stress, Mn-SOD activity is increased in various plants like chickpea pea and liquorice plant (Gómez et al. 1999; Eyidogan and Oz 2005; Pan et al. 2006), while its protein content was maintained during the salt treatment in pea plants (Martí et al. 2011; Camejo et al. 2013) or increased in *Arabidopsis* (Jiang et al. 2007). The fact that increased antioxidant activity can prevent the oxidative damage accompanying salt stress was evident in overexpressing Mn-SOD plants in different transgenic plants which showed increased salt tolerance (Wang et al. 2004, 2005). Together with Mn-SOD activity, the activities of Cu/Zn-SOD, Fe-SOD, CAT, and POD were significantly higher in transgenic *Arabidopsis* than in wild-type plants (Wang et al. 2004). Another example of collaboration among antioxidative proteins is the observed coordinated increase in wheat AOX and Mn-SOD proteins which could prevent the over-reduction of the mitochondrial ubiquinone pool, decreasing the content of $O_2^{\bullet-}$ in this organelle (Jacoby et al. 2010). Similarly, in pea mitochondria, an increased expression and protein components of the ascorbate–glutathione cycle, mainly APX and MDHAR, were enhanced under salinity conditions (Gómez et al. 1999). The importance of the subcellular compartments in the response of these enzymes to salt stress was evidenced by these authors showing that the pattern of changes was found different in pea chloroplast isoenzymes. In this organelle, APX behaved differently in thylakoids and stroma in response to NaCl, with a significant increase of stromal APX, whereas the thylakoidal activity was significantly and progressively lost (Gómez et al. 2004). Moreover, GR and DHAR but not MDHAR activities were significantly enhanced. All these inductions in the different cell compartments are thought to contribute to preventing ROS formation and oxidative damage, giving rise to the tolerance to salt stress in the studied pea cultivar. In sweet potato, the expression of mitochondrial APX together with cytosolic and chloroplast isoforms was tissue specific and dependent on salt stress duration (Lin and Pu 2010). Other authors indicated a higher expression and activities of mitochondrial APX, MHAR, DHAR, and GPX in a salt-tolerant variety of *Lycopersicon pennellii* plants (Mittova et al. 2003). Likewise, the overexpression of MDHAR in transgenic tobacco increased the tolerance against salt and osmotic stresses (Eltayeb et al. 2006) and the overexpression of DHAR increased salt tolerance in *Arabidopsis* (Ushimaru et al. 2006), revealing the importance of these antioxidant components in the tolerance to salinity conditions.

Less information exists related to the involvement of mitochondrial antioxidants ASC and GSH in the salt stress response. Mitochondria are the organelles responsible for ascorbate biosynthesis and although glutathione synthesis is not mitochondrial, glutathione also shows highly compartment-specific changes in plants during stress situations, indicating important subcellular roles for this antioxidant in plant defenses (Zechmann 2014). To sense the oxidative stress, the reduced/oxidized balance of ASC and GSH is crucial for the cell response (Mullineaux and Rausch 2005; Foyer and Noctor 2009, 2011), taking into account their known influence on gene expression (Munné-Bosch et al. 2013). Abiotic stress usually decreases the content of these antioxidants, and balance mechanisms exist to maintain ASC- and GSH-dependent processes in specific compartments (Foyer and Noctor 2011). Therefore, during stress situations, subcellular changes in the glutathione contents may reflect the occurrence of compartment-specific oxidative stress. Moreover, it seems that high and stable levels of this antioxidant in the mitochondria are essential for cell development and survival especially in stress situations in which glutathione is depleted. However, in mitochondria, information about ASC and GSH contents and redox state is scarce and their accurate role in these organelles under abiotic stress is not well established (Lázaro et al. 2013).

The role of Prx system in plant mitochondria in response to abiotic stress is not widely reported in the literature. Nevertheless, there are data that allow the attribution to this system of a redox sensing and signal transduction function (Finkemeier et al. 2005; Dietz 2011; Martí et al. 2011; Lázaro et al. 2013). There are results that suggest a role for mitochondrial PrxIIF in antioxidant response and as a sensitive marker for stress. Transcript and protein levels of this PrxIIF were upregulated in leaves of pea, but not in roots, under salinity and cadmium stress (Barranco-Medina et al. 2007) both causing oxidative stress in sensitive species (Sandalio et al. 2001). However, this contrasts with the results obtained by Horling et al. (2002, 2003) where no changes were observed in PrxIIF mRNA levels in *Arabidopsis* leaves under salt, light, ozone, and hydrogen peroxide treatment. These partially discrepant results in *Arabidopsis* and pea could be attributed to the higher sensitivity of pea (Metwally et al. 2005). On the other hand, in *Arabidopsis* PrxIIF knockout mutants, PrxIIF seems not to be essential for plant survival, and lack of this enzyme in *Arabidopsis* PrxIIF knockout mutants might be compensated by increased mitochondrial ascorbate peroxidase activity or by the presence of the ascorbate–glutathione cycle (Finkemeier et al. 2005; Lázaro et al. 2013). In the response to salt stress, pea mitochondrial PrxIIF was S-nitrosylated over a long time

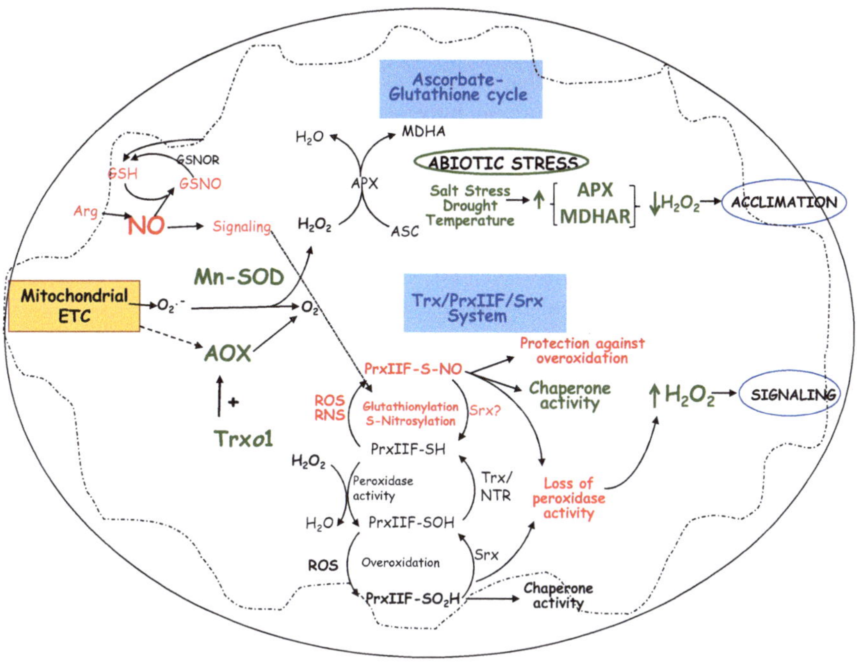

Fig. 4 Interaction between antioxidant and redox systems in ROS and RNS signaling in plant mitochondria under salinity. Superoxide radicals ($O_2^{\bullet -}$) produced by the mitochondrial ETC are dismutated to H_2O_2 by Mn-SOD. H_2O_2 is reduced by the ascorbate–glutathione cycle and the Trx/PrxIIF/Srx system (*in blue*). APX and PrxIIF scavenge H_2O_2 throughout the cycles indicated in Figs. 1 and 3. Under oxidative stress during salinity, AOX activity is induced, potentially decreasing ROS generation by ETC. In parallel, APX and MDHAR activities increased, which produces a decrease in the concentration of H_2O_2 allowing acclimation (*in green*). Also PrxIIF-SOH can be overoxidized to the sulfinic form (PrxIIF-SO_2H) gaining a chaperone activity and losing its peroxidase activity, increasing H_2O_2 concentration in the mitochondria allowing signaling. PrxIIF-SO_2H can be regenerated to the reduced form by the action of Srx and Trx. NO generated in the mitochondria during stress allows signaling in addition of forming GSNO by reduction with GSH from ascorbate–glutathione cycle (*in red*). NO also reacts with $O_2^{\bullet -}$ from the ETC, to form $ONOO^-$ that bursts nitrosative stress. Under nitrosative stress, PrxIIF-SH can be glutathionylated or S-nitrosylated in order to protect the enzyme against overoxidation and to gain chaperone activity, losing the peroxidase activity and allowing the signaling by H_2O_2 (*in red*). These posttranslational modifications could be reverted to PrxIIF-SH by the Srx activity. A similar pattern of induction of the AOX, Mn-SOD, ASC–GSH components, and Prx system has been reported under drought and extreme temperatures (see text)

period, possibly responding to increased NO under long-term salt stress (Camejo et al. 2013). It has been shown that salt stress modifies the S-nitrosylation level of some proteins in *Arabidopsis* (Fares et al. 2011). Subsequently, we have demonstrated that recombinant PrxIIF is S-nitrosylated by in vitro GSNO treatments, resulting in a conformational change and functional switch. Thus, PrxIIF activity was reduced acquiring a new transnitrosylase activity, which could probably function as a protective mechanism under conditions inducing oxidative and nitrosative stress (Camejo et al. 2015) (Fig. 4).

The involvement of mitochondrial TR in oxidative stress has been shown in yeast and mammalian (Lopert et al. 2012; Greetham et al. 2013). Serrato et al. (2004) have observed that *Arabidopsis* chloroplast TR knockout mutant lines showed growth inhibition and hypersensitivity to salt stress, drought, and methyl viologen. To our knowledge, there is no information about the response of mitochondrial TR to salinity as well as of Srx, probably due to its more recent discovery in plants (Biteau et al. 2003). *Arabidopsis* mutant lines lacking Srx are highly sensitive to methyl viologen, a drug that produce severe oxidative stress. In contrast to this observation, Rey et al. (2007) have observed that in vivo grown *Arabidopsis* Srx knockout mutant lines exhibited less oxidative damage than Wt when subjected to photoxidative treatment. This discrepancy could be related to the nature and intensity of the stress conditions. High levels of ROS in the mitochondria overoxidize PrxIIF to the sulfinic inactive form, and Srx, by controlling the reversion of this form, could regulate the signaling process. The absence of Srx in knockout lines of *Arabidopsis* produced an accumulation of inactive sulfinic forms of chloroplast 2Cys Prx and mitochondrial PrxIIF. Besides, the deletion of Srx yielded into more sensitive plants against high concentration of hydrogen peroxide when compared with wild-type plants (Iglesias-Baena et al. 2010, 2011). Therefore, Srx protects chloroplast and mitochondria from the oxidative stress regenerating the inactive sulfinic Prxs. As a summary, a model of response of the antioxidant and redox systems in mitochondria allowing acclimation and signaling under salinity is presented in Fig. 4.

5.2 Mitochondrial Response Under Drought

Drought stress in plants has severe effects, including changes inside the cell, with the result of reduced vegetative growth and cell division accompanied by decreased productivity. Changes in gene expression levels and activation of the enzymes involved in the production and scavenging of ROS as a resulting of drought stress situation have also been widely described (Mahajan and Tuteja 2005; Flexas et al. 2006). Moreover, plants subjected to drought stress suffer a disruption of cellular homeostasis, with inevitable consequences for the functioning of mitochondria (Flexas et al. 2005). These effects included their ability to adjust energy status to cope with the adverse conditions, although chloroplasts are more affected than mitochondria (Taylor et al. 2009). The response of leaf respiration to water stress varied depending on the duration and severity of the drought. In most cases studied (2/3) at whole-plant level, the respiration rate decreased with water stress (see review in Atkin and Macherel 2009). In other studies (1/3), respiration was unaffected but this did not mean that drought had no effect on mitochondrial metabolism. Ribas-Carbó et al. (2005a) found that the partitioning between CP and AP pathways was affected by severe stress in favor to AOX, which increased its activity but not its protein abundance. This could be the result of a greater electron flow through complex I and non-phosphorylating alternative dehydrogenases.

Drought stress also enhanced AOX activity as a percentage of total electron flow in tobacco (Galle et al. 2010). Changes in gene expression have been shown in *Arabidopsis* under drought, with an increase in *AOX1* accompanied by decreases in the transcription of the antioxidants PrxIIF, Trx*o*1, Grx, and Trx x (Mittler et al. 2004) (Fig. 4). In contrast, drought decreased leaf AOX transcript in *Medicago* (Filippou et al. 2011). An increase in AOX protein or capacity in response to drought has also been shown in wheat leaves (Bartoli et al. 2004; Vassileva et al. 2009), while no increase in AOX protein was reported for soybean leaves (Ribas-Carbó et al. 2005a) and drought. Also, the loss of function of mitochondrial AOX1a in *Arabidopsis* mutant plants led to a higher sensitivity to a combination of drought and moderate light stress (Giraud et al. 2008). These mutants, as we described above, showed changes in transcript abundance of other mitochondrial and chloroplast antioxidant enzymes, when growing under drought and salinity stress conditions (Giraud et al. 2008). Another mitochondrial protein involved in the response to drought is Mn-SOD. Its gene expression responded positively but decreased after rehydration in wheat seedlings (Wu et al. 1999), and the increase was greater in tolerant than in susceptible maize plants (Shiriga et al. 2014), suggesting that this enzyme may protect tolerant genotypes upon drought stress (Fig. 4). Moreover, Mn-SOD expression can improve drought tolerance in rice: transgenic plants overexpressing a pea Mn-SOD in rice chloroplasts exhibited less injury, measured by net photosynthetic rate, when treated with PEG, and showed reduced electrolyte leakage compared with wild-type plants (Wang et al. 2005). However, no change in its activity has been reported in other species such as pea plants under drought or after recovery (Moran et al. 1994). Other enzymes involved in the response to drought are the components of the ASC–GSH cycle. GR has been co-localized in chloroplasts and pea mitochondria and it is involved in the response of different species to drought. In pea plants, *GOR1* transcript levels remained constant during the drought phase although an increase was observed upon rewatering (Stevens et al. 1997). However, Sharma and Dubey (2005) described a significant increase in GR activity in drought-stressed *O. sativa* seedlings and also the enzymes involved in regeneration of ASC such as MDHAR and DHAR were higher in the stressed seedlings. Moreover, DHAR overexpression increased drought and ozone stress tolerance in tobacco (Eltayeb et al. 2006) (Fig. 4). Other examples of the involvement of the antioxidant system in the response to water deficit are evidenced in maize in which a mild water stress induced by a PEG treatment led to an increase in the activity of several antioxidant enzymes in mitochondria, including Mn-SOD, GR, APX, ASC, and GSH, during the first 12 h, after which they decreased (Tan et al. 2011). Similarly, in white clover, mitochondrial GR and Mn-SOD activities were found as the main ROS scavengers in mild water stress conditions (Chang-Quan and Rui-Chang 2008). All these imply the cooperation of these defense mechanisms against mild water stress detoxifying ROS at the site of production. In agreement, a decrease in APX protein content in tolerant *Amaranth* plants is probably involved in the observed increase in H_2O_2 which may act as signaling molecule for the stress response (Huerta-Ocampo et al. 2009). Similarly, in tolerant wheat, direct exposure to severe drought without previous acclimation produced high oxidative stress mediated by a poor response of the mitochondrial antioxidant system (Selote and Khanna-Chopra 2010), although a period of drought acclimation prior to the stress induced the upregulation of SOD, APX, and ASC–GSH cycle components, limiting the accumulation of H_2O_2 and lipid damages (Selote and Khanna-Chopra 2010). Also changes in ascorbate and glutathione contents are a commonly observed stress response of plants during drought (Hernández et al. 2013; Chan et al. 2013). A recent study of drought in vitamin C (*vitc2-1*)- and glutathione (*pad2-1*)-deficient *Arabidopsis* mutants revealed a decrease of glutathione in the nuclei and mitochondria of both mutants, respectively, early in the drought stress, while ascorbate remained unchanged. This suggests that GSH acts as a signal for drought stress from roots to leaves (Koffler et al. 2014). As regards ASC, using two maize varieties differing in their constitutive content Bartoli et al. (2005) showed that the enzyme responsible for the last step in ASC synthesis, GalLDH, was induced by drought in a cultivar-specific manner while not affecting the ASC content in wheat or the regenerating ASC–GSH cycle enzymes. These authors pointed to DHAR and GR, which were higher in the variety containing higher levels of ASC, together with the ETC, as the key components controlling the ascorbate content in the cell and, in this way, possibly influencing the drought stress response.

Changes in the mitochondrial Prx system under drought are not widely described. In *Vitis vinifera*, peroxiredoxin *vvprxIIF* targeted to mitochondria has been shown to be the most responsive gene among the peroxiredoxins to water stress, and it might be involved in drought tolerance through H_2O_2 signaling (Vidigal et al. 2013). Also PrxIIF protein increased under water deficit in *Arabidopsis* in mature but not in young leaves (Gama et al. 2007). However, to our knowledge, the response of mitochondrial Trx*o*, Srx, or TR under this condition is not reported.

All the changes described allow to propose mitochondria as target organelle for improving plant performance under water stress (Tan et al. 2011) and point mitochondria as a key organelle involved in the response and tolerance to salt and drought stresses through the coordinated function of its antioxidative defense metabolism.

5.3 Mitochondrial Response Under Extreme Temperatures

Plants are exposed to changing temperature in natural conditions. Both cold and heat stress induce significant changes in photosynthesis, membrane stability, and respiration and, as those occurring under salinity and drought conditions, constrain agricultural productivity (Camejo et al. 2006; Hasanuzzaman et al. 2013). Low temperature is probably the most studied abiotic stress in relation to AOX respiration, and, as in the case of salinity and drought, it seems that the response of AOX to temperature is tissue and species specific and depends on the severity and length of the stress. In many species, a sharp increase in AOX transcript and/or protein has been widely reported under this condition (Armstrong et al. 2008), although in other species, such as soybean, no change was observed (González-Meler et al. 1999). Using isotope discrimination, instantaneous (minutes to hours) changes in temperature had little impact on the relative partitioning of electrons between the CP and AOX (MacFarlane et al. 2009). However, longer treatments in *Arabidopsis* produced an increase in AOX activity with no accumulation of protein (Armstrong et al. 2008), while other studies have shown an accumulation of protein at low temperatures (Fiorani et al. 2005; Campbell et al. 2007; Umbach et al. 2009). Also in chilling-sensitive maize, a short-term cold treatment increased the AOX activity with a decreased CP (Ribas-Carbó et al. 2000). The effect of temperature has also been described in field-grown alpine grasses (Searle and Turnbull 2011), which presented higher AOX protein levels in the cold months. When transferred to cold, plants usually exhibited a strong increase in *AOX1a* transcript and protein but also an increase in transcript level of ROS-scavenging enzymes. Also AOX amount is important for the acclimation: plants with strong suppression of *AOX* transferred to cold presented a higher accumulation of these ROS-scavenging enzymes' transcripts than Wt plants, and, as a consequence, mutant plants suffered less oxidative damage (Fig. 4). In this way, cold stress affected *AOX* expression, although this did not always involve an increase in AOX respiration but probably gathered signaling pathways in mitochondria to respond at least at antioxidant level. The role of Mn-SOD in cold stress is observed in experiments targeting this enzyme from *Nicotiana* into maize chloroplasts, with resulting foliar tolerance to the oxidative stress induced by the unfavorable situation (Van Breusegem et al. 1999). Also the Mn-SOD gene was induced in both winter and spring plants when grown in cold-acclimating conditions in controlled environments and in the field (Wu et al. 1999). This similar behavior to drought conditions has been thought to be related to the cellular dehydration occurring in the cold acclimation step (Malone 1993). During freezing stress, overexpression of Mn-SOD in alfalfa plants protected them by minimizing ROS radical production with a rapid regrowth of these plants compared to mutant plants lacking a functional Mn-SOD gene (McKersie et al. 1996). In sacred lotus germinating embryos, chilling stress strongly increased the expression of Mn-SOD gene but was strongly inhibited under heat shock (Li et al. 2009). High temperature causes cell damage through increased ROS levels due to a decline in antioxidant activity in the stressed tissues (Mittler et al. 2004), and tolerance is based in part on an increase in this protective system (Jha et al. 2014).Under a heat shock treatment of two tomato varieties with different thermotolerance, the sensitive variety cv. Amalia presented a higher decrease in Mn-SOD activity than the tolerant one Nagcarlan, pointing to this isoenzyme as implicated in the tolerance of tomato to heat treatments (Camejo et al. 2006) (Fig. 4). It was also shown that an acclimation period to moderate temperatures in tomato cv. Amalia plants induced the Mn-SOD activity when plants were subjected to the heat shock treatment (Camejo et al. 2007).

The role of ASC–GSH cycle enzymes in cold or heat stress is evident from increases in MDHAR activity which contributed to chilling tolerance in tomato fruits (Stevens et al. 2008). Changes in the GR isoform population as a result of chilling stress have been reported in *Arabidopsis* and pea plants as well as in cucumbers, in which the enhancement of total activity was due to the preferential induction of some of the isoforms (Kubo et al. 1993; Edwards et al. 1994; Lee and Lee 2000). In cucumber plants under heat stress, the activities of mitochondrial, cytosolic, and chloroplastic APX isoenzymes increased after an initial decline (Song et al. 2005) (Fig. 4). Also, studies on heat-acclimated turfgrass have revealed reduced ROS production probably due to enhanced ascorbate and glutathione synthesis (Xu et al. 2006).

Related to redox proteins, little information is reported about their response to changes in temperature. As an example, mitochondrial PrxIIF transcript and protein levels have been shown to increase in leaves of pea and *Arabidopsis* plants under cold stress (Barranco-Medina et al. 2007; Gama et al. 2007). In rice, a Trx*h* OsTrx23 which negatively regulates the activity of kinases OsMPK3 and OsMPK6 is known to be induced after chilling stress (Xie et al. 2009). This induction is coincident with the decline in *OsMPK3* expression and activity. H_2O_2 produced under the stress conditions has been implicated in the negative regulation of the activity and function of this Trx*h*, in this way decreasing its inhibitory action on the kinases. The overexpression of chloroplastic AtTDX in *Arabidopsis* plants conferred more resistance to heat shock stress (Sanz-Barrio et al. 2012) but the role of Trx*o*1 under extreme temperature stress situations is not reported. Related to Srx, an induction of the chloroplast transcript level has been reported in *Arabidopsis* plants responding to cold stress treatment (Liu et al. 2006).

However, although the signaling molecules involved in this response are under investigation, research on this line is imperative to try to enhance antioxidant production to better cope with unfavorable temperature conditions (Dat et al. 1998; Wahid et al. 2007).

6 Conclusions and Prospectives

A major challenge in studying stress responses in plants is to understand how different signaling pathways are activated by similar stimuli at subcellular levels. It has been suggested that, under abiotic and biotic stress conditions, cellular fate depends on several factors, among which ROS and NO play a pivotal role. In this context, timing, kinds, and amounts of ROS production in mitochondria as well as the interactions of ROS/NO are all important factors in determining cellular response. The involvement of the cross talk between ROS/NO in the regulation of protein activity and function and the effect on signaling affecting gene transcription in response to different stress begin to be elucidated. However, we know little about the actual flux of $O_2^{\bullet-}$ and NO within mitochondria in vivo, about how their changes under different physiological and stress conditions, or about their quantitative importance relative to other sources of ROS/NO under these circumstances. More work is required to develop techniques to measure mitochondrial ROS and

NO production in vivo. Such methods are essential to understand the role of mitochondrial $O_2^{\bullet -}$, H_2O_2, and NO production in redox signaling and in processes driving oxidative damage under normal and stress conditions.

An understanding of the subcellular GSH and ASC changes can be critical to our knowledge of redox homeostasis in plant cells under physiological and stress conditions. Such information can be also useful to clarify the importance of both antioxidants as protection mechanisms during abiotic stress and as redox signals.

In this context, there is a need to progress on the molecular identification of the GSH transporters. Despite their importance in glutathione homeostasis, the identification and characterization of mitochondrial glutathione transporters in plants is still not clear and at times controversial. In order to achieve further progress in this field, research on the combination and progression of current methods available for the detection of subcellular glutathione will be also necessary to obtain a combined measurement of the actual GSH concentration and the redox state in each cell compartment. Future research on ASC should be similarly focused on.

Moreover, an induced expression of mitochondrial antioxidant defense genes is usually crucial in determining the ability of a plant to cope with abiotic stress although the molecular mechanisms involved in the regulation of this induction remain unrevealed. In mitochondria, AOX is an important factor in maintaining redox balance and in alleviating ROS formation particularly during stress. APX in the ascorbate–glutathione cycle and PrxII F in the Trx/Prx/Srx system contribute to the redox control by means of the reduction of hydrogen peroxide produced by the Mn-SOD activity. However, changes at a transcript level did not usually correlate well with changes in protein responsive to stress, and posttranscriptional mechanisms are believed to play an important role in defining the mitochondrial stress response. While there has been important progress on the signaling role of redox state of thiol in Trxs and Prxs through posttranslational modifications like protein oxidation, S-glutathionylation, and S-nitrosylation/denitrosylation, many challenges remain about the thiol specificity, when and how they are coordinated to allow specific proteins to respond, and their repercussion in the regulation of target proteins and/or specific genes (Fig. 4). Likewise, as far as we know, the specific function, protein–protein interaction, and redox-network implication for the mitochondrial Trxs are far from being elucidated under abiotic stress. Further studies seeking to identify functional targets for Trx*o*1 in the mitochondria are needed to learn more about new physiological roles of this Trx*o*1 in plant cell. Thus, the interaction between both systems, the ASC–GSH cycle and the Trx–Prx–Srx, needs to be investigated in depth as well as that relative to the functional links between AOX and Trx*o*1. All these studies may help to unravel the role of ROS, RNS, and mitochondrial antioxidant proteins as redox sensors and in the signaling process under normal and stress conditions.

Acknowledgments This work was supported by MICINN (BFU2014-52452-P) and Séneca Foundation, Murcia, Spain (04553/GERM/06). The authors apologize to the scientists that are not cited because of space limitation and thank Phillip Thomas for correction of the written English in the manuscript.

References

Abogadallah GM (2010) Antioxidative defense under salt stress. Plant Signal Behav 5:369–374

Agrawal GK, Bourguignon J, Rolland N, Ephritikhine G, Ferro M, Jaquinod M, Alexiou KG, Chardot T, Chakraborty N, Jolivet P, Doonan JH, Rakwal R (2011) Plant organelle proteomics: collaborating for optimal cell function. Mass Spectrom Rev 30:772–853

Andronis EA, Roubelakis-Angelakis KA (2010) Short-term salinity stress in tobacco leads to the onset of animal-like PCD hallmarks in planta in contrast to long-term stress. Planta 231:437–448

Armstrong AF, Badger MR, Day DA, Barthet MM, Smith PMC, Millar AH, Whelan J, Atkin OK (2008) Dynamic changes in the mitochondrial electron transport chain underpinning cold acclimation of leaf respiration. Plant Cell Environ 31:1156–1169

Arrigoni O, Dipierro S, Borranccino G (1981) Ascorbate free radical reductase: a key enzyme of the ascorbic acid system. FEBS Lett 125:242–244

Asada K (1999) The water-water cycle in chloroplast: scavenging of active oxygen and dissipation of excess photons. Annu Rev Plant Physiol Plant Mol Biol 50:601–639

Ashraf M (2009) Biotechnological approach of improving plant salt tolerance using antioxidants as markers. Biotechnol Adv 27:84–93

Atkin OK, Macherel D (2009) The crucial role of plant mitochondria in orchestrating drought tolerance. Ann Bot 103:581–597

Bachhawat AK, Thakur A, Kaur J, Zulkifli M (2013) Glutathione transporters. Biochim Biophys Acta 1830:3154–3164

Barranco-Medina S, López-Jaramillo FJ, Bernier-Villamor L, Sevilla F, Lázaro JJ (2006) Cloning, overexpression, purification and preliminary crystallographic studies of a mitochondrial type II peroxiredoxin from *Pisum sativum*. Acta Crystallogr F 62:696–698

Barranco-Medina S, Krell T, Finkemeier I, Sevilla F, Lazaro JJ, Dietz KJ (2007) Biochemical and molecular characterization of the mitochondrial peroxiredoxin PsPrxII F from *Pisum sativum*. Plant Physiol Biochem 45:729–739

Barranco-Medina S, Krell T, Bernier-Villamor L, Sevilla F, Lázaro JJ, Dietz KJ (2008a) Hexameric oligomerization of mitochondrial peroxiredoxin PrxIIF and formation of an ultrahigh affinity complex with its electron donor thioredoxin Trx-o. J Exp Bot 59:3259–3269

Barranco-Medina S, Kakorin S, Lázaro JJ, Dietz KJ (2008b) Thermodynamics of the dimmer-decamer transition of reduced human and plant 2-Cys peroxiredoxin. Biochemistry 47:7196–7204

Bartoli CG, Pastori GM, Foyer CH (2000) Ascorbate biosynthesis in mitochondria is linked to the electron transport chain between complexes III and IV. Plant Physiol 123:335–344

Bartoli CG, Gómez F, Martínez DE, Guiamet JJ (2004) Mitochondria are the main target for oxidative damage in leaves of wheat (*Triticum aestivum* L.). J Exp Bot 55:1663–1669

Bartoli CG, Guiamet JJ, Kiddle G, Pastori GM, Cagno R, Theodoulou FL, Foyer CH (2005) Ascorbate content of wheat leaves is not determined by maximal L-galactono-1,4-lactone dehydrogenase (GalLDH) activity under drought stress. Plant Cell Environ 28:1073–1081

Beer SM, Taylor ER, Brown SE, Dahm CC, Costa NJ, Runswick MJ, Murphy MP (2004) Glutaredoxin 2 catalyzes the reversible oxidation and glutathionylation of mitochondrial membrane thiol proteins: implications for mitochondrial redox regulation and antioxidant defense. J Biol Chem 279:47939–47951

Biteau B, Labarre J, Toledano MB (2003) ATP-dependent reduction of cysteine-sulphinic acid by *S. cerevisiae* sulphiredoxin. Nature 425:980–984

Boyer JS (1982) Plant productivity and environment. Science 218:443–448

Camejo D, Jiménez A, Alarcón JJ, Torres W, Gómez JM, Sevilla F (2006) Changes in photosynthetic parameters and antioxidant activities during heat shock treatment in tomato plants. Funct Plant Biol 33:177–187

Camejo D, Martí MC, Nicolás E, Alarcón JJ, Jiménez A, Sevilla F (2007) Response of superoxide dismutase isoenzymes in tomato plants (*Lycopersicon esculentum*) during thermo-acclimation of the photosynthetic apparatus. Physiol Planta 131:367–2007

Camejo D, Romero-Puertas MC, Rodríguez-Serrano M, Sandalio LM, Lázaro JJ, Jiménez A, Sevilla F (2013) Salinity-induced changes in S-nitrosylation of pea mitochondrial proteins. J Proteom 79:87–99

Camejo D, Ortiz-Espín A, Lázaro JJ, Romero-Puertas MC, Lázaro-Payo A, Sevilla F, Jiménez A (2015) Functional and structural changes in plant mitochondrial PrxIIF caused by NO. J Proteom 119:112–125

Campbell C, Atkinson L, Zaragoza-Castells J, Lundmark M, Atkin O, Hurry V (2007) Acclimation of photosynthesis and respiration is asynchronous in response to changes in temperature regardless of plant functional group. New Phytol 176:375–389

Chan KX, Wirtz M, Phua SY, Estavillo GM, Pogson BJ (2013) Balancing metabolites in drought: the sulphur assimilation conundrum. Trends Plant Sci 18:18–29

Chang-Quan W, Rui-Chang L (2008) Enhancement of superoxide dismutase activity in the leaves of white clover (*Trifolium repens* L.) in response to polyethylene glycol-induced water stress. Acta Physiol Planta 30:841–847

Chew O, Whelan J, Millar AH (2003) Molecular definition of the ascorbate-glutathione cycle in Arabidopsis mitochondria reveals dual targeting of antioxidant defenses in plants. J Biol Chem 278:46869–46877

Clifton R, Millar AH, Whelan J (2006) Alternative oxidases in Arabidopsis: a comparative analysis of differential expression in the gene family provides new insights into function of non-phosphorylating bypasses. Biophys Biochem Acta 1757:730–741

Collins Y, Chouchani ET, James AM, Menger KE, Cochemé HM, Murphy MP (2012) Mitochondrial redox signaling at a glance. J Cell Sci 125:801–806

Corpas FJ, Alché JD, Barroso JB (2013) Current overview of S-nitrosoglutathione (GSNO) in higher plants. Front Plant Sci 4

Cramer GR, Urano K, Delrot S, Pezzotti M, Shinozaki K (2011) Effects of abiotic stress on plants: a systems biology perspective. BMC Plant Biol 11:163

Cvetkovska M, Vanlerberghe GC (2012) Alternative oxidase modulates leaf mitochondrial concentrations of superoxide and nitric oxide. New Phytol 195:32–39

Dat J, Foyer C, Scott I (1998) Change in salicylic acid and antioxidants during induced thermo tolerance in mustard seedlings. Plant Physiol 118:1455–1461

Day DA, Krab K, Lambers H, Moore AL, Siedow JN, Wagner AM, Wiskich JT (1996) The cyanide-resistant oxidase: to inhibit or not to inhibit, that is the question. Plant Physiol 110:1–2

del Río LA (2011) Peroxisomes as a cellular source of reactive nitrogen species signal molecules. Arch Biochem Biophys 506:1–11

del Río LA, Pastori GM, Palma JM, Sandalio LM, Sevilla F, Corpas FJ, Jiménez A, López-Huertas E, Hernández JA (1998) The activated oxygen role of peroxisomes in senescence. Plant Physiol 116:1195–1200

del Río LA, Corpas FJ, Sandalio LM, Palma JM, Gómez M, Barroso JB (2002) Reactive oxygen species, antioxidant systems and nitric oxide in peroxisomes. J Exp Bot 53:1255–1272

Dietz KJ (2011) Peroxiredoxins in plants and cyanobacteria. Antioxid Redox Signal 15:1129–1159

Edwards EA, Rawsthorne S, Mullineux PM (1990) Subcellular distribution of multiple forms of glutathione reductase in leaves of pea (*Pisum sativum* L.). Planta 180:278–284

Edwards E, Enard C, Creissen GP, Mullineaux PM (1994) Synthesis and properties of glutathione reductase in stressed peas. Planta 192:137–143

Elhafez D, Murcha MW, Clifton R, Soole KL, Day DA, Whelan J (2005) Characterisation of mitochondrial alternative NAD(P)H dehydrogenases in Arabidopsis: intraorganelle location and expression. Plant Cell Physiol 47:43–54

Eltayeb AE, Kawano N, Badawi GH, Kaminaka H, Sanekata T, Morishima I, Shibahara T, Inanaga S, Tanaka K (2006) Enhanced tolerance to ozone and drought stresses in transgenic tobacco overexpressing dehydroascorbate reductase in cytosol. Physiol Planta 127:57–65

Eyidogan F, Oz MT (2005) Effect of salinity on antioxidant responses of chickpea seedlings. Acta Physiol Plant 29:485–493

Fares A, Rossignol M, Peltier JB (2011) Proteomics investigation of endogenous S-nitrosylation in Arabidopsis. Biochem Biophys Res Commun 416:331–336

Farmer EE, Mueller MJ (2013) ROS-mediated lipid peroxidation and RES-activated signaling. Annu Rev Plant Biol 64:429–450

Fernández-García N, Martí MC, Jiménez A, Sevilla F, Olmos E (2009) Sub-cellular distribution of glutathione in an Arabidopsis mutant (*vtc1*) deficient in ascorbate. J Plant Physiol 166:2004–2012

Filipovic MR, Stanic D, Raicevic S, Spasic M, Niketic V (2007) Consequences of MnSOD interactions with nitric oxide: nitric oxide dismutation and the generation of peroxynitrite and hydrogen peroxide. Free Radic Res 4:62–72

Filippou P, Antoniou C, Fotopoulos V (2011) Effect of drought and rewatering on the cellular status and antioxidant response of *Medicago truncatula* plants. Plant Signal Behav 6:270–277

Finkemeier I, Goodman M, Lankemeyer P, Kandlbinder A, Sweetlove LJ, Dietz KJ (2005) The mitochondrial type II peroxiredoxin F is essential for redox homeostasis and root growth of *Arabidopsis thaliana* under stress. J Biol Chem 280:12168–12180

Fiorani F, Umbach AL, Siedow JN (2005) The alternative oxidase of plant mitochondria is involved in the acclimation of shoot growth at low temperature. A study of Arabidopsis AOX1a transgenic plants. Plant Physiol 139:1795–1805

Flexas J, Galmes J, Ribas-Carbo M, Medrano H (2005) The effects of water stress on plant respiration. In: Lambers H, Ribas-Carbó M (eds) Plant respiration: from cell to ecosystem. Springer, Berlin

Flexas J, Bota J, Galmes J, Medrano H, Ribas-Carbo M (2006) Keeping a positive carbon balance under adverse conditions: responses of photosynthesis and respiration to water stress. Physiol Planta 127:343–352

Florez-Sarasa ID, Bouma TJ, Medrano H, Azcon-Bieto J, Ribas-Carbó M (2007) Contribution of the cytochrome and alternative pathways to growth respiration and maintenance respiration in *Arabidopsis thaliana*. Physiol Planta 129:143–151

Foyer CH, Halliwell B (1976) Presence of glutathione and glutathione reductase in chloroplasts: a proposed role in ascorbic acid metabolism. Planta 133:21–25

Foyer CH, Noctor G (2009) Redox regulation in photosynthetic organisms: signaling, acclimation, and practical implications. Antioxid Redox Signal 11:861–905

Foyer CH, Noctor G (2011) Ascorbate and glutathione: the heart of the redox hub. Plant Physiol 155:2–18

Foyer CH, Noctor G (2013) Redox signaling in plants. Antioxid Redox Signal 18:2087–2090

Fröhlich A, Durner J (2011) The hunt for plant nitric oxide synthase (NOS): is one really needed? Plant Sci 181:401–440

Galle A, Florez-Sarasa I, Thameur A, de Paepe R, Flexas J, Ribas-Carbó M (2010) Effects of drought stress and subsequent rewatering on photosynthetic and respiratory pathways in *Nicotiana sylvestris* wild type and the mitochondrial complex I-deficient CMSII mutant. J Exp Bot 61:765–775

Gama O, Keech F, Eymery F, Finkemeier I, Gelhaye E, Gardeström P, Dietz KJ, Rey P, Jacquot JP, Rouhier N (2007) The mitochondrial type II peroxiredoxin from poplar. Physiol Plant 129:196–206

Gelhaye E, Rouhier N, Gerard J, Jolivet Y, Gualberto J, Navrot N, Ohlsson P, Wingsle G, Hirasawa M, Knaff DB, Wang H, Dizengremei P, Meyer Y, Jacquot JP (2004) A specific form of thioredoxin h occurs in plant mitochondria and regulates the alternative oxidase. Proc Natl Acad Sci USA 101:14545–14550

Gelhaye E, Rouhier N, Navrot N, Jaquot JP (2005) The plant thioredoxin system. Cell Mol Life Sci 62:24–35

Giraud E, Ho LHM, Clifton R, Carroll A, Estavillo G, Tan YF, Howell KA, Ivanova A, Pogson BJ, Millar AH, Whelan J (2008) The absence of ALTERNATIVE OXIDASE1a in Arabidopsis results in acute sensitivity to combined light and drought stress. Plant Physiol 147:595–610

Gómez JM, Hernández JA, Jiménez A, del Río LA, Sevilla F (1999) Differential response of antioxidative enzymes of chloroplasts and mitochondria to long-term NaCl stress of pea plants. Free Radic Res 31:S11–S18

Gómez JM, Jiménez A, Olmos E, Sevilla F (2004) Location and effects of long-term NaCl stress on superoxide dismutase and ascorbate peroxidase isoenzymes of pea (*Pisum sativum* cv. Puget) chloroplasts. J Exp Bot 55:119–130

González-Meler MA, Ribas-Carbó M, Giles L, Siedow JN (1999) The effect of growth and measurement temperature on the activity of the alternative respiratory pathway. Plant Physiol 120:765–772

Greetham D, Kritsiligkou P, Watkins RH, Carter Z, Parkin J, Grant CM (2013) Oxidation of the yeast mitochondrial thioredoxin promotes cell death. Antioxid Redox Sign 18:376–385

Gueta-Dahan Y, Yaniv Z, Zilinskas BA, Ben-Hayyim G (1997) Salt and oxidative stress: similar and specific responses and their relation to salt tolerance in citrus. Planta 203:460–469

Gupta KJ, Igamberdiev AU, Manjunatha G, Segu S, Moran JF, Neelawarne B, Bauwe H, Kaiser WM (2011) The emerging roles of nitric oxide (NO) in plant mitochondria. Plant Sci 181:520–526

Guy RD, Vanlerberghe GC (2005) Partitioning of respiratory electrons in the dark in leaves of transgenic tobacco with modified levels of alternative oxidase. Physiol Planta 125:171–180

Hasanuzzaman M, Vahar K, Alam MM, Roychowdhury R, Fujita M (2013) Physiological, biochemical, and molecular mechanisms of heat stress tolerance in plants. Int J Mol Sci 14:9643–9684

Hasegawa PM, Bressan RA, Zhu JK, Bohnert HJ (2000) Plant cellular and molecular responses to high salinity. Annu Rev Plant Biol 51:463–499

Hernández JA, Corpas FJ, Gómez M, del Río LA, Sevilla F (1993) Salt-induced oxidative stress mediated by activated oxygen species in pea leaf mitochondria. Physiol Planta 89:103–110

Hernández JA, Jiménez A, Mullineaux PM, Sevilla F (2000) Tolerance of pea (*Pisum sativum* L.) to long-term salt stress is associated with induction of antioxidant defences. Plant Cell Environ 23:853–862

Hernández JA, Ferrer MA, Jiménez A, Barceló AR, Sevilla F (2001) Antioxidant systems and $O_2^{\cdot-}/H_2O_2$ production in the apoplast of pea leaves. Its relation with salt-induced necrotic lesions in minor veins. Plant Physiol 127:817–831

Hernández I, Cela J, Alegre L (2013) Antioxidant defenses against drought stress. In: Aroca R (ed) Plant responses to drought stress. Springer, Berlin

Horling F, König J, Dietz KJ (2002) Type II peroxiredoxin C, a member of the peroxiredoxin family of *Arabidopsis thaliana*: its expression and activity in comparison with other peroxiredoxins. Plant Physiol Biochem 40:491–499

Horling F, Lamkemeyer P, König J, Finkemeier I, Kandlbinder A, Baier M, Dietz KJ (2003) Divergent light, ascorbate, and oxidative stress-dependent regulation of expression of the peroxiredoxin gene family in Arabidopsis. Plant Physiol 131:317–325

Huerta-Ocampo JA, Briones-Cerecero EP, Mendoza-Hernández G, De León-Rodríguez A, Barba de la Rosa AP (2009) Proteomic analysis of Amaranth (*Amaranthus hypochondriacus* L.) leaves under drought stress. Int J Plant Sci 170:990–998

Igamberdiev AU, Ratcliffe RG, Gupta KJ (2014) Plant mitochondria: source and target for nitric oxide. Mitochondrion 19:329–333

Iglesias-Baena I, Barranco-Medina S, Lázaro-Payo A, López-Jaramillo FJ, Sevilla F, Lázaro JJ (2010) Characterization of plant sulfiredoxin and role of sulphinic form of 2-Cys peroxiredoxin. J Exp Bot 6:1509–1521

Iglesias-Baena I, Barranco-Medina S, Sevilla F, Lázaro JJ (2011) The dual targeted plant sulfiredoxin retroreduces the sulfinic form of atypical mitochondrial peroxiredoxin. Plant Physiol 155:944–955

Jacoby RP, Millar AH, Taylor NL (2010) Wheat mitochondrial proteomes provide new links between antioxidant defense and plant salinity tolerance. J Proteom Res 9:6595–6604

Jensen PK (1966) Antimycin-insensitive oxidation of succinate and reduced nicotinamide-adenine dinucleotide in electron-transport particles. I. pH dependency and hydrogen peroxide formation. Biochim Biophys Acta 122:157–166

Jha UC, Bohra A, Singh NP (2014) Heat stress in crop plants: its nature, impacts and integrated breeding strategies to improve heat tolerance. Plant Breed 133:679–701

Jiang Y, Yang B, Harris NS, Deyholos MK (2007) Comparative proteomic analysis of NaCl stress-responsive proteins in Arabidopsis roots. J Exp Bot 58:3591–3607

Jiménez A, Hernández JA, del Río LA, Sevilla F (1997) Evidence for the presence of the ascorbate-glutathione cycle in mitochondria and peroxisomes of pea leaves. Plant Physiol 114:275–284

Jiménez A, Hernández JA, Ros Barceló A, Sandalio LM, Del Río LA, Sevilla F (1998a) Characterization of mitochondrial and peroxisomal ascorbate peroxidase of pea leaves. Physiol Planta 104:687–692

Jiménez A, Hernández JA, Pastori G, del Río LA, Sevilla F (1998b) Role of the ascorbate-glutathione cycle of mitochondria and peroxisomes in the senescence of pea leaves. Plant Physiol 118:1327–1335

Kerchev PI, Pellny TK, Vivancos PD, Kiddle G, Hedden P, Driscoll S, Vanacker H, Verrier P, Hancock RD, Foyer CH (2011) The transcription factor ABI4 Is required for the ascorbic acid-dependent regulation of growth and regulation of jasmonate-dependent defense signaling pathways in Arabidopsis. Plant Cell 23:3319–3334

Kido Y, Shiba T, Inaoka DK, Sakamoto K, Nara T, Aoki T, Honma T, Inoue M, Matsuoka S, Moore A, Harada S, Kita K (2010) Crystallization and preliminary crystallographic analysis of cyanide-insensitive alternative oxidase from *Trypanosoma brucei*. Acta Crystallogr F Struct Biol Cryst Commun 66:275–278

Kocsy G, Tari I, Vankova R, Zechmann B, Gulyas Z, Poor P, Galiba G (2013) Redox control of plant growth and development. Plant Sci 211:77–91

Koffler BE, Luschin-Ebengreuth N, Stabentheiner E, Müller M, Zechman B (2014) Compartment specific response of antioxidants to drought stress in Arabidopsis. Plant Sci 227:133–144

Koprivova A, Mugford ST, Kopriva S (2010) Arabidopsis root growth dependence on glutathione is linked to auxin transport. Plant Cell Rep 29:1157–1167

Kubo A, Sano A, Saji H, Tanaka K, Kondo N, Tanaka K (1993) Primary structure and properties of glutathione reductase from *Arabidopsis thaliana*. Plant Cell Physiol 34:1259–1266

Laloi C, Rayapuram N, Chartier Y, Grienenberger JM, Bonnard G, Meyer Y (2001) Identification and characterization of a mitochondrial thioredoxin system in plants. Proc Natl Acad Sci USA 98:14144–14149

Lambers H, Robinson SA, Ribas-Carbo M (2005) Regulation of respiration in vivo. In: Lambers H, Ribas-Carbo M (eds) Plant respiration: from cell to ecosystem, vol 18, Advances in photosynthesis and respiration series. Springer, Amsterdam

Lázaro JJ, Jiménez A, Camejo D, Iglesias-Baena I, Martí MC, Lázaro-Payo A, Barranco-Medina S, Sevilla F (2013) Dissecting the integrative antioxidant and redox systems in plant mitochondria. Effect of stress and *S*-nitrosylation. Front Plant Sci 4:460

Lee DH, Lee CB (2000) Chilling stress-induced changes of antioxidant enzymes in the leaves of cucumber: in gel enzyme activity assays. Plant Sci 159:75–85

Leitner M, Vandelle E, Gaupels F, Bellin D, Delledonne M (2009) NO signals in the haze. Nitric oxide signalling in plant defence. Curr Opin Plant Biol 12:451–458

Li W, Qi L, Lin X, Chen H, Ma Z, Wu K, Huang S (2009) The expression of manganese superoxide dismutase gene from *Nelumbo nucifera* responds strongly to chilling and oxidative stresses. J Integr Plant Biol 51:279–286

Lin KH, Pu SF (2010) Tissue- and genotype-specific ascorbate peroxidase expression in sweet potato in response to salt stress. Biol Plant 54:664–670

Liu XP, Liu XY, Zhang J, Xia ZL, Liu X, Qin HJ, Wang DW (2006) Molecular and functional characterization of sulfiredoxin homologs from higher plants. Cell Res 16:287–296

Lobell DB, Schlenker W, Costa-Roberts J (2011) Climate trends and global crop production since 1980. Science 333:616–620

Lopert P, Day BJ, Patel M (2012) Thioredoxin reductase deficiency potentiates oxidative stress, mitochondrial dysfunction and cell death in dopaminergic cells. PLOS One 7:e50683

Maathuis FJM, Ahmad I, Patishtan J (2014) Regulation of Na^+ fluxes in plants. Front Plant Sci 5:467

MacFarlane C, Hansen LD, Florez-Sarasa I, Ribas-Carbo M (2009) Plant mitochondria electron partitioning is independent of short-term temperature changes. Plant Cell Environ 32:585–591

Mahajan S, Tuteja N (2005) Cold, salinity and drought stresses: an overview. Arch Biochem Biophys 444:139–158

Malone M (1993) Rapid inhibition of leaf growth by root cooling in wheat: kinetics and mechanism. J Exp Bot 44:1663–1669

Marques AT, Santos SP, Rosa MG, Rodrigues MAA, Abreu JA, Frazao C, Romao CV (2014) Expression, purification and crystallization of MnSOD from *Arabidopsis thaliana*. Acta Crystallogr 70:669–672

Martí MC, Olmos E, Calvete JJ, Díaz I, Barranco-Medina S, Whelan J, Lázaro JJ, Sevilla F, Jiménez A (2009) Mitochondrial and nuclear localization of a novel pea thioredoxin: identification of its mitochondrial target proteins. Plant Physiol 150:646–657

Martí MC, Florez-Sarasa, Camejo D, Ribas-Carbó M, Lázaro JJ, Sevilla F, Jiménez A (2011) Response of the mitochondrial antioxidant redox system and respiration to salinity in pea plants. J Exp Bot 62:3863–3874

Martí MC, Florez-Sarasa I, Camejo D, Pallol B, Ortiz A, Ribas-Carbó M, Jiménez A, Sevilla F (2012) Response of mitochondrial antioxidant system and respiratory pathways to reactive nitrogen species in pea leaves. Physiol Planta 147:194–206

Maxwell DP, Wang Y, McIntosh L (1999) The alternative oxidase lowers mitochondrial reactive oxygen production in plant cells. Proc Natl Acad Sci USA 96:8271–8276

McDonald AE, Vanlerberghe GC (2006) Origins, evolutionary history, and taxonomic distribution of alternative oxidase and plastoquinol terminal oxidase. Comp Biochem Physiol Part D1:357–364

McDonald AE, Vanlerberghe GC, Staples JF (2009) Alternative oxidase in animals: unique characteristics and taxonomic distribution. J Exp Biol 212:2627–2634

McKersie BD, Bowley SR, Harjanto E, Leprince O (1996) Water-deficit tolerance and field performance of transgenic Alfalfa overexpressing superoxide dismutase. Plant Physiol 111:1177–1181

Metwally A, Safronova VI, Belimov AA, Dietz KJ (2005) Genotypic variation of the response to cadmium toxicity in *Pisum sativum* L. J Exp Bot 56:167–178

Meyer Y, Belin C, Delorme-Hinoux V, Reichheld JP, Riondet C, Meyer Y (2012) Thioredoxin and glutaredoxin systems in plants: molecular mechanisms, crosstalks, and functional significance. Antioxid Redox Signal 17:1124–1160

Millar AH, Mittova V, Kiddle G, Heazlewood JL, Bartoli CG, Theodoulou FL, Foyer CH (2003) Control of ascorbate synthesis by respiration and its implications for stress responses. Plant Physiol 133:443–447

Millar AH, Whelan J, Soole KL, Day DA (2011) Organization and regulation of mitochondrial respiration in plants. Annu Rev Plant Biol 62:79–104

Millenaar FF, Lambers H (2003) The alternative oxidase: in vivo regulation and function. Plant Biol 5:2–15

Mittler R, Vanderauwera S, Gollery M, Van Breusegem F (2004) The reactive oxygen gene network of plants. Trends Plant Sci 9:490–498

Mittler R, Vanderauwera S, Suzuki N, Miller G, Tognetti VB, Vandepoele K, Gollery M, Shulaev V, Van BF (2011) ROS signaling: the new wave? Trends Plant Sci 16:300–309

Mittova V, Tal M, Volokita M, Guy M (2003) Up-regulation of the leaf mitochondrial and peroxisomal antioxidative systems in response to salt-induced oxidative stress in the wild salt-tolerant tomato species *Lycopersicon pennellii*. Plant Cell Environ 26:845–856

Moller IM (2001) Plant mitochondria and oxidative stress: electron transport, NADPH turnover and metabolism of reactive oxygen species. Annu Rev Plant Physiol Plant Mol Biol 52:561–591

Moore AL, Shiba T, Young L, Harada S, Kita K, Ito K (2013) Unraveling the heater: new insights into the structure of the alternative oxidase. Annu Rev Plant Biol 64:637–663

Moran JF, Becana M, Iturbe-Ormaetxe I, Frechillam S, Klucas RV, Aparicio-Tejo P (1994) Drought induces oxidative stress in pea plants. Planta 194:346–352

Morgan MJ, Lehmann M, Schwarzländer M, Baxter CJ, Sienkiewicz-Porzucek A, Williams TCR, Schauer N, Fernie AR, Fricker MD, Ratcliffe RG, Sweetlove LJ, Finkemeier I (2008) Decrease in manganese superoxide dismutase leads to reduced root growth and affects tricarboxylic acid cycle flux and mitochondrial redox homeostasis. Plant Physiol 147:101–114

Mullineaux PM, Rausch T (2005) Glutathione, photosynthesis and the redox regulation of stress-responsive gene expression. Photosynth Res 86:459–474

Munné-Bosch S, Queval G, Foyer CH (2013) The impact of global change factors on redox signaling underpinning stress tolerance. Plant Physiol 161:5–19

Munns R (2002) Comparative physiology of salt and water stress. Plant Cell Environ 25:239–250

Murphy MP (2009) How mitochondria produce reactive oxygen species. Biochem J 417:1–13

Noctor G, De Paepe R, Foyer CH (2007) Mitochondrial redox biology and homeostasis in plants. Trends Plant Sci 12:125–134

Noctor G, Mhamdi A, Chaouch S, Han Y, Neukermans J, Marquez-Garcia B, Queval G, Foyer CH (2012) Glutathione in plants: an integrated overview. Plant Cell Environ 35:454–484

Olmos E, Hernández JA, Sevilla F, Hellín E (1994) Induction of several antioxidant enzymes in the selection of a salt-tolerant cell-line of *Pisum sativum*. J Plant Physiol 144:594–598

Pan Y, Wu LJ, Yu ZL (2006) Effect of salt and drought stress on antioxidant enzymes activities and SOD isoenzymes of liquorice (*Glycyrrhiza uralensis* Fisch). Plant Growth Regul 49:157–165

Preuss ML, Cameron JC, Berg RH, Jez JM (2014) Immunolocalization of glutathione biosynthesis enzymes in *Arabidopsis thaliana*. Plant Physiol Biochem 75:9–13

Pulido P, Cazalis R, Cejudo FJ (2009) An antioxidant redox system in the nucleus of wheat seed cells suffering oxidative stress. Plant J 57:132–145

Rasmusson AG, Moller IM (1991) NAD(P)H dehydrogenases on the inner surface of the inner mitochondrial membrane studied using inside out submitochondrial particles. Physiol Planta 83:357–365

Rasmusson AG, Wallström SV (2010) Involvement of mitochondria in the control of plant cell NAD(P)H reduction levels. Biochem Soc Trans 38:661–666

Rasmusson AG, Fernie AR, van Dongen JT (2009) Alternative oxidase: a defence against metabolic fluctuations? Physiol Planta 137:371–382

Rautenkranz A, Li L, Machler F, Martinoia E, Oertli JJ (1994) Transport of ascorbic and dehydroascorbic acids across protoplast and vacuole membranes isolated from barley (*Hordeum vulgare* L. cv Gerbel) leaves. Plant Physiol 106:187–193

Rey P, Becuwe N, Barrault MB, Rumeau D, Havaux M, Biteau B, Toledano MB (2007) The *Arabidopsis thaliana* sulfiredoxin is a plastidic cysteine-sulfinic acid reductase involved in the photooxidative stress response. Plant J 49:505–514

Rhoads DM, Subbaiah CC (2007) Mitochondrial retrograde regulation in plants. Mitochondrion 7:177–194

Ribas-Carbó M, Berry JA, Yakir D, Giles L, Robinson SA, Lennon AM, Siedow JN (1995) Electron partitioning between the cytochrome and alternative pathways in plant mitochondria. Plant Physiol 109:829–837

Ribas-Carbó M, Lennon AM, Robinson SA, Giles L, Berry J, Siedow JN (1997) The regulation of the electron partitioning between the cytochrome and alternative pathways in soybean cotyledon and root mitochondria. Plant Physiol 113:903–911

Ribas-Carbó M, Aroca R, Gonzàlez-Meler MA, Irigoyen JJ, Sánchez-Díaz M (2000) The electron partitioning between the cytochrome and alternative respiratory pathways during chilling recovery in two cultivars of maize differing in chilling sensitivity. Plant Physiol 122:199–204

Ribas-Carbó M, Robinson SA, Giles L (2005a) The application of the oxygen-isotope technique to assess respiratory pathway partitioning. In: Lambers H, Ribas-Carbó M (eds) Advances in photosynthesis and respiration. Plant respiration: from cell to ecosystem. Springer, Amsterdam

Ribas-Carbó M, Taylor NL, Giles L, Busquets S, Finnegan PM, Day DA, Lambers H, Medrano H, Berry JA, Flexas J (2005b) Effects of water stress on respiration in soybean leaves. Plant Physiol 139:466–473

Rich PR, Bonner WD Jr (1978) The sites of superoxide anion generation in higher plant mitochondria. Arch Biochem Biophys 188:206–213

Rodríguez-Serrano M, Romero-Puertas MC, Pazmiño DM, Testillano PS, Risueño MC, Del Río LA, Sandalio LM (2009) Cellular response of pea plants to cadmium toxicity: cross talk between reactive oxygen species, nitric oxide, and calcium. Plant Physiol 150:229–243

Romero-Puertas MC, Rodríguez-Serrano M, Sandalio LM (2013) Protein S-nitrosylation in plants under abiotic stress: an overview. Front Plant Sci 20:373

Sandalio LM, Dalurzo HC, Gómez M, Romero-Puertas MC, del Río LA (2001) Cadmium-induced changes in the growth and oxidative metabolism of pea plants. J Exp Bot 52:2115–2126

Sanz-Barrio R, Fernández-San Millán A, Carballeda J, Corral-Martínez P, Seguí-Simarro JM, Farran I (2012) Chaperone-like properties of tobacco plastid thioredoxins f and m. J Exp Bot 63:365–379

Schmitt FJ, Renger G, Friedrich T, Kreslavksi VD, Zharmukhadmedov SK, Los DA, Kuznetsov VV, Allakhverdiev SI (2014) Reactive oxygen species: re-evaluation of generation, monitoring and role in stress-signaling in phototrophic organisms. Biochim Biophys Acta 1837:385–848

Schnaubelt D, Schulz P, Hannah MA, Yocgo RE, Foyer CH (2013) A phenomics approach to the analysis of the influence of glutathione on leaf area and abiotic stress tolerance in *Arabidopsis thaliana*. Front Plant Sci 4:416

Searle SY, Turnbull MH (2011) Seasonal variation of leaf respiration and the alternative pathway in field-grown Populus × canadensis. Physiol Planta 141:332–342

Selote DS, Khanna-Chopra R (2010) Antioxidant response of wheat roots to drought acclimation. Protoplasma 245:153–163

Serrato AJ, Pérez-Ruiz JM, Spínola MC, Cejudo FJ (2004) A novel NADPH thioredoxin reductase, localized in the chloroplast, which deficiency causes hypersensitivity to abiotic stress in *Arabidopsis thaliana*. J Biol Chem 279:43821–43827

Sevilla F, López-Gorgé J, del Río LA (1982) Characterization of a manganese superoxide dismutase from the higher plant *Pisum sativum*. Plant Physiol 70:1321–1326

Sevilla F, Camejo D, Ortiz-Espín A, Calderón A, Lázaro JJ, Jiménez A (2015) Thioredoxin/peroxiredoxin/sulfiredoxin system: current overview on its redox function in plants and regulation by ROS and RNS. J Exp Bot. doi:10.1093/jxb/erv146

Sharma P, Dubey RS (2005) Modulation of nitrate reductase activity in rice seedlings under aluminium toxicity and water stress: role of osmolytes as enzyme protectant. J Plant Physiol 162:854–864

Shiriga K, Sharma R, Kumar K, Yadav SK, Hossain F, Thirunavukkarasu N (2014) Expression pattern of superoxide dismutase under drought stress in maize. Int J Innov Res Sci Eng Technol 3:11333–11337

Song XS, Hu WH, Mao WH, Ogweno JO, Zhou YH, Yu JQ (2005) Response of ascorbate peroxidase isoenzymes and ascorbate regeneration system to abiotic stresses in *Cucumis sativus* L. Plant Physiol Biochem 43:1082–1088

Stevens RG, Creissen GP, Mullineaux PM (1997) Cloning and characterisation of a cytosolic glutathione reductase cDNA from pea (*Pisum sativum* L.) and its expression in response to stress. Plant Mol Biol 35:641–654

Stevens R, Page D, Gouble B, Garchery C, Zamir D, Causse M (2008) Tomato fruit ascorbic acid content is linked with monodehydroascorbate reductase activity and tolerance to chilling stress. Plant Cell Environ 31:1086–1096

Stoimenova M, Igamberdiev AU, Gupta KJ, Hill RD (2007) Nitrite-driven anaerobic ATP synthesis in barley and rice root mitochondria. Planta 226:465–474

Sweetlove LJ, Heazlewood JL, Herald V, Holtzapffel R, Day DA, Leaver CJ, Millar AH (2002) The impact of oxidative stress on Arabidopsis mitochondria. Plant J 32:891–904

Szarka A, Horemans N, Kovacs Z, Gróf P, Mayer M, Bánhegyi G (2007) Dehydroascorbate reduction in plant mitochondria is coupled to the respiratory electron transfer chain. Physiol Planta 129:225–232

Takahama U (2004) Oxidation of vacuolar and apoplastic phenolic substrates by peroxidase: physiological significance of the oxidation reactions. Phytochem Rev 3:207–219

Talla S, Riazunnisa K, Padmavathi L, Sunil B, Rajsheel P, Raghavendra AS (2011) Ascorbic acid is a key participant during the interactions between chloroplasts and mitochondria to optimize photosynthesis and protect against photoinhibition. J Biosci 36:163–173

Tan YF, O'Toole N, Taylor NL, Millar AH (2010) Divalent metal ions in plant mitochondria and their role in interactions with proteins and oxidative stress-induced damage to respiratory function. Plant Physiol 152:747–761

Tan M, Lu J, Zhang A, Hu B, Zhu X, Li W (2011) The distribution and cooperation of antioxidant (iso)enzymes and antioxidants in different subcellular compartments in maize leaves during water stress. J Plant Grow Regul 30:255–271

Taylor NL, Tan YF, Jacoby RP, Millar AH (2009) Abiotic environmental stress induced changes in the *Arabidopsis thaliana* chloroplast, mitochondria and peroxisome proteomes. J Proteom 72:367–378

Trono D, Flagella Z, Laus MN, Di Fonzo N, Pastore D (2004) The uncoupling protein and the potassium channel are activated by hyperosmotic stress in mitochondria from durum wheat seedlings. Plant Cell Environ 27:437–448

Turrens JF (2003) Mitochondrial formation of reactive oxygen species. J Physiol (Lond) 552:335–344

Umbach AL, Fiorani F, Siedow JN (2005) Characterization of transformed Arabidopsis with altered alternative oxidase levels and analysis of effects on reactive oxygen species in tissue. Plant Physiol 139:1806–1820

Umbach AL, Lacey EP, Richter SJ (2009) Temperature-sensitive alternative oxidase protein content and its relationship to floral reflectance in natural *Plantago lanceolata* populations. New Phytol 181:662–671

Ushimaru T, Nakagawa T, Fujioka Y, Daicho K, Naito M, Yamauchi Y, Nonaka H, Amako K, Yamawaki K, Murata N (2006) Transgenic Arabidopsis plants expressing the rice dehydroascorbate reductase gene are resistant to salt stress. J Plant Physiol 163:1179–1184

Van Breusegem F, Slooten L, Stassart JM, Botterman J, Monees T, Van Montagu M, Inze D (1999) Effects of overproduction of tobacco MnSOD in maize chloroplasts on foliar tolerance to cold and oxidative stress. J Exp Bot 50:71–78

Vanlerberghe GC (2013) Alternative oxidase: a mitochondrial respiratory pathway to maintain metabolic and signaling homeostasis during abiotic and biotic stress in plants. Int J Mol Sci 14:6805–6847

Vassileva V, Simova-Stoilova L, Demirevska K, Feller U (2009) Variety-specific response of wheat (*Triticum aestivum* L.) leaf mitochondria to drought stress. J Plant Res 122:445–454

Vidal G, Ribas-Carbó M, Garmier M, Dubertret G, Rasmusson AG, Mathieu C, Foyer CH, De Paepe R (2007) Lack of respiratory chain complex I impairs alternative oxidase engagement and modulates redox signalling during elicitor-induced cell death in tobacco. Plant Cell 19:640–655

Vidigal P, Carvalho R, Amancio S, Carvalho L (2013) Peroxiredoxins are involved in two independent signalling pathways in the abiotic stress protection in *Vitis vinifera*. Biol Planta 57:675–683

Wahid A, Gelani S, Ashraf M, Foolad MR (2007) Heat tolerance in plants: an overview. Environ Exp Bot 61:199–223

Wang YH, Ying Y, Vhen J, Wang X (2004) Transgenic Arabidopsis overexpressing Mn-SOD enhanced salt-tolerance. Plant Sci 167:671–677

Wang FZ, Wang QB, Kwon SY, Kwak SS, Su WA (2005) Enhanced drought tolerance of transgenic rice plants expressing a pea manganese superoxide dismutase. J Plant Physiol 162:465–472

Wilson ID, Neill SJ, Hancock JT (2008) Nitric oxide synthesis and signalling in plants. Plant Cell Environ 31:622–631

Woodson JD, Chory J (2008) Coordination of gene expression between organellar and nuclear genomes. Nat Rev Genet 9:383–395

Wu G, Wilen RW, Robertson AJ, Gusta LV (1999) Isolation, chromosomal localization, and differential expression of mitochondrial manganese superoxide dismutase and chloroplastic copper/zinc superoxide dismutase genes in wheat. Plant Physiol 120:513–520

Xie G, Kato H, Sasaki K, Imai R (2009) A cold-induced thioredoxin h of rice, OsTrx23, negatively regulates kinase activities of OsMPK3 and OsMPK6 *in vitro*. FEBS Lett 583:2734–2738

Xu S, Li J, Zhang X, Wei H, Cui L (2006) Effects of heat acclimation pretreatment on changes of membrane lipid peroxidation, antioxidant metabolites, and ultrastructure of chloroplasts in two cool-season turfgrass species under heat stress. Environ Exp Bot 56:274–285

Yu M, Lamattina L, Spoel SH, Loake GJ (2014) Nitric oxide function in plant biology: a redox cue in deconvolution. New Phytol 202:1142–1156

Zaffagnini M, Bedhomme M, Marchand CH, Morisse S, Trost P, Lemaire SD (2012) Redox regulation in photosynthetic organisms: focus on glutathionylation. Antioxid Redox Signal 16:567–586

Zechmann B (2014) Compartment-specific importance of glutathione during abiotic and biotic stress. Front Plant Sci 5:566

Zechmann B, Müller M (2010) Subcellular compartmentation of glutathione in dicotyledonous plants. Protoplasma 246:15–24

Zechmann B, Stumpe M, Mauch F (2011) Immunocytochemical determination of the subcellular distribution of ascorbate in plants. Planta 233:1–12

ROS as Key Players of Abiotic Stress Responses in Plants

Nobuhiro Suzuki

Contents

Abstract Despite their toxic potential, reactive oxygen species (ROS) play an integral role as signaling molecules in the regulation of a broad range of biological processes such as growth, development, and responses to biotic and/or abiotic stimuli in plants. To some extent, various functions of ROS signaling are attributed to differences in the regulatory mechanisms of respiratory burst oxidase homologs (RBOHs) that are involved in a multitude of different signal transduction pathways activated in assorted tissue and cell types under fluctuating environmental conditions. To acclimate or survive under abiotic stress conditions, plants possess powerful strategies involving systemic signaling, retrograde signaling, and programmed cell death (PCD), in which ROS signals are integrated with other pathways to generate highly coordinated signaling networks. In this chapter, beneficial roles of ROS as signaling molecules in the regulation of abiotic stress responses in plants will be addressed.

N. Suzuki (✉)
Department of Materials and Life Sciences, Faculty of Science and Technology, Sophia University, 7-1 Kioi-cho, Chiyoda-ku, 102-8554 Tokyo, Japan
e-mail: n-suzuki-cs6@sophia.ac.jp

D.K. Gupta et al. (eds.), *Reactive Oxygen Species and Oxidative Damage in Plants Under Stress*, DOI 10.1007/978-3-319-20421-5_3

Keywords Abiotic stress • Hormones • NADPH oxidase • Programmed cell death • Redox signaling • ROS signal • Systemic signaling • Retrograde signaling

Abbreviations

ABA	Abscisic acid
ABI4	ABA-insensitive 4
ACC	1-Aminocyclopropane-1-carboxylic acid
ANAC017	A membrane-bound NAC 017
AOX	Alternative oxidase
APX	Ascorbate peroxidase
BCL2	B-cell lymphoma 2
BR	Brassinosteroid
CDPKs	Calcium-dependent protein kinases
CTR1	Constitutive triple response 1
DPI	Diphenyleneiodonium
EDS1	Enhanced disease susceptibility 1
EIN2	Ethylene-insensitive 2
EEE	Excess excitation energy
GUN1	Genomes uncoupled 1
IAA	Indole-3-acetic acid
JA	Jasmonic acid
LSD1	Lesion simulating disease 1
MAPK	Mitogen-activated protein kinase
NPQ	Non-photochemical quenching
MeJA	Methyl jasmonate
OST1	Open stomata 1
PA	Phosphatidic acid
PCD	Programmed cell death
PAD4	Phytoalexin-deficient 4
PRL1	Pleiotropic response locus 1
PQ	Plastoquinone
RBOH	Respiratory burst oxidase homolog
ROS	Reactive oxygen species
SA	Salicylic acid
SAA	Systemic acquired acclimation
SAR	Systemic acquired resistance
SID2	Salicylic acid induction deficient 2
TPC1	Two-pore channel 1
VPE	Vacuolar processing enzymes

1 Introduction

The reactive oxygen species (ROS) signaling network is highly conserved among aerobic organisms and controls a broad range of biological processes such as growth, development, and responses to biotic and/or abiotic stimuli (Mittler et al. 2011). Although early researches related to ROS metabolism focused on their potential toxicity and the different mechanisms to scavenge them, more recent studies have focused on the roles of ROS as signaling molecules. Why did plants acquire the ability to utilize ROS as signaling molecules during the evolutionary process? The existence of different types of ROS might be an advantage for the fine-tuning of complex signaling networks in cells, because coordinated production of ROS with different properties could at least partially contribute to generation of various signals. Indeed, specificity of ROS signaling was previously indicated by the finding that different sets of genes were upregulated in response to different types of ROS (Gadjev et al. 2006; Suzuki et al. 2011). In addition, almost any changes in cellular homeostasis could lead to alteration of redox state in particular organelles, followed by changes in cellular level of ROS. ROS can be therefore integrated with the activities of many other signaling components such as hormones, Ca^{2+}, and kinases and involved in a number of pathways underlying different biological outcomes (Petrov and Van Breusegem 2012). Mobility of ROS, especially H_2O_2, is also an advantage to act as a signaling molecule. H_2O_2, a relatively long-lived and small ROS, readily penetrates cell membranes and is freely diffusible between cells (Bienert et al. 2006). Diffusion of H_2O_2 that is facilitated by plasma membrane aquaporins could also influence the efficiency and direction of H_2O_2 signaling between cells (Bienert et al. 2007; Dynowski et al. 2008).

To utilize ROS as signaling molecules, nontoxic levels of ROS must be maintained in a delicate balance between production involving enzymatic reactions and the unavoidable production during basic cellular processes and the metabolic counter-process involving ROS scavenging pathways (Mittler et al. 2004). Another possible advantage for using ROS as signaling molecules is the specific localization of their production/scavenging mechanisms in different organelles, cell types, and tissues (Mittler et al. 2004; Miller et al. 2009b; Suzuki et al. 2011). Local increase in ROS production can be limited to particular locations in the cell or plant. Therefore, level of ROS throughout the cell or plant as well as spatial coordination of ROS signals can be tightly regulated. In plants, many antioxidant mechanisms that function in different organelles modulate cellular redox state by detoxifying excess ROS produced during basic biological processes such as photosynthesis and respiration (Miller et al. 2009b). In addition, NADPH oxidases, respiratory burst oxidase homologs (RBOHs), deliberately produce ROS in different cell types, tissues, or stages (Suzuki et al. 2011). The tight regulation of RBOH protein activity makes these enzymes good candidates for the fine-tuning of ROS production in terms of amplitude, duration, and localization in plants (Marino et al. 2012; Gilroy et al. 2014) (Fig. 1).

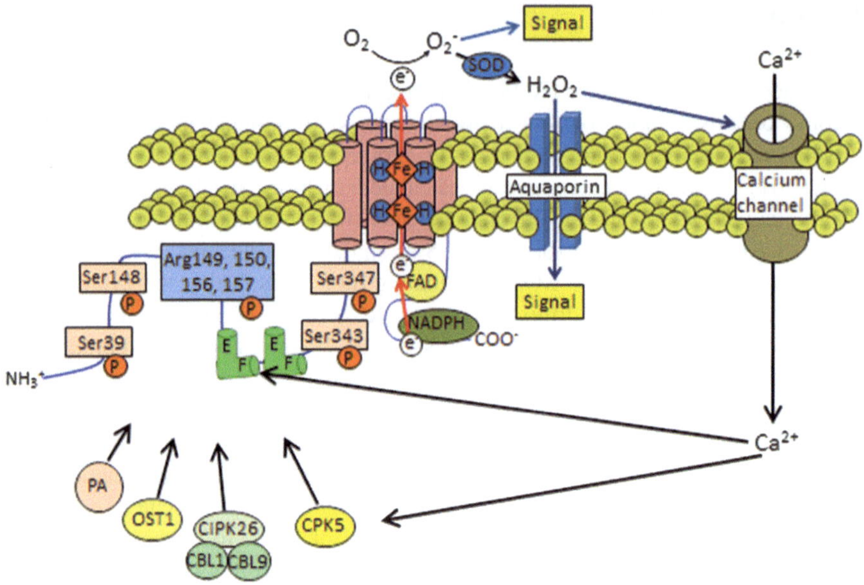

Fig. 1 Regulatory mechanisms of RBOH proteins in *Arabidopsis*. ROS can activate or suppress calcium channels to control the release of Ca^{2+} into cytosol. Ca^{2+} can directly or indirectly regulate the production of ROS by respiratory burst homolog (RBOH) proteins resulting in the generation of superoxide radicals that are dismutated to H_2O_2 spontaneously or via superoxide dismutase (SOD). H_2O_2 diffused into cytosol through aquaporin then functions as a signaling molecule. RBOHs have cytosolic FAD and NADPH-binding domains in the C-terminal region and six conserved transmembrane-spanning domains (*pink cylinders*). The N-terminal domain contains two EF-hand motifs and phosphorylation target sites that are important for activity of RBOHs. Binding of Ca^{2+} to EF-hand motifs is required for activation of RBOHs (Ogasawara et al. 2008; Drerup et al. 2013). Several kinases including OST1 (Sirichandra et al. 2009), CPK5 (Dubiella et al. 2013), and CBL1/9–CIPK26 complexes (Drerup et al. 2013) activate RBOH proteins by phosphorylation of the target sites (Ser or Arg residues). In addition, phosphatidic acid (PA) produced via function of phospholipase Dα1 (PLDα1) was also shown to activate RBOHD (Zhang et al. 2009b). *Abbreviations*: *CBL1* calcineurin B-like protein 1, *CBL9* calcineurin B-like protein 9, *CIPK26* calcineurin B-like interacting protein 26, *CPK5* calmodulin domain protein kinase 5, *OST1* open stomata 1, *FAD* flavin adenine dinucleotide, *NAD* nicotinamide adenine dinucleotide, *PA* phosphatidic acid

Being sessile organisms, plants evolved sophisticated strategies to acclimate to or survive under fluctuating environmental conditions. These strategies involve systemic signaling, retrograde signaling, and programmed cell death (PCD), in which ROS play beneficial roles as signaling molecules (Woodson and Chory 2008; Baxter et al. 2014; Szechynska-Hebda and Karpinski 2013; Petrov et al. 2015). Systemic signaling mechanisms evolutionally improved the ability of plants to alert all remote and unstressed tissues of the plant to the existence of a biotic or abiotic threat and to trigger the activation of resistance or acclimation pathways in these tissues (Suzuki et al. 2013; Szechynska-Hebda et al. 2010). Recent studies have highlighted the importance of rapid systemic responses for acclimation of plants to abiotic stimuli, focusing on temporal–spatial coordination between several players of complex network of cell-to-cell communication, active propagation of ROS wave and calcium wave, hormones, and electric signals (Miller et al. 2009b; Mittler et al. 2011; Baxter et al. 2014; Gilroy et al. 2014). Signaling networks between the organelles employ ROS as second messenger. Alterations in redox state and ROS metabolism in the chloroplast and mitochondria are sources for retrograde signals to regulate nuclear gene expression, which play an important role in the acclimation of plants to environmental stimuli (Rhoads and Subbaiah 2007; Pogson et al. 2008; Woodson and Chory 2008; Shapiguzov et al. 2012; Szechynska-Hebda and Karpinski 2013). In addition, chloroplastic ROS production and photosynthetic functions were recently shown to be regulated by cues perceived by cell wall or apoplastic spaces (Padmanabhan and Dinesh-Kumar 2010), suggesting that chloroplastic retrograde signaling might be a part of a large signaling network involving various cues generated from different organelles, and ROS might be an important mediator that integrates these different signals. PCD is one of the essential strategies for survival of plants that are subjected to severe biotic or abiotic stresses, because only cells that are destined to die can be eliminated to prevent the spreading of damage to the neighboring cells and maintain appropriate metabolic status in the rest of cells and tissues (Petrov et al. 2015). It is an active and genetically controlled process in which cells are selectively eliminated in multistep fashion through the involvement of high concentrations of ROS and specific proteases and nuclease (Petrov et al. 2015).

In this chapter, beneficial roles of ROS as signaling molecules in the regulation of abiotic stress responses of plants will be addressed, especially in the context of ROS-producing mechanisms, integration of ROS signals with other pathways, and plant-specific mechanisms that underlie response of plants to abiotic stimuli.

2 ROS-Generating Pathways and Their Regulatory Mechanisms in Plants

In plants, NADPH oxidases, respiratory burst oxidase homologs (RBOHs), are responsible for the production of ROS that act as important signaling molecules (Torres and Dangl 2005; Suzuki et al. 2011). In *Arabidopsis*, RBOHs constitute a multigenic family comprised of ten genes (i.e., AtRBOHA–AtRBOHJ). Very specific function of each RBOH isoform was implicated by the finding that expression of genes encoding each RBOH protein, except for RBOHD, is restricted to one or two organs (Suzuki et al. 2011; Marino et al. 2012). Indeed, several studies have revealed that different plant RBOHs are involved in a multitude of different signaling pathways underlying growth, development, and response to biotic and abiotic stress (Torres et al. 2005; McInnis et al. 2006; Monshausen et al. 2007; Jammes et al. 2009; Nishimura and Dangl 2010; Miller et al. 2009b; Suzuki et al. 2011). In addition, a previous study revealed distinctive expression profiles of nine RBOHs in rice in response to various abiotic stimuli, which demonstrated their linked but diverse functions (Wang et al. 2013).

Plant RBOHs consist of C-terminal region containing cytosolic FAD and NADPH-binding domains, six conserved transmembrane-spanning domains, and N-terminal extension containing two Ca^{2+}-binding EF-hand motifs and phosphorylation target sites that are important for their activity (Kobayashi et al. 2007; Oda et al. 2010; Kimura et al. 2012; Drerup et al. 2013). Previous studies in *Arabidopsis* have revealed several regulatory mechanisms of RBOH proteins, which involve protein phosphorylation, Ca^{2+}, calcium-dependent protein kinases (CDPKs), and phospholipase Dα1 (PLDα1) (Lin et al. 2009; Monshausen et al. 2009; Zhang et al. 2009b; Jakubowicz et al. 2010; Dubiella et al. 2013; Drerup et al. 2013). Mechanical stimulation of plant tissue can induce an increase in cytosolic Ca^{2+} via an influx from the apoplast across the plasma membrane (Monshausen et al. 2009). The increased Ca^{2+} then enhances RBOHC-dependent ROS production followed by the activation of positive feedback loop between Ca^{2+} and RBOHC to regulate root hair development (Monshausen et al. 2007, 2009; Takeda et al. 2008). Ca^{2+} binding and phosphorylation synergistically activate RBOHD and RBOHF in *Arabidopsis* (Ogasawara et al. 2008; Kimura et al. 2012). A Ca^{2+} increase in the cytosol and conformational changes in EF-hand motifs by Ca^{2+} binding were found to be necessary for the activation of RBOHD (Ogasawara et al. 2008). PLDα1 and its lipid product phosphatidic acid (PA) play an essential role in abscisic acid (ABA)-induced production of ROS in guard cells via the function of RBOHD and RBOHF (Zhang et al. 2009b). ABA-dependent stomatal movement involves binding of PA to Arg residues 149, 150, 156, and 157 in RBOHD and phosphorylation of Ser13 and Ser174 in RBOHF by OPEN STOMATA 1 (OST1) kinase (Sirichandra et al. 2009). These findings indicate integration between RBOHD and RBOHF in the regulation of ABA-dependent stomatal closure. The coordination between PA and OST1 however, still needs to be addressed in future studies. In a recent study, systemic acquired resistance to pathogen was shown to involve phosphorylation of RBOHD by calcium-dependent protein kinase 5 (CPK5) and H_2O_2 production (Dubiella et al. 2013), supporting the hypothesis that Ca^{2+}-dependent ROS production is required for the propagation of the ROS wave over long distances (Miller et al. 2009b). In addition, a recent finding demonstrated that the activity of RBOHF is regulated by direct Ca^{2+} binding to its EF-hands and Ca^{2+}-dependent phosphorylation by CBL1/9–CIPK26 complexes (Drerup et al. 2013). Taken together, these findings indicate that the diverse functions of RBOH proteins in plants might be, at least partially, attributed to differences in regulatory mechanisms.

RBOH proteins are not the only source of ROS in plant cells. Numerous pathways for ROS production exist in plants and include photosynthesis, respiration, glycolate oxidase, oxalate oxidase, xanthine oxidase, amine oxidase, excited chlorophyll, fatty acid oxidation, and peroxidases (Mittler 2002). These pathways were also found to play important roles in the response of plants to abiotic stresses. For example, oxalate oxidase was shown to be involved in ROS production in root cells during drought stress (Voothuluru and Sharp 2013). In addition, recent studies uncovered a role for peroxidase-dependent ROS in the regulation of root growth and response to potassium deficiency (Kim et al. 2010; Jia 2011; Kwasniewski et al. 2013). Interestingly, ROS production by peroxidases might not be functionally equivalent to ROS generated by RBOH proteins (Daudi et al. 2012; Wrzaczek et al. 2013). This hypothesis can be supported by the finding that stomatal closure and ROS burst induced by a yeast elicitor were not inhibited in *rbohD* and *rbohF* mutants in *Arabidopsis* (Khokon et al. 2010). Functional differences between RBOH proteins and peroxidases may be at least partially attributed to differences in the types of ROS generated via functions of these enzymes. Superoxide ($O_2^{\bullet -}$), generated by RBOH proteins, can activate specific signaling pathways distinct from those activated by H_2O_2 (Suzuki et al. 2011). Another possibility is that diverse functions between these different types of enzymes might be due to differences in their respective reductants. RBOH proteins utilize NADPH as a reductant for the generation of $O_2^{\bullet -}$. In contrast, different chemicals or compounds including phenols, organic acids, and auxin have been suggested as candidate reductants employed in the peroxidase-dependent generation of H_2O_2 (O'Brien et al. 2012). Pathways involving these different reductants could be integrated with ROS signals activated via the different functions of these enzymes.

3 Involvement of ROS in the Regulation of Systemic Acquired Acclimation to Abiotic Stress

Recent findings highlight the significance of cell-to-cell communication mediating long-distance systemic signaling in plants. Plants evolved sophisticated acclimation and defense mechanisms that can be activated in the tissue(s) locally exposed to biotic or abiotic stimuli, as well as in distal portions not directly exposed to these stimuli. These mechanisms play an important role in preventing further damage to the entire plant when part of tissues is exposed to biotic or abiotic stimuli. The activation of defense or acclimation mechanisms in systemic or non-challenged tissues is termed systemic acquired resistance (SAR) or systemic acquired acclimation (SAA), respectively (Karpinski et al. 1999; Rossel et al. 2007; Carr et al. 2010; Szechynska-Hebda et al. 2010; Dempsey and Klessig 2012; Spoel and Dong 2012; Shah and Zeier 2013; Baxter et al. 2014). A recent study revealed the existence of an H_2O_2-dependent long-distance signal induced by various abiotic stimuli (Miller et al. 2009b). RBOHD was shown to be required for the initiation and self-propagation of a rapid cell-to-cell signal transduction that is dependent upon H_2O_2 accumulation in the apoplast to generate a "ROS wave" (yellow arrows in Fig. 2) (Mittler et al. 2011). In addition, more recent study demonstrated the significance of the ROS wave in the SAA of plants to heat or high-light stresses

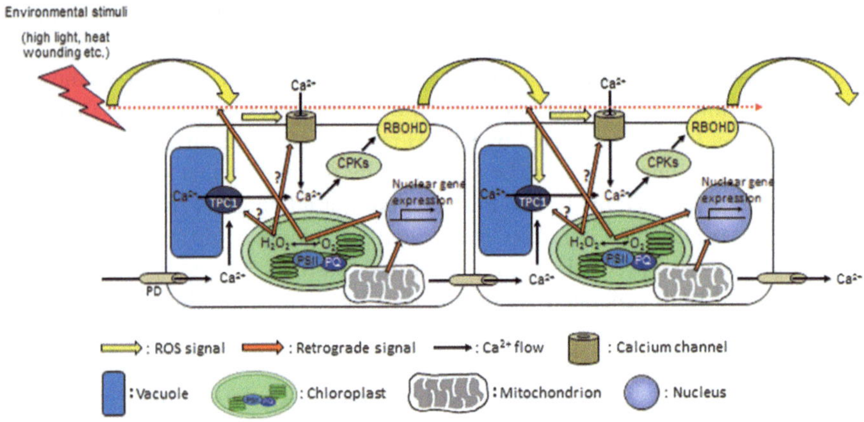

Fig. 2 Cellular pathways that regulate ROS-dependent systemic signaling and retrograde signaling. H_2O_2 generated via function of RBOHD is required for rapid signal propagation from cell to cell in response to different environmental stimuli, such as high light, heat, or wounding (Miller et al. 2009). The ROS and Ca^{2+} waves in cells are integrated via the function of respiratory burst homolog (RBOH) proteins, Ca^{2+}-dependent protein kinases, and calcium channels such as two-pore channel (TPC) 1. Changes in the redox state of PSII and the PQ pools in the chloroplast might activate electric signaling (red dotted arrow) at the plasma membrane involved in systemic responses (Szechynska-Hebda et al. 2010). The electric signal that is dependent upon the chloroplast redox state is also accompanied by ROS generation, implicating a cross talk between the RBOHD-dependent signal, electric signals on the plasma membrane, and ROS/redox signaling from chloroplast. Mitochondrial retrograde signaling (Rhoads and Subbaiah 2007; Pogson et al. 2008; Woodson and Chory 2008) could also play a key role in these responses as part of a cross talk network that senses different metabolic/environmental states and activates rapid systemic signaling. In addition, ROS produced via function of RBOH proteins or basic biological processes in different cellular components might induce different PCD pathways depending on abiotic stimuli. *Abbreviations*: *PD* plasmodesmata, *CPK* Ca^{2+}-dependent protein kinase

(Suzuki et al. 2013). The SAA of plants to abiotic stress is mediated by temporal–spatial interactions of the ROS wave, which function as a general priming signal, with stress-specific hormone or amino acid signals activated in systemic tissues. Calcium wave was recently shown to function as pivotal element of the systemic communication machinery (black arrows in Fig. 2) (Choi et al. 2014; Gilroy et al. 2014). In response to local stimulation with salt stress in the root tip, a wave of increased cytosolic Ca^{2+} level moves systemically through the plant paralleling with the ROS wave. The Ca^{2+} wave that travels at ~400 μm/s can spread through the root system and be transmitted to the aerial part of the plant. Application of the Ca^{2+} channel blocker lanthanum (La^{3+}) inhibited systemic induction of marker gene expression associated with ROS as well as Ca^{2+} wave, implying the link between ROS and Ca^{2+} wave. In addition, plant deficient in two-pore channel 1 (TPC1), a vacuolar ion channel, exhibited disruption of the propagation of the systemic Ca^{2+} wave.

4 Temporal Coordination Between ROS and Other Signals in the Regulation of Systemic Signaling in Plants

In response to changes in environmental conditions, early signaling events including increased Ca^{2+} levels in the cytosol, activation of MAPKs, accumulation of hormones, and production of ROS can all occur within seconds or minutes following application of abiotic stimuli (Benschop et al. 2007; Finka et al. 2012; Miller et al. 2009b). Early systemic responses of plants to abiotic stimuli have been previously described (Karpinski et al. 1999; Rossel et al. 2007; Muhlenbock et al. 2008; Szechynska-Hebda et al. 2010; Gordon et al. 2012). Studies employing transgenic plants expressing a luciferase reporter gene under the control of an APX1, APX2, or ZAT10 promoter demonstrated the activation of acclamatory responses within 5–20 min following application of high light both in leaves locally exposed to the stimuli and in distal tissues that were not directly exposed to the stimuli (Karpinski et al. 1999; Rossel et al. 2007; Szechynska-Hebda et al. 2010). Systemic responses to high light were shown to be associated with redox changes in the plastoquinone (PQ) pool, increased production of ROS and ethylene, reduction of maximal photochemical efficiency and non-photochemical quenching (NPQ), and changes in extracellular electric potential (Karpinski et al. 1999; Rossel et al. 2007; Szechynska-Hebda et al. 2010). In addition, amino acids involved in photorespiratory machinery, such as glycine, serine, and glycerate, rapidly accumulate in leaves directly exposed to high light within 60 s as well as in systemic tissues of plants at 15 and 45 min (Suzuki et al. 2013). Rapid local responses of these metabolites to high light were altered in the mutant lacking cytosolic APX1, demonstrating the involvement of H_2O_2 scavenging in this process.

In *Arabidopsis*, elevated levels of jasmonic acid (JA) accumulate in damaged tissues as well as undamaged systemic leaves within 30 s to 5 min in response to mechanical wounding (Glauser et al. 2009; Koo et al. 2009). The velocity of this long-distance signal leading to synthesis of JA in systemic tissues was 3.4–4.5 cm/min (Koo et al. 2009; Glauser et al. 2009). RBOHD-dependent long-distance signal is also a rapid auto-propagating systemic signal that travels at the rate of approximately 8.4 cm/min and is induced by various abiotic stimuli including mechanical wounding (Miller et al. 2009b). In addition, the potential involvement of electric signals that propagate with similar rates was also implicated in RBOHD-triggered rapid systemic signaling during wounding (Zimmermann et al. 2009; Mittler et al. 2011; Suzuki and Mittler 2012). These findings implicate the integration of JA and mobile signals such as ROS and electric signals. Peroxisomes that are responsible for the production of both H_2O_2 and JA might be a candidate of a key player to integrate JA and mobile ROS signals (Leon 2013). H_2O_2 is produced during the oxidation of glycolate to glyoxylic acid in photorespiratory processes (Mittler et al. 2004). JA is synthesized through the octadecanoid pathway involving the translocation of lipid intermediates from the chloroplast to the cytosol and later on into peroxisomes (Leon 2013). JA synthesized in the peroxisomes is transported to the cytosol, and JA-isoleucine conjugate, the bioactive form of the hormone, is then produced. Long-term responses to fluctuating environmental conditions regulate phenotypic changes such as growth, development, and survival of cells. Exposure of mature leaves to changes in light conditions and atmospheric CO_2 induces alterations in photosynthetic rate and tolerance to high light in new developing leaves not directly exposed to these environmental changes (Coupe et al. 2006; Araya et al. 2008; Jiang et al. 2012). Although alterations in photosynthetic rate and response to high light implicate ROS and redox signaling in the systemic regulation of long-term responses in new developing leaves (Muhlenbock et al. 2008; Li et al. 2009; Mittler et al. 2011), links between ROS signaling and these responses are still not uncovered.

Previous studies demonstrated that the biphasic production of ROS consists of a primary phase that occurs within minutes and a secondary phase that occurs within hours/days (Nishimura and Dangl 2010; Soares et al. 2009; Kunihiro et al. 2011; Mittler et al. 2011). For example, mechanical wounding induced an initial burst of $O_2^{\bullet -}$ within 3 min followed by later production of $O_2^{\bullet -}$ and H_2O_2 after 6 h (Soares et al. 2009). Inhibition of early phase of ROS production by an NADPH oxidase inhibitor suppresses later production of $O_2^{\bullet -}$ and accumulation of wound response proteins, indicating that an initial burst of ROS is required for the later phase of ROS production which regulates downstream acclamatory responses of plants to stress stimuli. In addition, a recent study suggests that these two phases of the ROS burst are linked via the ROS wave that communicates the initial ROS burst in the local tissue to the systemic tissue via a cell-to-cell relay mechanism (Miller et al. 2009b; Suzuki et al. 2013).

5 Spatial Coordination Between ROS and Other Signals in the Regulation of Systemic Signaling in Plants

To some extent, signals generated in plants during SAA are similar in local and systemic tissues. Rossel et al. (2007) compared the transcriptomes of local leaves, directly exposed to high light, and systemic leaves, not directly challenged by the stimulus. More than 70 % of the transcripts upregulated in local leaves in response to high light were also altered in their expression in systemic leaves, suggesting that similar signals exist between local and systemic tissues during SAA to high light. Similarities between local and systemic responses to high light is also supported by findings that alterations in ROS and redox signals and accumulation of amino acids associated with the photorespiratory pathway occurred both in local and systemic tissues (Muhlenbock et al. 2008; Szechynska-Hebda et al. 2010; Miller et al. 2009b). In addition, local application of heat or cold stimuli also can induce similar stress response proteins or transcripts in both local and systemic tissues (Gorsuch et al. 2010; Suzuki et al. 2013). In particular, induction of heat-responsive proteins in systemic tissue was shown to be RBOHD dependent (Suzuki et al. 2013).

Although signals generated in local and systemic tissues showed considerable overlap, previous studies have also demonstrated differences in alterations of transcripts or metabolites between these types of tissues. For example, ethylene accumulated both in local and systemic tissue in response to local application of high light; nevertheless, the signal regulated by EIN2 was shown to be required for induction of APX2 only in systemic tissues that are not directly subjected to high light (Muhlenbock et al. 2008). In addition, SID2 delays induction of APX2 only in leaves directly exposed to high light. These findings suggest that specific patterns of APX2 expression in local and systemic tissue might be regulated by the coordination between ethylene and salicylic acid (SA) signaling during SAA to high light. Moreover, spatial diversity in high-light responses between different leaves during SAA was also demonstrated by the findings that local high-light treatment resulted in the different expression levels of transcripts associated with regulation of ROS and redox signals depending on leaf position (Gordon et al. 2012).

How are signals generated in local and systemic tissues linked? The ROS wave may play a key role in propagating signals from local tissues to systemic tissues. The initial burst of ROS in a local group of plant cells triggers a cascade of cell-to-cell communication events that carries a systemic signal over long distances throughout different tissues of the plant (Miller et al. 2009b). Szechynska-Hebda et al. (2010) uncovered the pattern of systemic changes in NPQ, H_2O_2 concentration, and APX1 expression during SAA response of plants to high light. Wavelike patterns of APX1 expression in systemic tissue of plants correlate positively with H_2O_2 accumulation but negatively with NPQ (Szechynska-Hebda et al. 2010; Karpinski et al. 2013). The activation of systemic signals by local application of high light was recently shown to be accompanied by plasma membrane electrical signals in a light wavelength-specific manner (Szechynska-Hebda et al. 2010). In addition, the RBOHD-dependent ROS wave is associated with the generation and/or propagation of systemic potential variations (Suzuki et al. 2013). These finding suggest a link between electric signals in plants and ROS production.

6 Integration of ROS Signals with Other Signals

ROS signaling is integrated with various other signals including Ca^{2+} signaling, protein kinases, redox responses, and hormone signals. One good example of signaling networks associated with ROS signaling is MAPK cascade (Mittler et al. 2011; Petrov and Van Breusegem 2012). MPK3, MPK4, and MPK6 can all be activated by ROS and abiotic stresses, but different MKKs might transmit the signal depending on different stimuli (Jaspers and Kangasjarvi 2010). *Arabidopsis* overexpressing MKK2 exhibited constitutive MPK4 and MPK6 activity and resulted in increased tolerance of transgenic plans to salt and cold stress (Teige et al. 2004; Taj et al. 2010; Ismail et al. 2014). In contrast, overexpression of MKK9 that activates MPK3/MPK6 resulted in enhanced sensitivity of the transgenic plants to salt stress (Xu et al. 2008). In addition, activation of MPK6 by MKK3 was shown to be not required for salt stress response in plants (Takahashi et al. 2007). MEKK1 was suggested to be specifically required for the activation of MPK4 by H_2O_2 (Nakagami et al. 2006), and the signal between MEKK1 and MPK4 is mediated by MKK1 and MKK2 (Qiu et al. 2008). The MEKK1–MKK1/2–MPK4 pathway might play integral roles to regulate transcription factors that are highly responsive to ROS-generating conditions (Pitzschke et al. 2009). MPK8 could be a negative regulator of ROS wave (Marino et al. 2012). MPK8 which is activated by phosphorylation and direct binding of CaM in a Ca^{2+}-dependent manner has been shown to negatively regulate ROS production via control of RBOHD (Takahashi et al. 2011). Various forms of abiotic stress result in increased production of ROS which can be liked to signals caused by changes in the regulation of plant hormones (Fujita et al. 2006). Ethylene biosynthesis was found to be modulated by positive regulation via RBOH proteins and negative regulation via CTR1 (constitutive triple response 1) (Jakubowicz et al. 2010). In *Arabidopsis*, CTR1 can be inhibited by phosphatidic acid (PA) that positively enhances activation of RBOHD and RBOHF (Jakubowicz et al. 2010). Previous studies revealed the involvement of ethylene in the regulation of SAA to high light induced by local high-light application (Muhlenbock et al. 2008; Karpinski et al. 2013). In response to high light, alterations in the redox state of the PQ pool can initiate a signal that induces production of 1-aminocyclopropane-1-carboxylate (ACC, the immediate precursor of ethylene), ROS, and the expression of ethylene-regulated genes (Muhlenbock et al. 2008). Increased ROS production results in bleaching of leaves and programmed cell death that relies on regulation of ethylene-insensitive 2 (EIN2) by lesion stimulating disease 1 (LSD1) (Muhlenbock et al. 2008; Karpinski et al. 2013).

Involvement of brassinosteroid (BR) signaling in ROS-dependent stress responses was also supported by previous findings. For example, exogenous BR treatments resulted in enhanced tolerance to oxidative stress accompanied by induction of H_2O_2 production in apoplast and expression of RBOH, MPK1, and MPK3 (Xia et al. 2009). More recent studies demonstrated the involvement of BR in SAA to high light in cucumber (Xia et al. 2009, 2011; Li et al. 2013a). Although BRs are not directly involved in long-distance signaling, they affect other signals such as auxins and polyamines (Li et al. 2013a). A recent study indicated the involvement of auxin in SAA response of plants to high light. Large portions of the transcripts that are altered in their expression in the distal leaves overlap with auxin-responsive transcripts (Gordon et al. 2012), indicating a connection between SAA and developmental processes mediated by auxin. Integration between ethylene and BRs during SAA response to HL needs to be elucidated in future studies.

ROS can affect auxin biosynthesis, transport, metabolism, and signaling underlying regulation of stress responses in plants (Krishnamurthy and Rathinasabapathi 2013b). The integration of RBOH functions with auxin has been evidenced by the analyses of the various RBOH expressions following the exogenous auxin (IAA: indole-3-acetic acid) treatment in *Arabidopsis*. It was observed that RBOHD was highly activated in response to auxin (Peer et al. 2013). Auxin transport mutant *aux1* exhibited higher sensitivity to arsenite compared to WT plants, and H_2O_2

production was inhibited in *aux1* under this stress condition (Krishnamurthy and Rathinasabapathi 2013a). These results indicate that auxin transport plays positive role in induction of ROS production and protection of plants against arsenite. In addition, the auxin signaling mutant *tir afb2* showed reduced production of H_2O_2 and $O_2^{\bullet -}$ and increased activity of catalase and ascorbate peroxidase accompanied by enhanced tolerance to salt stress (Iglesias et al. 2010).

ABA is involved in a broad range of biological functions, and its integration with ROS has been evidenced in many reports (Ma et al. 2012; Sagi et al. 2004; Drerup et al. 2013; Kwak et al. 2003). For example, RBOHD and RBOHF function together to regulate stomatal closure, seed germination, root elongation, and Na^+/K^+ homeostasis under salt stress (Ma et al. 2012; Kwak et al. 2003). Overexpression of 9-*cis*-epoxycarotenoid in tobacco, an enzyme of ABA synthesis, enhanced tolerance of transgenic plants to drought and salt stress accompanied by increased ABA-induced production of ROS via the function of RBOHs (Zhang et al. 2009a). In addition, mild salt stress triggers biphasic changes in ROS production in *Arabidopsis* and maize, and RBOHD in *Arabidopsis* is required for ROS production following mild salt stress (Lin et al. 2009; Xie et al. 2011). Moreover, SAA of plants to heat stress was shown to be correlated with activation of the ROS wave and transient accumulation of ABA in systemic tissues, and these responses were suppressed in a mutant lacking RBOHD (Suzuki et al. 2013). The SAA response to heat stress was also attenuated in mutants deficient in ABA signaling. These results indicate that temporal–spatial interactions between RBOHD-dependent ROS and ABA mediate SAA to heat stress (Suzuki et al. 2013). Moreover, ABA and SA treatment have been shown to result in transient increases in H_2O_2 production which induces tolerance to heat, salt, high-light, and oxidative stress (Xia et al. 2009).

Integration of SA or JA signaling with ABA underlying stomatal closure has been addressed in a recent review (Song et al. 2014). ABA-deficient mutant *aba2-1* failed to close stomata in response to exogenous SA treatment, whereas guard cells of SA-deficient mutants *sid2* and *NahG* responded to ABA (Zeng and He 2010; Montillet and Hirt 2013; Song et al. 2014), indicating that SA signaling functions upstream of ABA signaling. SA-induced stomatal closure was inhibited by DPI, an NADPH oxidase inhibitor, and plants deficient in RBOHD exhibited defect in stomatal response to exogenous SA treatment (Kalachova et al. 2013). In contrast, several lines of other evidences suggested that SA mediates ROS production, not via NADPH oxidases, but rather via a peroxidase-catalyzed reaction (Mori et al. 2001; Song et al. 2014). The reduced stomatal apertures in SA-accumulating *siz1* mutant were rescued by the application of peroxidase inhibitors, but not by DPI (Miura et al. 2013). In addition, SA induces stomatal closure accompanied by extracellular ROS production mediated by peroxidase (Khokon et al. 2011). Mechanisms that regulate integration of SA signals to ROS-generating pathways need to be elucidated in future works. It was suggested that there is an overlap in the signals associated with stomatal closure between JA and ABA and that many common components including ROS production and Ca^{2+} oscillation are shared between these hormone signalings (Santino et al. 2013). *Arabidopsis* RBOHD and RBOHF were shown to be involved in the expression of methyl jasmonate (MeJA) response genes regulated by MYC2 transcription factor (Maruta et al. 2012).

7 Involvement of ROS in the Regulation of Retrograde Signaling

Under stress conditions, changes in the redox state of the chloroplast and mitochondria, ROS-producing organelles are the source of retrograde signaling that play crucial roles in stress acclimation of plants (Orange arrows in Fig. 2) (Suzuki et al. 2012; Choudhury et al. 2013). Three different processes in *Arabidopsis* were shown to induce chloroplast to nucleus signaling that alter the expression of nuclear genes, depending on the presence of GUN1 in the chloroplast and ABI4 in the nucleus: (1) accumulation of the chlorophyll biosynthesis intermediate Mg-protoporphyrin IX (Mg-Proto IX) and its methylester (Mg-Proto IX-ME), (2) inhibition of plastid gene expression (PGE) at the protein translation stage, and (3) changes in the redox state of the photosynthetic electron transfer (PET) chain (Koussevitzky et al. 2007; Woodson and Chory 2008).

Signals from the chloroplast to nuclei, modulated by changes in cellular redox state depending on the light intensity, might be an important process to respond to fluctuating light conditions (Szechynska-Hebda and Karpinski 2013). Under low light intensity, the transfer of signal from chloroplast to nucleus followed by changes in nuclear gene expression contributes to the adjustment of morphology of the leaves and cells, and number and structure of chloroplasts (Oelze et al. 2012). Light use efficiency is optimized at least partially by modulating stoichiometry of photosystems, light-harvesting antenna size, composition of the stromal enzymes, and activity of antioxidant systems (Muhlenbock et al. 2008; Pfannschmidt 2010; Foyer and Noctor 2011; Ruckle et al. 2012; Szechynska-Hebda and Karpinski 2013). Low light intensity can promote oxidation of PQ and reduction of thioredoxin, whereas high light intensity oppositely affects redox state in PQ pool and thioredoxin (Muhlenbock et al. 2008). Under high light intensity, PQ redox state and non-photochemical quenching (NPQ), the process to dissipate excess energy by heat, were shown to be potential regulators of retrograde signals that are required for at least regulation of APX1 and APX2 expression (Szechynska-Hebda et al. 2010). Imbalance between ATP and NADPH synthesis might be also an initiator of retrograde signaling (Szechynska-Hebda and Karpinski 2013). High proportion of NAD(P)H/NAD(P), ATP/ADP, and ATP/NADPH ratio could lead to inactivation of the photosynthetic electron transport chain components, resulting in production of $^1O_2$ and $O_2^{\bullet-}$ followed by deregulation of chloroplast metabolism and altered expression of stress response genes (Szechynska-Hebda and Karpinski 2013). In addition, excess excitation energy (EEE) also induces the changes in cellular ROS homeostasis and affects the expression of nuclear genes that control SAR and SAA (Rossel et al. 2007; Szechynska-Hebda et al. 2010). Progression of excess energy-induced PCD initiated by redox changes in PQ pool is regulated by the coordination between LSD1, EDS1, PAD4, and EIN2 that are associated with ethylene signaling and ROS production.

Upon illumination, chloroplast precursor, protochlorophyllide (PChlide), Mg-Proto IX, and Mg-Proto IX-ME produce $^1O_2$ that can act as a signaling molecule from chloroplast to nuclei (Tripathy and Oelmuller 2012). The $^1O_2$-dependent retrograde signaling has been extensively studied using the *Arabidopsis flu* mutant that massively accumulates PChlide and $^1O_2$ and exhibits growth retardation and cell death under constant dark/light cycle (op den Camp et al. 2003; Apel and Hirt 2004; Wagner et al. 2004; Laloi et al. 2007; Lee et al. 2007). Transcriptome analysis revealed a set of genes that can be specifically activated by $^1O_2$, not by H_2O_2 or $O_2^{\bullet-}$ (op den Camp et al. 2003; Gadjev et al. 2006; Suzuki et al. 2011), suggesting that chloroplast to nucleus retrograde signaling is at least partially regulated by $^1O_2$-specific signaling. Two plastid-localized proteins EXECUTER1 and EXECUTER2 are required for $^1O_2$-dependent gene regulation and cell death (Lee et al. 2007). Moderate light exposure induces acclimation mechanisms that protect plants against more severe high-light stress, and $^1O_2$ production and EXECUTER-dependent signal can be activated during this acclamatory process (Zhang et al. 2014), indicating the significance of $^1O_2$-dependent retrograde signaling in the regulation of light acclimation of plants. In addition, nuclear topoisomerase VI and oxygenation derivatives of linoleic acid, the prominent polyunsaturated fatty acid of chloroplast membrane lipid, might also play an important role to integrate $^1O_2$-dependent signal with regulation of nuclear gene expression (op den Camp et al. 2003; Tripathy and Oelmuller 2012). The oxidation of linoleic acid was shown to be not directly caused by $^1O_2$, but enzymatic reactions regulated by $^1O_2$ signals (op den Camp et al. 2003; Tripathy and Oelmuller 2012). Furthermore, $^1O_2$-linked cell death activator (soldat8) that encodes SIGMA6 factor of the plastid RNA polymerase was identified as specific suppressor of $^1O_2$-dependent stress responses in *flu* mutant (Coll et al. 2009). The other protein pleiotropic response locus 1 (PRL1) also affects the expression of $^1O_2$ response genes in *Arabidopsis* (Baruah et al. 2009).

In a recent study, integration between ABI4 and redox metabolism was investigated using the ascorbic acid-deficient mutants, *vtc1* and *vtc2* (Kerchev et al. 2011). The transcriptome signatures of *abi4*, *vtc1*, and *vtc2* mutants extensively overlap with large number of transcription factors and other regulatory genes (Kerchev et al. 2011). In addition, ABA and JA signaling synergistically function to regulate the growth of plants through ABI4 in ascorbic acid-dependent manner (Kerchev et al. 2011). Although it is still not clearly understood, H_2O_2 produced in chloroplasts was also implicated in retrograde signaling (Shapiguzov et al. 2012). H_2O_2-dependent retrograde signaling might be a combination of passive diffusion of H_2O_2 with indirect pathways involving ABA signaling (Mullineaux and Karpinski 2002; Galvez-Valdivieso and Mullineaux 2010). H_2O_2 might not be, however, the signaling molecule that directly affects nuclear gene expression; rather, redox-sensitive components such as oxidized proteins or peptides might mediate the signal transfer from the chloroplast to nucleus (Moller and Sweetlove 2010; Sierla et al. 2013).

Although mitochondrial retrograde signaling is poorly understood compared with the chloroplast to nucleus retrograde signaling, key regulators of mitochondrial retrograde signaling have been identified in recent studies. A membrane-bound NAC transcription factor, ANAC017, was found to be a regulator of

AOX1a, a marker gene of mitochondrial retrograde signaling (Ng et al. 2013). ANAC017 mediates H_2O_2-induced alterations in transcript abundance. The *abi4* mutants are insensitive to transcriptional derepression of AOX1a by the mitochondrial complex I inhibitor rotenone, indicating a role for ABI4 in redox regulation and mitochondria to nucleus retrograde signaling (Giraud et al. 2009). This work demonstrated the integral role of ABI4 to mediate mitochondrial and chloroplast retrograde signaling pathways, and perhaps it is the convergence point for mitochondria–plastid–nucleus coordination.

8 Programmed Cell Death Regulated by ROS Under Abiotic Stress

A low dose of ROS acts as signaling molecules that mediate at least part of stress responses, but they induce programmed cell death (PCD) at higher concentrations. PCD is a genetically controlled process in which only cells that are destined to die are selectively destroyed in a multistep fashion via the functions of specific proteases and nuclease (Petrov et al. 2015). Thus, no damage to the neighboring cells is inflicted. In addition, metabolism in the rest of cells and tissues can be appropriately adjusted to acclimate to or survive under stressed conditions by this process. Here, stress-specific pathways that positively regulate PCD will be mainly discussed. More detail of PCD under abiotic stress has been addressed in more extensive review (Petrov et al. 2015).

Severe drought enhances ROS production mainly due to decreased CO_2 fixation accompanied by increased leakage of electron to O_2, which may result in induction of PCD (Gechev et al. 2012). ROS production is inhibited by suppression of chlorophyll synthesis and photosynthetic activity, and modification of sucrose metabolisms under drought (Petrov et al. 2015; Liu et al. 2013). Although plants possess mechanisms to prevent unnecessary PCD, leaf senescence executed by ROS-triggered PCD acts as an important adaptation process of plants to drought (Petrov et al. 2015). ROS-triggered PCD under drought was shown to be regulated by coordination between ABA and cytokinin (Munn-Bosch and Alegre 2004). PCD characterized by degradation of organelles, increased size of the vacuole, and plasmalemma collapse can be induced in root tip meristems under drought (Duan et al. 2010). This adaptive mechanism in the root might be a strategy of plants to enhance lateral root growth. Reorientation of polyamine metabolism, as well as stomatal closure, is a drought response mechanism in which ABA plays pivotal roles. In grapevine, ABA induces accumulation of polyamine that is metabolized by amine oxidases, and H_2O_2 can be produced as a by-product of this reaction (Toumi et al. 2010). H_2O_2 produced in this reaction might be a regulator of further stress response or induction of PCD. PCD acts as an important process to protect plants against flooding stress. Under this stress condition, aerenchyma is formed by PCD in selected cells to produce air channels in roots (Gunawardena et al. 2001). Significance of ROS in the formation of aerenchyma has been indicated by the finding that exogenous application of H_2O_2 stimulated formation of aerenchyma in rice (Steffens et al. 2011). In addition, expression of RBOHD was highly enhanced during water logging conditions in maize, suggesting that RBOHD-dependent production of H_2O_2 play a key role in the formation of aerenchyma (Rajhi et al. 2011). Furthermore, mitochondrial dysfunction and accumulation of metal ion was also shown to increase ROS production under flooding stress (Shabala et al. 2014).

Although involvement of $O_2^{\bullet -}$ produced via the function of NADPH oxidase is implicated in PCD pathway under salt and osmotic stress, signals activated by ionic salt and nonionic osmotic stress inducer (i.e., sorbitol) did not show extensive overlap (Monetti et al. 2014; Petrov et al. 2015), indicating the specificity of the mechanism regulating PCD under these stresses. Specific feature of salt-induced PCD is characterized by ion disequilibrium that may be due to invasion of Na^+ into the cytosol accompanied by a decrease in K^+ (Kim et al. 2014). Hydroxyl radicals in cells might affect K^+/Na^+ ratio and regulate enzymes involved in PCD (Demidchik et al. 2010). In addition, an anti-apoptotic protein BCL2 can regulate vacuolar processing enzymes (VPE) by modulating ion fluxes in rice (Kim et al. 2014). Overexpression of BCL2 exhibited enhanced PCD symptoms accompanied by significantly reduced K^+ efflux and represses the expression of VPEs. These findings suggest that maintenance of proper Na^+/K^+ ratios could be a key process to enhance salt stress tolerance in plants (Huh et al. 2002; Teakle and Tyerman 2010).

Mitochondria are responsible for the regulation of PCD induced during heat stress (Vacca et al. 2004). Functionally active cytochrome c can be released from the mitochondria in a ROS-dependent manner, and caspase-like protease that induces PCD is then activated (Vacca et al. 2004). Proline that was implicated in mitochondrial ROS metabolism (Miller et al. 2009a) might be a specific component to cause cell death during heat stress. Although it acts as an osmolyte and ROS detoxifier under salt and osmotic stresses, it can negatively impact on cells under heat stress (Lv et al. 2011). In addition, exogenous application of ascorbate or glutathione (i.e., unspecific antioxidants) induced PCD in *Arabidopsis* cells under heat stress. On the other hand, catalase that specifically detoxifies H_2O_2 suppressed PCD. These finding suggest that H_2O_2 functions as PCD-inducing signal, but other types of ROS might act as negative regulator of PCD during heat stress (Doyle and McCabe 2010). Furthermore, MPK6 was found to be a component of PCD under heat stress. Enhanced accumulation of ROS and cytosolic Ca^{2+} can activate NPK6 that functions upstream to hydrolases and proteases associated with PCD (Li et al. 2012).

Low temperature is also able to induce PCD in plants when combined with elevated light intensity. LSD1, a negative regulator of PCD, interacts with catalase, and its deficiency can enhance low-temperature-induced PCD in plants (Huang et al. 2010; Li et al. 2013b). LSD1 and EDS1 proteins antagonistically regulate the acclimation of plants to UV-C stress. Plants deficient in LSD1 exhibited enhanced PCD following UV-C treatment. On the other hand, knocking out of EDS1 repressed UV-C-induced PCD (Wituszynska et al. 2015). The UV-C response associated with LSD1 and EDS1 is regulated by the modulation of ROS homeostasis. In response to high dose of UV-B, a defense program involving SA, JA, ethylene, and senescence-associated processes is activated to prevent oxidative damage in plants (Bandurska et al. 2013). These findings indicate the cross talks in the pathways associated with PCD between low-temperature, high-light, UV-B and UV-C, and pathogen responses.

9 Conclusions

ROS play an integral role in the regulation of numerous responses to abiotic stresses in plants. The complexity in ROS responses to various environmental stimuli might be, at least partially, attributed to different regulatory mechanisms of ROS production via basic biological processes and NADPH oxidases (RBOHs) that function in an array of organelles, tissue types, and developmental stages under various environmental conditions. A key mechanism in coordinating the complex spatial and temporal responses in plants is the cascade of cell-to-cell communication events that result in the formation of a wave of ROS production and increase in cytosolic Ca^{2+} that rapidly propagates throughout the different tissues of the plant. Networks of ROS/redox signaling in the chloroplast and mitochondria contribute to a delicate balance of homeostasis within each organelle, as well as to cross talk between different cellular components by regulating important biological pathways such as nuclear gene expression and energy metabolism under stress conditions. Functions of chloroplasts and mitochondria might be the key mediators for signal transductions in the cell because they are involved in the regulation of important pathways including systemic signaling, retrograde signaling, and PCD under stress conditions. In addition, there is a great overlap in the regulatory mechanisms between RBOHD-dependent systemic response, chloroplast/mitochondrial retrograde signaling and PCD, suggesting a cross talk between these pathways. Detailed mechanisms to coordinate or integrate these different ROS-dependent pathways need to be elucidated in future works. In addition, we also need to address how different pathways switched depending on the different environmental stimuli. This question could be answered by studying integration between ROS signaling and sensors of different stimuli.

Acknowledgment This work was supported by funding from Sophia University in Japan.

References

Apel K, Hirt H (2004) Reactive oxygen species: metabolism, oxidative stress, and signal transduction. Annu Rev Plant Biol 55:373–399

Araya T, Noguchi K, Terashima I (2008) Manipulation of light and CO_2 environments of the primary leaves of bean (*Phaseolus vulgaris* L.) affects photosynthesis in both the primary and the first trifoliate leaves: involvement of systemic regulation. Plant Cell Environ 31:50–61

Bandurska H, Niedziela J, Chadzinikolau T (2013) Separate and combined responses to water deficit and UV-B radiation. Plant Sci Int J Exp Plant Biol 213:98–105
Baruah A, Simkova K, Hincha DK, Apel K, Laloi C (2009) Modulation of O-mediated retrograde signaling by the PLEIOTROPIC RESPONSE LOCUS 1 (PRL1) protein, a central integrator of stress and energy signaling. Plant J Cell Mol Biol 60:22–32
Baxter A, Mittler R, Suzuki N (2014) ROS as key players in plant stress signalling. J Exp Bot 65:1229–1240
Benschop JJ, Mohammed S, O'Flaherty M, Heck AJ, Slijper M, Menke FL (2007) Quantitative phosphoproteomics of early elicitor signaling in Arabidopsis. Mol Cell Proteomics 6:1198–1214
Bienert GP, Schjoerring JK, Jahn TP (2006) Membrane transport of hydrogen peroxide. Biochim Biophys Acta 1758:994–1003
Bienert GP, Moller AL, Kristiansen KA, Schulz A, Moller IM, Schjoerring JK, Jahn TP (2007) Specific aquaporins facilitate the diffusion of hydrogen peroxide across membranes. J Biol Chem 282:1183–1192
Carr JP, Lewsey MG, Palukaitis P (2010) Signaling in induced resistance. Adv Virus Res 76:57–121
Choi WG, Toyota M, Kim SH, Hilleary R, Gilroy S (2014) Salt stress-induced Ca^{2+} waves are associated with rapid, long-distance root-to-shoot signaling in plants. Proc Natl Acad Sci USA 111:6497–6502
Choudhury S, Panda P, Sahoo L, Panda SK (2013) Reactive oxygen species signaling in plants under abiotic stress. Plant Signal Behav 8:e23681
Coll NS, Danon A, Meurer J, Cho WK, Apel K (2009) Characterization of soldat8, a suppressor of singlet oxygen-induced cell death in Arabidopsis seedlings. Plant Cell Physiol 50:707–718
Coupe SA, Palmer BG, Lake JA, Overy SA, Oxborough K, Woodward FI, Gray JE, Quick WP (2006) Systemic signalling of environmental cues in Arabidopsis leaves. J Exp Bot 57:329–341
Daudi A, Cheng Z, O'Brien JA, Mammarella N, Khan S, Ausubel FM, Bolwell GP (2012) The apoplastic oxidative burst peroxidase in Arabidopsis is a major component of pattern-triggered immunity. Plant Cell 24:275–287
Demidchik V, Cuin TA, Svistunenko D, Smith SJ, Miller AJ, Shabala S, Sokolik A, Yurin V (2010) Arabidopsis root K^{+}-efflux conductance activated by hydroxyl radicals: single-channel properties, genetic basis and involvement in stress-induced cell death. J Cell Sci 123:1468–1479
Dempsey DA, Klessig DF (2012) SOS - too many signals for systemic acquired resistance? Trends Plant Sci 17:538–545
Doyle SM, McCabe PF (2010) Type and cellular location of reactive oxygen species determine activation or suppression of programmed cell death in Arabidopsis suspension cultures. Plant Signal Behav 5:467–468
Drerup MM, Schlucking K, Hashimoto K, Manishankar P, Steinhorst L, Kuchitsu K, Kudla J (2013) The Calcineurin B-Like calcium sensors CBL1 and CBL9 together with their interacting protein kinase CIPK26 regulate the Arabidopsis NADPH oxidase RBOHF. Mol Plant 6:559–569
Duan Y, Zhang W, Li B, Wang Y, Li K, Sodmergen, Han C, Zhang Y, Li X (2010) An endoplasmic reticulum response pathway mediates programmed cell death of root tip induced by water stress in Arabidopsis. New Phytol 186:681–695
Dubiella U, Seybold H, Durian G, Komander E, Lassig R, Witte CP, Schulze WX, Romeis T (2013) Calcium-dependent protein kinase/NADPH oxidase activation circuit is required for rapid defense signal propagation. Proc Natl Acad Sci USA 110:8744–8749
Dynowski M, Schaaf G, Loque D, Moran O, Ludewig U (2008) Plant plasma membrane water channels conduct the signalling molecule H_2O_2. Biochem J 414:53–61
Finka A, Cuendet AF, Maathuis FJ, Saidi Y, Goloubinoff P (2012) Plasma membrane cyclic nucleotide gated calcium channels control land plant thermal sensing and acquired thermotolerance. Plant Cell 24:3333–3348

Foyer CH, Noctor G (2011) Ascorbate and glutathione: the heart of the redox hub. Plant Physiol 155:2–18
Fujita M, Fujita Y, Noutoshi Y, Takahashi F, Narusaka Y, Yamaguchi-Shinozaki K, Shinozaki K (2006) Crosstalk between abiotic and biotic stress responses: a current view from the points of convergence in the stress signaling networks. Curr Opin Plant Biol 9:436–442
Gadjev I, Vanderauwera S, Gechev TS, Laloi C, Minkov IN, Shulaev V, Apel K, Inze D, Mittler R, Van Breusegem F (2006) Transcriptomic footprints disclose specificity of reactive oxygen species signaling in Arabidopsis. Plant Physiol 141:436–445
Galvez-Valdivieso G, Mullineaux PM (2010) The role of reactive oxygen species in signalling from chloroplasts to the nucleus. Physiol Plant 138:430–439
Gechev TS, Dinakar C, Benina M, Toneva V, Bartels D (2012) Molecular mechanisms of desiccation tolerance in resurrection plants. Cell Mol Life Sci 69:3175–3186
Gilroy S, Suzuki N, Miller G, Choi WG, Toyota M, Devireddy AR, Mittler R (2014) A tidal wave of signals: calcium and ROS at the forefront of rapid systemic signaling. Trends Plant Sci 19:623–630
Giraud E, Van Aken O, Ho LH, Whelan J (2009) The transcription factor ABI4 is a regulator of mitochondrial retrograde expression of ALTERNATIVE OXIDASE1a. Plant Physiol 150:1286–1296
Glauser G, Dubugnon L, Mousavi SA, Rudaz S, Wolfender JL, Farmer EE (2009) Velocity estimates for signal propagation leading to systemic jasmonic acid accumulation in wounded Arabidopsis. J Biol Chem 284:34506–34513
Gordon MJ, Carmody M, Albrecht V, Pogson B (2012) Systemic and local responses to repeated HL stress-induced retrograde signaling in Arabidopsis. Front Plant Sci 3:303
Gorsuch PA, Sargeant AW, Penfield SD, Quick WP, Atkin OK (2010) Systemic low temperature signaling in Arabidopsis. Plant Cell Physiol 51:1488–1498
Gunawardena A, Pearce DM, Jackson MB, Hawes CR, Evans DE (2001) Characterisation of programmed cell death during aerenchyma formation induced by ethylene or hypoxia in roots of maize (*Zea mays* L.). Planta 212:205–214
Huang X, Li Y, Zhang X, Zuo J, Yang S (2010) The Arabidopsis LSD1 gene plays an important role in the regulation of low temperature-dependent cell death. New Phytol 187:301–312
Huh GH, Damsz B, Matsumoto TK, Reddy MP, Rus AM, Ibeas JI, Narasimhan ML, Bressan RA, Hasegawa PM (2002) Salt causes ion disequilibrium-induced programmed cell death in yeast and plants. Plant J Cell Mol Biol 29:649–659
Iglesias MJ, Terrile MC, Bartoli CG, D'Ippolito S, Casalongue CA (2010) Auxin signaling participates in the adaptative response against oxidative stress and salinity by interacting with redox metabolism in Arabidopsis. Plant Mol Biol 74:215–222
Ismail A, Takeda S, Nick P (2014) Life and death under salt stress: same players, different timing? J Exp Bot 65:2963–2979
Jakubowicz M, Galganska H, Nowak W, Sadowski J (2010) Exogenously induced expression of ethylene biosynthesis, ethylene perception, phospholipase D, and Rboh-oxidase genes in broccoli seedlings. J Exp Bot 61:3475–3491
Jammes F, Song C, Shin D, Munemasa S, Takeda K, Gu D, Cho D, Lee S, Giordo R, Sritubtim S, Leonhardt N, Ellis BE, Murata Y, Kwak JM (2009) MAP kinases MPK9 and MPK12 are preferentially expressed in guard cells and positively regulate ROS-mediated ABA signaling. Proc Natl Acad Sci USA 106:20520–20525
Jaspers P, Kangasjarvi J (2010) Reactive oxygen species in abiotic stress signaling. Physiol Plant 138:405–413
Jia L (2011) Is reactive oxygen species (ROS) the underlying factor for inhibited root growth in Osspr1? Plant Signal Behav 6:1024–1025
Jiang C, Belfield EJ, Mithani A, Visscher A, Ragoussis J, Mott R, Smith JA, Harberd NP (2012) ROS-mediated vascular homeostatic control of root-to-shoot soil Na delivery in Arabidopsis. EMBO J 31:4359–4370

Kalachova T, Iakovenko O, Kretinin S, Kravets V (2013) Involvement of phospholipase D and NADPH-oxidase in salicylic acid signaling cascade. Plant Physiol Biochem 66:127–133
Karpinski S, Reynolds H, Karpinska B, Wingsle G, Creissen G, Mullineaux P (1999) Systemic signaling and acclimation in response to excess excitation energy in Arabidopsis. Science 284:654–657
Karpinski S, Szechynska-Hebda M, Wituszynska W, Burdiak P (2013) Light acclimation, retrograde signalling, cell death and immune defences in plants. Plant Cell Environ 36:736–744
Kerchev PI, Pellny TK, Vivancos PD, Kiddle G, Hedden P, Driscoll S, Vanacker H, Verrier P, Hancock RD, Foyer CH (2011) The transcription factor ABI4 Is required for the ascorbic acid-dependent regulation of growth and regulation of jasmonate-dependent defense signaling pathways in Arabidopsis. Plant Cell 23:3319–3334
Khokon MA, Hossain MA, Munemasa S, Uraji M, Nakamura Y, Mori IC, Murata Y (2010) Yeast elicitor-induced stomatal closure and peroxidase-mediated ROS production in Arabidopsis. Plant Cell Physiol 51:1915–1921
Khokon AR, Okuma E, Hossain MA, Munemasa S, Uraji M, Nakamura Y, Mori IC, Murata Y (2011) Involvement of extracellular oxidative burst in salicylic acid-induced stomatal closure in Arabidopsis. Plant Cell Environ 34:434–443
Kim MJ, Ciani S, Schachtman DP (2010) A peroxidase contributes to ROS production during Arabidopsis root response to potassium deficiency. Mol Plant 3:420–427
Kim Y, Wang M, Bai Y, Zeng Z, Guo F, Han N, Bian H, Wang J, Pan J, Zhu M (2014) Bcl-2 suppresses activation of VPEs by inhibiting cytosolic Ca(2)(+) level with elevated K(+) efflux in NaCl-induced PCD in rice. Plant Physiol Biochem 80:168–175
Kimura S, Kaya H, Kawarazaki T, Hiraoka G, Senzaki E, Michikawa M, Kuchitsu K (2012) Protein phosphorylation is a prerequisite for the Ca^{2+}-dependent activation of Arabidopsis NADPH oxidases and may function as a trigger for the positive feedback regulation of Ca^{2+} and reactive oxygen species. Biochim Biophys Acta Mol Cell Res 1823:398–405
Kobayashi M, Ohura I, Kawakita K, Yokota N, Fujiwara M, Shimamoto K, Doke N, Yoshioka H (2007) Calcium-dependent protein kinases regulate the production of reactive oxygen species by potato NADPH oxidase. Plant Cell 19:1065–1080
Koo AJ, Gao X, Jones AD, Howe GA (2009) A rapid wound signal activates the systemic synthesis of bioactive jasmonates in Arabidopsis. Plant J Cell Mol Biol 59:974–986
Koussevitzky S, Nott A, Mockler TC, Hong F, Sachetto-Martins G, Surpin M, Lim J, Mittler R, Chory J (2007) Signals from chloroplasts converge to regulate nuclear gene expression. Science 316:715–719
Krishnamurthy A, Rathinasabapathi B (2013a) Auxin and its transport play a role in plant tolerance to arsenite-induced oxidative stress in *Arabidopsis thaliana*. Plant Cell Environ 36:1838–1849
Krishnamurthy A, Rathinasabapathi B (2013b) Oxidative stress tolerance in plants: novel interplay between auxin and reactive oxygen species signaling. Plant Signal Behav. doi:10.4161/psb.25761
Kunihiro S, Hiramatsu T, Kawano T (2011) Involvement of salicylic acid signal transduction in aluminum-responsive oxidative burst in *Arabidopsis thaliana* cell suspension culture. Plant Signal Behav 6:611–616
Kwak JM, Mori IC, Pei ZM, Leonhardt N, Torres MA, Dangl JL, Bloom RE, Bodde S, Jones JD, Schroeder JI (2003) NADPH oxidase AtrbohD and AtrbohF genes function in ROS-dependent ABA signaling in Arabidopsis. EMBO J 22:2623–2633
Kwasniewski M, Chwialkowska K, Kwasniewska J, Kusak J, Siwinski K, Szarejko I (2013) Accumulation of peroxidase-related reactive oxygen species in trichoblasts correlates with root hair initiation in barley. J Plant Physiol 170:185–195
Laloi C, Stachowiak M, Pers-Kamczyc E, Warzych E, Murgia I, Apel K (2007) Cross-talk between singlet oxygen- and hydrogen peroxide-dependent signaling of stress responses in *Arabidopsis thaliana*. Proc Natl Acad Sci USA 104:672–677

Lee KP, Kim C, Landgraf F, Apel K (2007) EXECUTER1- and EXECUTER2-dependent transfer of stress-related signals from the plastid to the nucleus of *Arabidopsis thaliana*. Proc Natl Acad Sci USA 104:10270–10275
Leon J (2013) Role of plant peroxisomes in the production of jasmonic acid-based signals. Subcell Biochem 69:299–313
Li Z, Wakao S, Fischer BB, Niyogi KK (2009) Sensing and responding to excess light. Annu Rev Plant Biol 60:239–260
Li Z, Yue H, Xing D (2012) MAP Kinase 6-mediated activation of vacuolar processing enzyme modulates heat shock-induced programmed cell death in Arabidopsis. New Phytol 195:85–96
Li P, Chen L, Zhou Y, Xia X, Shi K, Chen Z, Yu J (2013a) Brassinosteroids-induced systemic stress tolerance was associated with increased transcripts of several defence-related genes in the phloem in. PLoS One 8:e66582
Li Y, Chen L, Mu J, Zuo J (2013b) LESION SIMULATING DISEASE1 interacts with catalases to regulate hypersensitive cell death in Arabidopsis. Plant Physiol 163:1059–1070
Lin F, Ding HD, Wang JX, Zhang H, Zhang AY, Zhang Y, Tan MP, Dong W, Jiang MY (2009) Positive feedback regulation of maize NADPH oxidase by mitogen-activated protein kinase cascade in abscisic acid signalling. J Exp Bot 60:3221–3238
Liu YH, Offler CE, Ruan YL (2013) Regulation of fruit and seed response to heat and drought by sugars as nutrients and signals. Front Plant Sci 4:282
Lv WT, Lin B, Zhang M, Hua XJ (2011) Proline accumulation is inhibitory to Arabidopsis seedlings during heat stress. Plant Physiol 156:1921–1933
Ma L, Zhang H, Sun L, Jiao Y, Zhang G, Miao C, Hao F (2012) NADPH oxidase AtrbohD and AtrbohF function in ROS-dependent regulation of Na(+)/K(+)homeostasis in Arabidopsis under salt stress. J Exp Bot 63:305–317
Marino D, Dunand C, Puppo A, Pauly N (2012) A burst of plant NADPH oxidases. Trends Plant Sci 17:9–15
Maruta T, Inoue T, Noshi M, Tamoi M, Yabuta Y, Yoshimura K, Ishikawa T, Shigeoka S (2012) Cytosolic ascorbate peroxidase 1 protects organelles against oxidative stress by wounding- and jasmonate-induced H(2)O(2) in Arabidopsis plants. Biochim Biophys Acta 1820:1901–1907
McInnis SM, Desikan R, Hancock JT, Hiscock SJ (2006) Production of reactive oxygen species and reactive nitrogen species by angiosperm stigmas and pollen: potential signalling crosstalk? New Phytol 172:221–228
Miller G, Honig A, Stein H, Suzuki N, Mittler R, Zilberstein A (2009a) Unraveling delta1-pyrroline-5-carboxylate-proline cycle in plants by uncoupled expression of proline oxidation enzymes. J Biol Chem 284:26482–26492
Miller G, Schlauch K, Tam R, Cortes D, Torres MA, Shulaev V, Dangl JL, Mittler R (2009b) The plant NADPH oxidase RBOHD mediates rapid systemic signaling in response to diverse stimuli. Sci Signal 2:ra45
Mittler R (2002) Oxidative stress, antioxidants and stress tolerance. Trends Plant Sci 7:405–410
Mittler R, Vanderauwera S, Gollery M, Van Breusegem F (2004) Reactive oxygen gene network of plants. Trends Plant Sci 9:490–498
Mittler R, Vanderauwera S, Suzuki N, Miller G, Tognetti VB, Vandepoele K, Gollery M, Shulaev V, Van Breusegem F (2011) ROS signaling: the new wave? Trends Plant Sci 16:300–309
Miura K, Okamoto H, Okuma E, Shiba H, Kamada H, Hasegawa PM, Murata Y (2013) SIZ1 deficiency causes reduced stomatal aperture and enhanced drought tolerance via controlling salicylic acid-induced accumulation of reactive oxygen species in Arabidopsis. Plant J 73:91–104
Moller IM, Sweetlove LJ (2010) ROS signalling–specificity is required. Trends Plant Sci 15:370–374
Monetti E, Kadono T, Tran D, Azzarello E, Arbelet-Bonnin D, Biligui B, Briand J, Kawano T, Mancuso S, Bouteau F (2014) Deciphering early events involved in hyperosmotic stress-induced programmed cell death in tobacco BY-2 cells. J Exp Bot 65:1361–1375

Monshausen GB, Bibikova TN, Messerli MA, Shi C, Gilroy S (2007) Oscillations in extracellular pH and reactive oxygen species modulate tip growth of Arabidopsis root hairs. Proc Natl Acad Sci USA 104:20996–21001

Monshausen GB, Bibikova TN, Weisenseel MH, Gilroy S (2009) Ca^{2+} regulates reactive oxygen species production and pH during mechanosensing in Arabidopsis roots. Plant Cell 21:2341–2356

Montillet JL, Hirt H (2013) New checkpoints in stomatal defense. Trends Plant Sci 18:295–297

Mori IC, Pinontoan R, Kawano T, Muto S (2001) Involvement of superoxide generation in salicylic acid-induced stomatal closure in *Vicia faba*. Plant Cell Physiol 42:1383–1388

Muhlenbock P, Szechynska-Hebda M, Plaszczyca M, Baudo M, Mullineaux PM, Parker JE, Karpinska B, Karpinski S (2008) Chloroplast signaling and LESION SIMULATING DISEASE1 regulate crosstalk between light acclimation and immunity in Arabidopsis. Plant Cell 20:2339–2356

Mullineaux P, Karpinski S (2002) Signal transduction in response to excess light: getting out of the chloroplast. Curr Opin Plant Biol 5:43–48

Munn-Bosch S, Alegre L (2004) Die and let live: leaf senescence contributes to plant survival under drought stress. Funct Plant Biol 31:203–216

Nakagami H, Soukupova H, Schikora A, Zarsky V, Hirt H (2006) A mitogen-activated protein kinase kinase kinase mediates reactive oxygen species homeostasis in Arabidopsis. J Biol Chem 281:38697–38704

Ng S, Ivanova A, Duncan O, Law SR, Van Aken O, De Clercq I, Wang Y, Carrie C, Xu L, Kmiec B, Walker H, Van Breusegem F, Whelan J, Giraud E (2013) A membrane-bound NAC transcription factor, ANAC017, mediates mitochondrial retrograde signaling in Arabidopsis. Plant Cell 25:3450–3471

Nishimura MT, Dangl JL (2010) Arabidopsis and the plant immune system. Plant J Cell Mol Biol 61:1053–1066

O'Brien JA, Daudi A, Butt VS, Bolwell GP (2012) Reactive oxygen species and their role in plant defence and cell wall metabolism. Planta 236:765–779

Oda T, Hashimoto H, Kuwabara N, Akashi S, Hayashi K, Kojima C, Wong HL, Kawasaki T, Shimamoto K, Sato M, Shimizu T (2010) Structure of the N-terminal regulatory domain of a plant NADPH oxidase and its functional implications. J Biol Chem 285:1435–1445

Oelze ML, Vogel MO, Alsharafa K, Kahmann U, Viehhauser A, Maurino VG, Dietz KJ (2012) Efficient acclimation of the chloroplast antioxidant defence of *Arabidopsis thaliana* leaves in response to a 10- or 100-fold light increment and the possible involvement of retrograde signals. J Exp Bot 63:1297–1313

Ogasawara Y, Kaya H, Hiraoka G, Yumoto F, Kimura S, Kadota Y, Hishinuma H, Senzaki E, Yamagoe S, Nagata K, Nara M, Suzuki K, Tanokura M, Kuchitsu K (2008) Synergistic activation of the Arabidopsis NADPH oxidase AtrbohD by Ca^{2+} and phosphorylation. J Biol Chem 283:8885–8892

op den Camp RG, Przybyla D, Ochsenbein C, Laloi C, Kim C, Danon A, Wagner D, Hideg E, Gobel C, Feussner I, Nater M, Apel K (2003) Rapid induction of distinct stress responses after the release of singlet oxygen in Arabidopsis. Plant Cell 15:2320–2332

Padmanabhan MS, Dinesh-Kumar SP (2010) All hands on deck-the role of chloroplasts, endoplasmic reticulum, and the nucleus in driving plant innate immunity. Mol Plant Microbe Interact 23:1368–1380

Peer WA, Cheng Y, Murphy AS (2013) Evidence of oxidative attenuation of auxin signalling. J Exp Bot 64:2629–2639

Petrov VD, Van Breusegem F (2012) Hydrogen peroxide-a central hub for information flow in plant cells. AoB Plants 2012:pls014

Petrov V, Hille J, Mueller-Roeber B, Gechev TS (2015) ROS-mediated abiotic stress-induced programmed cell death in plants. Front Plant Sci 6:69

Pfannschmidt T (2010) Plastidial retrograde signalling–a true "plastid factor" or just metabolite signatures? Trends Plant Sci 15:427–435

Pitzschke A, Djamei A, Bitton F, Hirt H (2009) A major role of the MEKK1-MKK1/2-MPK4 pathway in ROS signalling. Mol Plant 2:120–137
Pogson BJ, Woo NS, Forster B, Small ID (2008) Plastid signalling to the nucleus and beyond. Trends Plant Sci 13:602–609
Qiu JL, Zhou L, Yun BW, Nielsen HB, Fiil BK, Petersen K, Mackinlay J, Loake GJ, Mundy J, Morris PC (2008) Arabidopsis mitogen-activated protein kinase kinases MKK1 and MKK2 have overlapping functions in defense signaling mediated by MEKK1, MPK4, and MKS1. Plant Physiol 148:212–222
Rajhi I, Yamauchi T, Takahashi H, Nishiuchi S, Shiono K, Watanabe R, Mliki A, Nagamura Y, Tsutsumi N, Nishizawa NK, Nakazono M (2011) Identification of genes expressed in maize root cortical cells during lysigenous aerenchyma formation using laser microdissection and microarray analyses. New Phytol 190:351–368
Rhoads DM, Subbaiah CC (2007) Mitochondrial retrograde regulation in plants. Mitochondrion 7:177–194
Rossel JB, Wilson PB, Hussain D, Woo NS, Gordon MJ, Mewett OP, Howell KA, Whelan J, Kazan K, Pogson BJ (2007) Systemic and intracellular responses to photooxidative stress in Arabidopsis. Plant Cell 19:4091–4110
Ruckle ME, Burgoon LD, Lawrence LA, Sinkler CA, Larkin RM (2012) Plastids are major regulators of light signaling in Arabidopsis. Plant Physiol 159:366–390
Sagi M, Davydov O, Orazova S, Yesbergenova Z, Ophir R, Stratmann JW, Fluhr R (2004) Plant respiratory burst oxidase homologs impinge on wound responsiveness and development in *Lycopersicon esculentum*. Plant Cell 16:616–628
Santino A, Taurino M, De Domenico S, Bonsegna S, Poltronieri P, Pastor V, Flors V (2013) Jasmonate signaling in plant development and defense response to multiple (a)biotic stresses. Plant Cell Rep 32:1085–1098
Shabala S, Shabala L, Barcelo J, Poschenrieder C (2014) Membrane transporters mediating root signalling and adaptive responses to oxygen deprivation and soil flooding. Plant Cell Environ 37:2216–2233
Shah J, Zeier J (2013) Long-distance communication and signal amplification in systemic acquired resistance. Front Plant Sci 4:30
Shapiguzov A, Vainonen JP, Wrzaczek M, Kangasjarvi J (2012) ROS-talk-how the apoplast, the chloroplast, and the nucleus get the message through. Front Plant Sci 3:292
Sierla M, Rahikainen M, Salojarvi J, Kangasjarvi J, Kangasjarvi S (2013) Apoplastic and chloroplastic redox signaling networks in plant stress responses. Antioxid Redox Signal 18:2220–2239
Sirichandra C, Gu D, Hu HC, Davanture M, Lee S, Djaoui M, Valot B, Zivy M, Leung J, Merlot S, Kwak JM (2009) Phosphorylation of the Arabidopsis AtrbohF NADPH oxidase by OST1 protein kinase. FEBS Lett 583:2982–2986
Soares NC, Francisco R, Vielba JM, Ricardo CP, Jackson PA (2009) Associating wound-related changes in the apoplast proteome of Medicago with early steps in the ROS signal-transduction pathway. J Proteome Res 8:2298–2309
Song Y, Miao Y, Song CP (2014) Behind the scenes: the roles of reactive oxygen species in guard cells. New Phytol 201:1121–1140
Spoel SH, Dong X (2012) How do plants achieve immunity? Defence without specialized immune cells. Nat Rev Immunol 12:89–100
Steffens B, Geske T, Sauter M (2011) Aerenchyma formation in the rice stem and its promotion by H_2O_2. New Phytol 190:369–378
Suzuki N, Mittler R (2012) Reactive oxygen species-dependent wound responses in animals and plants. Free Radic Biol Med 53:2269–2276
Suzuki N, Miller G, Morales J, Shulaev V, Torres MA, Mittler R (2011) Respiratory burst oxidases: the engines of ROS signaling. Curr Opin Plant Biol 14:691–699
Suzuki N, Koussevitzky S, Mittler R, Miller G (2012) ROS and redox signalling in the response of plants to abiotic stress. Plant Cell Environ 35:259–270

Suzuki N, Miller G, Salazar C, Mondal HA, Shulaev E, Cortes DF, Shuman JL, Luo X, Shah J, Schlauch K, Shulaev V, Mittler R (2013) Temporal-spatial interaction between reactive oxygen species and abscisic acid regulates rapid systemic acclimation in plants. Plant Cell 25:3553–3569

Szechynska-Hebda M, Karpinski S (2013) Light intensity-dependent retrograde signalling in higher plants. J Plant Physiol 170:1501–1516

Szechynska-Hebda M, Kruk J, Gorecka M, Karpinska B, Karpinski S (2010) Evidence for light wavelength-specific photoelectrophysiological signaling and memory of excess light episodes in Arabidopsis. Plant Cell 22:2201–2218

Taj G, Agarwal P, Grant M, Kumar A (2010) MAPK machinery in plants: recognition and response to different stresses through multiple signal transduction pathways. Plant Signal Behav 5:1370–1378

Takahashi F, Yoshida R, Ichimura K, Mizoguchi T, Seo S, Yonezawa M, Maruyama K, Yamaguchi-Shinozaki K, Shinozaki K (2007) The mitogen-activated protein kinase cascade MKK3-MPK6 is an important part of the jasmonate signal transduction pathway in Arabidopsis. Plant Cell 19:805–818

Takahashi F, Mizoguchi T, Yoshida R, Ichimura K, Shinozaki K (2011) Calmodulin-dependent activation of MAP kinase for ROS homeostasis in Arabidopsis. Mol Cell 41:649–660

Takeda S, Gapper C, Kaya H, Bell E, Kuchitsu K, Dolan L (2008) Local positive feedback regulation determines cell shape in root hair cells. Science 319:1241–1244

Teakle NL, Tyerman SD (2010) Mechanisms of Cl(-) transport contributing to salt tolerance. Plant Cell Environ 33:566–589

Teige M, Scheikl E, Eulgem T, Doczi R, Ichimura K, Shinozaki K, Dangl JL, Hirt H (2004) The MKK2 pathway mediates cold and salt stress signaling in Arabidopsis. Mol Cell 15:141–152

Torres MA, Dangl JL (2005) Functions of the respiratory burst oxidase in biotic interactions, abiotic stress and development. Curr Opin Plant Biol 8:397–403

Torres MA, Jones JD, Dangl JL (2005) Pathogen-induced, NADPH oxidase-derived reactive oxygen intermediates suppress spread of cell death in *Arabidopsis thaliana*. Nat Genet 37:1130–1134

Toumi I, Moschou PN, Paschalidis KA, Bouamama B, Ben Salem-Fnayou A, Ghorbel AW, Mliki A, Roubelakis-Angelakis KA (2010) Abscisic acid signals reorientation of polyamine metabolism to orchestrate stress responses via the polyamine exodus pathway in grapevine. J Plant Physiol 167:519–525

Tripathy BC, Oelmuller R (2012) Reactive oxygen species generation and signaling in plants. Plant Signal Behav 7:1621–1633

Vacca RA, de Pinto MC, Valenti D, Passarella S, Marra E, De Gara L (2004) Production of reactive oxygen species, alteration of cytosolic ascorbate peroxidase, and impairment of mitochondrial metabolism are early events in heat shock-induced programmed cell death in tobacco Bright-Yellow 2 cells. Plant Physiol 134:1100–1112

Voothuluru P, Sharp RE (2013) Apoplastic hydrogen peroxide in the growth zone of the maize primary root under water stress. I. Increased levels are specific to the apical region of growth maintenance. J Exp Bot 64:1223–1233

Wagner D, Przybyla D, Op den Camp R, Kim C, Landgraf F, Lee KP, Wursch M, Laloi C, Nater M, Hideg E, Apel K (2004) The genetic basis of singlet oxygen-induced stress responses of *Arabidopsis thaliana*. Science 306:1183–1185

Wang GF, Li WQ, Li WY, Wu GL, Zhou CY, Chen KM (2013) Characterization of rice NADPH oxidase genes and their expression under various environmental conditions. Int J Mol Sci 14:9440–9458

Wituszynska W, Szechynska-Hebda M, Sobczak M, Rusaczonek A, Kozlowska-Makulska A, Witon D, Karpinski S (2015) Lesion simulating disease 1 and enhanced disease susceptibility 1 differentially regulate UV-C-induced photooxidative stress signalling and programmed cell death in *Arabidopsis thaliana*. Plant Cell Environ 38:315–330

Woodson JD, Chory J (2008) Coordination of gene expression between organellar and nuclear genomes. Nat Rev Genet 9:383–395

Wrzaczek M, Brosche M, Kangasjarvi J (2013) ROS signaling loops-production, perception, regulation. Curr Opin Plant Biol 16:575–582

Xia XJ, Wang YJ, Zhou YH, Tao Y, Mao WH, Shi K, Asami T, Chen Z, Yu JQ (2009) Reactive oxygen species are involved in brassinosteroid-induced stress tolerance in cucumber. Plant Physiol 150:801–814

Xia XJ, Zhou YH, Ding J, Shi K, Asami T, Chen Z, Yu JQ (2011) Induction of systemic stress tolerance by brassinosteroid in *Cucumis sativus*. New Phytol 191:706–720

Xie YJ, Xu S, Han B, Wu MZ, Yuan XX, Han Y, Gu Q, Xu DK, Yang Q, Shen WB (2011) Evidence of Arabidopsis salt acclimation induced by up-regulation of HY1 and the regulatory role of RbohD-derived reactive oxygen species synthesis. Plant J Cell Mol Biol 66:280–292

Xu J, Li Y, Wang Y, Liu H, Lei L, Yang H, Liu G, Ren D (2008) Activation of MAPK kinase 9 induces ethylene and camalexin biosynthesis and enhances sensitivity to salt stress in Arabidopsis. J Biol Chem 283:26996–27006

Zeng W, He SY (2010) A prominent role of the flagellin receptor FLAGELLIN-SENSING2 in mediating stomatal response to *Pseudomonas syringae* pv tomato DC3000 in Arabidopsis. Plant Physiol 153:1188–1198

Zhang HJ, Fang Q, Zhang ZG, Wang YC, Zheng XB (2009a) The role of respiratory burst oxidase homologues in elicitor-induced stomatal closure and hypersensitive response in *Nicotiana benthamiana*. J Exp Bot 60:3109–3122

Zhang Y, Zhu H, Zhang Q, Li M, Yan M, Wang R, Wang L, Welti R, Zhang W, Wang X (2009b) Phospholipase dalpha1 and phosphatidic acid regulate NADPH oxidase activity and production of reactive oxygen species in ABA-mediated stomatal closure in Arabidopsis. Plant Cell 21:2357–2377

Zhang S, Apel K, Kim C (2014) Singlet oxygen-mediated and EXECUTER-dependent signalling and acclimation of *Arabidopsis thaliana* exposed to light stress. Philos Trans R Soc Lond B Biol Sci 369:20130227

Zimmermann MR, Maischak H, Mithofer A, Boland W, Felle HH (2009) System potentials, a novel electrical long-distance apoplastic signal in plants, induced by wounding. Plant Physiol 149:1593–1600

Redox Regulation and Antioxidant Defence During Abiotic Stress: What Have We Learned from *Arabidopsis* and Its Relatives?

Baris Uzilday, Rengin Ozgur, A. Hediye Sekmen, and Ismail Turkan

Contents

Abstract Abiotic stress conditions are a global constraint that affects plant growth and crop yield worldwide, and this phenomenon is expected to be increased in the forthcoming future due to global climate change. *Arabidopsis thaliana* is the model organism for plant science since the early 1990s, and its genome has been known for more than a decade. Studies conducted with *Arabidopsis* created a foundation that could be transferred and used in its close relatives to similarity of genetic sequences. Up to now, studies on *A. thaliana* gave deep insight into different abiotic stress tolerance mechanisms. However, *A. thaliana* is not a stress-tolerant plant species. Therefore some of the stress tolerance mechanisms that are used by its

B. Uzilday • R. Ozgur • A.H. Sekmen • I. Turkan (✉)
Department of Biology, Faculty of Science, Ege University, 35100 Bornova, Izmir, Turkey
e-mail: ismail.turkan@ege.edu.tr

D.K. Gupta et al. (eds.), *Reactive Oxygen Species and Oxidative Damage in Plants Under Stress*, DOI 10.1007/978-3-319-20421-5_4

stress-tolerant relatives might not even be observed in *Arabidopsis*. This chapter focused on reactive oxygen species (ROS) production during environmental stress and antioxidant defence systems activated against it in *A. thaliana* and its close relatives such as *Thellungiella* sp., *A. halleri, Thlaspi* sp., *Lepidium sativum* and *Arabis paniculata*.

Keywords *Arabidopsis*-related model species (ARMS) • *Arabidopsis* • Antioxidant defence • Drought stress • Heavy metal stress • Redox regulation • Salinity • Oxidative stress

1 Introduction

Arabidopsis thaliana is an invaluable tool for plant biologists since it has been used as a model organism. Its short life cycle, small size, low number of chromosomes, small genome and prolific seed production through self-pollination made *Arabidopsis* a useful tool for genetic studies. Since the production of the first induced mutant with X-ray in 1947 and the development of different protocols for transformation in 1980s and publishing of its full genome in 2000, *A. thaliana* played a pivotal role in the development of plant biology (Koornneef and Meinke 2010). Most of our detailed knowledge about key processes such as plant growth, development, reproduction and response to environment was gained through *A. thaliana*, its mutants, and transgenic lines. Global climate change and rapidly increasing human population brings a new challenge to the field of plant biology: producing enough food to feed everyone even under adverse environmental conditions. For this aim, a better understanding of stress tolerance mechanisms is needed to cultivate marginal lands such as arid or saline areas and to sustain plant production during the climate change. Even though it is an indispensable genetics tool, *A. thaliana* is not a stress-tolerant plant species and some of the stress tolerance mechanisms that are used by its stress-tolerant relatives might not even be observed in *Arabidopsis*. Considering this, some researchers started searching for alternative hosts that evolved to tolerate stress factors at hand (Amtmann et al. 2005). While doing so, selection of plant species that are close relatives of *A. thaliana* provided researchers with the chance to exploit gene sequences and genetics tools available from *Arabidopsis* due to high genetic sequence similarity. In these days, efforts in this area have started to bear its fruits. Model species such as *Thellungiella* as a salt-tolerant model and *A. halleri*, *Arabis* sp. and *Thlaspi* sp. as heavy metal-tolerant models have been very useful tools to investigate tolerance mechanisms to these stresses. In this chapter, we have summarised and discussed studies that are related to redox regulation and antioxidant defence in *Arabidopsis* and its close relative *Arabidopsis*-related model species (ARMS) under abiotic stress such as salinity, drought and heavy metals. For this, first, we will introduce mechanisms of ROS production, and then we will summarise enzymatic and nonenzymatic antioxidant defence mechanisms. Following this, studies in *Arabidopsis* and its relatives in stressed conditions will be discussed.

2 What is ROS and How it is Produced in Plant Cell?

Reactive oxygen species (ROS), which can also be named as active oxygen species (AOS) or reactive oxygen intermediates (ROI), are formed in all aerobic organisms from bacteria to mammalian cells (Bose et al. 2014). They are inevitable by-products of the aerobic metabolism. There are basically four forms of ROS within the cell: singlet oxygen ($^1O_2$), superoxide anion radical ($O_2^{\bullet-}$), hydrogen peroxide (H_2O_2) and hydroxyl radical ($HO^{\bullet}$). The main sites of ROS production in a plant cell are mitochondria, chloroplast, peroxisomes, plasma membrane and apoplast. ROS play a dual role: they are involved in cellular signalling on one hand and on the other hand as toxic products that accumulate under different environmental stress conditions such as salinity, drought, etc. To utilise ROS as signalling molecules, nontoxic levels of ROS must be maintained by the control of ROS production and the metabolic counterprocess involving ROS-scavenging pathways and the expression of ROS-regulated genes (Mittler et al. 2004; Lai et al. 2012). On the other hand, uncontrolled accumulation of ROS causes protein denaturation, lipid peroxidation in cellular membrane, oxidation of DNA and RNA and carbohydrate oxidation and also negatively affects enzymatic activity (Scandalios 1993; Noctor and Foyer 1998).

Under normal conditions, plants continuously produce ROS that are localised in different cellular compartments as by-products of various metabolic pathways. While the main sources of ROS production are chloroplast and peroxisomes in the light (Foyer and Noctor 2003), in darkness, it is mitochondria (Moller 2001). Chloroplasts and peroxisomes produce 20-fold more ROS than mitochondria in the light (Rhoads et al. 2006). However, in the dark and in nongreen tissues, mitochondria are the major sources of ROS. In addition, ROS also can be produced at the apoplastic space by plant NADPH oxidases, cell wall-associated peroxidases (POXs) and oxalate oxidases (Kawano 2003), but reliable quantification of apoplastic ROS accumulation is still unavailable (Wrzaczek et al. 2013).

2.1 *Chloroplasts*

In chloroplasts, the main sites of ROS production are PSI and PSII reaction centres. Under abiotic stresses such as drought and salt stress, decreased availability of CO_2 as a result of limiting gas exchange enhances the production of ROS in chloroplasts. Under normal conditions, the electron flow from excited photosystem centres is directed to $NADP^+$ to produce NADPH; however, when electron transfer chain (ETC) is overloaded, a part of this flow can be diverted from ferredoxin to O_2, resulting in the production of superoxide anion radical ($O_2^{\bullet -}$) through the Mehler reaction. On the other hand, $^1O_2$ is formed in PSII by energy transfer from triplet excited state chlorophyll, which is produced by excess excitation by light, to $^3O_2$, mainly under high light intensities (Laloi et al. 2004). $^1O_2$ can diffuse distances of only hundred nanometers and react with the first available biomolecule. Moreover, it can cause light-induced activity loss in PSII, triggering cell death (Wagner et al. 2004; Krieger-Liszkay et al. 2008). The production of $^1O_2$ can be prevented by the Mehler reaction due to its ability to relax ETC (Asada and Takahashi 1987). Hence, the production of $O_2^{\bullet -}$ might act as an alternative sink for the consumption of electrons in the ETC, and this is called the 'water–water cycle'. The major site of $O_2^{\bullet -}$ production is thylakoid membrane-bound primary antenna pigments. H_2O_2 is the most stable ROS as compared to other free radicals. While the half-lives of $^1O_2$, $O_2^{\bullet -}$ and $OH^\bullet$ are 2–4 μs, the half-life of H_2O_2 is close to a minute (Bhattachrjee 2005; Pitzschke et al. 2006). Among ROS, only H_2O_2 can diffuse some distance from its site of production. Excess accumulation of H_2O_2 in plant cell leads to the occurrence of oxidative stress, leading to necrosis or even programmed cell death (PCD) (Quan et al. 2008). Moreover, it may inactivate enzymes by oxidising thiol groups (Gill and Tuteja 2010). Interestingly, at low concentrations, H_2O_2 can act as a signal molecule, which triggers tolerance to abiotic and biotic stress and is a key regulator in some physiological processes such as growth and development (Foreman et al. 2003), stomatal movement (Bright et al. 2006) and senescence (Peng et al. 2005).

2.2 *Mitochondria*

In mitochondria, under abiotic stress, electrons can be transferred from complex I and complex III to O_2 due to overreduction of electron acceptors, which results in the production of $O_2^{\bullet -}$. $O_2^{\bullet -}$ is reduced by SOD activity to H_2O_2. This H_2O_2 can react with reduced Fe^{2+} and Cu^+ to produce highly toxic $OH^\bullet$, which are lethal (Rhoads et al. 2006; Sweetlove and Foyer 2004). Therefore, in mitochondria, the main sites of ROS production are complex I and complex III. ETC system in mitochondria may prevent this overreduction of electron carriers by means of energy-dissipating systems. For example, plants utilise an alternative oxidase pathway (AOX), which is an alternative to the cytochrome pathway in mitochondria. Electrons are taken from the ubiquinone pool and transferred to oxygen to form H_2O in alternative pathway, and energy is dissipated as heat (Millar et al. 2011). In this way, this system prevents ROS generation. Therefore, mitochondria may play a central role in cell adaptation to abiotic stresses, which are known to induce oxidative stress at cellular level.

2.3 Peroxisomes

$O_2^{\bullet-}$ radicals generated as a consequence of their normal metabolism are produced in peroxisomes. In peroxisomes, one of the main sites of ROS production is organelle matrix where xanthine oxidase (XOD) catalyses the oxidation of xanthine and hypoxanthine to uric acid (Corpas et al. 2001). Apart from mitochondria and plastids, peroxisome membranes have sites for ETC composed of a flavoprotein NADH and cytochrome b. In peroxisome membranes, $O_2^{\bullet-}$ is produced by peroxisome ETC (del Río et al. 2002). H_2O_2 is produced by the main metabolic processes such as β-oxidation of the fatty acids, photorespiratory glycolate oxidase reaction, enzymatic reactions of flavin oxidases and the disproportionation of $O_2^{\bullet-}$ radicals. Similar to chloroplast and mitochondria, peroxisome-generated ROS especially H_2O_2 at low levels also act as mediators in intracellular signalling (Masters 2001). In addition, in peroxisomes, transition metal ions like Fe^{2+} and Cu^{2+} that can catalyse the formation of hydroxyl radical ($OH^{\bullet}$) in the Fenton reaction ($Fe^{2+} + H_2O_2 \rightarrow Fe^{3+} + OH^- + OH$) are also abundant. Excess formation of this ROS can lead to lipid peroxidation, followed by damage to peroxisomal membrane and ultimately resulting in loss of peroxisomal functions (Yokota and Oda 2001). Therefore, peroxisomes are a source of oxidative stress, but it is a compartment where controlled production and scavenging of ROS (mostly due to photorespiration) simultaneously occur.

2.4 Other Sources of ROS Production

ROS are also produced at the apoplastic space and plasma membrane. This production is stimulated by process like ozone fumigation and during pathogen attack. H_2O_2 can be produced in apoplast by pH-dependent cell wall peroxidases, germin-like oxalate oxidases, amine oxidases and glycolate oxidases (Kawano 2003). Moreover, in plasma membranes, other sources of this radical are NADPH-dependent oxidases (NADPH oxidases, known in plants as respiratory burst oxidase homologues [RBOH]) (Baxter et al. 2014). Also, oxalate oxidase was shown to be involved in ROS production in root cells during drought stress (Voothuluru and Sharp 2013), whereas glycolate oxidase (GOX) has been shown to play a role in nonhost pathogen defence in *Arabidopsis* and tobacco (Rojas and Mysore 2012; Rojas et al. 2012).

3 Antioxidant Defence Mechanism

Plants possess an array of antioxidant defence enzymes to cope with detrimental effects of ROS. These enzymes work in coordination to scavenge $O_2^{\bullet-}$, H_2O_2 and other free radicals. Among different ROS mentioned above, $OH^{\bullet}$ and $^1O_2$ cannot be scavenged by enzymatic antioxidants, and their detoxification is purely dependent on nonenzymatic low-molecular-weight antioxidants. Although mechanisms underlying the production of ROS in plants under stress are universal, based on plant species, different antioxidant defence strategies are involved to scavenge ROS. The mechanism of different types of enzymatic and nonenzymatic antioxidants will be briefly discussed with special emphasis on *Arabidopsis*. Our aim of focusing discussion in *Arabidopsis* is that although most of the enzymes were first identified in other species such as spinach, maize or barley, molecular mechanisms underlying their activities, their regulation and their interactions with other cellular components were first studied and elucidated in *Arabidopsis* due to genetic and molecular tools available.

3.1 Superoxide Dismutase

Among antioxidant enzymes, SOD has a unique role: it is the only enzyme that can scavenge $O_2^{\bullet-}$ and convert it to H_2O_2. Therefore, cellular defence against $O_2^{\bullet-}$, which is the main ROS produced during photosynthesis and respiration, is dependent on SOD activity. As it can be seen, SOD activity is the first step in the detoxification of the $O_2^{\bullet-}$, and hence, it has been named as the 'first line of defence' against oxidative stress. Delicate balance between SOD and other antioxidant enzymes that scavenge H_2O_2 such as APX, CAT, peroxidases and peroxiredoxins needs to be maintained in the cell for full detoxification of ROS. If SOD activity exceeds the capacity of H_2O_2-scavenging enzymes, H_2O_2 can accumulate and exert its toxic effects. On the other hand, insufficient SOD activity leads to the accumulation of $O_2^{\bullet-}$, which can also cause oxidative damage.

It is known that phospholipid membranes are not permeable to $O_2^{\bullet-}$ due to its negative charge (Takahashi and Asada 1983). Therefore $O_2^{\bullet-}$ that is produced in a compartment needs to be scavenged in the same organelle. For this reason, plants have a wide arsenal of SOD isoforms that work in different cellular compartments. Three types of SOD have been identified in plants that differ by their cofactor in the active site (Cu and Zn, Fe and Mn). These enzymes are located in different compartments of the cell and scavenge $O_2^{\bullet-}$ in their specific locations. FeSOD is believed to be the most ancient isoform due to the abundance of iron in soluble Fe (II) form at that time (Bannister et al. 1991). As the levels of O_2 in the environment increased and minerals were oxidised, Fe (II) become less available and Mn (III) was utilised in the active site of SOD (MnSOD). Similarly, increasing O_2 concentration in the atmosphere caused the conversion of insoluble Cu (I) to soluble

Cu (II). At this phase, Cu (II) began to be used at the active site of SOD (Cu/ZnSOD) (Alscher et al. 2002).

FeSOD is localised in chloroplasts, while MnSOD is localised in mitochondria. Cu/ZnSOD can be found in different compartments such as chloroplasts, cytosol and peroxisomes. In *Arabidopsis*, there are three genes encoding FeSOD (*FSD1*, *FSD2*, *FSD3*), three genes encoding Cu/ZnSOD (*CSD1*, *CSD2*, *CSD3*) and one gene encoding MnSOD (*MSD1*). Among these *CSD* genes, *CSD1* and *CSD2* encode cytoplasmic and chloroplastic isoforms, respectively, whereas *CSD3* encodes the peroxisomal Cu/ZnSOD isoform. It has been shown that *FSD1* expression is under circadian regulation in response to a diurnal rhythm, while other isoforms of SOD respond to oxidative stress (Kliebenstein et al. 1998).

3.2 Catalase

Catalase was first described in 1900 by Oscar Loew, and it is the first antioxidant enzyme to be discovered. This enzyme was named as 'catalase' by the author because of 'its catalytic action on hydrogen peroxide' (Loew 1900). Catalase dismutates two molecules of H_2O_2 to water and O_2. All known eukaryotic catalase are haem containing tetramer proteins, and each polypeptide in a tetramer contains a haem prosthetic group (Regelsberger et al. 2002). Investigations showed that this enzyme is encoded by three genes in all angiosperms species studied till date. Products of all three genes encode 492 amino acid polypeptides, which show a high similarity between each other. In *Arabidopsis*, *CAT1* gene is mainly expressed in pollen and seeds, while CAT2 is expressed in photosynthetic tissues. *CAT3* is expressed in vascular tissues and leaves. Among these genes, *CAT2* follows a circadian rhythm similar to photosynthetic enzymes, while *CAT3* follows an opposite one. Although there are three genes encoding CAT in *Arabidopsis*, seven different bands can be detected in native activity gels (Hu et al. 2010). This is due to heterotetrameric nature of the enzyme. According to this, if three CAT genes encode a heterotetrameric protein, there can be 15 possible distinct isoforms (McClung 1997). However, fewer than the possible number of catalase heterotetramers might indicate a specific interaction between different subunits, eventually limiting the number of isoforms (McClung 1997).

In plants, CAT is mainly found in peroxisomes and is responsible for scavenging of photorespiratory H_2O_2 and H_2O_2 produced by β-oxidation of fatty acids. In *Arabidopsis*, leaf CAT activity is reduced by 80 and 20 % in *cat2* and *cat3* knockout plants, respectively, while there is no evident decrease in *cat3* knockouts. Moreover, double *cat2 cat1* and *cat2 cat3* knockout plants have similar CAT activities to that of *cat2* knockout plants (Bueso et al. 2007). When grown at irradiances between 50 and 200 μmol m^{-2} s^{-1}, *cat2* plants showed a dwarf phenotype, which became more severe with the increasing light intensity. This phenotype under increasing light intensity was caused by redox perturbation due to insufficient scavenging of H_2O_2 (Queval et al. 2007). Contrary to *cat2*, *cat1* and *cat3* plants did not show any obvious leaf phenotype, which was also consistent with the reduction in leaf CAT activity (Hu et al. 2010). In addition, Hu et al. (2010) showed that *cat2* phenotype could be complemented with CAT2 or CAT3 protein expressed under *CAT2* promoter, indicating that the function of CAT2 is rather regulated at transcriptional level. Previously it was reported that calmodulin (CaM) binds to CAT3 in Ca^{2+}-dependent manner and regulates its activity, indicating an interaction between Ca^{2+} signalling and the regulation of CAT activity (Yang and Poovaiah 2002).

3.3 Ascorbate Peroxidase and Other Ascorbate–Glutathione Cycle Enzymes

Ascorbate peroxidase (APX) is a Class I peroxidase that exclusively uses ascorbate (AsA) as its electron donor to scavenge H_2O_2. During this reaction, two AsA molecules are oxidised to yield two monodehydroascorbate (MDHA) molecules and H_2O_2 is converted to water. MDHA formed is spontaneously disproportionated to AsA and dehydroascorbate (DHA). MDHA can also be converted directly to AsA via the action of NADPH-dependent enzyme monodehydroascorbate reductase (MDHAR). On the other hand, oxidised DHA can be regenerated back to AsA by dehydroascorbate reductase (DHAR), which utilises GSH as a source of reducing power. Oxidised GSH (GSSG) is then regenerated back to GSH by glutathione reductase (GR) with NADPH. In conclusion, APX supported by an efficient AsA–GSH cycle for the regeneration of electron donor AsA prevents the accumulation of toxic levels of H_2O_2 in the plant cell. This regeneration mechanism is also important for sustainable activity of APX enzyme. When the concentration of AsA is lower than 20 μM in the medium, APX activity was rapidly and irreversibly lost. This inactivation occurred in 30 s for chloroplastic APX isoforms, while it was approximately an hour for cytosolic APX isoforms (De Leonardis et al. 2000).

In *Arabidopsis*, there are at least nine different isoforms of APX (Mittler et al. 2004). The chloroplast contains at least three isoforms, which are located in different parts of the chloroplasts such as lumen (*APX4*), stroma (*sAPX*) and thylakoids (*tAPX*). Stromal APX can also be targeted to mitochondria (Chew et al. 2003). In some plant species, stromal and thylakoid isoforms of APX are encoded by a single gene, and different isoforms are a product of alternative splicing. However, in *Arabidopsis*, there are two separate genes for each isoform (Shigeoka et al. 2002). It is known that many photosynthetic enzymes can be regulated via oxidation and reduction of their thiol groups (Dietz and Pfannschmidt 2011); therefore fine-tuning of H_2O_2 concentrations in chloroplasts is vital and APX plays a major role in this regulation. In contrast to a variety of chloroplastic APX isoforms, cytoplasm of *Arabidopsis* only contains one APX isoform (*APX1*) under normal conditions, and an additional isoform (*APX2*) was induced under high light stress (Karpinski et al. 1999). In addition, one isoform of APX (*APX3*) was reported in peroxisomes, site that is known to be a major source of H_2O_2 in plant cells. Similar to APX, isoforms of AsA-GSH cycle enzymes are also diverse. There are five *MDHAR*, five *DHAR* and two *GR* genes in *Arabidopsis*, which are dispersed to cytosol, chloroplasts and mitochondrial compartments. Like stromal APX, products of *MDHAR1* and *DHAR1* and *GR2* genes are also dual targeted to chloroplasts and mitochondria.

3.4 Glutathione Peroxidase

Glutathione peroxidise (GPX) is the general name for the family of enzymes that catalyse the reduction of H_2O_2, lipid peroxides and organic peroxides by utilising GSH as electron donor (Ursini et al. 1997). The global phylogenetic analysis of GPX between vertebrates, invertebrates, bacteria, fungi and plants showed that plant GPXs form an independent cluster (Margis et al. 2008). An ancestral gene led to the origin of all plant GPX genes, and major duplication events of this gene to generate known GPXs occurred before the divergence of monocots and dicots. There are 8 genes encoding GPX in *Arabidopsis* (Mittler et al. 2004). Among these *GPX1*, *GPX6* and *GPX7* are predicted to be localised in chloroplasts, while *GPX3* is in mitochondria and *GPX5* is in endoplasmic reticulum. Besides these organellar isoforms, there are also two additional cytoplasmic isoforms *GPX4* and *GPX8* (Mittler et al. 2004). These isoforms show spatiotemporal expression pattern. Among these, *GPX1* is strongly expressed in leaves, stems and flowers; however, it is not expressed in roots indicating a role for defence against photosynthetic ROS (Milla et al. 2003). Milla et al. (2003) also showed that these genes were differentially induced under abiotic stresses such as salinity, osmotic stress, cold and high temperature. Under salinity, *GPX1*, *GPX2*, *GPX 5* and *GPX 6* were induced, while under osmotic stress, all of the *GPX* genes were induced. However, *GPX1* was the only gene to respond to high temperature stress. Recently, Passaia et al. (2014) showed that *GPX*s can regulate root phenotype of *A. thaliana* plants. By using different *gpx* mutants, they demonstrated that *gpx1*, *gpx4*, *gpx5*, *gpx7* and *gpx8* mutants had a greater lateral root density, when compared to wild type, while, *gpx2* and *gpx3* mutants had lower lateral root densities.

3.5 Peroxiredoxins

Peroxiredoxins (Prxs) are a family of thiol-specific reductases or peroxidases that catalyse the reduction of a range of different peroxides such as H_2O_2, peroxynitrites and alkyl hydroperoxides to water, nitrite or corresponding alcohol (Hofmann et al. 2002). They are nonheme-containing peroxidases that require an external electron donor (Rouhier and Jacquot 2002). When Prxs reduce a substrate and become oxidised, they lose their enzymatic activity, and for the next catalytic cycle, they need to be regenerated (Dietz et al. 2002). It was shown that for many Prxs, thioredoxins (TRXs) serve as electron donor to regenerate the enzyme as well as other dithiols such as glutathione (Dietz et al. 2002; Finkemeier et al. 2005). Four out of six subclasses of Prx have been reported in plants, and these are 1-Cys Prx, 2-Cys Prx, type II Prx and PrxQ (Dietz et al. 2002). 1-Cys Prx contains a single catalytic cysteine, and it gets its name from this single cysteine. 2-Cys Prx is a dimeric enzyme with two cysteine residues in each polypeptide, and in plants, they are localised exclusively in chloroplasts. Similarly, type II Prx contains two conserved cysteines, but these Prxs are localised in other subcellular compartments such as chloroplasts, mitochondria and cytosol (Finkemeier et al. 2005; Bréhélin et al. 2003). The fourth type of Prxs is PrxQ, which are monomeric enzymes that contain two cysteines separated by four amino acids (Rouhier and Jacquot 2002). In *Arabidopsis,* there are ten genes that encode Prx genes (Dietz et al. 2002). These encode one 1-Cys Prx, two 2-Cys Prxs (2-Cys Prxs A and B), one PrxQ and six type II Prxs (type II Prxs A–F). Baier et al. (2000) demonstrated that antisense suppression of *Arabidopsis* 2-Cys Prx leads to increased oxidation of Asc pool. In response to this, stromal and thylakoid APX and MDHAR expressions were increased. Induction of H_2O_2-scavenging mechanism by APX under decreased 2-Cys Prx expression and oxidation of Asc pool under these circumstances suggested that 2-Cys Prx plays a major role in the detoxification of H_2O_2 in chloroplasts. Although 2-Cys Prx is not efficient as APX in terms of H_2O_2 reduction, it can detoxify other hydroperoxides such as tert-butyl hydroperoxide and cumene hydroperoxide. By this way, Prx can combat with lipid peroxidation, which is in agreement with thylakoid localisation of this enzyme where most of the lipid peroxides are produced in chloroplasts (König et al. 2002).

3.6 Nonenzymatic Antioxidants

There are many nonenzymatic antioxidants such as ascorbate, glutathione, tocopherols, carotenoids, polyphenols like flavonoids, and some compatible solutes such as proline. As compared with enzymatic antioxidants, nonenzymatic antioxidants have higher potential to avoid the toxic effects of ROS. Among them, AsA and GSH are the most important and well-studied compounds that mediate the redox status of the cell under strict control.

Ascorbate is one of the most important molecules that directly scavenge H_2O_2 and is the electron donor of APX (Foyer and Noctor 2003). GSH is a part of AsA–GSH cycle and plays crucial role in diverse physiological processes like plant growth, sulphate transport, signal transduction, cell cycle, cell death and senescence (Kranner et al. 2006). Moreover, it is responsible for the removal of excess H_2O_2 and a strong partner of redox regulatory system (Mullineaux and Rausch 2005). Detailed information of AsA–GSH cycle compounds and enzymes of relatives of *Arabidopsis* will be discussed under environmental stress sections in this chapter.

α-Tocopherol prevents the membranes from lipid peroxidation by scavenging lipid peroxyl radicals and also $^1O_2$ in chloroplasts (Munne-Bosch 2005). In the presence of AsA, tocopherols can be regenerated and continue to scavenge the lipid peroxyls. For instance, ascorbate-deficient *vtc1* mutant of *Arabidopsis* showed α-tocopherol deficit under stressed conditions due to the absence of ascorbate in chloroplasts (Munné-Bosch and Alegre 2002). Overexpression of tocopherol biosynthesis genes (*HPPD*, *VTE1*, *VTE2*, *VTE3*, *VTE4*) in *A. thaliana* enhanced the expressions of genes that encode APX, DHAR and MDHAR enzymes and altered AsA and GSH pool in the cell (Li et al. 2010). Recently, an excellent review was published on the relation between stress tolerance and tocopherols, which can be used for further information (Miret and Munné-Bosch 2015).

Carotenoids are another group of low-molecular-weight antioxidants playing a central role in scavenging $^1O_2$ in chloroplasts and preventing damage of $^1O_2$ and peroxyl radicals in the cell. There are many side products of the oxidation of β-carotene such as β-cyclocitral, which are accumulated as a consequence of environmental stress. β-Cyclocitral was recently identified to enhance the expressions of $^1O_2$-responsive genes but not H_2O_2-responsive genes in *Arabidopsis*. These findings suggest that oxidation products of nonenzymatic antioxidants might act as signalling molecules in the transduction of $^1O_2$ signalling in plants (Ramel et al. 2012). This phenomenon was only documented in *Arabidopsis*, and it was not investigated in any other related model plants which are more tolerant to abiotic stress. It might be interesting to test if this signalling type also exists in other stress-tolerant relatives of *Arabidopsis*.

Besides their effects on water balance in the cell, there are many studies suggesting that compatible solutes such as proline can act as nonenzymatic antioxidants, which can scavenge $HO^{\bullet}$ (Szabados and Savouré 2010). *Arabidopsis* overexpressing Δ(1)-pyrroline-5-carboxylate synthetase under hsp 17.6II promoter enhanced proline biosynthesis after heat induction, which also resulted in increase in the activity of SOD, GPX and CAT (Lv et al. 2011). Kant et al. (2006) showed that salt tolerance of *Thellungiella halophila* is highly related to proline metabolism. Exogenous proline application to cultured cells of *Thellungiella* before H_2O_2 application alleviated the damage of oxidative stress, demonstrating the protective role of proline in *Thellungiella* (Soshinkova et al. 2013). Also, drought stress increased the proline content and activities of antioxidant enzymes in *B. juncea* (Fariduddin et al. 2009). Lee et al. (2012) examined the proline contents of 16 *Thellungiella* accessions and compared them with the data obtained from 54 *A. thaliana* accessions under low temperature (4 °C, 14 days) and found that proline content was higher in *Thellungiella* species.

4 ROS Formation and Antioxidant Defence Under Abiotic Stress

4.1 Salt Stress

In the last 20 years, understanding of the mechanism underlying plant tolerance and adaptation to salt stress has been based on comparative analyses between halophytic and glycophytic (salt-sensitive) species, rather than genetic studies that were only conducted in halophytes. Moreover, multigenes involved in salt tolerance have been revealed by mutational approaches derived from salt-sensitive *A. thaliana*. However, *A. thaliana* may not be suitable to study salinity tolerance due to its sensitivity to this stress. Therefore, although it is a genetic model, *A. thaliana* may not provide sufficient information about the mechanism of salt tolerance. This situation has raised the idea of investigating halophytic species that are closely related to *A. thaliana* for this kind of research. In this part, responses of several species, which are close relatives of *A. thaliana*, to salt stress will be discussed.

Thellungiella parvula, a close relative of *Arabidopsis*, can withstand salt concentrations up to 600 mM, while *A. thaliana* (genetic model, glycophyte) shows significant damage and even plant death at 50 mM (Orsini et al. 2010; Uzilday et al. 2015). While 50 mM NaCl treatment did not negatively influence growth of *T. parvula*, 300 mM (Ghars et al. 2008; Uzilday et al. 2015) and 500 mM NaCl (Kant et al. 2006) significantly decreased it. These results showed that 50 mM NaCl could be counted as a 'nonstress' condition for *T. parvula*, while higher salt concentrations become stressful as reported by Uzilday et al. (2015) (Table 1). Recently, *Thellungiella parvula* is renamed as *Eutrema parvulum*, but for the sake of recognition and wide use of *Thellungiella* name in the literature, we preferred to use the genus name *Thellungiella* in this chapter.

Salt stress can significantly decrease maximum quantum yield of PSII (Fv/Fm), which is an indicator of damage to the photosynthetic machinery (Lim et al. 2007; Kim et al. 2010). However, Fv/Fm of *T. parvula* was not altered even at 300 mM NaCl. This result suggested that there is no any apparent damage in photosynthetic machinery of *T. parvula* (Uzilday et al. 2015). On the other hand, M'rah et al. (2007) found that 50 mM NaCl did not affect the photosynthetic activity of *Thellungiella halophila* due to decreased stomatal conductance and unchanged Rubisco pool. But 100 and 200 mM NaCl decreased photosynthetic activities in both *T. halophila* (halophyte) and *A. thaliana* (glycophyte). However, it was observed that the detrimental effect of 200 mM NaCl in *T. halophila* was less than *A. thaliana* as evidenced by lower contents of lipid peroxidation (M'rah et al. 2007).

M'rah et al. (2007) determined also the expression level of CDSP32, which is a plastidic thioredoxin-like protein, in the abovementioned study. CDSP32 plays a role in limiting the accumulation of alkyl hydroperoxides in chloroplast and is accepted as one of the components of the defence system against oxidative damage. It was found that its expression level increased in both *T. halophila* and *A. thaliana*

Table 1 Antioxidant enzyme activities in close relatives of *Arabidopsis* under stress conditions

Species	Reference	Treatment	SOD	CAT	POX	APX	GR	MDHAR	DHAR
Thellungiella halophila	M'rah et al. (2006)	NaCl 100 mM	–	+					
Thlaspi caerulescens	Wójcik et al. (2006)	Cd 25 μM	0	–	–	–			
Thlaspi caerulescens	Wójcik et al. (2006)	Cd 500 μM	–	–	0	0			
Thlaspi caerulescens	Wójcik et al. (2006)	Zn 500 μM	+	+	–	+			
Thlaspi caerulescens	Wójcik et al. (2006)	Zn 2000 μM	+	–	+	–			
Arabis paniculata	Qiu et al. (2008)	Cd 178 μM	+	+		+	+		
Thellungiella parvula	Uzilday et al. (2015)	NaCl 300 μM	+		+	+	+	+	+
Lepidium sativum	Gill et al. (2012)	Cd 100 μM	+	+		+	+		

SOD Superoxide dismutase, *CAT* Catalase, *APX* Ascorbate peroxidase, *GR* Glutathione reductase, *MDHAR* Malondialdehyde reductase, *DHAR* Dehydroascorbate reductases. +: increase, –: decrease, 0: no change under salt stress. However, the highest expression level was observed in *T. halophila*. These results suggest that salt tolerance of *T. halophila* might be associated with the upregulated protection systems such as CDSP32. Lipid peroxidation is widely used as a marker of oxidative stress (Mittler 2002). Similar to those of other halophytes such as *Crithmum maritimum* (Ben Amor et al. 2005) and *Cakile maritima* (Ellouzi et al. 2011), lipid peroxidation levels of *T. parvula* and *T. halophila* were not altered by salinity (Uzilday et al. 2015; M'rah et al. 2006). Uzilday et al. (2015) and M'rah et al. (2007) reported that salinity tolerance of *T. parvula* and *T. halophila* might be closely related with the increased capacity of antioxidant system to scavenge ROS. As compared with lipid peroxidation level in whole cell of *T. parvula*, peroxidation level in chloroplast isolated from 300 mM NaCl-treated plants increased with salinity (Uzilday et al. 2015), but, even so, this situation did not affect the rate of linear electron flow in chloroplasts (Stepien and Johnson 2009) due to the upregulation of redox regulatory enzymes such as TRX–Prx (Uzilday et al. 2015).

Several studies have demonstrated that salt-tolerant species increase their antioxidant enzyme activities and antioxidant contents in response to salt stress, whereas salt-sensitive species failed (Olmos et al. 1994; Jbir et al. 2001; Shalata et al. 2001). M'rah et al. (2006) also investigated the effects of NaCl on the activities of antioxidant enzymes of *T. halophila*. It was found that SOD activity in leaves of salt-treated *T. halophila* decreased, which was probably associated with the decrease of Cu/Zn SOD isoform. Moreover, the activities of CAT increased only in 50 and 100 mM NaCl-treated plants, but it did not change with 200 mM NaCl. Similarly, guaiacol peroxidase activity was also not changed under salt stress (M'rah et al. 2006). These results showed that antioxidant system of *T. halophila* was not induced with salinity as expected. On the other hand, Uzilday et al. (2015) determined changes in antioxidant defence in both chloroplast and whole cell of *T. parvula* under salinity. In contrast to *T. halophila*, the activities of water–water cycle enzymes (SOD, APX, MDHAR, DHAR and GR) and expressions of genes encoding these enzymes in chloroplast of *T. parvula* were significantly induced by salt stress, associated with increase in $O_2^{\bullet-}$ production caused by overload of electron transport system. This increase in the activities of water–water cycle enzymes might cause the relaxation of electron transport chain in the chloroplast of *T. parvula*. All the above results suggest that *T. parvula* was able to protect the photosynthetic machinery with the efficient use of water–water cycle as reported by Uzilday et al. (2015). Similar to chloroplast, the activities of APX, GR, MDHAR and DHAR were increased in whole cell of *T. parvula* indicating the importance Asada–Halliwell–Foyer enzymes in preventing salt stress-induced oxidative stress (Uzilday et al. 2015).

Besides the activities of water–water cycle enzymes, Uzilday et al. (2015) also determined the expression levels of some of key chloroplastic redox components such as ferredoxin (*FED*), ferredoxin thioredoxin reductase (*FTR*), NADPH thioredoxin reductase (*NTRC*), some thioredoxins (*TRX*) and four types of peroxiredoxins (*Prxs*) in *T. parvula*. The relative expression of *TRX-X*, *Prxs*, *FTR*, *NTRC*, *2CPA* and *2CPB* reached their highest level in 200 mM NaCl-treated

T. parvula. TRXs can reduce Prxs, which scavenge H_2O_2 and detoxify alkyl hydroperoxides and peroxynitrite. They are also involved in regulating the activity of enzymes involved in carbon metabolism. Therefore, induction of TRX–Prx system under salt stress prevented any change in the rate of linear electron transport in chloroplast of *T. parvula*. As opposed to *T. parvula* and other halophytes (such as *Spartina alterniflora* (Baisakh et al. 2008) and *Nitraria sphaerocarpa* (Chen et al. 2012)), in *Arabidopsis*, the capacity of TRX–Prx system was decreased with salinity implying decrease in linear electron flow in chloroplasts.

Perez-Ruiz et al. (2006) reported that NTRC is an alternative system for chloroplast protection against oxidative damage. Moreover, Perez-Ruiz et al. (2006) observed damage to photosynthetic apparatus of *Arabidopsis ntrc*-knockout mutants due to overload of ETC. On the other hand, Uzilday et al. (2015) found that in *T. parvula*, similar to FTR expression, NTRC expression was also induced by salinity implying protection of photosynthetic apparatus. These results revealed that *T. parvula* was able to protect its chloroplast from salinity-induced oxidative stress by the induction of redox regulatory components (Uzilday et al. 2015).

Under salt stress, proline (Pro) accumulation leads to the reduction of osmotic potential of the cell, which results in the attraction of water into the cell and maintenance of turgor (Cechin et al. 2006). In addition to this, Pro might act as a compatible solute that can protect membrane structure and conformation of proteins as well as scavenge ROS to regulate cellular redox status (Verslues and Bray 2006). Hence, proline may act as a regulatory or signalling molecule to activate multiple stress response mechanisms that are part of the adaptation process (Claussen 2005). Pro accumulation was studied also in some halophytes, which are closely related to *Arabidopsis* such as *T. parvula* and *T. halophila* under salt stress. Pro accumulation increased in *T. parvula* and *T. halophila* under salt stress as determined by Uzilday et al. (2015) and M'rah et al. (2006), respectively. To understand this increase in proline accumulation with salinity, Uzilday et al. (2015) determined expressions of genes related to proline biosynthesis (*P5CS1* and *P5CS2*) and its degradation (*PRODH1* and *PRODH2*) in *T. parvula*. It was found that in *T. parvula* under salt stress, the expression of *P5CS1* was increased, while expressions of *PRODH1* and *PRODH2* were decreased, which results in the accumulation of Pro. However, expression of *P5CDH*, which converts P5C to glutamate, was increased to degrade P5C which is a toxic compound and can cause tissue damage. Uzilday et al. (2015) reported that in *T. parvula*, Pro takes the role of both being an osmotic regulator to maintain turgor of the cell and an antioxidant to prevent ROS induced damage.

Pro accumulation in response of *T. halophila* to salt stress was determined by M'rah et al. (2007). It was found that while Pro accumulation increased in both *A. thaliana* and *T. halophila* under moderate salinity (50 and 75 mM NaCl), at NaCl concentrations exceeding 300 mM, additional increase in Pro content was not observed. These results suggested that Pro is not effective in imparting tolerance to salt stress of *T. halophila* as indicated in *T. parvula*. On the other hand, Ghars et al. (2012) also investigated the relationship between lipid signalling enzymes and Pro level in *T. halophila* and found that Pro accumulation was prevented in phospholipase C (PLC) inhibitor-treated *T. halophila* under salt stress. In contrast,

Table 2 The proteomic studies that identified changed protein profile in antioxidant defence of close relatives of *Arabidopsis* under abiotic and oxidative stress conditions

Species	Stress	ROS-related proteins	References
Thellungiella halophila	50–150 mM NaCl	APX, DHAR, Prx, GST, APX1	Pang et al. (2010)
Thellungiella halophila	300 mM NaCl	GST, TPX, Cu/ZnSOD	Zhou et al. (2010)
Thellungiella halophila	0–200–400–600 mM NaCl	NADPH dehydrogenase, MnSOD	Wang et al. (2013)
Thellungiella halophila	4 °C Low temperature	Prx, dehydrin, 2-Cys Prx, aldo-keto reductase	Gao et al. (2009)
Arabidopsis halleri	1 mM Cd, 10 mM Zn	GST, FeSOD, MDHAR	Farinati et al. (2009)
Thlaspi caerulescens	2 μm Zn	APX, Prx, Cu/ZnSOD, DHAR	Tuomainen et al. (2006)
Brassica juncea	H_2O_2	GST, DHAR, TRX	Alvarez et al. (2011)

SOD Superoxide dismutase, *CAT* Catalase, *APX* Ascorbate peroxidase, *MDHAR* Malondialdehyde reductase, *Prx* peroxiredoxin, *TRX* thioredoxin, *GST* glutathione-S-transferases

under control conditions, this inhibitor induced Pro accumulation. These results suggested that in the absence of stress, PLC negatively regulates proline accumulation. In contrast to PLC, the phospholipase D (PLD) inhibitor inhibited Pro biosynthesis under control conditions. These results indicated that PLD positively controls Pro accumulation in *T. halophila* via Ca^{2+} signalling under severe stress.

In addition, 100 and 150 mM NaCl treatments increased the carotenoids and proline levels of *B. juncea*. Antioxidant defence system was also induced by salinity such as activities of SOD, POX, CAT, GR and APX (Ahmad 2010). The salinity tolerance of *B. juncea* showed differences between varieties. Salinity increased proline accumulation in *B. juncea* var. Bio902, and SOD, APX and CAT activities of this cultivar were higher than *B. juncea* var. Urvaslu (Mittal et al. 2012).

In a comparative proteomic study between *T. halophila* and *A. thaliana*, it was found that 5 of 32 and 1 of 79 proteins were identified as related with antioxidant defence system under salt stress (Pang et al. 2010) (Table 2). In *Thellungiella* roots, differential expression of phosphoproteome was reported under less than 300 mM NaCl (Zhou et al. 2010). Some of the identified proteins also include GST, TRX and Cu/ZnSOD. Wang et al. (2013) also compared the proteomic profiles of *T. halophila* which were subjected to different salt concentrations (0, 200, 400, 600 mM NaCl) and found that salinity changed the expressions of antioxidant defence system elements such as NADPH dehydrogenase and MnSOD.

4.2 Drought

Drought is a global constraint that affects plant growth. According to IPCC forecasts, frequency and duration of drought will increase in the forthcoming future due to global warming. These future scenarios urged plant scientist and agronomists to develop drought-resistant crops and to better understand the mechanisms underlying drought tolerance.

There are several reports on the involvement of antioxidant defence as an important component of drought stress response. As also mentioned above, limitation of gas exchange due to stomatal closure impairs the balance between light reactions and carbon reactions of photosynthesis and causes a disturbance in the metabolism. Especially under high light conditions, load on chloroplastic electron transfer chain is increased, and this causes excess production of ROS. Particularly in C_3 plants, photorespiration is one of the major sources of ROS (H_2O_2) due to decreased availability of CO_2. If plant cannot cope with this increased ROS production, damage to biological molecules is an inevitable phenomenon.

Considering this, some antioxidant genes were overexpressed (OE) in plants to confer drought stress tolerance. For example, overexpression of different *APX* isoforms (Murgia et al. 2004; Yan et al. 2003; Badawi et al. 2004a, b; Lu et al. 2007), *MDHAR* (Eltayeb et al. 2007), *DHAR* (Ushimaru et al. 2006), *MnSOD* (Wang et al. 2004) and *Cu/ZnSOD* (Badawi et al. 2004a, b) conferred tolerance to drought as compared to wild types. In addition, Xi et al. (2010) overexpressed *MnSOD* and *CAT* (single OE and double OE plants) under control of seed-specific promoter in *A. thaliana* and observed drought tolerance during seedling stage. Increased expression of both enzymes conferred stress tolerance by decreasing ROS levels and preventing oxidative damage.

As expected, loss of the function of antioxidant genes resulted in plants to be more sensitive to drought. *gpx3* knockout *Arabidopsis* exhibited a higher rate of water loss under drought stress, higher sensitivity to H_2O_2 treatment during seed germination and seedling development (Miao et al. 2006). Also, AsA-deficient *Arabidopsis* plants (*vtc1*) exhibited stress sensitivity due to insufficient ROS scavenging (Huang et al. 2005). Davletova et al. (2005) showed that the absence of cytosolic *APX1* resulted in altered H_2O_2-scavenging system, accumulation of ROS and oxidation of proteins during photooxidative stress. However, interestingly, *apx1* plants grow better than WT plants under osmotic stress, although it was more sensitive to photooxidative stress (Miller et al. 2007). When *apx1* mutants were compared to double *cytosolic apx1/thylakoid apx* mutants, again it showed that double mutant plants developed a different cellular response that results in late flowering, low protein oxidation under high light and enhanced accumulation of anthocyanin (Miller et al. 2007). These findings suggest that plant antioxidant defence system is a dynamic process and there is some degree of redundancy and overlapping signals can induce different responses.

Most of the drought-related molecular studies are conducted with *A. thaliana*. Interestingly, there is no such model to investigate drought tolerance mechanisms such as *Thellungiella* or *A. halleri* that is used for salinity or heavy metal tolerance. However, researchers focused on comparatively investigating drought tolerance mechanisms between different *A. thaliana* ecotypes. Unfortunately none of these studies tried to elucidate ROS defence and redox regulation mechanisms under drought in detail. In this part, some reports on *A. thaliana* drought responses with a special focus on antioxidant defence and redox regulation will be discussed.

While some of the studies investigated plant responses to drought at whole plant level, Jung (2004) investigated drought responses of young and mature leaves of *A. thaliana*. In this study, it was shown that both young and mature leaves exhibited an increase in nonphotochemical quenching (NPQ), which aids in dissipating excess light as heat and reduced electron load on chloroplastic electron transport chain. Increase in NPQ is a mechanism that is used by both young and mature leaves to avoid ROS production in chloroplasts. In addition, although nonenzymatic antioxidants increased in both young and mature leaves, increase in enzymatic antioxidant including SOD, CAT, total POX and GR were only observed in mature leaves but not young leaves. These findings suggest that tolerance mechanisms are diverse even in young and mature leaves of the same plant.

In another study, Koffler et al. (2014) investigated compartment-specific response of antioxidants to drought stress in *Arabidopsis*. They showed that in early stages of drought (4–7 days of water withholding), Asc content does not decrease in mitochondria, chloroplasts, cytosol or nuclei and it increased in peroxisomes and vacuole. Asc content reduced only after severe drought (7–10 days after stress) in these compartments, and it is still at levels as the start of the experiment in vacuoles. Contrastingly, a significant decrease in GSH was observed starting at the 4th day of drought treatment. These results indicate that GSH can serve as an early signal for drought stress. Especially decrease of glutathione in the nuclei can regulate gene expression through redox regulation of transcription factors.

Ozfidan et al. (2012) investigated regulation of antioxidant defence in *A. thaliana* under osmotic stress and ABA treatment. As it is widely known, ABA is a regulator of stress responses in plants and takes part in development, germination and stomatal closure. In this study, the authors showed that osmotic stress increased activities of SOD, CAT and APX, but these inductions were more prominent in the groups that were treated with osmotic stress with the addition of ABA. H_2O_2 levels were lower in plants treated with osmotic stress + ABA, when compared with only osmotic stress-treated plants, demonstrating the effect of ABA on antioxidant defence system.

For a full picture of transcriptomic response of antioxidant response of *Arabidopsis*, Noctor et al. (2014) defined 406 *Arabidopsis* genes that encode core antioxidant and redox homeostasis genes. Of these, probes for 302 genes were available in ATH1 microarray chips. Using data from two recent drought experiments that was available through Genevestigator, the authors investigated the changes of these 302 genes under drought and observed that only 15 genes respond to the same trend (increase or decrease) in both experiments with a twofold cutoff. Among all, only ROS-producing enzyme was increased, i.e., annexin-1, which has a peroxidase activity, while none of the NADPH oxidases were increased.

Similarly, CAT2 expression and expression of cytosolic and chloroplastic SODs did not change. A clear response was observed in a gene that encodes ascorbate oxidase, *Senescence-Related Gene 1*. The expression of two TRX-linked genes and subunit of a chloroplastic 2-Cys Prx was decreased. In addition, two glutaredoxin genes (GRX) were observed to be responsive to drought in both experiments, which are known to interact with transcription factors (Zander et al. 2012). Also inductions in several GSTs were observed. Among these GSTs, GSTU4 was observed to be induced in response to drought, salt, ABA and osmotic stress. As it can be seen, the accumulation of available data makes it possible to dissect and extract more detailed information related to events under stressed conditions.

4.3 Temperature Stress

Another important stress factor is low and high temperatures such as cold, freezing and heat shock. These stress factors can also cause the induction of ROS production. Heat stress causes lipid peroxidation which is a marker of oxidative stress (Larkindale and Knight 2002). Griffith et al. (2007) investigated the freezing temperature effects on Yukon ecotype of *Thellungiella halophila*, which is native to subarctic Canada. It has genetic potential as model species whose genome had been published, and it naturally resides in extreme cold temperatures such as −18 °C. In their study, Griffith et al. (2007) found that seed germination and viability of Yukon ecotype of *T. halophila* were resistant to cold temperature and it was ascribed as freezing plant, which continues to live in extreme cold under the formation of ice in its tissues.

Under chilling stress (4 °C), GSH content of *T. halophila* was measured as two times higher than that of *A. thaliana*, but GSSG content of *A. thaliana* was slightly higher than *T. halophila*. In the same study, low-temperature marker genes were induced by chilling conditions in both species which are *COR15A*, *COR47*, *RD29A*, *RD29B*, *LEA*, *ELIP2* and *ADH1* (Benina et al. 2013).

Proteomic analysis of *T. halophila* under low temperature showed that 16 % of total protein profile was related to plant defence. Among them, four proteins were involved in chloroplastic redox balance, and also two nonchloroplastic ROS-scavenging proteins were elevated by cold stress (Gao et al. 2009) (Table 2).

As discussed below, heavy metal accumulation alters membrane integrity and increases ROS. Taulavuori et al. (2005) suggested that together with freezing, this membrane damage by heavy metals led to severe membrane damage as a result of ROS accumulation (Taulavuori et al. 2005).

4.4 Heavy Metal Stress

Heavy metals are a group of elements in the periodic table that especially refer to cadmium (Cd), zinc (Zn), nickel (Ni), mercury (Hg) and others. These metals are the main environmental pollution factors around old and current industrial areas. Some of the heavy metals such as Cu are essential for plants, but excess levels of these essential elements and nonessential elements such as Cd can be toxic after a certain threshold. It is important to determine the toxicity levels of these elements in plants and to evaluate their effects on plant growth and development. Heavy metal stress transforms the metabolism of plants by affecting enzyme activities and obstructing membrane integrity. Heavy metal stress also disturbs ROS balance in the cell and induces plant-detoxifying mechanisms such as antioxidant machinery. For example, excess accumulation of Cd leads to the production of ROS and inevitably oxidative stress (Gill and Tuteja 2010; Foyer and Noctor 2003).

Cho and Seo (2005) demonstrated that Cd-resistant *A. thaliana* had lower H_2O_2 and TBARS content as compared to wild type. Resistant type had greater activities of SOD, GPX, APX and GR as compared to wild type. The authors suggest that decreased H_2O_2 accumulation of resistant type enhanced Cd tolerance. Similarly, Maksymiec and Krupa (2006) investigated the effects of 100 μM Cu and Cd on antioxidant defence system of *A. thaliana*. This study revealed that Cd treatment increased the ROS and MDA levels and the activities of SOD, CAT and APX, while Cu treatment enhanced the ROS and MDA levels and APX activity, but it decreased the activities of SOD and CAT. In another study, in *A. thaliana*, 10 μM Cd treatment increased ROS and MDA levels and also enhanced the activities of SOD, CAT, APX, GR and GSH content. Moreover, 5 μM Cu elevated ROS, MDA and GSH contents, while it decreased the activities of CAT, APX and GR in contrast to Cd treatments (Cuypers et al. 2011).

There are many *Arabidopsis*-related heavy metal hyperaccumulators such as *Arabis paniculata*, *Thlaspi* sp. and *A. halleri*. Many researchers exploited the use of these plant species for the elucidation of heavy metal stress tolerance due to their close relation to *A. thaliana*. Similarity between genome sequences of these plants and *A. thaliana* makes it possible to use available sequence data and molecular tools for this research.

Qiu et al. (2008) investigated the effects of Cd treatments in the range from 0 μM to 178 μM (0, 22, 44, 89, 178 μM Cd) on antioxidant system of Cd hyperaccumulator *A. paniculata*. They found that 22–89 μM Cd decreased the lipid peroxidation and activities of SOD, GPX and APX in roots, whereas these concentrations did not have any effects in leaves. While in roots, 22–89 μM Cd inhibited the oxidative stress, 178 μM Cd enhanced the accumulation of ROS in *A. paniculata* (Qiu et al. 2008) (Table 1).

T. geosingense is a hyperaccumulator of Ni and it can tolerate Zn, Co and Cd. It is a well-known metal hyperaccumulator which is related to model plant *A. thaliana*. The accumulation of glutathione (GSH) in *T. geosingense* increased the oxidative stress tolerance capacity of the cell and enhanced the tolerance to Ni,

Co and Zn by the activity of mitochondrial serine acetyltransferases (SATm) (Freeman and Salt 2007).

A. halleri can tolerate Zn, Cd, Pb and can hyperaccumulate Cd and Zn (Farinati et al. 2009). According to proteomic studies, 1 mM Cd and 10 mM Zn treatments enhanced the levels of several stress response proteins including oxidative stress-related GST and FeSOD in *A. halleri* (Farinati et al. 2009) (Table 2). In their transcriptomic study, Chiang et al. (2006) revealed that expressions of Asc–GSH cycle genes such as APX, MDHAR4 and POX were higher in *A. halleri* than that of *A. thaliana*. They also found that the activities of APX and POX were induced in *A. halleri*, when compared to *A. thaliana*. Under paraquat and Cd treatments, *A. halleri* was able to efficiently scavenge the ROS, but this was not the case for *A. thaliana*. The higher antioxidant activities of *A. halleri* resulted in heavy metal tolerance by decreased levels of ROS (Chiang et al. 2006).

Phytochelatins (PCs), which bind metals in the cytosol and then segregate them to vacuole, are formed by the polymerisation of GSH. γ-EC synthetase and GSH synthetase are two enzymes that are responsible for the production of GSH from glutamate. Expressions of these GSH synthesis genes were highly induced by Cd or Cu application to *A. thaliana* (Xiang and Oliver 1998).

The content of Ni, an essential micronutrient, can reach up to 10 μg g^{-1} dry weight in plants grown in normal soil, while generally it is toxic over this concentration. On the other hand, in Ni-rich soils, there are many plant species that can accumulate Ni up to 1000 μg g^{-1} dry weight. Till date, approximately 400 species were characterised as heavy metal hyperaccumulators (Assunção et al. 2003). In the late 1800s, *Thlaspi* was identified as growing in Zn-rich soils, and even now, it is widely used as a model species in heavy metal tolerance studies. *T. caerulescens* is a well-known species, which is a hyperaccumulator of Ni (up to 4700 μg g^{-1} dry weight), Zn (30,000 μg g^{-1} dry weight) and Cd (14,000 μg g^{-1} dry weight).

A comparative study between Ni hyperaccumulators (*T. geoesingense*, *T. oxyceras*, *T. rosulare*) and nonaccumulators (*T. perfoliatum*, *T. arvense*, *A. thaliana*) exhibited that GSH concentration was highly related with hyperaccumulation capability (Freeman et al. 2004). Serine acetyltransferase (SAT) is another key regulator enzyme in Cys and glutathione synthesis in plants. Enhanced activity of SAT showed a significant correlation with tolerance to oxidative stress in hyperaccumulator *T. geosingense*, which also caused elevated tolerance to Ni. The same study also revealed that when *TgSAT* was overexpressed in *A. thaliana*, it led to Ni tolerance in transgenic lines as compared to wild type (Freeman et al. 2004). These results prove the importance of genetic diversity of *Arabidopsis*-related model plants and give us the opportunity to engineer these characters to *A. thaliana* or other crop plants.

In a comparative study between hyperaccumulator *T. caerulescens* and nonaccumulator *Nicotiana tabacum*, *T. caerulescens* showed higher antioxidant defence capacity than that of *N. tabacum* under Cd. For instance, *T. caerulescens* had higher basal levels of CAT and SOD activities, and after Cd treatment, their activities were increased. Lipid peroxidation was elevated by Cd treatments in both species; however, it was 83 % higher in *N. tabacum* as compared to *T. caerulescens*.

T. caerulescens controlled H_2O_2 levels by CAT activity, which indicated the pivotal role of this antioxidant enzyme in heavy metal stress response (Boominathan and Doran 2003).

Craciun et al. (2006) crossed *A. halleri* and *A. lyrata* to produce genotypes that have different degrees of heavy metal tolerance. They ascribed the transcriptomic differences between Cd-tolerant and Cd-sensitive genotypes and found that genes involved in cellular detoxification and DNA repair were upregulated in Cd-tolerant genotypes.

Lepidium sativum is another Cd-tolerant plant which is related to *Arabidopsis* and belongs to *Brassicaceae*. While high concentrations of Cd (100 mg Cd kg^{-1} soil) inhibited the photosynthetic rate, growth, and decreased N content and nitrate reductase activity, it increased the activities of SOD, CAT, GR, APX and GSH levels. Moreover, antioxidant defence mechanism of *L. sativum* was efficiently working at low Cd concentrations, and as a result of this, no effects of low Cd concentrations on *L. sativum* had been reported by Gill et al. (2012) (Table 1).

Although it is an essential element, Cu as well as Cd are also considered as a heavy metal stressor for plants and have impacts on antioxidant defence system. For example, excess Cu (5, 25, 50, 100 μM) application to *A. thaliana* for 1, 3 and 7 days influenced the antioxidant defence system (Drążkiewicz et al. 2003). This study revealed that responses of antioxidant defence system of *Arabidopsis* against Cu depend on both the duration of treatment and Cu concentration used. Cu affected the activities of DHAR, MDHAR, GR, APX and levels of DHA, AsA, GSH and GSSG (Drążkiewicz et al. 2003). Similar to this, it was shown that the activity of another GSH-related enzyme GST was induced by Cu (Skórzyńska-Polit et al. 2010). In the same study, it was also shown that in *Arabidopsis*, membrane damage by Cu is higher than that of Cd application, which was evidenced by electrolyte leakage.

Proteomic approaches are widely used to understand the metal tolerance of hyperaccumulator *T. caerulescens*. *T. caerulescens* is a widely used plant for heavy metal stress studies because its genome is 87 % identical to *Arabidopsis* (Peer et al. 2003) (Table 2). In a study where three accessions of *T. caerulescens* were used to compare proteomic profiles under Zn and Cd treatments, ROS-scavenging proteins such as Cu/Zn SOD, APX, DHAR and Prx were identified under Zn treatment (Tuomainen et al. 2006). In an experiment, 500, 1000, 1250 and 2000 μM Zn and 25, 200 and 500 μM Cd were applied to *T. caerulescens* for 14 days by Wójcik et al. (2006) (Table 1). Elevated Zn increased the SOD and POX activities but decreased the APX and CAT activities, while Cd application increased the APX and POX activities and decreased the SOD activity.

B. juncea is a member of Brassica family related to *A. thaliana* and is an important oil plant worldwide. Under toxic levels of Zn, activities of CAT, POX, APX, MDHAR, DHAR and GR enzymes were increased in *B. juncea* and AsA and GSH contents were enhanced (Prasad et al. 1999). *B. juncea* is also an accumulator of Cd, and for this attribute, it was proposed for bioremediation for contaminated lands. It was found that under high levels of Cd, it had high antioxidant capacity due to increased levels of GSH (Seth et al. 2008). Markovska et al. (2009) reported that

100 μM Cd induced the H_2O_2 levels but not MDA and enhanced the activities of APX, MDHAR and GR in *B. juncea*. Moreover, Mobin and Khan (2007) compared the antioxidant capacity of two different cultivars of *B. juncea* under excess Cd. In this study, it was shown that two cultivars had different responses to Cd and increased activities of CAT, APX and GR played roles in the alleviation of Cd stress. Mohamed et al. (2012) demonstrated difference between antioxidant responses of roots and shoots of *B. juncea* to Cd stress and showed that roots could tolerate more Cd than shoots, which can be associated with GSH levels. The effects of Cu stress on antioxidant system was also investigated in *B. juncea*, and it was found that Cu application decreased the content of GSH and excess Cu (200 μm) increased the activities of CAT and APX (Devi and Prasad 2005). A comparative study on antioxidant defence systems of *B. juncea* and *B. napus* under Cd stress showed that MDA levels of *B. napus* were increased by stress and the activities of SOD, CAT, APX and GR were decreased upon Cd stress, whereas antioxidant defence system was not changed in *B. juncea* that showed higher tolerance to Cd than *B. napus* (Nouairi et al. 2009).

All the studies mentioned above indicated that there is a strong correlation between antioxidant capacity and heavy metal stress tolerance. Hyperaccumulator plants usually display high amounts of antioxidants within the cell either enzymatically or nonenzymatically, and they are capable of conserving the membrane integrity by reducing reactive oxygen species.

5 Conclusions

As it can be seen from the studies above, relatives of *A. thaliana* can give us insights to improve performance of crop plants when cultivated in the marginal areas such as arid, saline and heavy metal-contaminated soils. In this stage, researchers are trying to understand how basic mechanisms that are well known in *A. thaliana* work in its stress-tolerant relatives. The next step will be to analyse the differences between sensitive and tolerant mechanisms and then transfer this knowledge to application via crop plants.

Due to the availability of comprehensive microarray chips, most of our knowledge at transcriptomic level depends on experiments that use *A. thaliana*. However, recent developments in non-Sanger next-generation DNA sequencing technologies such as 454 and Illumina sequencing make it possible to investigate transcriptomic of any species that is interested. Combined with sequence and gene annotation data of *Arabidopsis*, these developments can make it possible to unravel yet to be discovered tolerance mechanisms in its relatives.

References

Ahmad P (2010) Growth and antioxidant responses in mustard (*Brassica juncea* L.) plants subjected to combined effect of gibberellic acid and salinity. Arch Agron Soil Sci 56:575–588

Alscher RG, Erturk N, Heath LS (2002) Role of superoxide dismutases (SODs) in controlling oxidative stress in plants. J Exp Bot 53:1331–1341

Alvarez S, Galant A, Jez JM, Hicks LM (2011) Redox-regulatory mechanisms induced by oxidative stress in *Brassica juncea* roots monitored by 2-DE proteomics. Proteomics 11: 1346–1350

Amor NB, Hamed KB, Debez A, Grignon C, Abdelly C (2005) Physiological and antioxidant responses of the perennial halophyte Crithmummaritimum to salinity. Plant Sci 168(4): 889–899

Amtmann A, Bohnert HJ, Bressan RA (2005) Abiotic stress and plant genome evolution. Search for new models. Plant Physiol 138:127–130

Asada K, Takahashi M (1987) Production and scavenging of active oxygen in chloroplasts. In: Kyle DJ, Osmond CB, Arntzen CJ (eds) Photoinhibition. Elsevier, Amsterdam

Assunção AG, Bookum WM, Nelissen HJ, Vooijs R, Schat H, Ernst WH (2003) Differential metalspecific tolerance and accumulation patterns among *Thlaspi caerulescens* populations originating from different soil types. New Phytol 159(2):411–419

Badawi GH, Kawano N, Yamauchi Y, Shimada E, Sasaki R, Kubo A, Tanaka K (2004a) Over-expression of ascorbate peroxidase in tobacco chloroplasts enhances the tolerance to salt stress and water deficit. Physiol Plant 121:231–238

Badawi GH, Yamauchi Y, Shimada E, Sasaki R, Kawano N, Tanaka K, Tanaka K (2004b) Enhanced tolerance to salt stress and water deficit by overexpressing superoxide dismutase in tobacco (*Nicotiana tabacum*) chloroplasts. Plant Sci 166:919–928

Baier M, Noctor G, Foyer CH, Dietz KJ (2000) Antisense suppression of 2-cysteine peroxiredoxin in Arabidopsis specifically enhances the activities and expression of enzymes associated with ascorbate metabolism but not glutathione metabolism. Plant Physiol 124:823–832

Baisakh N, Subudhi PK, Varadwaj P (2008) Primary responses to salt stress in a halophyte, smooth cordgrass (*Spartina alterniflora* Loisel.). Funct Integr Genom 8:287–300

Bannister WH, Bannister JV, Barra D, Bond J, Bossa F (1991) Evolutionary aspects of superoxide dismutase: the copper/zinc enzyme. Free Radic Res 12:349–361

Baxter A, Mittler R, Suzuki N (2014) ROS as key players in plant stress signalling. J Exp Bot 65: 1229–1240

Benina M, Obata T, Mehterov N, Ivanov I, Petrov V, Toneva V, Ferniw AR, Gechev TS (2013) Comparative metabolic profiling of *Haberlea rhodopensis*, *Thellungiella halophyla*, and *Arabidopsis thaliana* exposed to low temperature. Front Plant Sci 4:499

Bhattachrjee S (2005) Reactive oxygen species and oxidative burst: roles in stress, senescence and signal transduction in plant. Curr Sci 89:1113–1121

Boominathan R, Doran PM (2003) Cadmium tolerance and antioxidative defenses in hairy roots of the cadmium hyperaccumulator, *Thlaspi caerulescens*. Biotechnol Bioeng 83:158–167

Bose J, Rodrigo Moreno A, Shabala S (2014) ROS homeostasis in halophytes in the context of salinity stress tolerance. J Exp Bot 65:1241–1257

Bréhélin C, Meyer EH, de Souris JP, Bonnard G, Meyer Y (2003) Resemblance and dissemblance of Arabidopsis type II peroxiredoxins: similar sequences for divergent gene expression, protein localization, and activity. Plant Physiol 132:2045–2057

Bright J, Desikan R, Hancock JT, Weir IS, Neill SJ (2006) ABA-induced NO generation and stomatal closure in Arabidopsis are dependent on H_2O_2 synthesis. Plant J 45:113–122

Bueso E, Alejandro S, Carbonell P, Perez-Amador MA, Fayos J, Bellés JM, Serrano R (2007) The lithium tolerance of the Arabidopsis cat2 mutant reveals a cross-talk between oxidative stress and ethylene. Plant J 52:1052–1065

Cechin I, Rossi SC, Oliveira VC, Fumis TF (2006) Photosynthetic responses and proline content of mature and young leaves of sunflower plants under water deficit. Photosynthetica 44:143–146

Chen J, Cheng T, Wang P, Liu W, Xiao J, Hu X, Jiang Z, Zhang S, Shi J (2012) Salinity-induced changes in protein expression in the halophytic plant *Nitraria sphaerocarpa*. J Proteomics 75:5226–5243

Chew O, Whelan J, Millar AH (2003) Molecular definition of the ascorbate-glutathione cycle in Arabidopsis mitochondria reveals dual targeting of antioxidant defenses in plants. J Biol Chem 278:46869–46877

Chiang HC, Lo JC, Yeh KC (2006) Genes associated with heavy metal tolerance and accumulation in Zn/Cd hyperaccumulator *Arabidopsis halleri*: a genomic survey with cDNA microarray. Environ Sci Technol 40:6792–6798

Cho UH, Seo NH (2005) Oxidative stress in *Arabidopsis thaliana* exposed to cadmium is due to hydrogen peroxide accumulation. Plant Sci 168:113–120

Claussen W (2005) Proline as a measure of stress in tomato plants. Plant Sci 168:241–248

Corpas FJ, Barroso JB, del Río LA (2001) Peroxisomes as a source of reactive oxygen species and nitric oxide signal molecules in plant cells. Trends Plant Sci 6:145–150

Craciun AR, Courbot M, Bourgis F, Salis P, Saumitou-Laprade P, Verbruggen N (2006) Comparative cDNA-AFLP analysis of Cd-tolerant and-sensitive genotypes derived from crosses between the Cd hyperaccumulator *Arabidopsis halleri* and *Arabidopsis lyrata* ssp. petraea. J Exp Bot 57:2967–2983

Cuypers A, Karen S, Jos R, Kelly O, Els K, Tony R, Nele H, Nathalie V, Suzy VS, Frank VB, Yves G, Jan C, Jaco V (2011) The cellular redox state as a modulator in cadmium and copper responses in Arabidopsis thaliana seedlings. J Plant Physiol 168:309–316

Davletova S, Rizhsky L, Liang H, Shengqiang Z, Oliver DJ, Coutu J, Mittler R (2005) Cytosolic ascorbate peroxidase 1 is a central component of the reactive oxygen gene network of Arabidopsis. Plant Cell Online 17:268–281

De Leonardis S, Dipierro N, Dipierro S (2000) Purification and characterization of an ascorbate peroxidase from potato tuber mitochondria. Plant Physiol Biochem 38:773–779

del Río LA, Corpas FJ, Sandalio LM, Palma JM, Gómez M, Barroso JB (2002) Reactive oxygen species, antioxidant systems and nitric oxide in peroxisomes. J Exp Bot 53:1255–1272

Devi SR, Prasad MNV (2005) Antioxidant capacity of *Brassica juncea* plants exposed to elevated levels of copper. Russ J Plant Physiol 52:205–208

Dietz KJ, Horling F, König J, Baier M (2002) The function of the chloroplast 2-cysteine peroxiredoxin in peroxide detoxification and its regulation. J Exp Bot 53:1321–1329

Dietz KJ, Pfannschmidt T (2011) Novel regulators in photosynthetic redox control of plant metabolism and gene expression. Plant Physiol 155:1477–1485

Drążkiewicz M, Skórzyńska-Polit E, Krupa Z (2003) Response of the ascorbate–glutathione cycle to excess copper in *Arabidopsis thaliana* (L.). Plant Sci 164:195–202

Ellouzi H, Ben Hamed K, Cela J, Munné-Bosch S, Abdelly C (2011) Early effects of salt stress on the physiological and oxidative status of Cakile maritima (halophyte) and *Arabidopsis thaliana* (glycophyte). Physiol Planta 142:128–143

Eltayeb AE, Kawano N, Badawi GH, Kaminaka H, Sanekata T, Shibahara T, Tanaka K (2007) Overexpression of monodehydroascorbate reductase in transgenic tobacco confers enhanced tolerance to ozone, salt and polyethylene glycol stresses. Planta 225:1255–1264

Fariduddin Q, Khanam S, Hasan SA, Ali B, Hayat S, Ahmad A (2009) Effect of 28-homobrassinolide on the drought stress-induced changes in photosynthesis and antioxidant system of *Brassica juncea* L. Acta Physiol Plant 31:889–897

Farinati S, DalCorso G, Bona E, Corbella M, Lampis S, Cecconi D, Furini A (2009) Proteomic analysis of *Arabidopsis halleri* shoots in response to the heavy metals cadmium and zinc and rhizosphere microorganisms. Proteomics 9:4837–4850

Finkemeier I, Goodman M, Lamkemeyer P, Kandlbinder A, Sweetlove LJ, Dietz KJ (2005) The mitochondrial type II peroxiredoxin F is essential for redox homeostasis and root growth of *Arabidopsis thaliana* under stress. J Biol Chem 280:12168–12180

Foreman J, Demidchik V, Bothwell JH, Mylona P, Miedema H, Torres MA, Linstead P, Costa S, Brownlee C, Jones JD, Davies JM, Dolan L (2003) Reactive oxygen species produced by NADPH oxidase regulate plant cell growth. Nature 422:442–446

Foyer CH, Noctor G (2003) Redox sensing and signaling associated with reactive oxygen in chloroplasts, peroxisomes and mitochondria. Physiol Plant 119:355–364

Freeman JL, Persans MW, Nieman K, Albrecht C, Peer W, Pickering IJ, Salt DE (2004) Increased glutathione biosynthesis plays a role in nickel tolerance in *Thlaspi* nickel hyperaccumulators. Plant Cell Online 16:2176–2191

Freeman JL, Salt DE (2007) The metal tolerance profile of *Thlaspi goesingense* is mimicked in *Arabidopsis thaliana* heterologously expressing serine acetyl-transferase. BMC Plant Biol 7:63

Gao F, Zhou Y, Zhu W, Li X, Fan L, Zhang G (2009) Proteomic analysis of cold stress-responsive proteins in Thellungiella rosette leaves. Planta 230:1033–1046

Ghars MA, Parre E, Debez A, Bordenave M, Richard L, Leport C, Bouchereau A, Savoure A, Abdelly C (2008) Comparative salt tolerance analysis between *Arabidopsis thaliana* and *Thellungiella halophila*, with special emphasis on K^+/Na^+ selectivity and proline accumulation. J Plant Physiol 165:588–599

Ghars MA, Richard L, Lefebvre-DeVos D, Leprince AS, Parre E, Bordenave M, Abdelly C, Savoure A (2012) Phospholipases C and D modulate proline accumulation in *Thellungiella halophila*/salsuginea differently according to the severity of salt or hyperosmotic stress. Plant Cell Physiol 53:183–192

Gill F, Tuteja N (2010) Reactive oxygen species and antioxidant machinery in abiotic stress tolerance in crop plants. Plant Physiol Biochem 48:909–930

Gill SS, Khan NA, Tuteja N (2012) Cadmium at high dose perturbs growth, photosynthesis and nitrogen metabolism while at low dose it up regulates sulfur assimilation and antioxidant machinery in garden cress (*Lepidium sativum* L.). Plant Sci 182:112–120

Griffith M, Timonin M, Wong AC, Gray GR, Akhter SR, Saldanha M, Rogers MA, Weretilnyk EA, Moffatt B (2007) Thellungiella: an Arabidopsis-related model plant adapted to cold temperatures. Plant Cell Environ 30:529–538

Hofmann B, Hecht HJ, Flohé L (2002) Peroxiredoxins. Biol Chem 383:347–364

Hu YQ, Liu S, Yuan HM, Li J, Yan DW, Zhang JF, Lu YT (2010) Functional comparison of catalase genes in the elimination of photorespiratory H_2O_2 using promoter-and 3′-untranslated region exchange experiments in the Arabidopsis cat2 photorespiratory mutant. Plant Cell Environ 33:1656–1670

Huang C, He W, Guo J, Chang X, Su P, Zhang L (2005) Increased sensitivity to salt stress in an ascorbate-deficient Arabidopsis mutant. J Exp Bot 56:3041–3049

Jbir N, Chaïbi W, Ammar S, Jemmali A, Ayadi A (2001) Root growth and lignification of two wheat species differing in their sensitivity to NaCl, in response to salt stress. Comptes Rendus de l'Académie des Sciences-Series III-Sciences de la Vie 324(9):863–868

Jung S (2004) Variation in antioxidant metabolism of young and mature leaves of *Arabidopsis thaliana* subjected to drought. Plant Sci 166:459–466

Kant S, Kant P, Raveh E, Barak S (2006) Evidence that differential gene expression between the halophyte, *Thellungiella halophila*, and *Arabidopsis thaliana* is responsible for higher levels of the compatible osmolyte proline and tight control of Na^+ uptake in *T. halophila*. Plant Cell Environ 29:1220–1234

Karpinski S, Reynolds H, Karpinska B, Wingsle G, Creissen G, Mullineaux P (1999) Systemic signaling and acclimation in response to excess excitation energy in Arabidopsis. Science 284:654–657

Kawano T (2003) Roles of the reactive oxygen species-generating peroxidase reactions in plant defense and growth induction. Plant Cell Rep 21:829–837

Kim KH, Alam I, Lee KW, Sharmim SA, Kwak SS, Lee SY, Lee BH (2010) Enhanced tolerance of transgenic tall fescue plants overexpressing 2-Cys peroxiredoxin against methyl viologen and heat stresses. Biotechnol Lett 32:571–576

Kliebenstein DJ, Monde RA, Last RL (1998) Superoxide dismutase in Arabidopsis: an eclectic enzyme family with disparate regulation and protein localization. Plant Physiol 118:637–650
Koffler BE, Luschin-Ebengreuth N, Stabentheiner E, Müller M, Zechmann B (2014) Compartment specific response of antioxidants to drought stress in Arabidopsis. Plant Sci 227:133–144
König J, Baier M, Horling F, Kahmann U, Harris G, Schürmann P, Dietz KJ (2002) The plant-specific function of 2-Cys peroxiredoxin-mediated detoxification of peroxides in the redox-hierarchy of photosynthetic electron flux. Proc Natl Acad Sci USA 99:5738–5743
Koornneef M, Meinke D (2010) The development of Arabidopsis as a model plant. Plant J 61: 909–921
Kranner I, Birtić S, Anderson KM, Pritchard HW (2006) Glutathione half-cell reduction potential: a universal stress marker and modulator of programmed cell death? Free Radic Biol Med 40:2155–2165
Lai AG, Doherty CJ, Mueller-Roeber B, Kay SA, Schippers JH, Dijkwel PP (2012) CIRCADIAN CLOCK-ASSOCIATED 1 regulates ROS homeostasis and oxidative stress responses. Proc Natl Acad Sci USA 109:17129–17134
Laloi C, Apel K, Danon A (2004) Reactive oxygen signalling: the latest news. Curr Opin Plant Biol 7:323–328
Larkindale J, Knight MR (2002) Protection against heat stress-induced oxidative damage in Arabidopsis involves calcium, abscisic acid, ethylene, and salicylic acid. Plant Physiol 128:682–695
Lee YP, Babakov A, de Boer B, Zuther E, Hincha DK (2012) Comparison of freezing tolerance, compatible solutes and polyamines in geographically diverse collections of *Thellungiella* sp. and *Arabidopsis thaliana* accessions. BMC Plant Biol 12:131
Li Y, Zhou Y, Wang Z, Sun X, Tang K (2010) Engineering tocopherol biosynthetic pathway in Arabidopsis leaves and its effect on antioxidant metabolism. Plant Sci 178:312–320
Lim S, Kim YH, Kim SH, Kwon SY, Lee HS, Kim JS, Cho KY, Paek KY, Kwak SS (2007) Enhanced tolerance of transgenic sweet potato plants that express both CuZnSOD and APX in chloroplasts to methyl viologen-mediated oxidative stress and chilling. Mol Breed 19:227–239
Liszkay K, Fufezan C, Trebst A (2008) Singlet oxygen production in photosystem II and related protection mechanism. Photosynth Res 98:551–564
Loew O (1900) A new enzyme of general occurrence in organisms. Science 11:701–702
Lu Z, Liu D, Liu S (2007) Two rice cytosolic ascorbate peroxidases differentially improve salt tolerance in transgenic Arabidopsis. Plant Cell Rep 26:1909–1917
Lv WT, Lin B, Zhang M, Hua XJ (2011) Proline accumulation is inhibitory to Arabidopsis seedlings during heat stress. Plant Physiol 156:1921–1933
M'rah S, Ouerghi Z, Berthomieu C, Havaux M, Jungas C, Hajii M, Grignon C, Lachaal M (2006) Effects of NaCl on the growth, ion accumulation and photosynthetic parameters of *Thellungiella halophila*. J Plant Physiol 163:1022–1031
M'rah S, Ouerghi Z, Eymery F, Rey P, Hajii M, Grignon C, Lachaal M (2007) Efficiency of biochemical protection against toxic effects of accumulated salt differentiates *Thellungiella halophila* from *Arabidopsis thaliana*. J Plant Physiol 164:375–384
Maksymiec W, Krupa Z (2006) The effects of short-term exposition to Cd, excess Cu ions and jasmonate on oxidative stress appearing in *Arabidopsis thaliana*. Environ Exp Bot 57:187–194
Margis R, Dunand C, Teixeira FK, Margis-Pinheiro M (2008) Glutathione peroxidase family–an evolutionary overview. FEBS J 275:3959–3970
Markovska YK, Gorinova NI, Nedkovska MP, Miteva KM (2009) Cadmium-induced oxidative damage and antioxidant responses in *Brassica juncea* plants. Biol Plant 53:151–154
Masters CJ (2001) Cellular signalling: the role of the peroxisome. Cell Signal 8:197–208
McClung CR (1997) Regulation of catalases in Arabidopsis. Free Radic Biol Med 23:489–496
Miao Y, Lu D, Wang P, Wang XC, Chen J, Miao C, Song CP (2006) An Arabidopsis glutathione peroxidase functions as both a redox transducer and a scavenger in abscisic acid and drought stress responses. Plant Cell Online 18:2749–2766

Milla MAR, Maurer A, Huete AR, Gustafson JP (2003) Glutathione peroxidase genes in Arabidopsis are ubiquitous and regulated by abiotic stresses through diverse signaling pathways. Plant J 36:602–615

Millar AH, Whelan J, Soole KL, Day DA (2011) Organization and regulation of mitochondrial respiration in plants. Annu Rev Plant Biol 62:79–104

Miller G, Suzuki N, Rizhsky L, Hegie A, Koussevitzky S, Mittler R (2007) Double mutants deficient in cytosolic and thylakoid ascorbate peroxidase reveal a complex mode of interaction between reactive oxygen species, plant development, and response to abiotic stresses. Plant Physiol 144:1777–1785

Miret JA, Munné-Bosch S (2015) Redox signaling and stress tolerance in plants: a focus on vitamin E. Ann New York Acad Sci. doi:10.1111/nyas.12639

Mittal S, Kumari N, Sharma V (2012) Differential response of salt stress on *Brassica juncea*: photosynthetic performance, pigment, proline, D1 and antioxidant enzymes. Plant Physiol Biochem 54:17–26

Mittler R (2002) Oxidative stress, antioxidants and stress tolerance. Trends Plant Sci 7(9):405–410

Mittler R, Vanderauwera S, Gollery M, Van Breusegem F (2004) Reactive oxygen gene network of plants. Trends Plant Sci 9:490–498

Mobin M, Khan NA (2007) Photosynthetic activity, pigment composition and antioxidative response of two mustard (*Brassica juncea*) cultivars differing in photosynthetic capacity subjected to cadmium stress. J Plant Physiol 164:601–610

Mohamed AA, Castagna A, Ranieri A, di Toppi LS (2012) Cadmium tolerance in *Brassica juncea* roots and shoots is affected by antioxidant status and phytochelatin biosynthesis. Plant Physiol Biochem 57:15–22

Moller IM (2001) Plant mitochondria and oxidative stress: electron transport, NADPH turnover, and metabolism of reactive oxygen species. Annu Rev Plant Physiol Mol Biol 52:561–591

Mullineaux PM, Rausch T (2005) Glutathione, photosynthesis and the redox regulation of stressresponsive gene expression. Photosynth Res 86(3):459–474

Munné-Bosch S, Alegre L (2002) The function of tocopherols and tocotrienols in plants. Crit Rev Plant Sci 21:31–57

Munne-Bosch S (2005) The role of α-tocopherol in plant stress tolerance. J Plant Physiol 162: 743–748

Murgia I, Tarantino D, Vannini C, Bracale M, Carravieri S, Soave C (2004) *Arabidopsis thaliana* plants overexpressing thylakoidal ascorbate peroxidase show increased resistance to Paraquat-induced photooxidative stress and to nitric oxide-induced cell death. Plant J 38:940–953

Noctor G, Foyer CH (1998) Ascorbate and glutathione: keeping active oxygen under control. Annu Rev Plant Physiol Plant Mol Biol 49:249–279

Noctor G, Mhamdi A, Foyer CH (2014) The roles of reactive oxygen metabolism in drought: not so cut and dried. Plant Physiol 164:1636–1648

Nouairi I, Ammar WB, Youssef NB, Miled DDB, Ghorbal MH, Zarrouk M (2009) Antioxidant defense system in leaves of Indian mustard (*Brassica juncea*) and rape (*Brassica napus*) under cadmium stress. Acta Physiol Plant 31:237–247

Olmos E, Hernandez JA, Sevilla F, Hellin E (1994) Induction of several antioxidant enzymes in the selection of a salt-tolerant cell line of *Pisum sativum*. J Plant Physiol 144:594–598

Orsini F, D'Urzo MP, Inan G, Serra S, Oh DH, Mickelbart MV, Consiglio F, Li X, Jeong JC, Yun DJ, Bohnert HJ, Bressan RA, Maggio A (2010) A comparative study of salt tolerance parameters in 11 wild relatives of *Arabidopsis thaliana*. J Exp Bot 61:3787–3798

Ozfidan C, Turkan I, Sekmen AH, Seckin B (2012) Abscisic acid-regulated responses of aba2-1 under osmotic stress: the abscisic acid-inducible antioxidant defence system and reactive oxygen species production. Plant Biol 14:337–346

Pang Q, Chen S, Dai S, Chen Y, Wang Y, Yan X (2010) Comparative proteomics of salt tolerance in *Arabidopsis thaliana* and *Thellungiella halophila*. J Proteome Res 9:2584–2599

Passaia G, Queval G, Bai J, Margis-Pinheiro M, Foyer CH (2014) The effects of redox controls mediated by glutathione peroxidases on root architecture in *Arabidopsis thaliana*. J Exp Bot 65:1403–1413
Peer WA, Mamoudian M, Lahner B, Reeves RD, Murphy AS, Salt DE (2003) Identifying model metal hyperaccumulating plants: germplasm analysis of 20 Brassicaceae accessions from a wide geographical area. New Phytol 159:421–430
Peng CL, Ou ZY, Liu N, Lin GZ (2005) Response to high temperature in flag leaves of super high-yielding rice Pei'ai 64S/E32 and Liangyoupeijiu. Rice Sci 12:179–186
Perez-Ruiz JM, Spinola MC, Kirchsteiger K, Moreno J, Sahraway M, Cejudo FJ (2006) Rice NTRC is a high-efficiency redox system for chloroplast protection against oxidative damage. Plant Cell 18:2356–2368
Pitzschke A, Forzani C, Hirt H (2006) Reactive oxygen species signaling in plants. Antioxid Redox Signal 8:1757–1764
Prasad KVSK, Saradhi PP, Sharmila P (1999) Concerted action of antioxidant enzymes and curtailed growth under zinc toxicity in *Brassica juncea*. Environ Exp Bot 42:1–10
Qiu RL, Zhao X, Tang YT, Yu FM, Hu PJ (2008) Antioxidative response to Cd in a newly discovered cadmium hyperaccumulator *Arabis paniculata* F. Chemosphere 74:6–12
Quan LJ, Zhang B, Shi WW, Li HY (2008) Hydrogen peroxide in plants: a versatile molecule of the reactive oxygen species network. J Integr Plant Biol 50:2–18
Queval G, Issakidis-Bourguet E, Hoeberichts FA, Vandorpe M, Gakière B, Vanacker H, Noctor G (2007) Conditional oxidative stress responses in the Arabidopsis photorespiratory mutant cat2 demonstrate that redox state is a key modulator of daylength-dependent gene expression, and define photoperiod as a crucial factor in the regulation of H_2O_2-induced cell death. Plant J 52:640–657
Ramel F, Birtic S, Ginies C, Soubigou-Taconnat L, Triantaphylidès C, Havaux M (2012) Carotenoid oxidation products are stress signals that mediate gene responses to singlet oxygen in plants. Proc Natl Acad Sci USA 109:5535–5540
Regelsberger G, Jakopitsch C, Plasser L, Schwaiger H, Furtmüller PG, Peschek GA, Obinger C (2002) Occurrence and biochemistry of hydroperoxidases in oxygenic phototrophic prokaryotes (cyanobacteria). Plant Physiol Biochem 40:479–490
Rhoads DM, Umbach AL, Subbaiah CC, Siedow JN (2006) Mitochondrial reactive oxygen species. Contribution to oxidative stress and interorganellar signaling. Plant Physiol 141: 357–366
Rojas C, Mysore KS (2012) Glycolate oxidase is an alternative source for H_2O_2 production during plant defense responses and functions independently from NADPH oxidase. Plant Signal Behav 7:752–755
Rojas CM, Senthil-Kumar M, Wang K, Ryu CM, Kaundal A, Mysore KS (2012) Glycolate oxidase modulates reactive oxygen species-mediated signal transduction during nonhost resistance in *Nicotiana benthamiana* and Arabidopsis. Plant Cell 24:336–352
Rouhier N, Jacquot JP (2002) Plant peroxiredoxins: alternative hydroperoxide scavenging enzymes. Photosynth Res 74:259–268
Scandalios JG (1993) Oxygen stress and superoxide dismutases. Plant Physiol 101:7–12
Seth CS, Chaturvedi PK, Misra V (2008) The role of phytochelatins and antioxidants in tolerance to Cd accumulation in *Brassica juncea* L. Ecotoxicol Environ Saf 71:76–85
Shalata A, Mittova V, Volokita M, Guy M, Tal M (2001) Response of the cultivated tomato and its wild salt-tolerant relative *Lycopersicon pennellii* to salt-dependent oxidative stress: the root antioxidative system. Physiol Plant 112:487–494
Shigeoka S, Ishikawa T, Tamoi M, Miyagawa Y, Takeda T, Yabuta Y, Yoshimura K (2002) Regulation and function of ascorbate peroxidase isoenzymes. J Exp Bot 53:1305–1319
Skórzyńska-Polit E, Drążkiewicz M, Krupa Z (2010) Lipid peroxidation and antioxidative response in *Arabidopsis thaliana* exposed to cadmium and copper. Acta Physiol Plant 32: 169–175

Soshinkova TN, Radyukina NL, Korolkova DV, Nosov AV (2013) Proline and functioning of the antioxidant system in *Thellungiella salsuginea* plants and cultured cells subjected to oxidative stress. Russ J Plant Physiol 60:41–54

Stepien P, Johnson GN (2009) Contrasting responses of photosynthesis to salt stress in the glycophyte Arabidopsis and the halophyte Thellungiella: role of the plastid terminal oxidase as an alternative electron sink. Plant Physiol 149:1154–1165

Sweetlove LJ, Foyer CH (2004) Roles for reactive oxygen species and antioxidants in plant mitochondria. In: Day DA, Millar AH, Whelan J (eds) Plant mitochondria: from genome to function, advances in photosynthesis and respiration, vol 1. Kluwer Academic, Dordrecht

Szabados L, Savouré A (2010) Proline: a multifunctional amino acid. Trends Plant Sci 15:89–97

Takahashi MA, Asada K (1983) Superoxide anion permeability of phospholipid membranes and chloroplast thylakoids. Arch Biochem Biophys 226:558–566

Taulavuori K, Prasad MNV, Taulavuori E, Laine K (2005) Metal stress consequences on frost hardiness of plants at northern high latitudes: a review and hypothesis. Environ Pollut 135: 209–220

Tuomainen MH, Nunan N, Lehesranta SJ, Tervahauta AI, Hassinen VH, Schat H, Kärenlampi SO (2006) Multivariate analysis of protein profiles of metal hyperaccumulator *Thlaspi caerulescens* accessions. Proteomics 6:3696–3706

Ursini F, Maiorin M, Roveri A (1997) Phospholipid hydroperoxide glutathione peroxidase (PHGPx): more than an antioxidant enzyme? Biomed Environ Sci 10:327–332

Ushimaru T, Nakagawa T, Fujioka Y, Daicho K, Naito M, Yamauchi Y, Murata N (2006) Transgenic Arabidopsis plants expressing the rice dehydroascorbate reductase gene are resistant to salt stress. J Plant Physiol 163:1179–1184

Uzilday B, Ozgur R, Sekmen AH, Yildiztugay E, Turkan I (2015) Changes in the alternative electron sinks and antioxidant defence in chloroplasts of the extreme halophyte *Eutrema parvulum* (*Thellungiella parvula*) under salinity. Ann Bot 115:449–463

Verslues PE, Bray EA (2006) Role of abscisic acid (ABA) and *Arabidopsis thaliana* ABA-insensitive loci in low water potential-induced ABA and proline accumulation. J Exp Bot 57:201–212

Voothuluru P, Sharp RE (2013) Apoplastic hydrogen peroxide in the growth zone of the maize primary root under water stress. I. Increased levels are specific to the apical region of growth maintenance. J Exp Bot 64:1223–1233

Wagner D, Przybyla D, den Camp R, Kim C, Landgraf F, Lee KP, Wursch M, Laloi C, Nater M, Hideg E, Apel K (2004) The genetic basis of singlet oxygen induced stress responses of *Arabidopsis thaliana*. Science 306:1183–1185

Wang X, Chang L, Wang B, Wang D, Li P, Wang L, Guo A (2013) Comparative proteomics of *Thellungiella halophila* leaves from plants subjected to salinity reveals the importance of chloroplastic starch and soluble sugars in halophyte salt tolerance. Mol Cell Proteomics 12:2174–2195

Wang Y, Ying Y, Chen J, Wang X (2004) Transgenic Arabidopsis overexpressing Mn-SOD enhanced salt-tolerance. Plant Sci 167:671–677

Wójcik M, Skórzyńska-Polit E, Tukiendorf A (2006) Organic acids accumulation and antioxidant enzyme activities in *Thlaspi caerulescens* under Zn and Cd stress. Plant Growth Regul 48: 145–155

Wrzaczek M, Brosché M, Kangasjarvi J (2013) ROS signaling loops - production, perception, regulation. Curr Opin Plant Biol 16:575–582

Xi DM, Liu WS, Yang GD, Wu CA, Zheng CC (2010) Seed-specific overexpression of antioxidant genes in Arabidopsis enhances oxidative stress tolerance during germination and early seedling growth. Plant Biotechnol J 8:796–806

Xiang C, Oliver DJ (1998) Glutathione metabolic genes coordinately respond to heavy metals and jasmonic acid in Arabidopsis. Plant Cell Online 10:1539–1550

Yan J, Wang J, Tissue D, Holaday AS, Allen R, Zhang H (2003) Photosynthesis and seed production under water-deficit conditions in transgenic tobacco plants that overexpress an ascorbate peroxidase gene. Crop Sci 43:1477–1483

Yang T, Poovaiah BW (2002) Hydrogen peroxide homeostasis: activation of plant catalase by calcium/calmodulin. Proc Natl Acad Sci USA 99:4097–4102

Yokota T, Oda HD (2001) The role of 15-lipoxygenase in disruption of the peroxisomal membrane and in programmed degradation of peroxisomes in normal rat liver. J Histochem Cytochem 49:613–622

Zander M, Chen S, Imkampe J, Thurow C, Gatz C (2012) Repression of the *Arabidopsis thaliana* jasmonic acid/ethylene-induced defense pathway by TGA-interacting glutaredoxins depends on their C-terminal ALWL motif. Mol Plant 5:831–840

Zhou Y, Gao F, Li X, Zhang J, Zhang G (2010) Alterations in phosphoproteome under salt stress in Thellungiella roots. Chin Sci Bull 55:3673–3679

ROS Signaling: Relevance with Site of Production and Metabolism of ROS

Rup Kumar Kar

Contents

Abstract One of the inevitable consequences of aerobic metabolism is the production of ROS in biological organisms including plants. Accordingly, plants have evolved antioxidant system (antioxidant enzymes and antioxidant molecules) to protect the cell components from oxidative threat. Usually cells accumulate ROS to a fatal level due to imbalance between generation and scavenging under stress or pathogenic attack. Recent observations led to an idea of involvement of ROS in signaling for plant growth and development. Although common sites of ROS generation are chloroplasts, mitochondria, and other organelles, ROS produced by plasma membrane localized NADPH oxidase (Rboh) in extracellular space has been implicated to participate in signaling process. Calcium, being most important signal molecule, has a cross talk with Rboh through a positive feedback loop that forms the basis of ROS-driven signaling network. Such signaling most often works behind plant growth and developmental processes like seed germination, root growth, stomatal regulation, and stress tolerance. Recent researches establish a ROS wave with an integration of Ca^{2+} signal that may operate for long distance signaling in plants. This may help to explain the event of systemic signaling in case of systemic acquired resistance (SAR) during pathogen attack and systemic acquired acclimation (SAA), which is achieved by gradual exposure to stress.

R.K. Kar (✉)
Plant Physiology & Biochemistry Laboratory, Department of Botany, Visva-Bharati University, Santiniketan 731235, West Bengal, India
e-mail: rupkumar.kar@visva-bharati.ac.in

D.K. Gupta et al. (eds.), *Reactive Oxygen Species and Oxidative Damage in Plants Under Stress*, DOI 10.1007/978-3-319-20421-5_5

Establishment of the possibilities of intracellular ROS signaling through vesicular trafficking and involvement in regulation of nuclear activities are on the way.

Keywords Antioxidant • Extracellular space • NADPH oxidase • Oxidative metabolism • Reactive oxygen species • ROS signaling • ROS wave

1 Introduction

Since the origin of life on the earth, it has been a challenge to perpetuate with the constantly changing environment. Accordingly, the life forms have undertaken great deal of modifications to cope up with newer milieu. Therefore, a long history of evolution includes diversity not only in shape, size, and morphology but also metabolism and behavior. Divergence of plants from animals has been a critical point during such evolution since a radical difference in respect of metabolism makes them two discrete opposite groups—autotrophic (producers) and heterotrophic (consumers), respectively. At the same time, plants, being sessile organisms, were more exposed to unavoidable adverse conditions that exerted an enormous selection pressure for suitable adaptations.

Another breakthrough in the history of metabolic evolution is the introduction of oxygen in the earth's outer layer, atmosphere, with the advent of oxygenic photosynthesis in cyanobacteria. Besides the advantages of living with O_2 (like energy efficient aerobic respiration, formation of ozone layer for UV filtration), generation of reactive oxygen species (ROS) as inevitable byproducts of aerobic metabolism (Halliwell 2006) has been presumed to be a curse to life, as ROS can attack major biomolecules viz. lipids, proteins, nucleic acids, and many others. But from the recent knowledge developed by extensive research for last one or two decades, it appears to be boon in a disguise. In recent years, role of ROS as signaling molecule for growth and development and also for defense has been mostly unraveled.

2 ROS: Types and Chemistry

Molecular oxygen as such mostly unreactive in ground state, by univalent reduction produces a series of free radicals and other chemical species through electron transfer reactions. Most important products are superoxide ($O_2^{\bullet -}$), that is further dismutated to H_2O_2 either spontaneously or enzymatically, and hydroxyl radical ($OH^{\bullet}$) (Gechev et al. 2006). Also energy transfer reactions lead to the formation of singlet oxygen ($^1O_2$) (Kim et al. 2008). Transition metals (Fe or Cu) having unpaired electrons react with O_2 to form $O_2^{\bullet -}$ frequently. These are also responsible for generation of $OH^{\bullet}$ from H_2O_2 by Fenton reaction or Haber–Weiss reaction

(Halliwell and Gutteridge 2000; Lee et al. 2007; Quan et al. 2008). Such different forms of reactive oxygen species (ROS) may be connected by a reaction cascade where $^1O_2$ or $O_2^{\bullet-}$ are sequentially reduced to H_2O_2 and $OH^{\bullet}$ (Halliwell and Gutteridge 2000). Besides, $O_2^{\bullet-}$ can react with $NO^{\bullet}$ to produce peroxynitrite ($OONO^-$) or may give rise to $HO_2^{\bullet}$ through protonation. Different ROS species are mostly short-lived and attack locally major cellular ingredients. Besides, other radicals like alkoxy radical ($RO^{\bullet}$), peroxy radical ($ROO^{\bullet}$), excited carbonyl (RO^*) that may be generated under different situations are all cytotoxic to plant cells (Vellosillo et al. 2010). $O_2^{\bullet-}$ and $OH^{\bullet}$ are highly unstable (half-life at the level of microseconds and nanoseconds, respectively) and cannot cross membrane, while H_2O_2, though not a free radical but a ROS, is relatively stable (half-life around 1 ms) (Møller et al. 2007) and can cross membranes through aquaporins. Therefore, under cellular conditions $O_2^{\bullet-}$ and $OH^{\bullet}$ cause damage locally, but H_2O_2 may move across cellular distances and thus act as signal molecules apart from damaging biomolecules.

3 ROS: Sites of Production

Reactive oxygen species are synthesized in plant cells through various reactions as a part of normal metabolism (Suzuki et al. 2011). But their rate of production aggravates under stress and pathogenic conditions, and such production is confined to particular cell compartments (Mittler 2002; Asada 2006; Navrot et al. 2007).

In green plants, chloroplast is the most important among the organelles in respect of ROS generation as O_2 is continuously synthesized and readily available inside the chloroplast because of photosynthetic electron transport activity (Tripathy and Oelmüller 2012). Thus, due to over-excitation of chlorophylls under stress excitation energy from triplet chlorophyll may be transferred to O_2 producing $^1O_2$ near the reactions centers of photosystem PSII. Besides, $O_2^{\bullet-}$ may also be formed at PSI via Mehler reaction (Karuppanapandian et al. 2011) or at PSII during electron transfer to O_2 through Q_A and Q_B (Das and Roychoudhury 2014). On the other hand, mitochondria, the major source of ROS in mammals, may generate $O_2^{\bullet-}$ and H_2O_2 by univalent reduction of O_2 near electron transport chain (ETC) in plants. Apart from these organelles, single membrane bound peroxisomes are the major sites of H_2O_2 production due to activities of flavin oxidases (del Río et al. 2006; Palma et al. 2009). Plant peroxisomes also produce NO during different metabolic reactions (Corpas et al. 2001). Cell wall and apoplastic space are active sites of ROS production and possibly play pivotal roles in defense and signaling. Cell wall localized peroxidase(s), diamine/polyamine oxidases, and oxalate oxidase produce H_2O_2 (Spiteller 2003; Higuchi 2006) that may, in turn, be metabolized to $OH^{\bullet}$ by the activity of class III peroxidases (Kärkönen and Kuchitsu 2015). Moreover, plasma membrane localized NADPH oxidase transfers electrons from NADPH on cytoplasmic side to O_2 producing $O_2^{\bullet-}$ in cell wall or apoplastic space. This superoxide is further dismutated to H_2O_2 which plays important roles in cell wall

Table 1 Different intracellular ROS production sites and associated ROS-action(s)

Site of production	Source(s) of ROS	ROS action(s)	Reference(s)
Chloroplasts	Photosynthetic electron transport chain, lipoxygenase	Induce chloroplast avoidance movement	Wen et al. (2008); Foyer and Noctor (2009)
Mitochondria	Complexes I and III of mitochondrial electron transport chain	Interorganellar signal transduction	Møller (2001); Sweetlove and Foyer (2004); Rhoads et al. (2006)
Peroxisome	Flavin oxidases (e.g., xanthine oxidase), glycolate oxidase, peroxisomal membrane polypeptides	Involved in leaf senescence, responses to some abiotic stresses	López-Huertas et al. (1999); Mittler et al. (2004); del Río et al. (2006)
Endoplasmic reticulum	NAD(P)H-dependent electron transport	Ca^{+2} transmission	Mittler (2002)
Plasma membrane	NADPH oxidase (NOX)	Defense in biotic–abiotic stresses	Kwak et al. (2003); Apel and Hirt (2004); Sagi and Fluhr (2006)
Apoplast	Class III Peroxidase, germin-like oxalate oxidase, polyamine oxidase	ABA-induced stomatal closure, modulation of wall extensibility	Mittler (2002); Müller et al. (2009)

modifications and pathogenic defense (Kärkönen and Kuchitsu 2015). A summarized chart showing the cellular sites for ROS production and possible action is given in Table 1.

4 Oxidative Metabolism and Antioxidant System

Plants growing under natural condition are always prone to be subjected to abiotic and biotic stresses. As a response to such stresses, plant cells invariably generate significant level of ROS leading to oxidative stress (Dangl and Jones 2001; Mittler et al. 2004; Gunes et al. 2008; Xiao et al. 2008). Plants thus developed the ability to control the oxidative load that finally results in stress tolerance (Cheeseman 2007). Stress-induced increase in ROS level can cause different degree of oxidation of cell components and a gross change in redox status. Thus, an oxidative outburst as a consequence of stress is reflected in the levels of ROS molecules like $O_2^{\bullet -}$, H_2O_2, and $OH^{\bullet}$, which are biochemically connected through metabolic reactions (Halliwell and Gutteridge 2000; Quan et al. 2008). However, plants have evolved defense system against such oxidative attack that may even occur under normal condition. Antioxidant system in plant comprises of enzymatic as well as nonenzymatic components. Superoxide dismutases (SODs) are considered as first line of defense that may become active in different cellular compartments (Alscher et al. 2002). Depending on the metal requirement, SODs are grouped under Fe SOD (chloroplasts), Mn SOD (mitochondria and peroxisomes), and Cu-Zn SOD (chloroplasts, cytosol, peroxisomes, and extracellular space), and all these are involved in dismutation of $O_2^{\bullet-}$ to H_2O_2 and O_2 (Mittler 2002; Alscher et al. 2002; Chen et al. 2010). Once H_2O_2 is formed, it may further be metabolized by the activities of several enzymes like catalase, ascorbate peroxidase, and glutathione peroxidase, the latter two enzymes requiring electron donor (ascorbate and glutathione, respectively) (Dat et al. 2000). Nonenzymatic antioxidants includes ascorbate and glutathione (GSH) as important members while other members like carotenoids, tocopherol, flavonoids, etc. also participate in ROS scavenging process (Apel and Hirt 2004). As major redox buffers ascorbate and GSH help in bringing down the oxidative load involving ascorbate–glutathione cycle (Noctor and Foyer 1998; Apel and Hirt 2004; Halliwell 2006). Therefore, a balance between ROS generation and ROS scavenging maintain the redox homeostasis, and a complex network of prooxidant and antioxidant system coordinates such balance (Quan et al. 2008). Often this balance is perturbed under stress, senescence, and pathogen attack leading to oxidative burst and plants having stress tolerance show upregulation of antioxidant system to reestablish the redox homeostasis (Kar 2011).

5 Role of ROS in Signaling

Although ROS have been initially envisaged as offenders causing damage to cell components under adverse conditions like stress and pathogen attack (Halliwell 2006), role of ROS as signaling system has emerged quite recently. Some properties of ROS that makes them likely to be signaling molecules are their compartment-specific rapid generation and scavenging, instability (very short half-life), and their oxidative reactions with other signal molecules and phytohormones that may integrate with the signaling cascade (Mittler et al. 2011; Kar 2011). But playing with potentially toxic molecules is dangerous unless there is some safeguard mechanism prevailing in the respective compartments. In fact, evolution of ROS scavenging mechanism occurred even before ROS has been used for signaling (Mittler et al. 2011). Therefore, it seems that the compartment chosen for ROS signaling must have tight regulation for combating oxidative effects. Moreover, effective signaling requires some sensor and related components in specific compartments that form a cascade ending with the final response that may involve gene expression (Apel and Hirt 2004). Reports are accumulating on ROS sensing by histidine kinases, MAPK cascade, inhibition of protein phosphatases, and activation of transcription factors (Apel and Hirt 2004). Besides, other redox molecules like thioredoxins, peroxiredoxins, glutaredoxins, and NADPH that constitute cellular redox networks also connect ROS for signaling (Dietz et al. 2010; Rouhier 2011). As a signaling component ROS may act downstream of hormones like ethylene and ABA, although there may be feedback loop where ROS may induce hormonal increase (e.g., ethylene synthesis may be induced by ROS). An interaction of hormones like GA, ABA, and ethylene with ROS has been demonstrated to control seed germination in *Vigna radiata* (Chaudhuri et al. 2013).

For last decade, ROS have been found to play key role in different plant growth and developmental processes like seed germination, root growth and root hair elongation, pollen tube growth, stomatal regulation, and gravitropic movement apart from stress and defense responses (Kar 2011; Müller et al. 2012; Baxter et al. 2013).

6 Signaling in Growth and Development

Plant growth depends on the cell division and cell elongation, the latter being the central event in case of seed germination, pollen tube formation, and root hair growth. Cell elongation growth is mostly driven by turgor and cell wall relaxation. ROS-mediated cell wall relaxation has been demonstrated in such process where ROS production is initiated by the activity of NADPH oxidase that generates $O_2^{\bullet-}$ in the apoplast (Singh et al. 2014). In fact, plant NADPH oxidases, known as respiratory burst oxidase homologues (rbohs), are homologues of catalytic domain of mammalian $gp91^{phox}$ (Keller et al. 1998) playing a key role in ROS production for signaling (Torres and Dangl 2005; Suzuki et al. 2011). That is why Rbohs have been aptly designated as engines of ROS signaling (Suzuki et al. 2011). Superoxide-producing plant NADPH oxidases or Rbohs are plasma membrane bound enzymes that contain EF-hand for Ca^{2+} binding and phosphorylation sites potentially controlled by cytosolic Ca^{2+} level (Ogasawara et al. 2008). On the other hand, ROS can induce increase in cytosolic Ca^{2+} level by activating hyperpolarization-activated Ca^{2+}-permeable cation channels (Pei et al. 2000), thus forming a positive feedback loop that help in amplifying the signal. Such ROS-Ca^{2+} interaction may form the basis of several ROS-mediated plant processes (Mori and Schroeder 2004). Further, extracellular $O_2^{\bullet-}$ thus formed by NADPH oxidase is immediately dismutated to H_2O_2 spontaneously or by the activity of SOD. Being relatively a stable form of ROS, H_2O_2 may enter into cytoplasm through aquaporins and become instrumental in causing elongation growth possibly through modulation of cytoskeleton including actin. Ca^{2+} ions may also have some role in this process, but less is known about this mechanism. However, a direct involvement of ROS ($OH^{\bullet}$) resulting from further metabolism of $O_2^{\bullet-}$ and H_2O_2 in the cell wall by cleaving polysaccharides (Müller et al. 2009) has also been implicated for seed germination and secondary root growth (Kranner et al. 2010; Roach and Kranner 2011; Singh et al. 2014). Apparently, apoplastic space is an ideal site for ROS signaling since this compartment has low redox buffering capacity (Noctor et al. 2002), thus less affecting signal strength.

7 Systemic Signaling and Acclimation

Recently, ROS are also proposed for systemic signaling thus constituting a cell-to-cell communication system that is involved in case of systemic acquired resistance (SAR) when pathogens invade the plant tissues and systemic acquired acclimation in case of exposure to different kinds of stress (Mittler et al. 2011; Baxter et al. 2013). During such systemic communication, ROS and Ca^{2+} interact with each other constituting a positive feedback loop, and this forms a wave of signal that can propagate from exposed tissues to systemic or non-challenged tissues (Gilroy et al. 2014). It is reported that such systemic signal, which is self-propagating, can move faster (8.4 cm min^{-1}), and this wave is essential for long distance signaling (Mittler et al. 2011). In this signaling process, NADPH oxidase initiates the ROS cascade being activated by stimuli followed by accumulation of H_2O_2 that can move to neighboring cells and activate Ca^{2+} influx channel on plasma membrane. Subsequently, cytosolic Ca^{2+} again activates Rboh by binding at EF-hands and phosphorylation to produce extracellular H_2O_2 that forms the ROS wave (Steinhorst and Kudla 2013).

Although the extracellular space has been emphasized for ROS signaling in most of research works based on the idea that this compartment has least interference with other components including antioxidants, arguments may be placed regarding any signaling role of intracellularly generated ROS like that are produced in chloroplasts, mitochondria, peroxisomes, and other organelles or compartments. Rhee (2006) first highlighted the possibility of intracellular signaling through H_2O_2. There are reports describing endosome-associated ROS generation by NADPH oxidase or ROS generation from endoplasmic reticulum and its intracellular compartments that suggest for intracellular communication system in response to stress (Park et al. 2003; Fedoroff 2006). ROS molecules have been demonstrated to be carried in vesicle and delivered to the destination by tightly regulated vesicular trafficking process in response to salt stress (Leshem et al. 2006; Leshem and Levine 2007). Therefore, a possibility of intracellular signaling by ROS through vesicular trafficking cannot be ruled out. Another important cell compartment is the nucleus where major signaling cascades end with modulation of nuclear functions like transcription and also epigenetic chromatin modifications. Extension of activities of ROS along with its cross talk with Ca^{2+} to the nucleus for signaling purpose has already been attempted (Mazars et al. 2010). Further research in near future will establish a complete array of ROS signaling in plant growth and development and interaction with environment. A working model including different subcellular sites of ROS production and possible downstream action towards signaling for plant growth and development and stress responses has been shown in the Fig. 1.

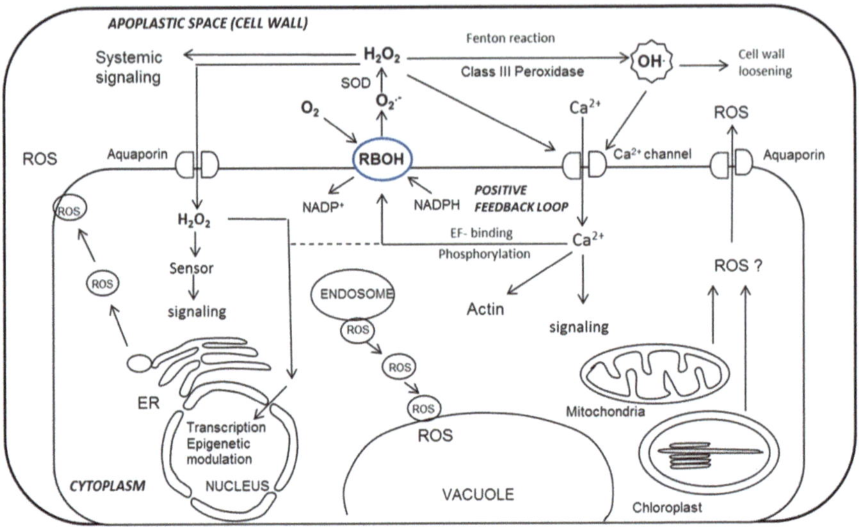

Fig. 1 An integrated model encrypting ROS production and metabolism at different subcellular sites and possible action towards signaling for plant responses. In the extracellular space (apoplast), the major site of ROS signaling, plasma membrane-located NADPH oxidase (RBOH) generates superoxide ($O_2^{\bullet -}$) that sequentially produces hydrogen peroxide (H_2O_2) and then hydroxyl radical ($OH^{\bullet}$), which directly causes cell wall loosening, required for cell elongation growth and/or activate hyperpolarization-activated calcium channels (HACC). Ca^{2+} influx into cytosol, in turn, activates RBOH though EF-binding and phosphorylation forming a positive feedback loop. Intracellular Ca^{2+} is also involved in interfering with actin and other signaling processes. Extracellular ROS (H_2O_2) may also enter into cytosol through aquaporins and lead to intracellular signaling through binding with redox censor in cytosol or move to nucleus along with Ca^{2+} to modulate nuclear activities. Besides, extracellular ROS (H_2O_2) may migrate to the neighboring cells and trigger the positive feedback loop with Ca^{2+} and sequentially propagate as systemic signal to far away cells. Mitochondria and chloroplasts, as intracellular sites, produce ROS that may participate in signaling. ROS may also be produced intracellularly in endoplasmic reticulum (ER) and endosomes and move for signaling either to apoplastic space or vacuole through vesicular trafficking

References

Alscher RG, Erturk N, Heath LS (2002) Role of superoxide dismutases (SODs) in controlling oxidative stress in plants. J Exp Bot 53:1331–1341

Apel K, Hirt H (2004) Reactive oxygen species: metabolism, oxidative stress, and signal transduction. Annu Rev Plant Biol 55:373–399

Asada K (2006) Production and scavenging of reactive oxygen species in chloroplasts and their functions. Plant Physiol 141:391–396

Baxter A, Mittler R, Suzuki N (2013) ROS as key players in plant stress signaling. J Exp Bot 65:1229–1240

Chaudhuri A, Singh KL, Kar RK (2013) Interaction of hormones with reactive oxygen species in regulating seed germination of *Vigna radiata* (L.) Wilczek. Plant Biochem Physiol 1:103

Cheeseman JM (2007) Hydrogen peroxide and plant stress: a challenging relationship. Plant Stress 1:4–15

Chen Q, Zhang M, Shen S (2010) Effect of salt on malondialdehyde and antioxidant enzymes in seedling roots of Jerusalem artichoke (*Helianthus tuberosus* L.). Acta Physiol Plant 33:273–278

Corpas FJ, Barroso JB, del Río LA (2001) Peroxisomes as a source of reactive oxygen species and nitric oxide signal molecules in plant cells. Trends Plant Sci 6:145–150

Dangl JL, Jones JDG (2001) Plant pathogens and integrated defence responses to infection. Nature 411:826–833

Das K, Roychoudhury A (2014) Reactive oxygen species (ROS) and response of antioxidants as ROS-scavengers during environmental stress in plants. Front Environ Sci 2:53

Dat J, Vandenabeele S, Vranová E, Van Montagu M, Inzé D, Van Breusegem F (2000) Dual action of the active oxygen species during plant stress responses. Cell Mol Life Sci 57:779–795

del Río LA, Sandalio LM, Corpas FJ, Palma JM, Barroso JB (2006) Reactive oxygen species and reactive nitrogen species in peroxisomes. Production, scavenging, and role in cell signaling. Plant Physiol 141:330–335

Dietz KJ, Jacquot JP, Harris G (2010) Hubs and bottlenecks in plant molecular signalling networks. New Phytol 188:919–938

Fedoroff N (2006) Redox regulatory mechanisms in cellular stress responses. Ann Bot 98:289–300

Foyer CH, Noctor G (2009) Redox regulation in photosynthetic organisms: signaling, acclimation, and practical implications. Antioxid Redox Signal 11:861–905

Gechev TS, Van Breusegem F, Stone JM, Denev I, Laloi C (2006) Reactive oxygen species as signals that modulate plant stress responses and programmed cell death. Bioessays 28:1091–1101

Gilroy S, Suzuki N, Miller G, Choi WG, Toyota M, Devireddy A, Mittler R (2014) A tidal wave of signals: calcium and ROS at the forefront of rapid systemic signaling. Trends Plant Sci 19:623–630

Gunes A, Pilbeam D, Inal A, Coban S (2008) Influence of silicon on sunflower cultivars under drought stress, I: Growth, antioxidant mechanisms and lipid peroxidation. Commun Soil Sci Plant Anal 39:1885–1903

Halliwell B (2006) Reactive species and antioxidants. Redox biology is a fundamental theme of aerobic life. Plant Physiol 141:312–322

Halliwell B, Gutteridge JMC (2000) Free radicals in biology and medicine, 4th edn. Oxford University Press, Oxford

Higuchi T (2006) Look back over the studies of lignin biochemistry. J Wood Sci 52:2–8

Kar RK (2011) Plant response to water stress: role of reactive oxygen species. Plant Signal Behav 6:1741–1745

Kärkönen A, Kuchitsu K (2015) Reactive oxygen species in cell wall metabolism and development in plants. Phytochemistry 112:22–32

Karuppanapandian T, Moon JC, Kim C, Manoharan K, Kim W (2011) Reactive oxygen species in plants: their generation, signal transduction, and scavenging mechanisms. Aust J Crop Sci 5:709–725

Keller T, Damude HG, Werner D, Doerner P, Dixon RA, Lamb C (1998) A plant homolog of the neutrophil NADPH oxidase gp91^{phox} subunit gene encodes a plasma membrane protein with Ca^{2+} binding motifs. Plant Cell 10:255–266

Kim C, Meskauskiene R, Apel K, Laloi C (2008) No single way to understand singlet oxygen signalling in plants. EMBO Rep 9:435–439

Kranner I, Roach T, Beckett RP, Whitaker C, Minibayeva M (2010) Extracellular production of reactive oxygen species during seed germination and early seedling growth in *Pisum sativum*. J Plant Physiol 167:805–811

Kwak JM, Mori IC, Pei Z-M, Leonhardt N, Torres MA, Dangl JL, Bloom RE, Bodde S, Jones JDG, Schroeder JI (2003) NADPH oxidase *AtrbohD* and *AtrbohF* genes function in ROS-dependent ABA signaling in *Arabidopsis*. EMBO J 22:2623–2633

Lee SH, Ahsan N, Lee KW, Kim DH, Lee DG, Kwak SS, Kwon SY, Kim TH, Lee BH (2007) Simultaneous overexpression of both CuZn superoxide dismutase and ascorbate peroxidase in transgenic tall fescue plants confers increased tolerance to a wide range of abiotic stresses. J Plant Physiol 164:1626–1638

Leshem Y, Levine A (2007) Intracellular ROS. What does it do there? Plant Signal Behav 2:155–156

Leshem Y, Melamed-Book N, Cagnac O, Ronen G, Nishri Y, Solomon M, Cohen G, Levine A (2006) Suppression of Arabidopsis vesicle-SNARE expression inhibited fusion of H_2O_2-containing vesicles with tonoplast and increased slat tolerance. Proc Natl Acad Sci U S A 103:18008–18013

López-Huertas E, Corpas FJ, Sandalio LM, del Río LA (1999) Characterization of membrane polypeptides from pea leaf peroxisomes involved in superoxide radical generation. Biochem J 337:531–536

Mazars C, Thuleau P, Lamotte O, Bourque S (2010) Cross-talk between ROS and calcium in regulation of nuclear activities. Mol Plant 3:706–718

Mittler R (2002) Oxidative stress, antioxidants and stress tolerance. Trends Plant Sci 7:405–410

Mittler R, Vanderauwera S, Gollery M, Van Breusegem F (2004) Reactive oxygen gene network of plants. Trends Plant Sci 9:490–498

Mittler R, Vanderauwera S, Suzuki N, Miller G, Tognetti VB, Vnadepoele K, Gollery M, Shulaev V, Breusegem FV (2011) ROS signaling: the new wave? Trends Plant Sci 16:300–309

Møller IM (2001) Plant mitochondria and oxidative stress: electron transport, NADPH turnover and metabolism of reactive oxygen species. Annu Rev Plant Physiol Plant Mol Biol 52:561–591

Møller IM, Jensen PE, Hansson A (2007) Oxidative modifications to cellular components in plants. Annu Rev Plant Biol 58:459–481

Mori IC, Schroeder JI (2004) Reactive oxygen species activation of plant Ca^{2+} channels. A signaling mechanism in polar growth, hormone transduction, stress signaling and hypothetically mechanotransduction. Plant Physiol 135:702–708

Müller K, Linkies A, Vreeburg RAM, Fry SC, Krieger-Liszkay A, Leubner-Metzger G (2009) *In vivo* cell wall loosening by hydroxyl radicals during cress seed germination and elongation growth. Plant Physiol 150:1855–1865

Müller K, Linkies A, Leubner-Metzger G, Kermode AR (2012) Role of a respiratory burst oxidase of *Lepidium sativum* (cress) seedlings in root development and auxin signaling. J Exp Bot 63:6325–6334

Navrot N, Roubier N, Gelbaye E, Jacquot JP (2007) Reactive oxygen species generation and antioxidant systems in plant mitochondria. Physiol Planta 129:185–195

Noctor G, Foyer CH (1998) Ascorbate and glutathione: keeping active oxygen under control. Annu Rev Plant Physiol Plant Mol Biol 49:249–279

Noctor G, Veljovic-Jovanovic S, Driscoll S, Novitskaya L, Foyer CH (2002) Drought and oxidative load in the leaves of C_3 plants: a predominant role for photorespiration? Ann Bot 89:841–850

Ogasawara Y, Kaya H, Hiraoka G, Yumoto F, Kimura S, Kadota Y, Hishinuma H, Senzaki E, Yamagoe S, Nagata K, Nara M, Suzuki K, Tanokura M, Kuchitsu K (2008) Synergistic activation of the *Arabidopsis* NADPH oxidase AtrbohD by Ca^{2+} and phosphorylation. J Biol Chem 283:8885–8892

Palma JM, Corpas FJ, del Río LA (2009) Proteome of plant peroxisomes: new perspectives on the role of these organelles in cell biology. Proteomics 9:2301–2312

Park KY, Jung JY, Park J, Hwang JU, Kim YW, Hwang I, Lee Y (2003) A role for phosphatidylinositol 3-phosphate in abscisic acid-induced reactive oxygen species generation in guard cells. Plant Physiol 132:92–98

Pei ZM, Murata Y, Benning G, Thomine S, Klüsener B, Allen GJ, Grill E, Schroeder JI (2000) Calcium channels activated by hydrogen peroxide mediate abscisic acid signaling in guard cells. Nature 406:731–734

Quan LJ, Zhang B, Shi WW, Li HY (2008) Hydrogen peroxide in plants: a versatile molecule of the reactive oxygen species network. J Integr Plant Biol 50:2–18

Rhee SG (2006) Cell signaling: H_2O_2, a necessary evil for cell signaling. Science 312:1882–1883

Rhoads DM, Umbach AL, Subbaiah CC, Siedow JN (2006) Mitochondrial reactive oxygen species. Contribution to oxidative stress and interorganellar signaling. Plant Physiol 141:357–366
Roach T, Kranner I (2011) Extracellular superoxide production associated with secondary root growth following desiccation of *Pisum sativum* seedlings. J Plant Physiol 168:1870–1873
Rouhier N (2011) Plant glutaredoxins: pivotal players in redox biology and iron-sulphur centre assembly. New Phytol 186:365–372
Sagi M, Fluhr R (2006) Production of reactive oxygen species by plant NADPH oxidases. Plant Physiol 141:336–340
Singh KL, Chaudhuri A, Kar RK (2014) Superoxide and its metabolism during germination and axis growth of *Vigna radiata* (L) Wilczek seeds. Plant Signal Behav 9:e29278
Spiteller G (2003) The relationship between changes in the cell wall, lipid peroxidation, proliferation, senescence and cell death. Physiol Plant 119:5–18
Steinhorst L, Kudla J (2013) Calcium and reactive oxygen species rule the waves of signaling. Plant Physiol 163:471–485
Suzuki N, Miller G, Morales J, Shulaev V, Torres MA, Mittler R (2011) Respiratory burst oxidases: the engines of ROS signaling. Curr Opin Plant Biol 14:691–699
Sweetlove LJ, Foyer CH (2004) Roles for reactive oxygen species and antioxidants in plant mitochondria. In: Day DA, Millar AH, Whelan J (eds) Plant mitochondria: from genome to function, vol 1, Advances in photosynthesis and respiration. Kluwer, Dordrecht
Torres MA, Dangl JL (2005) Functions of the respiratory burst oxidase in biotic interactions, abiotic stress and development. Curr Opin Plant Biol 8:397–403
Tripathy BC, Oelmüller R (2012) Reactive oxygen species generation and signaling in plants. Plant Signal Behav 7:1621–1633
Vellosillo T, Vicente J, Kulasekaran S, Hamberg M, Castresana C (2010) Emerging complexity in reactive oxygen species production and signaling during the response of plants to pathogens. Plant Physiol 154:444–448
Wen F, Xing D, Zhang L (2008) Hydrogen peroxide is involved in high blue light-induced chloroplast avoidance movements in *Arabidopsis*. J Exp Bot 59:2891–2901
Xiao X, Xu X, Yang F (2008) Adaptive responses to progressive drought stress in two *Populus cathayana* populations. Silva Fenn 42:705–719

Heavy Metal-Induced Oxidative Stress in Plants: Response of the Antioxidative System

Ivna Štolfa, Tanja Žuna Pfeiffer, Dubravka Špoljarić, Tihana Teklić, and Zdenko Lončarić

Contents

Abstract Heavy metals (HMs) are among the most important environmental pollutants, particularly in areas with strong anthropogenic pressure. For plants, high levels of HMs are extremely toxic since they may act in several different modes: by the direct inhibition of plant growth and biosynthetic pathways or through the production of reactive oxygen species (ROS). Certain metals generate ROS due to their involvement in redox reactions like Fenton and/or Haber–Weiss reactions, while metals without redox capacity enhance ROS production by reducing the antioxidant glutathione pool, activating calcium-dependent systems and

I. Štolfa (✉) • T.Ž. Pfeiffer • D. Špoljarić
Department of Biology, Josip Juraj Strossmayer University of Osijek, Cara Hadrijana 8/A, 31000 Osijek, Croatia
e-mail: istolfa@biologija.unios.hr

T. Teklić • Z. Lončarić
Faculty of Agriculture in Osijek, Josip Juraj Strossmayer University of Osijek, Ulica kralja Petra Svačića 1D, 31000 Osijek, Croatia

D.K. Gupta et al. (eds.), *Reactive Oxygen Species and Oxidative Damage in Plants Under Stress*, DOI 10.1007/978-3-319-20421-5_6

influencing iron-mediated processes. ROS production affects lipids, proteins, and DNA and consequently leads to cell death. In response, plants are equipped with complex enzymatic and nonenzymatic mechanisms involved in antioxidative defense to neutralize HM toxicity, and the main components of these mechanisms will be reviewed in this chapter.

Keywords Heavy metals • ROS • Antioxidants

1 Introduction

From a biological point of view, the term "heavy metal" refers to a series of metals that can be toxic for plants and animals even at very low concentrations. Some metals, such as copper (Cu), cobalt (Co), chromium (Cr), iron (Fe), nickel (Ni), zinc (Zn), manganese (Mn), and molybdenum (Mo), are essential elements for plant growth and metabolism, while others, such as cadmium (Cd), mercury (Hg), lead (Pb), selenium (Se), as well as metalloids like arsenium (As), have no known biological function (Rascio and Navari-Izzo 2011). However, it has been shown that carbonic anhydrase may bind Cd as a cofactor in the marine diatom *Thalassiosira weissflogii* (Lane et al. 2005).

Heavy metals (HMs) occur naturally in the Earth's crust and are found in rocks, soils, sediments, and marine and freshwater ecosystems. High levels of HMs enter in the environment from different anthropogenic activities such as mining and smelting, usage of phosphate fertilizers and pesticides, energy production from coal, metal-contaminated wastes, and sludge disposal and from various industrial processes such as plating, alloying, and production of NiCd batteries, pigments, and plastic (Alloway and Steinnes 1999). Generally, soil and water pollution by HMs is a serious environmental problem and in recent years, it has been reported from across the world (Ikenaka et al. 2010; Su et al. 2014; Naser 2013).

Both, essential and nonessential HMs may become extremely toxic for plants, if present in excess.

Higher plants take up metals from water and air by the shoots and leaves or from the soil and sediment by their roots, while algae use their thallus to accumulate metals from water (Greger 2004). Absorption by roots can be passive or active process. Passive uptake includes diffusion of ions from the soil into the root endodermis, while active uptake takes place against concentration gradient and requires metabolic energy. Although absorption differs between the metal ions, ions which are absorbed by the same mechanisms are likely to compete with each other (Alloway 2013). In addition to root absorption, plants can take up significant concentrations of different metal ions through foliar absorption. Absorption of atmospheric pollutants mainly depends on plant species, as well as epidermal and cuticle characteristics (Tomašević and Aničić 2010; Schreck et al. 2012). Once they have been absorbed through the roots or leaves, metal ions will be translocated into the xylem and then through the whole plant. In the xylem, metals could be chelated to different compounds such as histidine (His) (Krämer et al. 1996), nicotianamine (NA), citrate, malate, or oxalate (Senden et al. 1995). There is a considerable variation in the mobility of HM ions (Alrawiq et al. 2014). Mn, Zn, Cd, B, Mo, and Se have a high capability to be readily translocated. Ni, Co, and Cu are less mobile, while lowest mobility is typical for Cr, Pb, and Hg. Plants have highly specific mechanisms to translocate and store micronutrients as well as toxic elements. In some plant species, Ni and Cr were accumulated in their roots (Poniedziałek et al. 2005; Nematshahi et al. 2012), while in other plants, the highest concentration of HMs was found in stems and leaves (Rafati et al. 2011). Zn and Cu accumulation occurs mainly in the roots with poor translocation to the aboveground parts of the plants (Marques et al. 2007; Rivelli et al. 2012). Cd is accumulated mainly in the plant roots and to a smaller extent in other parts of the plant such as leaves, fruits, and grains (Grant et al. 1998). For different plant species, the most of the accumulated Pb is retained in the roots and only a small part can be translocated (Gichner et al. 2008; Gupta et al. 2009; Shahid et al. 2011). It has been shown that some hyperaccumulator plant species with specific detoxification mechanisms are capable of translocating higher concentrations of Pb to aboveground plant parts (Xiong 1998).

Manifested HM toxicity includes chlorosis, growth inhibition, and yield depression (Chaudri et al. 2000; Oancea et al. 2005; Sharma and Dubey 2005; Ebbs and Uchil 2008). For example, high levels of Cu may cause disruption of root cuticle, deformation of the root structure, and reduction of root hair proliferation (Sheldon and Menzies 2005). Excessive amount of Cr can not only inhibit seed germination, limit seedling growth (Akinci and Akinci 2010), and reduce root and shoot elongation but also can reduce chlorophyll (Chl) concentration (Amin et al. 2013). Exposure of algae to Cr (III) and Cr (VI) resulted in a significant inhibition of cell development after mitosis (Volland et al. 2012). Pb accumulation inhibits seed germination (Yang et al. 2010) and plant growth (Malar et al. 2014) and affects root system (Obroucheva et al. 1998; Fahr et al. 2013). Furthermore, high levels of Pb may cause reduction in: chlorophyll content (Zengin and Munzuroglu 2005), net photosynthetic and transpiration rate, stomatal conductance, and water use efficiency (Bharwana et al. 2013).

Algal growth inhibition is highly related to the amount of HM ions bound to the algal cell surface or present in intracellular spaces (Franklin et al. 2000, 2001; Ma et al. 2003) as well as to the chemical nature of the HMs (Tripathi and Gaur 2006).

According to their chemical and physical properties, HMs affect normal plant functioning in a variety of ways, including displacement of essential metal ions from biomolecules and blocking of essential functional groups in proteins and enzymes (Schützendübel and Polle 2002; Ali et al. 2013) and also distorting the integrity of cytoplasmic membrane (Janicka-Russak et al. 2008), adversely affecting processes such as respiration, photosynthesis, and enzymatic activities (Farid et al. 2013). In the case of algae, toxicity results from metal binding to sulphydryl groups of proteins and from the disruption of protein structure and displacement of essential metal ions (Arunakumara and Zhang 2008). Rise in HM levels is associated with the increased formation of reactive oxygen species (ROS). Highly reactive singlet oxygen $^1O_2$ is produced after spin conversion which usually occurs when triplet oxygen (O_2) absorbs sufficient energy. Transferring of one, two, or three electrons to O_2 results in the formation of superoxide radical ($^{\bullet}O_2^-$), hydrogen peroxide (H_2O_2), or a hydroxyl radical ($^{\bullet}OH$), respectively (Mittler 2002; Tripathy and Oelmüller 2012). Superoxide radical is a relatively stable and unreactive molecule. It is a reductant of the transition metal ions in the Haber–Weiss reaction to produce $^{\bullet}OH$ from H_2O_2. Furthermore, $^{\bullet}O_2^-$ oxidizes the quinols produced via disproportionation of $^{\bullet}OH$ and thiols to produce thiyl radicals which may initiate radical chain reactions. H_2O_2 in plant cells is mainly generated during the process of photosynthesis, photorespiration, and, to a lesser extent, cellular respiration (Bhattacharjee 2005). It can easily diffuse across the membranes, where it exerts specific functions at the subcellular level (Corpas et al. 2001; Neill et al. 2002; Mittler et al. 2004). It oxidizes thiols to the sulfenic acids, slowly reacts with cysteine in proteins, and also, inhibits CO_2 fixation in chloroplasts. H_2O_2 may also act as a signaling molecule that mediates resistance of plants to biotic and abiotic stresses but also in the process of growth and development (Ślesak et al. 2007). Hydroxyl radical reacts nonselectively with organic and inorganic molecules nearby its production site and does not diffuse across a long distance (Mano 2002). Certain metals such as Cu and Fe generate oxidative stress through their involvement in redox cycles like Fenton and/or Haber–Weiss reactions which result in an increased production of ROS (Moura et al. 2012). On the other hand, metals without redox capacity (Cd, Pb) enhance oxidative stress by reducing the antioxidant glutathione (GSH) pool, activating calcium-dependent systems, and influencing iron-mediated processes (Pinto et al. 2003). The majority of ROS are generated in the plasma membrane, apoplast, chloroplasts, mitochondria, peroxisomes, and glyoxysomes (Ahmad et al. 2008). Injury of the plasma membrane is one of the primary events of HM toxicity in plants. In the cell walls, high level of $^{\bullet}O_2^-$ and H_2O_2 occurs as a consequence of impaired function of oxidoreductive enzymes and electron transport chain leakage caused by HM uptake. Plasma membrane-bound NADPH oxidase is involved in HM-induced ROS generation (Hao et al. 2006). Radicals generated close to cell membranes may attack the unsaturated fatty acid side chains of membrane lipids which results with the formation of lipid hydroperoxides (Bestwick et al. 2001). From the apoplast, HM enters the symplast and bind to different molecules—proteins, nonenzymatic macromolecules, and other metabolites and may compete with essential cations. This occurs simultaneously in both, cytoplasm and cell organelles (Fodor 2002). Peroxisomes and glyoxysomes, single membrane organelles, contain enzymes involved in β-oxidation of fatty acids and C2 photorespiratory cycle. In these organelles,

H_2O_2 is produced via glycolate oxidase, while xanthine oxidase, urate oxidase, and NAD(P)H oxidases generate $^{\bullet}O_2^-$ (Perl-Treves and Perl 2002). The disruption of electron transfer by HM within the electron transport chain in mitochondria may also lead to an increase of ROS production (Keunen et al. 2011). Furthermore, HM accumulation may disrupt chloroplast ultrastructure, inhibit chlorophyll biosynthesis (Shukla et al. 2008), and interrupt O_2-evolving reactions of PSII and electron flow around PSI and PSII (Mallick and Mohn 2003). Thus, chloroplasts are very important source of ROS in plants under the metal-induced oxidative stress (Parmar et al. 2013; Wang et al. 2013).

In plants, resistance to toxic metal ions includes different mechanisms such as: exudation of complexing agents into the rhizosphere, binding in the cell wall, efflux of metal ions from the symplast, prevention of upward transport of metal ions into aboveground plant parts, complexation with various ligands in the symplast, transport of metal-ligand complexes into the vacuole, storage of metal ions in the vacuole by complexation with vacuolar ligands, and formation of metal-resistant enzymes (Gratão et al. 2006). Algal tolerance to HM is mainly dependent on the defense responses against the possible oxidative damages (Pinto et al. 2003), exudation capacity of chelating compounds, active efflux of metal ions by primary ATPase pumps, and reduced uptake (Gaur and Rai 2001). When exposed to HM-induced oxidative stress, plants will evolve enzymatic and nonenzymatic defense mechanisms (Sharma and Dietz 2009). These ROS-scavenging pathways are often species and tissue specific, as well as dependent on the metal used for the treatment and the stress intensity (Table 1). These comprise firstly the enzymatic pathways (Fig. 1):

(a) Ascorbate–glutathione cycle in chloroplasts, cytosol, mitochondria, apoplast, and peroxisomes which includes ascorbate peroxidase (APX), glutathione reductase (GR), monodehydroascorbate reductase (MDHAR), and dehydroascorbate reductase (DHAR);
(b) Water–water cycle in chloroplasts that includes superoxide dismutase (SOD);
(c) Catalase (CAT) in the peroxisomes.

Nonenzymatic antioxidant endpoints include substances such as ascorbate (AsA), phenolics (PHE), tocopherols and tocotrienols, sulfur-containing antioxidants including GSH, metallothioneins (MTs) and phytochelatins (PCs), soluble sugars, organic acids and amino acids, polyamines (PA), nicotianamine, phytates, carotenoids (Car), and phycobilines (Fig. 1). Many studies have indicated a positive relationship between the accumulation of these compounds during HM stress and plant stress tolerance but also a dual action of some antioxidants which also autoxidize, especially in the presence of metal ions, generating reactive substances and acting as prooxidants. Among the well-established and optimized antioxidant networks in plant cells induced by HMs, the most common and potent scavenging mechanisms are addressed in this chapter.

Table 1 Summary of the responses of antioxidative system to various HMs in different plant species

Species	Metal	Concentration	Duration	Organ	Enzyme	Nonenzymatic antioxidants	References
Ulva fasciata	Cd	0, 5, 10, 20, 50 μM	4 days	Thallus	MnSOD=, FeSOD↑, CAT↑, APX↑, GR↑	tAsA↑, tGSH↑,	Wu et al. (2009)
Scenedesmus sp.	Cu	2.5 μM	6 h	Whole algal cells	SOD↑, CAT↑, APX↑, GR↓	Pro↑, Car=	Tripathi et al. (2006)
		10 μM			SOD↑, CAT=, APX↓, GR↓	Pro↑, Car=	
	Zn	5 μM	6 h		SOD=, CAT=, APX=, GR=	Pro↑, Car=	
		25 μM			SOD↑, CAT=, APX=, GR↓	Pro↑, Car=	
Scenedesmus quadricauda	Cr (III)	1 μM	24 h	Whole algal cells	CAT=, APX↑, GR↑	PHE=, Thiols↑	Kováčik et al. (2015)
		10 μM			CAT↑, APX↑, GR=	PHE=, Thiols=	
	Cr (IV)	1 μM			CAT↑, APX=, GR=	PHE=, Thiols=	
		10 μM			CAT↑, APX↑, GR↓	PHE↓, Thiols↑	
Gracilaria manilaensis	Cu	2 mg/L	24 h	Whole algal cells	CAT↑, APX↑		Ahamad and Shuhanija (2013)
	Pb				CAT↑, APX↑		
Lemna minor	Cu	0, 0.05, 0.5, 5, 10 and 20 mg/L	4 days	Fronds	SOD↑, CAT↑↓, GPX↑↓	Car↓	Hou et al. (2007)
	Cd				SOD↑, CAT↑↓, GPX↑↓	Car↓	
Groenlandia densa	Ni	0, 0.05, 0.5, 5, 10, and 20 mg/L	7 days	Leaf	SOD↑, CAT↑↓, APX↑↓, GR↑↓	Pro↑	Yilmaz and Parlak (2011)
Maize	Cd	0, 0.3, 0.6 and 0.9 mM	8 days	Leaf	SOD↑, GPX↑↓, APX↑↓, GR↑↓	Car↓	Ekmekçi et al. (2008)

Arachis hypogaea L. seedlings	Cd	25, 50, and 100 μmol/L	10, 15, 20 and 25 days	Leaf Root	SOD↑, CAT↑, GPX↑, APX↑, GR↑	Pro↑	Dinakar et al. (2008)
Lepidium sativum L.	Cd	0, 25, 50 or 100 mg/kg soil	30 days	Leaf	SOD↑, CAT↑, APX↑, GR↑	tGSH↑	Gill et al. (2012)
Triticum aestivum	Cd	0.010, 0.101, 0.504, 0.971, 5.10, 9.68, 46.1, 92.3 μM	72 h	Root	SOD↓, CAT↓,		Dandan et al. (2011)
Cultured tobacco cells	Cd	100 μM	7 days	Whole cells	SOD↓, CAT↓, GPX↑	Pro↑, Betaine=	Islam et al. (2009)
Jatropha curcas L.	Pb	0.5, 1, 2, 3, 4 mM/kg soil	6 days	Leaves of seedlings	SOD=↑, CAT↑↓, GPX↑	Car↓	Shu et al. (2011)
Alyssum bertolonii	Ni	426 μM	21–28 days	Hairy roots	SOD↓, CAT↓, APX↓		Boominathan and Doran (2002)
Nicotiana tabacum					SOD=, CAT↓, APX↓		
Jatropha curcas L.	Ni	100, 200, 400, 800 μmol	7 days	Cotyledons	SOD↑↓, CAT↑↓, GPX↑↓		Yan et al. (2008)
Nasturtium officinale R. Br.	Ni	0, 1, 5, 10, 25 mg/L	1, 3, 5,7 days	Leaves	SOD↑↓, CAT↑↓, APX↑↓		Duman and Ozturk (2010)
				Roots	SOD↑↓, CAT↑↓, APX↑↓		
Triticum aestivum L. cv. 'Zyta'	Ni	100 μM	3, 6, 9 days	Leaves	SOD↓, CAT↓, APX↑, GPX↑		Gajewska and Skłodowska (2007)
Raphanus sativus L.	Cu	0.5 mM	2 days	Leaves	CAT↑, GPX↑	Pro=	Teklić et al. (2008b)
	Pb				CAT↑, GPX↑	Pro↑	
Saccharum spp.	Zn	65.0, 130 mg/L	30 days	Leaves	SOD↑, CAT↑, GPX↑	Car↑	Jain et al. (2010)

(continued)

Table 1 (continued)

Species	Metal	Concentration	Duration	Organ	Enzyme	Nonenzymatic antioxidants	References
Triticum aestivum	Zn	0.1, 1, 10, 100 mM	2, 4, 6, 6, 8 days	Leaves	SOD↓, CAT↓, GPX↓	Pro↑, Car↓, AsA↑, GSH↑↓	Panda et al. (2003)
	Cr				SOD↓, CAT↓, GPX↓	Pro↑, Car↓, AsA↑, GSH↑↓	
Lablab purpureus	Zn	100, 300, 600 μM	72 h	Leaves	APX↑, GR↑, GPX↑,	GSH↑↓, AsA↑, Pro↑, Put↑, Spd↑, Spm↓	D'Souza and Devaraj (2012)
				Roots	APX↓, GR↓, GPX=	GSH↑↓, AsA↑, Pro=, Put↑, Spd↑, Spm↓	
Hordeum vulgare L. cv. "Obzor"	Cu	15, 150 and 1500 μM	5 days	Leaves	SOD↑↓, CAT↑, GPX↑, APX↓↑,	AsA=↑	Demirevska-Kepova et al. (2004)

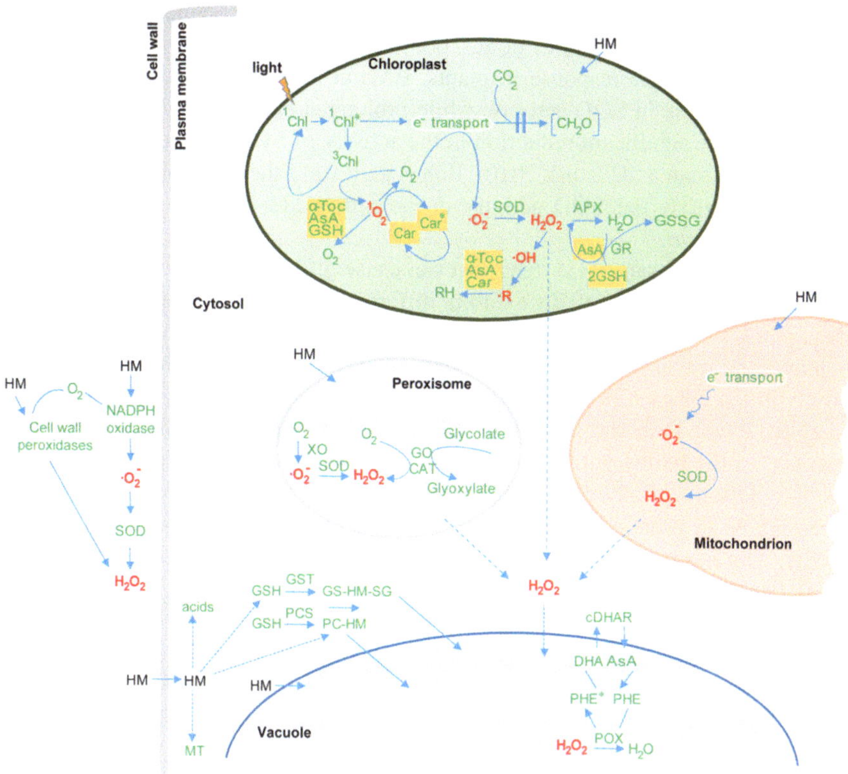

Fig. 1 General scheme of HM antioxidative mechanisms in plant cells

2 Antioxidative Enzymes

2.1 Superoxide Dismutase

Superoxide dismutase (SOD) is an essential component of plant antioxidative defense system as it dismutates two $^{\bullet}O_2^-$ radicals to H_2O_2 and O_2 and represents the first line of defense (Alscher et al. 2002). In this way, SOD maintains $^{\bullet}O_2^-$ at a steady state level, and it may contribute to a shift in the balance of free-radical metabolism towards H_2O_2 accumulation. In response to HM treatment, SOD activity can show induction of activity correlated with increased concentration of HMs, biphasic response—induction at low concentrations of HMs, and inhibition of activity at high concentrations of HMs (Table 1).

In algae, HM treatment (Cd, Cu, and Zn) caused the significant increase in SOD activity (Table 1). Data concerning the influence of HM treatment on SOD activity in plants are contradictory since both enzyme activation and inhibition can be observed (Table 1). Likely, SOD activity showed no contribution to the difference in Cd tolerance between the two groups of lines of the same species (Shah et al. 2001; Guo et al. 2007; Ci et al. 2009). The Cr, Pb, and Ni treatment causes biphasic SOD activity response in plants. HMs at low concentrations caused the significant increase in SOD activity, while prolonged exposure to high concentrations caused the significant reduction in the activity of SOD (Zou et al. 2009; Yan et al. 2008; Duman and Ozturk 2010). Bah et al. (2011) showed that Cd, Cr, and Pb treatments increase the SOD activity in the order Cd > Cr > Pb in seedlings of *Typha angustifolia*.

Noteworthy, the same HM treatment can cause differential response in different plant organs. An excess of Zn exhibits different response in leaves and roots of wheat plant, while it causes dose-dependent induction of SOD activity in roots and no changes in leaves (Li et al. 2013). Dixit et al. (2001), however, showed that SOD activity in Cd-exposed pea plants increased significantly in leaves, while it mostly remained lower than the control in the roots. Kumar et al. (2012) found that treatment of the barley with lower concentration of Ni caused a significant induction of SOD activity in both roots and leaves, while the SOD activity decreased with increasing Ni treatments only in leaves. The Pb treatment generally enhances SOD activity, which is more pronounced in roots than in shoots of rice seedlings (Verma and Dubey 2003).

In general, increase in SOD activity due to HM toxicity may be linked to an increase in $^{\bullet}O_2^-$ formation (Chongpraditnun et al. 1992) as well as to the de novo synthesis of enzyme protein (Cakmak and Horst 1991; Verma and Dubey 2003), which in turn may be associated with an induction of genes of SOD by superoxide-mediated signal transduction (Fatima and Ahmad 2005). Decrease in SOD activity caused by excess of HM can be a result of ROS overproduction and unspecific enzyme degradation (Filek et al. 2008) or the consequence of the binding of HM ions to the active center of the enzyme and replacement of elements such as Zn, Fe, and Mn, which are essential cofactors of SOD isozymes (Stroiński and Kozłowska 1997; Lopez-Millan et al. 2009).

Based on the metal cofactor used by the enzyme, SODs are classified into three groups: iron SOD (FeSOD) located in the chloroplast, manganese SOD (MnSOD) located in the mitochondrion and peroxisome, and copper–zinc SOD (CuZn SOD) located in chloroplast, cytosol, and extracellular space (Alscher et al. 2002). A completely distinct SOD class that contains Ni (NiSOD) was discovered in *Streptomyces* and cyanobacteria (Barondeau et al. 2004). The number of isoenzymes of each type of SOD varies greatly from plant to plant (Gratão et al. 2005). CuZnSOD proteins appear in both charophycean algae and land plants but other green algal groups lack them (De Jesus et al. 1989).

Different HMs can alter the SOD isoforms pattern in a different way in different plant species.

In green macroalga *Ulva fasciata*, MnSOD was not affected by Cd, but FeSOD was increased by Cd at higher concentrations, reflecting also SOD activity induction (Wu et al. 2009). In the marine dinoflagellate *Gonyaulax polyedra*, the activities of FeSOD and MnSOD, but not CuZnSOD, were induced by exposure to acute Cd (Okamoto and Colepicolo 1998). The Ni treatment caused increase in MnSOD activity and no changes in FeSOD activity in coffee cells, while no Cu/ZnSOD isoenzymes were detected (Gomes-Júnior et al. 2006a). On the contrary, Ni application can also result with a decrease in the total SOD activity in both leaves and roots of pea plants due to the inhibition of the activity of CuZn-containing isoenzyme (Gajewska and Skłodowska 2005). Similar results were found in pea plants treated with Cd, where CuZnSODs were the most sensitive forms to HM toxicity (Sandalio et al. 2001). This may be explained by the fact that Ni and Cd disturb Cu and Zn uptake by plants (Parida et al. 2003). In *Arabidopsis thaliana*, differences between the expressions of SOD isoenzymes were found between the roots and leaves. Smeets et al. (2008) found a significant increase in FSD1 (plastidic FeSOD) transcript level and small increase in MSD1 (mitochondrial MnSOD) transcript level, whereas reduction in CSD2 (plastidic CuZnSOD) transcript level in roots of *A. thaliana* due to Cd stress. On the other hand, transcript levels of CSD2, FSD1, and MSD1, as well as general SOD activity at the enzymatic level, are not increased in the leaves of *A. thaliana*.

In roots of *A. thaliana,* Cd and Cu treatment caused different response in gene expression level of SOD isoenzymes (Cuypers et al. 2011). In roots exposed to Cd stress, a slight increase in MSD1, strong induction in FSD1, and reductions in CSDs and FSD2-3 gene expression were noticed after exposure to Cd, while application of Cu resulted in an overall reduction in the gene expression of all superoxide scavenging enzymes except for CSD2. SOD-transcript levels are known to be posttranscriptionally regulated by miRNAs (Cuypers et al. 2011). MicroRNAs (miRNAs) are noncoding small RNAs, and they regulate gene transcription by targeting epigenetic modifications at the DNA level or act posttranscriptionally via RNA cleavage or translational repression (Mallory and Bouché 2008; Khraiwesh et al. 2010).

2.2 Antioxidative Enzymes That Remove H_2O_2

APX and CAT belong to two different classes of H_2O_2-scavenging enzymes because of their different affinities, with APX having a Km in the μM range and CAT in the mM range. Thus, while APX might be responsible for the fine modulation of ROS for signaling, CAT might be responsible for the removal of excess ROS during stress. Plants also contain plastidic and cytosolic isoenzymes of APX, which participate in the scavenging of H_2O_2 in the different compartments (Davletova et al. 2004).

2.2.1 Catalase

The main role of catalase (CAT), a heme-containing enzyme, is scavenging of hydrogen peroxide (H_2O_2) generated during different pathways under standard and stress condition (Mittler 2002; Foyer and Noctor 2005). CAT has a very low affinity for substrate (H_2O_2) compared to other peroxidases as for the actual reaction, the simultaneous entry of two molecules of H_2O_2 in the active site is required (Willkenes et al. 1997). Consequently, CAT is responsible for the gross removal and control of high H_2O_2 levels but is less suited for a fine tuning of sensitive redox balances with low H_2O_2 concentrations that may be important for regulatory mechanisms. In plant cells, CAT is primarily present in peroxisomes and plays a central role in maintaining the cellular redox balance (Corpas et al. 2001).

The contradiction between CAT activation and suppression in the presence of HM may depend on the element, its concentration, and the plant species (Cardoso et al. 2005).

In algae cells, CAT activity is generally induced or unchanged due to HM treatment (Table 1). HM treatment may cause biphasic response of CAT activity, it increases at moderate toxic level of HM and then decreases at high toxic level in different plant species (Table 1). It also may cause activation or reduction in CAT activity (Table 1).

Romero-Puertas et al. (2007) showed that Cd treatment causes reduction of CAT activity and its protein expression in pea leaves, while it causes induction of CAT transcripts. So they assumed that Cd promotes posttranslational modifications of this enzyme. On the other hand, the increase of CAT mRNA by the metal treatment could be induced by the higher H_2O_2 production that takes place in pea leaves in these conditions (Romero-Puertas et al. 2004). H_2O_2 is proven to be a signal molecule in cat1 induction in response to ABA and wounding in maize plants (Guan and Scandalios 2000, 2002). In *Brassica juncea,* four distinct CAT sequences have been cloned, and it has been shown that Cd exposure causes an increase of CAT3 transcript (Lang et al. 2005). Smeets et al. (2008) observed an increase in CAT transcription but no significant change in catalase enzymatic activity in the leaves. These discrepancies can be due to the presence of multiple allo- or isozymes. Thus, CAT activity was confirmed to closely correlate to Cd tolerance and can be an alternative biological indicator of Cd toxicity in wheat due to its high sensitivity to Cd, while it increases only in Cd-tolerant lines or species and decreases in sensitive ones (Ci et al. 2009; Li et al. 2011). In Ni hyperaccumulator *Thlaspi goesingense,* CAT activity also increased after Ni treatment (Freeman et al. 2004). Results in the study of Heidari and Saran (2011) also showed that at the highest concentration of heavy metals, the activity of CAT was higher in Cd treatment than Pb treatment.

HM treatment also showed different effect on CAT activity in plants due to the different plant organs. As an example, pea plants subjected to Ni stress generally exhibited no changes in CAT activity in leaves and roots (Gajewska and Skłodowska 2005). However, in barley, a significant increase in CAT activity was observed in Ni-treated leaves, while roots showed no significant changes (Kumar et al. 2012). Enhancement in the activity of CAT was recorded at low Cr(VI) stress in roots as well as in leaves of *Zea mays* seedlings, while after prolonged treatment with high concentration of Cr(VI), only CAT activity in roots was decreased (Zou et al. 2009). CAT activity was increased in all bioparts of heavy metal (Pb, Cr, and Cd)-stressed plant *Jatropha curcas*, while leaves showed the greatest increase in CAT activity (Devi Chinmayee et al. 2014).

In general, according to Pandey and Sharma (2002), decline in the activity of CAT might be attributed to the inhibition of synthesis of this iron-porphyrin enzyme and other oxidase proteins (Das et al. 1978), while some HMs have some similar characteristics to Fe and may interfere with its absorption and reduced the availability of Fe for heme biosynthesis (Siedlecka and Krupa 1999).

2.2.2 Peroxidases

Another enzyme class responsible for degrading H_2O_2 is the peroxidases (POX), which are capable of reducing H_2O_2 to H_2O with the concomitant oxidation of a series of substrates. POX isolated from plants are distinguishable from APX in both amino acid sequence and in physiological function. Peroxidases (POX) with large number of isoenzymatic forms participate in a variety of cellular functions such as growth, development, differentiation, senescence, auxin catabolism, and lignification having broad specificity for phenolic substrates (Passardi et al. 2004). Guaiacol peroxidase (GPOX) is present in vacuoles, the cell wall, cytosol, and extracellular spaces and consumes H_2O_2 to generate phenoxy compounds that are polymerized to produce cell wall components such as lignin (Mishra et al. 2006). Overall results indicate an enhancement in the activity of guaiacol peroxidase, suggesting that this enzyme serves as an intrinsic defense tool to resist HM-induced oxidative damage and may serve as a biochemical stress indicator for HM pollution (Van Assche and Clijsters 1990; Demirevska-Kepova et al. 2004; Radotic et al. 2000).

In the marine alga *Nannochloropsis oculata,* Cd treatment causes induction of GPX activity (Lee and Shin 2003). HM treatment can cause increase and decrease in GPX activity due to different HM and different plant species (Table 1). The GPX activity can also display biphasic responses due to HM treatment (Table 1) when low-level metal stress increased the activity, while the excess of HMs decreases the activity of GPX. Decline in the activity of GPOD might be due to the formation of protein complex with metals, to change the structure or integrity of proteins (Florence and Stauber 1986; Mohan and Hosetti 1997).

Van Assche et al. (1986) demonstrated that increase in GPX activity might be a result of de novo protein synthesis or the activation of enzymes already present in plant cells to diminish ROS deleterious effects. Enhanced POX activity reflects the modified mechanical properties of the cell wall and cell membrane integrity through participating in lignin biosynthesis which can be used in building up physical barrier against toxic HM (Hegedüs et al. 2001). Degenhardt and Gimmler (2000) revealed that cell walls isolated from the endodermal layer of *Z. mays* cultivated on metal-rich slag, including Cu and Zn, contain higher amounts of lignin than cell walls of control roots.

The toxicity of HM might not be the same at all stages of plant development (Hegedüs et al. 2001). The POD activity in roots of green or greening barley seedlings was not influenced by Cd concentration, but it increases in leaves treated with toxic concentrations of Cd (Hegedüs et al. 2001). Baccouch et al. (1998) showed that Ni treatment did not cause any changes in GPX isozyme profiles; the increased activity of the three GPX isoforms is the most likely explanation for the increase in GPX activity. Changes in isozyme profiles of peroxidases have been observed by others under HM stress conditions. In fact, Van Assche et al. (1986) showed the quantitative changes of GPX isozyme profile throughout Ni exposure.

2.2.3 Ascorbate–Glutathione Cycle

Ascorbate–glutathione cycle (AsA–GSH) operates in cytoplasm, mitochondria, peroxisomes, as well as chloroplasts, where it is the primary mechanism of H_2O_2 detoxification, and it appears to be an important defense mechanism against oxidative stress caused by HM (Cuypers et al. 2000, 2001). The reduction of H_2O_2 by AsA can occur directly or it can be catalyzed by ascorbate peroxidase (APX) (del Río et al. 2006). During this process, AsA is oxidized to monodehydroascorbate (MDH). The monodehydroascorbate formed can be directly reduced back to AsA by monodehydroascorbate reductase (MDHAR) or may first be converted to dehydroascorbate (DHA) and then reduced by dehydroascorbate reductase (DHAR). Efficiency of APX depends on rapid regeneration of AsA from DHA, which in turn is dependent upon the availability of GSH. To maintain homeostasis, GSH must be generated from oxidized glutathione (GSSG) by glutathione reductase (GR) at the expense of NADPH (Noctor and Foyer 1998). The enzymes of AsA–GSH cycle also play a significant role in scavenging H_2O_2 mainly in chloroplasts and in maintaining the redox status of the cell.

Cuypers et al. (2011) pointed out a central role for GSH in the regulation of the antioxidative gene expression under Cd stress. Also, a higher GSH/GSSG ratio is a marker of HM tolerance in hyperaccumulating ecotype of *Sedum alfredii,* which shows less ROS production than the non-hyperaccumulating ecotype under excess Cd (Tian et al. 2011). Cd treatment causes the enhancement of the transcript levels of genes responsible for GSH synthesis, which implies possible de novo synthesis of both enzymes, resulting in higher GSH production, although levels of GSH were lower in Cd-treated leaves (Semane et al. 2007). The decrease of GSH found in the study of Semane et al. (2007) coincides with the strongly increased phytochelatin (PC) level, which assumes an active complexation of free Cd in the leaves by PC. Although the Cd treatment causes the increase in GR activity, it still reduces the GSH/GSSG ratio in *A. thaliana* leaves. The same authors also showed reduction of AsA content, another important antioxidant in AsA–GSH cycle that correlates with the reduction of APX activity (Semane et al. 2007). One hypothesis to explain the reduction in reduced AsA is through a diminished biosynthesis as a consequence of the inhibition of the mitochondrial electron transport chain and increased levels of ROS after Cd exposure (Wang et al. 2004). The AsA biosynthetic pathway is linked to the mitochondrial enzyme, L-galactono-1,4-lactone dehydrogenase, a reaction coupled with cytochrome c oxidoreductase (complex I) from mitochondria (Millar et al. 2003; Smirnoff and Wheeler 2000). On the other hand, it is possible that regeneration of DHA to AsA by DHAR using GSH as an electron donor could not be maintained at a sufficient level because of the decreased GSH level under Cd stress (Semane et al. 2007). The activation of AsA–GSH cycle also depends upon duration of HM exposure. Schützendübel et al. (2001) gave the timetable of Cd-induced reactions of AsA–GSH cycle in roots. Initially, Cd causes a depletion of GSH, and it binds to thiol groups of GR (Creissen and Mullineaux 1995). The same inhibition mechanism may be possible for APX being sensitive to thiol reagents (Chen and Asada 1992). This causes inhibition of AsA–GSH cycle and an accumulation of H_2O_2 that acts as a signaling molecule in the induction of APX genes (Prasad et al. 1994; Karpinski et al. 1999). In contrast to APX gene expression, which is activated by H_2O_2, GR was not stimulated by H_2O_2 but by jasmonic acid (Xiang and Oliver 1998). The different time courses for the induction of APX and GR activities observed here also support the finding that these enzymes are regulated by different stimuli.

In algae, different HM treatment can cause different response of APX and GR activities (Table 1). The activities of APX and GR can also be enhanced, decreased, or unchanged by HM treatment or even showed biphasic response (Table 1).

Chou et al. (2012) found that the activities of APX and GR increased with increased Cd treatment duration in rice seedlings with no differences in expression of OsAPX and OsGR.

In contrast, APX2 expression is only induced during stress conditions and was clearly induced in roots of Cu and Cd (Cuypers et al. 2011). Smeets et al. (2008) showed the increase in APX activity and also enhanced level of APX1 transcript and DHAR in the Cd-treated leaves of *A. thaliana*. Interestingly, the GR1 gene was transcriptionally upregulated in both plant parts, while a significant decrease of the GR enzyme activity was observed. So, it is possible that Cd interacts with the translation mechanisms, disturbs the activity of GR, or even more likely influences the turnover of this enzyme.

Cd differentially altered GR activity and isoforms in leaves and roots of wheat plants. Wheat leaves did not show any change in their GR activity over time, whereas roots presented a remarkable increase after treatment (Yannarelli et al. 2007). Dixit et al. (2001) have showed in pea that GR was more activated in roots than in leaves when plants were treated with Cd. Moreover, GR activity did not show significant changes in pea leaves exposed to Cd stress (Sandalio et al. 2001). On the other hand, Cd treatment has decreased the activity of this enzyme in *Helianthus annuus* and *Ceratophyllum demersum* (Gallego et al. 1996; Aravind and Prasad 2005). Ni treatment strongly inhibited APX activity in maize root (Gajewska and Skłodowska 2005), which can disturb ROS homeostasis causing oxidative stress. In contrast to roots, APX activity in leaves of *Z. mays* plants increased in response to Ni treatment in Baccouch et al. (1998).

Cr treatment decreases the GSH and increases GSSG content in green gram leaves and induces the APX activities (Karuppanapandian et al. 2006). APX and GR activities were also upregulated upon exposure to Cr exceeding levels in cotton (Daud et al. 2014) and *Amaranthus viridis* (Liu et al. 2008).

D'Souza and Devaraj (2012) indicated that APX and GR activities in leaves of Hyacinth bean increase in the presence of Zn, and the early enhancement of APX activities suggests that the Zn-triggered antioxidative capacity mainly uses APX for the removal of H_2O_2. At the same time, GR activity declined with time in roots suggesting operation of AsA–GSH cycle only in leaves of Hyacinth bean (D'Souza and Devaraj 2012). Zn affects the AsA–GSH pathway in leaves and roots of *Phaseolus vulgaris*; nevertheless, the mechanism of action is different (Cuypers et al. 2001). In roots, an overall oxidation of metabolites and inhibition of APX and GR capacities were observed together with increased biosynthesis of AsA, which imposes the signal transport from roots to shoots. On the contrary, in primary leaves, the AsA–GSH pathway plays a prominent role in the defense against Zn stress. Nevertheless, the capacities of the four participating enzymes increased at various moments. Cu treatment, on the other hand, caused an increased capacity of the four enzymes (APX, MDHAR, DHAR, and GR) involved in the AsA–GSH cycle in the roots of *P. vulgaris* (Gupta et al. 1999). These results, therefore, clearly indicate that enhancement of the AsA–GSH cycle is part of the defense against oxidative stress that is imposed by Cu, shown also by Weckx and Clijsters (1996) in the primary leaves.

3 Nonenzymatic Antioxidants

3.1 Phenolics

Phenolics constitute very numerous groups of secondary plant metabolites. Through two most important ways of biosynthesis (shikimic and malic acid pathways), their production is indirectly stimulated by proline (Pro) which induces the synthesis of their precursors through pentose phosphate pathway (Hare and Cress 1997). These molecules, especially the complex flavonoid structures are among the most important plant antioxidants. Their functional diversity lies in their structural diversity with many biological functions, including multiple antioxidative mechanisms (Tahara 2007). Due to their structural properties, especially dihydroxy B-ring substitution, flavonoids perform antioxidant activities by directly scavenging $^{\bullet}OH$, HOCl, and $^1O_2$ and lipid peroxyl radicals, by inhibiting lipoxygenase and by metal chelation. For example, anthocyanins chelate metal ions enable their transportation to vacuole and even form complexes with other antioxidants, such as AsA, to prevent their oxidation (Sarma et al. 1997). In some cases, the resulting metal-flavonoid complexes are found to be more efficient ROS scavengers than the initial flavonoids (Malešev and Kuntić 2007). Flavonoids usually concentrate in plant vacuoles and are abundant in leaves, seeds, bark, flowers, and fruits. Efficient mechanisms have been recently identified for the transport of flavonoids from the endoplasmic reticulum, the site of their biosynthesis, to different cellular compartments. As summarized by Brunetti et al. (2013), the content of antioxidant flavonoids is high in leaf mesophyll cells, where they act in quenching of H_2O_2 and H_2O_2-generated $^{\bullet}OH$ in the nucleus, trapping of singlet oxygen in vivo in chloroplasts and their outer envelope membrane limiting the diffusion of ROS, and enhancing membrane rigidity. Also, flavonoids accumulate in vacuoles where they are capable of removing H_2O_2 freely diffusing out of other cellular compartments, when the activity of APX (or CAT) is strongly depressed, being of primary significance in plants suffering from severe stress conditions. Flavonoids can facilitate HM tolerance, which is seen in *A. thaliana* (Keilig and Ludwig-Müller 2009). Absorption profile of flavonoids showed a decrease in flavonoid contents in pea plants influenced by Cd under field conditions (Agrawal and Mishra 2009). Several classes of flavonoids can be found in microalgae and cyanobacteria, especially a large amount of phenolic acids, mostly salicylic, trans cinnamic, synaptic, chlorogenic, chimic, and caffeic acids (Miranda et al. 2001). Due to the high content of phenolic substances and their contribution to overall antioxidant activity in higher plants and algae, as well as a variety of beneficial biological properties of these compounds, they represent a good source of natural antioxidants (Hajimahmoodi et al. 2009).

3.2 Ascorbic Acid

The very important role of AsA in the plant antioxidative defense system was addressed earlier in this chapter in the context of AsA–GSH pathway which enables successive oxidations and reductions of AsA, GSH, and NADPH by the complex set of enzymes which operate together in the removal of ROS (Anjum et al. 2014). Hence, AsA is the most important reducing substrate for H_2O_2 detoxification in plants and exists in relatively high concentrations in leaves representing over 10 % of the total soluble carbohydrates (Noctor and Foyer 1998). Except for the high millimolar concentrations in chloroplasts and cytosol, AsA was found in mitochondria, peroxisomes, and vacuoles and also as one of the rare antioxidants in the apoplast (Gratão et al. 2005). Besides being a substrate for APX, AsA reacts directly with $^{\bullet}OH$, $^1O_2$, and $^{\bullet}O_2^-$ and also regenerates Car and tocopherols in thylakoid membranes (Mallick and Mohn 2000). Besides the total level of AsA in the cell, the ratio between its reduced and oxidized form maintained by associated enzymes plays an important role in the defense mechanisms from ROS and their products (Anjum et al. 2014). AsA content in plants and algae shows distinct trend under HM treatments (Table 1). Its content is found to decrease in different plant species (Rodríguez-Serrano et al. 2006; Romero-Puertas et al. 2007; Agrawal and Mishra 2009; Chen et al. 2010). In case of algal cells treated with HMs, slight increase in AsA pool of *Chlorella vulgaris* was observed in lower Cu concentrations, while a drastic reduction was evident at higher concentrations (Mallick 2004).

3.3 Tocopherols and Tocotrienols

Tocopherols and tocotrienols (group of vitamin E compounds) are denoted as lipid soluble membrane-associated antioxidants (Singh 2005). Tocotrienols are only found in certain plant families and almost exclusively in seeds and fruits, while tocopherols are more ubiquitous and accumulate in chloroplasts, vacuoles, and nuclei in plant leaves or in seeds within plastids and associated with cytoplasmic lipid bodies, while in green algae, they are present inside chloroplasts and mitochondria (Falk and Munné-Bosch 2010). As antioxidants tocopherols act as lipid protectors, membrane-stabilizing agents, inhibitors of lipid peroxidation, and ROS scavengers (Matringe et al. 2008), and their interplay with carotenoids in scavenging $^1O_2$-induced damage to photosystem is seen as the essential cell response in limiting photo-oxidative stress (Trebst et al. 2002). Out of four isomers found in plants, the greatest number of studies are focused on antioxidative properties of α-tocopherol (α-toc), which is the most potent scavenger of $^1O_2$, which is, like AsA, a relatively poor electron donor in physiological conditions that acts primarily by transfer of single hydrogen atoms (Kruk et al. 2005). Different methods were applied in studying tocopherol functions in plants, such as exogenous applications at the whole-plant or cellular level, studying variations in their content prior to and during the various physiological processes, and the use of mutants and transgenic lines with altered biosynthesis and levels of these compounds (Falk and Munné-Bosch 2010). In the case of heavy metal-induced stress, among other antioxidants, a significant increase of α-toc in *P. vulgaris* leaves was observed after 10-day exposure to HMs, decreasing in the following order Hg > Cd > Cu > Pb (Zengin and Munzuroglu 2005). Also, the increase of α-toc was observed after the exposure of *Solanum lycopersicum* (Hédiji et al. 2010) and *Vitis vinifera* cv. cell suspension cultures (Cetin et al. 2014) to Cd as well as Eridano *Populus* sp. clones to elevated Zn concentrations (Fernàndez-Martínez et al. 2014). It has been shown that oxidative stress activates the expression of genes responsible for the synthesis of tocopherols in higher plants (Tang et al. 2011). For example, overexpression of the γ-tocopherol methyl transferase gene from *A. thaliana* in transgenic *Brassica juncea* plants resulted in an over sixfold increase in the level of α-toc, induced by a variety of abiotic stresses (Yusuf and Sarin 2007). Moreover, α-toc-enriched transgenic *B. juncea* was found to increase the tocopherol levels which enhanced its tolerance against Cd-induced stress (Kumar et al. 2013). Also, some other transgenic plants, such as potato, with the reduced expression of some other proteins, e.g., manganese stabilizing protein, an important component of the PSII, were found to accumulate α-toc and other antioxidants as a consequence of Zn treatment (Gururani et al. 2013).

3.4 Amino Acids and Peptide Derivates

The importance of amino acids and peptide derivates accumulation in plant cells is especially pronounced under metal stress (reviewed by Gill et al. 2014). The main focus is attributed to Pro, but certain other amino acids and oligopeptides, i.e., glycinebetaine, polyamines (PAs), and nicotianamine, are also addressed (Sharma and Dietz 2006). Most of these compounds act as osmolytes which accumulate in response to water deficit conditions regulating the osmobalance which is often disturbed by HMs. Glycinebetaine and Pro are two major organic osmolytes with positive effects on enzyme and membrane integrity and important roles in osmotic adjustment in plants grown under stress conditions (Ashraf and Foolad 2007). While many studies have indicated a positive relationship between accumulation of these osmolytes and plant stress tolerance, some have proposed that the increase in their concentrations under stress is a product of and not an adaptive response to stress. Also, it is unclear if the Pro accumulation under HM stress is a sign of its role in metal detoxification or only an indirect response to metal-induced disturbances in plant water balance (Schat et al. 1997). However, besides osmoregulation, amino acids, particularly Pro, were shown to have many beneficial functions under metal stress, such as metal chelation and reducing metal uptake, $^1O_2$ and $^{\bullet}OH$ quenching, signaling, serving as a source of carbon and nitrogen, maintaining cytosolic pH and NAD(P)+/NAD(P)H ratio, and preventing denaturation of enzymes (Mehta and Gaur 1999; Sharma and Dietz 2006). Its accumulation can be achieved by increased synthesis or reduction of degradation which may be regulated by enzymes (Hong et al. 2000). *Lablab purpureus* plants exposed to Zn stress exhibited elevation in Pro levels in leaves and roots as a function of time (D'Souza and Devaraj 2012), which is also assumed to be connected with the role of Pro in osmotic stress during long-term metal exposure (Sharma et al. 1998). Accumulation of Pro was observed due to Cd treatment in different plant parts of *P. sativum* (Agrawal and Mishra 2009), as well as in carrot and radish roots (Chen et al. 2003). The same effect is seen in lettuce leaves of two cultivars after the administration of Cu in nutrient solution, as well as in soil (Teklić et al. 2008a). Also, the Pro concentration rises in the leaves of *V. faba* plants treated with Cd, Cu, Ni, Pb, or Zn, as well as in the stems of plants grown on Zn and Ni-contaminated soil. The concentration of Pro was higher in the leaves than in the stems of all treated plants, whereas the control plants had comparable amounts of this amino acid in both organs (Nadgórska-Socha et al. 2013). The content of Pro in radish plantlets treated with Cu or Pb differed depending on plant part (leaf or hypocotyl) and stress intensity (heavy metal content in growth medium and exposure duration) (Teklić et al. 2008b). This has been mainly linked to Ni hyperaccumulator plants (Sharma and Dietz 2006), and its content can be increased even 36-fold in the xylem of the hyperaccumulator *Alyssum lesbiacum* compared to nonaccumulator *Alyssum montanum* (Krämer et al. 1996). Another amino acid, NA, is a potent plant HM chelator involved in HM transport and HM tolerance in hyperaccumulating plants (Sharma and Dietz 2006).

Due to their effects on plant resistance against HMs or other abiotic stresses, exogenous applications of some amino acids, such as glycinebetaine and Pro to plants could lead to the significant increases in growth and final crop yield under the stress conditions (Ashraf and Foolad 2007). However, in some cases of metal toxicity, such as Cd application in barley roots, Pro accumulation is disturbed leading to water and oxidative stress responsible for the root growth inhibition (Tamás et al. 2008).

Some algae also respond to heavy metal toxicity by producing Pro (Table 1). Due to small size and specific uptake of metals in algal cells, Pro accumulation in unicellular microalgae was observed within a few hours, which differs from higher plants that require a longer period of exposure to metals (Mehta and Gaur 1999). *Chlorella vulgaris* cells accumulated high levels of Pro, with the highest content observed in maximum concentrations of Cu and Cr, followed by Ni and Zn treatment. Also, Pro pretreatments inhibited metal-induced lipid peroxidation in *C. vulgaris*. There is also a very important role for Pro in algal recovery from stress. Pro hyperaccumulation also seems to be an important strategy for overcoming oxidative stress induced by long-term exposure of *Scenedesmus* sp. to elevated levels of Cu and Zn (Tripathi et al. 2006).

PAs are low molecular-weight amines important in plant growth and development and are also involved in protection against environmental stresses (Gill and Tuteja 2010). Their hydrophilicity enables them to buffer intracellular pH and work as effective ROS-scavenging compounds, being mediators in protective reactions against different stresses (Grzesiak et al. 2013). Like with amino acids, the accumulation of these compounds in plants is one of the responses to water stress, as a consequence of HM influence or some other type of abiotic stresses. Their exact role in heavy metal defense is not clear, but a participation in the stabilization and protection of the membrane systems has been proposed (Sharma and Dietz 2006). It can be also related to their antioxidative properties in reducing the $^{\bullet}O_2^-$ accumulation when applied exogenously, like in tobacco protoplasts, which can be related to the inhibition of microsomal membrane NADPH oxidase, mostly by spermine (Spm) (Papadakis and Roubelakis-Angelakis 2005). The changes in PA levels vary among plant species and the stress duration. Individual PAs have different roles during stress response, and spermidine (Spd) could be implicated in enhancing HM tolerance, possibly by exerting an antioxidant activity and/or by metal chelation (D'Souza and Devaraj 2012). The most abundant of the three PAs, Spd, was found in twofold and onefold higher concentrations in leaves and roots of *L. purpureus* under Zn stress after 72 h at maximum concentration. It has been suggested that stress-tolerant plants increased exogenous PAs level to a greater extent than sensitive ones and, those PAs with a higher number of amino groups, i.e., Spm and Spd, are more effective in scavenging ROS, than diamine putrescine (Put) suggesting the involvement of amino groups in ROS scavenging (Kubiś 2008). In addition to common types of PAs, certain algae also have homospermidine, 1,3-diaminopropane, norspermidine, and norspermine, whose content varies due to the HM treatment (Agrawal et al. 1992).

3.5 Soluble Sugars

Soluble sugars, besides amino acids, present another important group of plant osmoregulators with antioxidant properties. An elevated level of total soluble sugars is important for energy production, stabilization of cellular membranes, maintenance of turgor, vitrification of the cytoplasm, and signaling, which may account for tolerance to heavy metal stress. Soluble sugars have different roles in events associated to metabolism of HM-stressed plants. Although soluble sugars have been linked to metabolic pathways that produce ROS, they can also have an important role in ROS-scavenging mechanisms. Sucrose and glucose either act as substrates for cellular respiration or as osmolytes to maintain cell homeostasis, while fructose is not related to osmoprotection and seems related to secondary metabolites synthesis (Rosa et al. 2009). Increased glucose levels can increase the production of NADPH (via the pentose phosphate pathway), that is, an important intermediate in the AsA–GSH cycle (Couée et al. 2006) as NADPH is the primary electron donor that assures an intracellular reduction status. Both glucose and sucrose levels have been shown to increase in some plant species treated with HMs. Both of these sugars also participate in signaling mechanisms. Van den Ende and Valluru (2009) suggest that sucrose might have a protective role against stress due to its capacity to scavenge ROS. Other sugars like raffinose and fructans are also reported to have a protective role as membranes against several stresses, namely freezing and drought stress (Van den Ende and Valluru 2009). Besides the benefits of endogenously derived sugars, exogenous application of trehalose in *Lemna gibba* also reduces the deleterious effects of Cd and Pb increasing the metal accumulation and is identified as a potential phytoremediation-enhancing agent (Duman et al. 2011; Duman 2011). Plant carbohydrate levels are often directly correlated to Pro synthesis, which was previously described in *L. purpureus* leaves and roots after Zn treatment (Sharma et al. 1998).

3.6 Thiols/Glutathione

These sulfur-containing antioxidants in plants are usually measured as total thiols, non-protein thiols, and GSH, as well as protein thiols i.e., PCs and MTs. As seen in algae (*Scenedesmus* sp.), the binding of metals to thiols can protect cell metabolism from metal toxicity only temporarily; however, during severe stress, depletion of thiol pool is known to make cells vulnerable to oxidative stress (Rijstenbil et al. 1994). Depletion of thiol content in *Scenedesmus* sp. after Cu or Zn treatments was significantly higher in longer treatments which can be related to high GSH demand needed for PC synthesis and was also followed by the inhibition of GR. Cd uptake is strongly connected to alterations in plant thiol content, having dual action. Elevated thiol content connected with lipid peroxidation and root growth inhibition was detected in all individual segments of barley root after Cd treatment (Tamás et al. 2008), as well as in *P. sativum* (Agrawal and Mishra 2009). On the other hand, 10 μM Cd decreased thiol content and induced oxidation of AsA and GSH indicating oxidative stress. Zn supplementation to Cd-treated plants can restore thiols, showing the protective role of Zn in modulating the redox status of the plant system through the antioxidant pathway for Cd-induced stress tolerance (Aravind and Prasad 2005). Thiols can play an important antioxidative role, protecting membrane lipids. Lipoic acid, both in its reduced and oxidized form, is reported to have antioxidative properties due to its direct scavenging of ROS. It is also able to chelate several metal ions that induces oxidative stress (Navari-Izzo et al. 2002) and thus can have an important role in cell protection. Glutathione is a low molecular-weight thiol tripeptide (Glu-Cys-Gly) that exists in plant cells in the reduced (GSH) and oxidized form (GSSG) with the typical GSH/GSSG ratio of 100:1 in glutathione homeostasis (Viehweger 2014). Being one of the major antioxidants, it is ubiquitous in plant tissues and active in many intracellular compartments (Noctor and Foyer 1998). Some of its most important functions include antioxidant activity in scavenging $^1O_2$ and H_2O_2, overall defense reactions against oxidative stress, and involvement in sulfur metabolism. Besides being a potent HM-ligand, it is a precursor of PCs. Upon HM exposure, GSH concentrations can decrease as a consequence of initiated PC biosynthesis. As it was mentioned in the previous chapter, GSH plays another key role in the plant antioxidative defense system by regenerating AsA via Asa–GSH pathway (Foyer and Halliwell 1976). In addition, GSH is a substrate for GPX and GST (glutathione peroxidase and glutathione-S-transferase) involved in the removal of ROS (Noctor et al. 2002). Exposure to heavy metals results in two different ways. In some cases of metal toxicity, the severe depletion of GSH is observed which is often strongly positively correlated with the inhibited activity of antioxidative enzymes, while in some cases, GSH content is found to increase (Table 1). A Cd-dependent reduction of glutathione (GSH) has been described in different tissues and plant species (Rodríguez-Serrano et al. 2006; Gomes-Júnior et al. 2006b). This reduction is often seen in total glutathione content, although the reduced form (GSH) is the most affected (Romero-Puertas et al. 2007). Moreover, exogenous GSH can modulate the antioxidant defense system against Cd stress in the two barley genotypes differing in Cd tolerance (Chen et al. 2010). Very large difference in antioxidant response related to GSH was also seen between the two species of green microalgae exposed to Cu. Among other measured endpoints, *Scenedesmus vacuolatus* exposed to Cu showed significantly higher GSH levels compared to *Chlorella kessleri* (Sabatini et al. 2009).

PCs are thiol-rich compounds formed in metal-exposed plants which are the main plant cellular mechanisms for HM detoxification and tolerance (Gupta et al. 2013). PCs are synthesized non-translationally in a transpeptidation reaction catalyzed by the enzyme phytochelatin synthase, a reaction needing GSH as a precursor. These peptides have chelating properties but also play other important roles in plant cells, including essential heavy-metal homeostasis, sulfur metabolism, and antioxidant action (Dietz et al. 1999; Cobbett 2000). Some in vitro analyses show that compared to GSH, PCs show stronger scavenger properties for H_2O_2 and O_2^- radicals (Tsuji et al. 2002). PCs have been most widely studied in plants, particularly in relation to Cd tolerance (Cobbett 2000). Some heavy metal hyperaccumulator plants show that their hypertolerance mechanisms to Cd accumulation are not dependant exclusively on PC-based sequestration (Schat et al. 2002), while in some hyperaccumulators like *Sedum alfredii,* it is probably related to GSH chelation (Sun et al. 2007; Gupta et al. 2010). The synthesis of PCs has also been reported in algal species exposed to high concentrations of metals such as Cu in *Scenedesmus bijugatus* (Nagalakshmi and Prasad 2001) and Cd in *Vaucheria* spp. (Skowroński et al. 1998). Scheidegger et al. (2011) suggested that PCs play a minor role in Pb detoxification in *Chlamydomonas reinhardtii* at environmentally relevant concentrations but are highly involved in Cu and Zn homeostasis. In several species of marine algae, PCs are synthesized even in untreated cultures, while elevated PC concentrations were induced also by very low but environmentally relevant concentrations of Cd (as low as 10^{-12} M free ion concn). Also, the treatment of *Stigeoclonium* spp. with individual metals (Cd, Pb, and Zn) induced higher content of PCs, opposed to metal mixture contained in the mining water (Pawlik-Skowrońska 2001). In certain algal species, such as *Dunaliella tertiolecta,* the Zn pretreatment induces the PCs synthesis-related tolerance toward other heavy metals (Cd, Hg, Cu, and Pb) and As and also increases tolerance toward oxidative stress caused by H_2O_2 or herbicide paraquat (Tsuji et al. 2002).

MTs are the second type of cysteine-rich, metal ion-binding proteins. They are special for their highly conserved cysteine arrangement, although in plants, MTs make a uniquely diverse family with many isoforms which make, based on the cysteine pattern, four different topologies (types 1–4 plant MTs) (Leszczyszyn et al. 2013). The biosynthesis of MTs is regulated at the transcriptional level, and MT genes can be differentially regulated under various HM stresses (Ahn et al. 2012). MTs are suggested to be involved in metal tolerance or homeostasis, by complexing ions (most commonly Cu(I), Zn(II), and Cd(II)) through the thiol groups of their cysteine residues in characteristic metal-thiolate clusters (Leszczyszyn et al. 2013). MTs are reported to be much more important in HM tolerance of certain plants than PCs (Mijovilovich et al. 2009). Except for their main function as metal chelators, novel investigations show that MTs are also involved in ROS removal, plasma membrane repair, and signaling, while the interplay between these functions needs to be fully characterized (Hassinen et al. 2011). Transgenic tobacco plants overexpressing GhMT3a-type protein showed increased tolerance against abiotic stresses, including HM treatments, compared with wild-type plants (Xue et al. 2009). Tsuji et al. (2002) suggested that MT2 in algae could play a role not only in detoxification of HMs but also in mitigation of oxidative stress. Morelli and Scarano (2004) found that metal-free MT3 was present in cell extracts of *Phaeodactylum tricornutum* treated with Cu, hypothesizing that this oxidized MT3 form participates in ROS scavenging.

3.7 Carotenoids and Phycobilins

Carotenoids (Car) function as accessory pigments in light absorption and as structural units of photosystem complexes serving also as photoprotective agents, quenchers of the excited states of Chlorophyll, and scavengers of $^{\bullet}O_2$ during photosynthesis, protecting the reaction center from photoinhibition (Yamamoto and Bassi 1996). The Car molecules become themselves excited but lack energy to form other ROS species what makes them very efficient antioxidants (Taiz and Zeiger 2002). Car contributes significantly to the total antioxidant capacity of both microalgae and plants. However, in most cases, HMs are known to target the pigment apparatus leading to Chl degradation followed by reduced Car content (Table 1), causing a marked effect on the entire metabolism of a plant. In some plants, like Cd-treated *Pisum sativum*, a decrease in total Chl and Car contents was observed in all treatments at all time points (Agrawal and Mishra 2009). Certain plants, like *P. vulgaris*, react adversely by increasing the levels of Car, with the strongest effect observed in plants exposed to Ni, followed by the sequence Co > Cr > Zn (Zengin 2013). In algae, such as *C. vulgaris*, a rising trend in Car content was observed after Cu treatment (Mallick 2004). Also, in some heterotrophically grown ROS-treated algal cultures, the employment of oxidative stress can be used to promote the formation of secondary carotenoids including astaxanthin (Ip and Chen 2005). Moreover, the biosynthesis of both primary pigments (chlorophylls and primary Car such as β-carotene and lutein) and secondary Car (astaxanthin, canthaxanthin, and adonixanthin) depends on the nature of applied ROS. In some cases, like in Cu and Zn-treated *Scenedesmus* sp. cells, Car content was unchanged during the treatments showing that some algal species use other mechanisms in protection from metal-induced oxidative stress (Tripathi et al. 2006), while in some red algae, Pb and Cu treatments significantly decreased total carotenoid content (Gouveia et al. 2013).

Phycobilins are open-chain tetrapyrroles that function as chromophores of light-harvesting chromoproteins called phycobiliproteins. Phycobiliproteins, such as phycoerythrin, phycoerythrocyanin, phycocyanin, and allophycocyanin, are highly absorbent, and fluorescent components of the photosynthetic light-harvesting complex (phycobilisome) found in cyanobacteria, rhodophytes, cryptomonads, prochlorophytes, and glaucocytophytes and their ratio in cells vary due to the chromatic adaptation (Larkum 2003). Their antioxidation activity was shown in cell-free extracts isolated from several species of cyanobacteria, which was mostly attributed to phycoerythrin as the main phycobiliprotein (Pumas et al. 2011). Moreover, phycocyanobilin, isolated from *Spirulina platensis*, prevented the peroxidation of linoleic acid (Hirata et al. 1999). However, it is indicated that some compounds such as phycocyanin exhibit pro-oxidant activity, generating $^{\bullet}OH$ in the light, while having radical scavenging activity in darkness attributed to phycobilin part of the compound (Zhou et al. 2005). Because of their linear tetrapyrrole structure, which resembles that of biliverdin and bilirubin, phycobilins are expected to have strong antioxidant properties (Wada et al. 2013).

The phycocyanobilin found in *S. platensis* cell extracts inhibited the initiation of lipid peroxidation and was proven to be responsible for majority of the antioxidative activity of phycocyanin with the potential as effective antioxidant agent (Hirata et al. 2000). Studies on *Anabaena flos-aquae* show that cellular proteins, including phycocyanin, represent the main targets of metal toxicity, which was more pronounced in the protein profile after Cu treatment under different temperature levels, compared to Cd treatment in the same temperature range (Surosz and Palinska 2005). Pb and Cu exposure significantly decreased the total phycobiliprotein content in rhodophyte *Gracilaria domingensis* (Gouveia et al. 2013), which was also observed for several cyanobacterial species, such as in *S. platensis* when exposed to Pb (Arunakumara and Zhang 2007) and after Ni or Pb exposure in *Oscillatoria* sp. and *Westiellopsis* sp., particularly in longer treatments. Phycocyanin and phycoerythrin were more susceptible to HM toxicity than allophycocyanin (Balakrishnan and Narayanan 2007).

4 Conclusion

The presented results reveal a high diversity in antioxidative responses not only among different plant or algal species but also among cultivars and different plant parts or algal strains. Therefore, the HM-induced oxidative stress is probably a general symptom in several species, but the antioxidative response is specific and dependent on genetic potential of individual cultivars or strains of given species and on HM concentration and/or a given period of exposure. Also, the key challenge in the current research of HM-induced oxidative stress in plants is to delineate the individual contribution of enzymatic and nonenzymatic biomolecules in the overall antioxidant capacity of plant organisms subjected to stress. In the last several decades, antioxidant systems including AsA, GSH, and their related enzymes which exist in relatively high content in cellular compartments shared a spotlight in the research field of HM-induced oxidative stress, and their roles have been examined in detail in the existing literature. However, other antioxidants which share a (potentially) minor role in the overall antioxidant capacity of plant cells are less studied, thus having great research potential.

References

Agrawal SB, Mishra S (2009) Effects of supplemental ultraviolet-B and cadmium on growth, antioxidants and yield of *Pisum sativum* L. Ecotoxicol Environ Saf 72:610–618

Agrawal SB, Agrawal M, Lee EH, Kramer GF, Pillai P (1992) Changes in polyamine and glutathione contents of a green algae, *Chlorogonium elongatum* (Dang) France exposed to mercur. Environ Exp Bot 32:145–151

Ahamad ZH, Shuhanija SN (2013) Physiological and biochemical responses of a Malaysian red alga, *Gracilaria manilaensis* treated with copper, lead and mercury. J Environ Res Dev 7:1246–1253

Ahmad P, Sarwat M, Sharma S (2008) Reactive oxygen species, antioxidants and signaling in plants. J Plant Biol 51:167–173

Ahn YO, Kim SH, Lee J, Kim HR, Lee HS, Kwak SS (2012) Three *Brassica rapa* metallothionein genes are differentially regulated under various stress conditions. Mol Biol Rep 39:2059–2067

Akinci IE, Akinci S (2010) Effect of chromium toxicity on germination and early seedling growth in melon (*Cucumis melo* L.). Afr J Biotechnol 9:4589–4594

Ali H, Khan E, Sajad MA (2013) Phytoremediation of heavy metals-concepts and applications. Chemosphere 91:869–881

Alloway BJ (2013) Sources of heavy metals and metalloids in soils. In: Alloway BJ (ed) Heavy metals in soils, trace metals and metalloids in soils and their bioavailability. Springer, London

Alloway BJ, Steinnes E (1999) Anthropogenic additions of cadmium to soils. In: McLaughlin MJ, Singh BR (eds) Cadmium in soils and plants, Developments in plants and soils sciences. Kluwer, Dordrecht

Alrawiq N, Khairiah J, Talib ML, Ismail BS, Anizan I (2014) Accumulation and translocation of heavy metals in soil and paddy plant samples collected from rice fields irrigated with recycled and non-recycled water in MADA Kedah, Malaysia. Int J ChemTech Res 6:2347–2356

Alscher RG, Erturk N, Heath LS (2002) Role of superoxide dismutases (SODs) in controlling oxidative stress in plants. J Exp Bot 53:1331–1341

Amin H, Arain BA, Amin F, Surhio MA (2013) Phytotoxicity of chromium on germination, growth and biochemical attributes of *Hibiscus esculentus* L. Am J Plant Sci 4:2431–2439

Anjum NA, Gill SS, Gill R, Hasanuzzaman M, Duarte AC, Pereira E, Ahmad I, Tuteja R, Tuteja N (2014) Metal/metalloid stress tolerance in plants: role of ascorbate, its redox couple, and associated enzymes. Protoplasma 251:1265–1283

Aravind P, Prasad MNV (2005) Modulation of cadmium-induced oxidative stress in *Ceratophyllum demersum* by zinc involves ascorbate-glutathione cycle and glutathione metabolism. Plant Physiol Biochem 43:107–116

Arunakumara KKIU, Zhang X (2007) Effect of Pb^{2+} on phycobiliprotein content of *Spirulina platensis*, an edible cyanobacterium. Trop Agric Res 19:150–159

Arunakumara KKIU, Zhang X (2008) Heavy metal bioaccumulation and toxicity with special reference to microalgae. J Ocean Univ Chin 7:25–30

Ashraf M, Foolad MR (2007) Roles of glycine betaine and proline in improving plant abiotic stress resistance. Environ Exp Bot 59:206–216

Baccouch S, Chaoui A, El Ferjani E (1998) Nickel-induced oxidative damage and antioxidant responses in *Zea mays* shoots. Plant Physiol Biochem 36:689–694

Bah AM, Dai H, Zhao J, Sun H, Cao F, Zhang G, Wu F (2011) Effects of cadmium, chromium and lead on growth, metal uptake and antioxidative capacity in *Typha angustifolia*. Biol Trace Elem Res 142:77–92

Balakrishnan CP, Narayanan CS (2007) Phytotoxicity of heavy metals nickel and lead on the cyanobacterial pigments. Seaweed Res Util 29:217–226

Barondeau DP, Kassmann CJ, Bruns CK, Tainer JA, Getzoff ED (2004) Nickel superoxide dismutase structure and mechanism. Biochemistry 43:8038–8047

Bestwick CS, Al A, Puri N, Mansfield JW (2001) Characterization of lipid peroxidation and changes to pro- and antioxidant enzyme activities during the hypersensitive reaction in lettuce (*Lactuca sativa* L.). Plant Sci 161:497–506

Bharwana SA, Ali S, Farooq MA, Abbas F, Iqbal N, Ahmad MSA, Shakoor MB (2013) Influence of lead stress on growth, photosynthesis and lead uptake in the seedlings of cotton. Int J Agron Plant Prod 4:2492–2501

Bhattacharjee S (2005) Reactive oxygen species and oxidative burst: roles in stress, senescence and signal transduction in plants. Curr Sci 89:1115–1121

Boominathan R, Doran PM (2002) Ni-induced oxidative stress in roots of the Ni hyperaccumulator, *Alyssum bertolonii*. New Phytol 156:205–215
Brunetti C, Di Ferdinando M, Fini A, Pollastri S, Tattini M (2013) Flavonoids as antioxidants and developmental regulators: relative significance in plants and humans. Int J Mol Sci 14:3540–3555
Cakmak I, Horst WJ (1991) Effect of aluminium on lipid peroxidation, superoxide dismutase, catalase and peroxidase activities in root tips of soybean (*Glycine max*). Physiol Planta 83:463–468
Cardoso PF, Gratão PL, Gomes-Junior RA, Medici LO, Azevedo RA (2005) Response of *Crotalaria juncea* to nickel exposure. Braz J Plant Physiol 17:267–272
Cetin ES, Babalik Z, Hallac-Turk F, Gokturk-Baydar N (2014) The effects of cadmium chloride on secondary metabolite production in *Vitis vinifera* cv. cell suspension cultures. Biol Res 47:47
Chaudri AM, Allain CM, Barbosa-Jefferson VL, Nicholson FA, Chambers BJ, McGrath SP (2000) A study of the impacts of Zn and Cu on two rhizobial species in soils of a long term field experiment. Plant Soil 22:167–179
Chen G, Asada K (1992) Inactivation of ascorbate peroxidase by thiols requires hydrogen peroxide. Plant Cell Physiol 33:117–123
Chen YX, He YF, Luo YM, Yu YL, Lin Q, Wong MH (2003) Physiological mechanism of plant roots exposed to cadmium. Chemosphere 50:789–793
Chen F, Wang F, Wu F, Mao W, Zhang G, Zhou M (2010) Modulation of exogenous glutathione in antioxidant defense system against Cd stress in the two barley genotypes differing in Cd tolerance. Plant Physiol Biochem 48:663–672
Chongpraditnun P, Mori S, Chino M (1992) Excess copper induces a cytosolic Cu, Zn-superoxide dismutase in soybean root. Plant Cell Physiol 33:239–244
Chou TS, Chao YY, Huei Kao C (2012) Involvement of hydrogen peroxide in heat shock- and cadmium-induced expression of ascorbate peroxidase and glutathione reductase in leaves of rice seedlings. J Plant Physiol 169:478–486
Ci D, Jiang D, Dai T, Jing Q, Cao W (2009) Effects of cadmium on plant growth and physiological traits in contrast wheat recombinant inbred lines differing in cadmium tolerance. Chemosphere 77:1620–1625
Cobbett CS (2000) Phytochelatins and their role in heavy metal detoxification. Plant Physiol 123:825–832
Corpas FJ, Barroso JB, del Río LA (2001) Peroxisomes as a source of reactive oxygen species and nitric oxide signal molecules in plant cells. Trends Plant Sci 6:145–150
Couée I, Sulmon C, Gouesbet G, El Amrani A (2006) Involvement of soluble sugars in reactive oxygen species balance and responses to oxidative stress in plants. J Exp Bot 57:449–459
Creissen GP, Mullineaux PM (1995) Cloning and characterization of glutathione reductase cDNAs and identification of two genes encoding the tobacco enzyme. Planta 197:422–425
Cuypers A, Vangronsveld J, Clijsters H (2000) Biphasic effect of copper on the ascorbate-glutathione pathway in primary leaves of *Phaseolus vulgaris* seedlings during the early stages of metal assimilation. Physiol Planta 110:512–517
Cuypers A, Vangronsveld J, Clijsters H (2001) The redox status of plant cells (AsA and GSH) is sensitive to zinc imposed oxidative stress in roots and primary leaves of *Phaseolus vulgaris*. Plant Physiol Biochem 39:657–664
Cuypers A, Smeets K, Ruytinx J, Opdenakker K, Keunen E, Remans T, Horemans N, Vanhoudt N, Van Sanden S, Van Belleghem F, Guisez Y, Colpaert J, Vangronsveld J (2011) The cellular redox state as a modulator in cadmium and copper responses in *Arabidopsis thaliana* seedlings. J Plant Physiol 168:309–316
D'Souza RM, Devaraj VR (2012) Induction of oxidative stress and antioxidative mechanisms in Hyacinth bean under zinc stress. Afr Crop Sci J 20:17–29
Dandan L, Dongmei Z, Peng W, Weng Nanyan W, Xiangdong Z (2011) Subcellular Cd distribution and its correlation with antioxidant enzymatic activities in wheat (*Triticum aestivum*) roots. Ecotoxicol Environ Saf 74:874–881

Das PK, Kar M, Mishra D (1978) Nickel nutrition of plants. 1. Effects of nickel on some oxidase activities during rice (*Oryza sativa* L.) seed germination. Z Pflanzenphysiol 90:225–233

Daud MK, Mei L, Variath MT, Ali S, Li C, Rafiq MT, Zhu SJ (2014) Chromium (VI) uptake and tolerance potential in cotton cultivars: effect on their root physiology, ultramorphology, and oxidative metabolism. Bio Med Res Int 2014:1–11

Davletova S, Rizhsky L, Liang H, Shengqiang D, Oliver D, Coutu J, Shulaev V, Schlauch K, Mittler R (2004) Cytosolic ascorbate peroxidase 1 is a central component of the reactive oxygen gene network of *Arabidopsis*. Plant Cell 17:268–281

De Jesus MD, Tabatabai F, Chapman DJ (1989) Taxonomic distribution of copper-zinc superoxide dismutase in green algae and its phylogenetic importance. J Phycol 25:767–772

Degenhardt B, Gimmler H (2000) Cell wall adaptations to multiple environmental stresses in maize roots. J Exp Bot 51:595–603

del Río LA, Scandalio LM, Corpas FJ, Palma JM, Barroso JM (2006) Reactive oxygen species and reactive nitrogen species in peroxisomes, production, scavenging, and role in cell signaling. Plant Physiol 141:330–335

Demirevska-Kepova K, Simova-Stoilova L, Stoyanova Z, Hölzer-Feller RU (2004) Biochemical changes in barley plants after excessive supply of copper and manganese. Environ Exp Bot 52:253–266

Devi Chinmayee M, Anu MS, Mahesh B, Mary Sheeba A, Mini I, Swapna TS (2014) A comparative study of heavy metal accumulation and antioxidant responses in *Jatropha curcas* L. J Environ Sci Toxicol Food Technol 8:58–67

Dietz KJ, Bair M, Kramer U (1999) Free radical and reactive oxygen species as mediators of heavy metal toxicity in plants. In: Prasad MNV, Hagemeyer J (eds) Heavy metal stress in plants from molecules to ecosystems. Springer, Berlin

Dinakar N, Nagajyothi PC, Suresh S, Udaykiran Y, Damodharam T (2008) Phytotoxicity of cadmium on protein, proline and antioxidant enzyme activities in growing *Arachis hypogaea* L. seedlings. J Environ Sci 20:199–206

Dixit V, Pandey V, Shyam R (2001) Differential antioxidative responses to cadmium in roots and leaves of pea (*Pisum sativum* L. cv. Azad). J Exp Bot 52:1101–1119

Duman F (2011) Effects of exogenous glycinebetaine and trehalose on lead accumulation in an aquatic plant (*Lemna gibba* L.). Int J Phytoremediation 13:492–497

Duman F, Ozturk F (2010) Nickel accumulation and its effect on biomass, protein content and antioxidative enzymes in roots and leaves of watercress (*Nasturtium officinale* R. Br.). J Environ Sci 22:526–532

Duman F, Aksoy A, Aydin Z, Temizgul R (2011) Effects of exogenous glycinebetaine and trehalose on cadmium accumulation and biological responses of an aquatic plant (*Lemna gibba* L.). Water Air Soil Pollut 217:545–556

Ebbs S, Uchil S (2008) Cadmium and zinc induced chlorosis in Indian mustard [*Brassica juncea* (L.) Czern] involves preferential loss of chlorophyll *b*. Photosynthetica 46:49–55

Ekmekçi Y, Tanyolaç D, Ayhan B (2008) Effects of cadmium on antioxidant enzyme and photosynthetic activities in leaves of two maize cultivars. J Plant Physiol 165:600–611

Fahr M, Laplaze L, Bendaou N, Hocher V, ElMzibri M, Bogusz D, Smouni A (2013) Effects of lead on root growth. Front Plant Sci 4:175

Falk J, Munné-Bosch S (2010) Tocochromanol functions in plants: antioxidation and beyond. J Exp Bot 61:1549–1566

Farid M, Shakoor MB, Ehsan S, Ali S, Zubair M, Hanif MA (2013) Morphological, physiological and biochemical responses of different plant species to Cd stress. Int J Chem Biochem Sci 3:53–60

Fatima RA, Ahmad M (2005) Certain antioxidant enzymes of *Allium cepa* as biomarkers for the detection of toxic heavy metals in wastewater. Sci Total Environ 346:256–273

Fernàndez-Martínez J, Zacchini M, Fernández-Marín B, García-Plazaola JI, Fleck I (2014) Gas-exchange, photo- and antioxidant protection, and metal accumulation in I-214 and Eridano *Populus* sp. clones subjected to elevated zinc concentrations. Environ Exp Bot 107:144–153

Filek M, Keskinen R, Hartikainen H, Szarejko I, Janiak A, Miszalski Z, Golda A (2008) The protective role of selenium in rape seedlings subjected to cadmium stress. J Plant Physiol 165:833–844
Florence TM, Stauber JL (1986) Toxicity of copper complexes to the marine diatom *Nitzschia closterium*. Aquat Toxicol 8:11–26
Fodor F (2002) Physiological responses of vascular plants to heavy metals. In: Prasad MN, Strzalka K (eds) Physiology and biochemistry of metal toxicity and tolerance in plants. Kluwer, Dordrecht
Foyer CH, Halliwell B (1976) The presence of glutathione and glutathione reductase in chloroplasts. A proposed role in ascorbic acid metabolism. Planta 133:21–25
Foyer CH, Noctor G (2005) Oxidant and antioxidant signalling in plants: a re-evaluation of the concept of oxidative stress in a physiological context. Plant Cell Environ 28:1056–1071
Franklin NM, Stauber JL, Markich SJ, Lim RP (2000) pH-dependent toxicity of copper and uranium to a tropical freshwater alga (*Chlorella* sp.). Aquat Toxicol 48:275–289
Franklin NM, Adams MS, Stauber JL, Lim RP (2001) Development of a rapid enzyme inhibition bioassay with marine and freshwater microalgae using flow cytometry. Arch Environ Contam Toxicol 40:469–480
Freeman JL, Persans MW, Nieman K, Albrecht C, Peer W, Pickering IJ, Salt DE (2004) Increased glutathione biosynthesis plays a role in nickel tolerance in *Thlaspi* nickel hyperaccumulators. Plant Cell 16:2176–2191
Gajewska E, Skłodowska M (2005) Antioxidative responses and proline level in leaves and roots of pea plants subjected to nickel stress. Acta Physiol Planta 27:329–339
Gajewska E, Skłodowska M (2007) Effect of nickel on ROS content and antioxidative enzyme activities in wheat leaves. Biometals 20:27–36
Gallego SM, Benavides MP, Tomaro ML (1996) Effect of heavy metal ion excess on sunflower leaves: evidence for involvement of oxidative stress. Plant Sci 121:151–159
Gaur JP, Rai LC (2001) Heavy metal tolerance in algae. In: Rai LC, Gaur JP (eds) Algal adaptation to environmental stresses: physiological, biochemical and molecular mechanisms. Springer, Berlin
Gichner T, Znidar I, Száková J (2008) Evaluation of DNA damage and mutagenicity induced by lead in tobacco plants. Mutat Res 652:186–190
Gill SS, Tuteja N (2010) Reactive oxygen species and antioxidant machinery in abiotic stress tolerance in crop plants. Plant Physiol Biochem 48:909–930
Gill SS, Khana NA, Tuteja N (2012) Cadmium at high dose perturbs growth, photosynthesis and nitrogen metabolism while at low dose it up regulates sulfur assimilation and antioxidant machinery in garden cress (*Lepidium sativum* L.). Plant Sci 182:112–120
Gill SS, Gill R, Anjum NA (2014) Target osmoprotectants for abiotic stress tolerance in crop plants-glycine betaine and proline. In: Anjum NA, Gill SS, Gill R (eds) Plant adaptation to environmental change: significance of amino acids and their derivatives. CAB, Wallingford, CT
Gomes-Júnior RA, Moldes CA, Delite FS, Gratão PL, Mazzafera P, Lea PJ, Azevedo RA (2006a) Nickel elicits a fast antioxidant response in *Coffea arabica* cells. Plant Physiol Biochem 44:420–429
Gomes-Júnior RA, Moldes CA, Delite FS, Pompeu GB, Gratão PL, Mazzafera P, Lea PJ, Azevedo RA (2006b) Antioxidant metabolism of coffee cell suspension cultures in response to cadmium. Chemosphere 65:1330–1337
Gouveia C, Kreusch M, Schmidt ÉC, Felix MR, Osorio LK, Pereira DT, dos Santos R, Ouriques LC, Martins Rde P, Latini A, Ramlov F, Carvalho TJ, Chow F, Maraschin M, Bouzon ZL (2013) The effects of lead and copper on the cellular architecture and metabolism of the red alga *Gracilaria domingensis*. Microsc Microanal 19:513–524
Grant CA, Buckley WT, Bailey LD, Selles F (1998) Cadmium accumulation in crops. Can J Plant Sci 78:1–17

Gratão PL, Polle A, Lea PJ, Azevedo RA (2005) Making the life of heavy metal-stressed plants a little easier. Funct Plant Biol 32:481–494
Gratão PL, Pompeu GB, Cardoso PF, Lea PJ, Azevedo RA (2006) Plant antioxidant responses to toxic elements. Curr Top Biochem Res 8:41–70
Greger M (2004) Metal availability, uptake, transport and accumulation in plants. In: Prasad MNV (ed) Heavy metal stress in plants: from biomolecules to ecosystems, 2nd edn. Springer, Berlin
Grzesiak M, Filek M, Barbasz A, Kreczmer B, Hartikainen H (2013) Relationships between polyamines, ethylene, osmoprotectants and antioxidant enzymes activities in wheat seedlings after short-term PEG- and NaCl-induced stresses. Plant Growth Regul 69:177–189
Guan LM, Scandalios JG (2000) Hydrogen peroxide-mediated catalase gene expression in response to wounding. Free Radic Biol Med 28:1182–1190
Guan LM, Scandalios JG (2002) Catalase gene expression in response to auxin-mediated developmental signals. Physiol Planta 114:288–295
Guo TR, Zhang GP, Zhang YH (2007) Physiological changes in barley plants under combined toxicity of aluminum, copper and cadmium. Colloids Surf B Biointerfaces 57:182–188
Gupta M, Cuypers A, Vangronsveld J, Clijsters H (1999) Copper affects the enzymes of the ascorbate-glutathione cycle and its related metabolites in the roots of *Phaseolus vulgaris*. Physiol Planta 106:262–267
Gupta DK, Nicoloso FT, Schetinger M, Rossato L, Pereira LB, Castro G, Srivastava S, Tripathi RD (2009) Antioxidant defense mechanism in hydroponically grown *Zea mays* seedlings under moderate lead stress. J Hazard Mater 172:479–484
Gupta DK, Huang HG, Yang XE, Razafindrabe BHN, Inouhe M (2010) The detoxification of lead in *Sedum alfredii* H. is not related to phytochelatins but the glutathione. J Hazard Mater 177:437–444
Gupta DK, Vandenhove H, Inouhe M (2013) Role of phytochelatins in heavy metal stress and detoxification mechanisms in plants. In: Gupta DK, Corpas FJ, Palma JM (eds) Heavy metal stress in plants. Springer, Berlin
Gururani MA, Upadhyaya CP, Strasser RJ, Yu JW, Park SW (2013) Evaluation of abiotic stress tolerance in transgenic potato plants with reduced expression of PSII manganese stabilizing protein. Plant Sci 198:7–16
Hajimahmoodi M, Ali Faramarzi M, Mohammadi N, Soltani N, Oveisi R, Nastaran M, Nafissi-Varcheh N (2009) Evaluation of antioxidant properties and total phenolic contents of some strains of microalgae. J Appl Phycol 22:43–50
Hao F, Wang X, Chen J (2006) Involvement of plasma-membrane NADPH oxidase in nickel-induced oxidative stress in roots of wheat seedlings. Plant Sci 170:151–158
Hare PD, Cress WA (1997) Metabolic implications of stress-induced proline accumulations in plants. Plant Growth Regul 21:79–102
Hassinen VH, Tervahauta AI, Schat H, Karenlampi SO (2011) Plant metallothioneins—metal chelators with ROS scavenging activity? Plant Biol 13:225–232
Hédiji H, Djebali W, Cabasson C, Maucourt M, Baldet P, Bertrand A, Zoghlami LB, Deborde C, Moing A, Brouquisse R, Chaïbi W, Gallusci P (2010) Effects of long-term cadmium exposure on growth and metabolomic profile of tomato plants. Ecotoxicol Environ Saf 73:1965–1974
Hegedüs A, Erdei S, Horváth G (2001) Comparative studies of H_2O_2 detoxifying enzymes in green and greening barley seedlings under cadmium stress. Plant Sci 160:1085–1093
Heidari M, Saran S (2011) Effects of lead and cadmium on seed germination, seedling growth and antioxidant enzymes activities of mustard (*Sinapis arvensis* L.). J Agric Biol Sci 6:6–11
Hirata T, Tanaka M, Ooike M, Tsunomura T, Sakaguchi M (1999) Radical scavenging activities of phycocyanobilin prepared from a cyanobacterium, *Spirulina platensis*. Fish Sci 65:971–972
Hirata T, Tanaka M, Ooike M, Tsunomura T, Sakaguchi M (2000) Antioxidant activities of phycocyanobilin prepared from *Spirulina platensis*. J App Phycol 12:435–439
Hong Z, Lakkineni K, Zhang Z, Verma DP (2000) Removal of feedback inhibition of delta(1)-pyrroline-5-carboxylate synthetase results in increased proline accumulation and protection of plants from osmotic stress. Plant Physiol 122:1129–1136

Hou W, Chen X, Song G, Wang Q, Chang CC (2007) Effects of copper and cadmium on heavy metal polluted waterbody restoration by duckweed (*Lemna minor*). Plant Physiol Biochem 45:62–69
Ikenaka Y, Nakayama SMM, Muzandu K, Choongo K, Teraoka H, Mizuno N, Ishizuka M (2010) Heavy metal contamination of soil and sediment in Zambia. Afr J Environ Sci Technol 4:729–739
Ip PF, Chen F (2005) Employment of reactive oxygen species to enhance astaxanthin formation in *Chlorella zofingiensis* in heterotrophic culture. Process Biochem 40:3491–3496
Islam MM, Hoque A, Okuma E, Banu NA, Shimoishi Y, Nakamura Y, Murata Y (2009) Exogenous proline and glycine betaine increase antioxidant enzyme activities and confer tolerance to cadmium stress in cultured tobacco cells. J Plant Physiol 166:1587–1597
Jain RS, Srivastava S, Solomon S, Shrivastava AK, Chandra A (2010) Impact of excess zinc on growth parameters, cell division, nutrient accumulation, photosynthetic pigments and oxidative stress of sugarcane (*Saccharum* spp.). Acta Physiol Plant 32:979–986
Janicka-Russak M, Kabala K, Burzyński M, Klobus G (2008) Response of plasma membrane H^+-ATPase to heavy metal stress in *Cucumis sativus* roots. J Exp Bot 59:3721–3728
Karpinski S, Reynolds H, Karpinska B, Wingsle G, Creissen G, Mullineaux P (1999) Systemic signaling and acclimation in response to excess excitation energy in *Arabidopsis*. Science 284:654–657
Karuppanapandian T, Sinha PB, Haniya AMK, Manoharan K (2006) Differential antioxidative responses of ascorbate-glutathione cycle enzymes metabolites to chromium stress in greengram (*Vigna radiata* L. Wilczek) leaves. J Plant Biol 49:440–447
Keilig K, Ludwig-Müller J (2009) Effect of flavonoids on heavy metal tolerance in *Arabidopsis thaliana* seedlings. Bot Stud 50:311–318
Keunen E, Remans T, Bohler S, Vangronsveld J, Cuypers A (2011) Metal-induced oxidative stress and plant mitochondria. Int J Mol Sci 12:6894–6918
Khraiwesh B, Arif MA, Seumel GI, Ossowski S, Weigel D, Reski R (2010) Transcriptional control of gene expression by microRNAs. Cell 140:111–122
Kováčik J, Babula P, Hedbavny J, Kryštofová O, Provaznik I (2015) Physiology and methodology of chromium toxicity using alga *Scenedesmus quadricauda* as model object. Chemosphere 120:23–30
Krämer U, Cotter-Howells JD, Charnock JM, Baker AJM, Smith AC (1996) Free histidine as a metal chelator in plants that accumulate nickel. Nature 379:635–638
Kruk J, Holländer-Czytko H, Oettmeier W, Trebst A (2005) Tocopherol as singlet oxygen scavenger in photosystem II. J Plant Physiol 162:749–757
Kubiś J (2008) Exogenous spermidine differentially alters activities of some scavenging system enzymes, H_2O_2 and superoxide radical levels in water-stressed cucumber leaves. J Plant Physiol 165:397–406
Kumar H, Sharma D, Kumar V (2012) Nickel-induced oxidative stress and role of antioxidant defence in barley roots and leaves. Int J Environ Biol 2:121–128
Kumar D, Yusuf MA, Singh P, Sardar M, Sarin NB (2013) Modulation of antioxidant machinery in α-tocopherol-enriched transgenic *Brassica juncea* plants tolerant to abiotic stress conditions. Protoplasma 250:1079–1089
Lane TW, Saito MA, George GN, Pickering IJ, Prince RC, Morel FMM (2005) A cadmium enzyme from a marine diatom. Nature 435:42
Lang M, Zhang Y, Chai T (2005) Identification of genes up-regulated in response to Cd exposure in *Brassica juncea* L. Gene 363:151–158
Larkum AWD (2003) Light harvesting systems in algae. In: Larkum AWD, Douglas SE, Raven JA (eds) Photosynthesis in algae. Kluwer, Dordrecht
Lee MY, Shin HW (2003) Cadmium-induced changes in antioxidant enzymes from the marine alga *Nannochloropsis oculata*. J Appl Phycol 15:13–19
Leszczyszyn OI, Imam HT, Blindauer CA (2013) Diversity and distribution of plant metallothioneins: a review of structure, properties and functions. Metallomics 5:1146–1169

Li D, Zhou D, Wang P, Weng N, Zhu X (2011) Subcellular Cd distribution and its correlation with antioxidant enzymatic activities in wheat (*Triticum aestivum*) roots. Ecotoxicol Environ Saf 74:874–881

Li X, Yang Y, Jia L, Chen H, Wei X (2013) Zinc-induced oxidative damage, antioxidant enzyme response and proline metabolism in roots and leaves of wheat plants. Ecotoxicol Environ Saf 89:150–157

Liu D, Zou J, Wang M, Jiang W (2008) Hexavalent chromium uptake and its effects on mineral uptake, antioxidant defence system and photosynthesis in *Amaranthus viridis* L. Bioresour Technol 99:2628–2636

Lopez-Millan A, Sagardoy R, Solanas M, Abadia A, Abadia J (2009) Cadmium toxicity in tomato (*Lycopersicon esculentum*) plants grown in hydroponics. J Environ Exp Bot 65:376–385

Ma M, Zhu W, Wang Z, Witkaamp GJ (2003) Accumulation, assimilation and growth inhibition of copper on fresh-water alga (*Scenedesmus subspicatus* 86.81 SAG) in the presence of EDTA and fulvic acid. Aquat Toxicol 63:221–228

Malar S, Vikram SS, Favas PJC, Perumal V (2014) Lead heavy metal toxicity induced changes on growth and antioxidative enzymes level in water hyacinths [*Eichhornia crassipes* (Mart.)]. Bot Stud 55:54

Malešev D, Kuntić V (2007) Investigation of metal-flavonoid chelates and the determination of flavonoids via metal-flavonoid complexing reactions. J Serb Chem Soc 72:921–939

Mallick N (2004) Copper-induced oxidative stress in the chlorophycean microalga *Chlorella vulgaris*: response of the antioxidant system. J Plant Physiol 161:591–597

Mallick N, Mohn FH (2000) Reactive oxygen species: response of algal cells. J Plant Physiol 157:183–193

Mallick N, Mohn FH (2003) Use of chlorophyll fluorescence in metal-stress research: a case study with the green microalga *Scenedesmus*. Ecotoxicol Environ Saf 55:64–69

Mallory AC, Bouché N (2008) MicroRNA-directed regulation: to cleave or not to cleave. Trends Plant Sci 13:359–367

Mano J (2002) Early events in environmental stresses in plants: induction mechanisms of oxidative stress. In: Inzé D (ed) Oxidative stress in plants. Taylor and Francis, London

Marques AP, Rangel AO, Castro PM (2007) Zinc accumulation in plant species indigenous to a Portuguese polluted site: relation with soil contamination. J Environ Qual 36:646–653

Matringe M, Ksas B, Rey P, Havaux M (2008) Tocotrienols, the unsaturated forms of vitamin E, can function as antioxidants and lipid protectors in tobacco leaves. Plant Physiol 147:764–778

Mehta SK, Gaur JP (1999) Heavy metal induced proline accumulation and its role in ameliorating metal toxicity in *Chlorella vulgaris*. New Phytol 143:253–259

Mijovilovich A, Leitenmaier B, Meyer-Klaucke W, Kroneck PMH, Gotz B, Kupper H (2009) Complexation and toxicity of copper in higher plants. II. Different mechanisms for copper versus cadmium detoxification in the copper-sensitive cadmium/zinc hyperaccumulator *Thlaspi caerulescens* (Ganges ecotype). Plant Physiol 151:715–731

Millar AH, Mittova V, Kiddle G, Heazlewood L, Bartoli CG, Theodoulou FL, Foyer CH (2003) Control of ascorbate synthesis by respiration and its implications for stress responses. Plant Physiol 133:443–447

Miranda MS, Sato S, Mancini-Filho J (2001) Antioxidant activity of the microalga *Chlorella vulgaris* cultured on special conditions. Boll Chim Farm 140:165–168

Mishra S, Srivastava S, Tripathi RD, Kumar R, Seth CS, Gupta DK (2006) Lead detoxification by coontail (*Ceratophyllum demersum* L.) involves induction of phytochelatins and antioxidant system in response to its accumulation. Chemosphere 65:1027–1039

Mittler R (2002) Oxidative stress, antioxidants and stress tolerance. Trends Plant Sci 7:405–410

Mittler R, Vanderauwera S, Gollery M, Van Breusegem F (2004) Reactive oxygen gene network of plants. Trends Plant Sci 9:490–498

Mohan BS, Hosetti BB (1997) Potential phytotoxicity of lead and cadmium to *Lemna minor* grown in sewage stabilization ponds. Environ Pollut 98:233–238

Morelli E, Scarano G (2004) Copper-induced changes of non-protein thiols and antioxidant enzymes in the marine microalga *Phaeodactylum tricornutum*. Plant Sci 167:289–296
Moura DJ, Péres VF, Jacques RA, Saffi J (2012) Heavy metal toxicity: oxidative stress parameters and DNA repair. In: Gupta DK, Sandalio LM (eds) Metal toxicity in plants: perception, signaling and remediation. Springer, Heidelberg
Nadgórska-Socha A, Kafel A, Kandziora-Ciupa M, Gospodarek J, Zawisza-Raszka A (2013) Accumulation of heavy metals and antioxidant responses in *Vicia faba* plants grown on monometallic contaminated soil. Environ Sci Pollut Res Int 20:1124–1134
Nagalakshmi N, Prasad MNV (2001) Responses of glutathione cycle enzymes and glutathione metabolism to copper stress in *Scenedesmus bijugatus*. Plant Sci 160:291–299
Naser AH (2013) Assessment and management of heavy metal pollution in the marine environment of the Arabian Gulf: a review. Mar Pollut Bull 72:6–13
Navari-Izzo F, Quartacci MF, Sgherri C (2002) Lipoic acid: a unique anti-oxidant in the detoxification of activated oxygen species. Plant Physiol Biochem 40:463–470
Neill JS, Desikan R, Clarke A, Hurst RD, Hanckok JT (2002) Hydrogen peroxide and nitric oxide as signaling molecules in plants. J Exp Bot 53:1237–1247
Nematshahi N, Lahouti M, Ganjeali A (2012) Accumulation of chromium and its effect on growth of (*Allium cepa* cv. Hybrid). Eur J Exp Biol 2:969–974
Noctor G, Foyer CH (1998) Ascorbate and glutathione: keeping active oxygen under control. Annu Rev Plant Physiol Plant Mol Biol 49:249–279
Noctor G, Gomez L, Vanacker H, Foyer CH (2002) Interactions between biosynthesis, compartmentation and transport in the control of glutathione homeostasis and signalling. J Exp Bot 53:1283–1304
Oancea S, Foca N, Airinei A (2005) Effects of heavy metals on plant growth and photosynthetic activity. Analele Univ. "Al. I. Cuza", Tom I, s, Biofizica, Fizica medicala si fizica mediului, pp 107–110
Obroucheva NV, Bystrova EI, Ivanov VB, Antipova OV, Seregin IV (1998) Root growth responses to lead in young maize seedlings. Plant Soil 200:55–61
Okamoto OK, Colepicolo P (1998) Response of superoxide dismutase to pollutant metal stress in the marine dinoflagellate *Gonyaulax polyedra*. Comp Biochem Physiol C Pharmacol Toxicol Endocrinol 119:67–73
Panda SK, Chaudhury I, Khan MH (2003) Heavy metals induce lipid peroxidation and affect antioxidants in wheat leaves. Biol Plant 46:289–294
Pandey N, Sharma CP (2002) Effect of heavy metals Co^{2+}, Ni^{2+} and Cd^{2+} on growth and metabolism of cabbage. Plant Sci 163:753–758
Papadakis AK, Roubelakis-Angelakis KA (2005) Polyamines inhibit NADPH oxidase-mediated superoxide generation and putrescine prevents programmed cell death induced by polyamine oxidase-generated hydrogen peroxide. Planta 220:826–837
Parida BK, Chhibba IM, Nayyar VK (2003) Influence of nickel-contaminated soils on fenugreek (*Trigonella corniculata* L.) growth and mineral composition. Sci Hortic 98:113–119
Parmar P, Kumari N, Sharma V (2013) Structural and functional alterations in photosynthetic apparatus of plants under cadmium stress. Bot Stud 54:45
Passardi F, Longet D, Penel C, Dunand C (2004) The class III peroxidase multigene family in rice and its evolution in land plants. Phytochemistry 65:1879–1893
Pawlik-Skowrońska B (2001) Phytochelatin production in freshwater algae *Stigeoclonium* in response to heavy metals contained in mining water; effects of some environmental factors. Aquat Toxicol 52:241–249
Perl-Treves R, Perl A (2002) Oxidative stress: an Introduction. In: Inzé D (ed) Oxidative stress in plants. Taylor and Francis, London
Pinto E, Sigaud-Kutner TCS, Leitão MAS, Okamoto OK, Morse D, Colepicolo P (2003) Heavy metal-induced oxidative stress in algae. J Phycol 39:1008–1018
Poniedziałek M, Sękara A, Ciura J, Jędrszczyk E (2005) Nickel and manganese accumulation and distribution in organs of nine crops. Folia Hortic 17:11–22

Prasad TK, Anderson MD, Martin BA, Steward CR (1994) Evidence for chilling-induced oxidative stress in maize seedlings and a regulatory role for hydrogen peroxide. Plant Cell 6:65–74
Pumas C, Vacharapiyasophon P, Peerapornpisal Y, Leelapornpisid P, Boonchum W, Ishii M, Khanongnuch C (2011) Thermostability of phycobiliproteins and antioxidant activity from four thermotolerant cyanobacteria. Phycol Res 59:166–174
Radotic K, Ducic T, Mutavdzic D (2000) Changes in peroxidase activity and isoenzymes in spruce needles after exposure to different concentrations of cadmium. Environ Exp Bot 44:105–113
Rafati M, Khorasani N, Moattar F, Shirvany A, Moraghebi F, Hosseinzadeh S (2011) Phytoremediation potential of *Populus alba* and *Morus alba* for cadmium, chromium and nickel absorption from polluted soil. Int J Environ Res 5:961–970
Rascio N, Navari-Izzo F (2011) Heavy metal hyperaccumulating plants: how and why do they do it? And what makes them so interesting? Plant Sci 180:169–181
Rijstenbil JW, Derksen JWM, Gerringa LJA, Poortvliet TCW, Sandee A, Van den Berg M, Van Drie J, Wijnholds JA (1994) Oxidative stress induced by copper: defense and damage in the marine planktonic diatom *Ditylum brightwellii* (Grunow) West, grown in continuous cultures with high and low zinc levels. Mar Biol 119:583–590
Rivelli AR, De Maria S, Puschenreiter M, Gherbin P (2012) Accumulation of cadmium, zinc, and copper by *Helianthus annuus* L.: impact on plant growth and uptake of nutritional elements. Int J Phytoremediation 14:320–334
Rodríguez-Serrano M, Romero-Puertas MC, Zabalza A, Corpas FJ, Gómez M, del Río LA, Sandalio LM (2006) Cadmium effect on oxidative metabolism of pea (*Pisum sativum* L.) roots. Imaging of reactive oxygen species and nitric oxide accumulation in vivo. Plant Cell Environ 29:1532–1544
Romero-Puertas MC, Rodríguez-Serrano M, Corpas F, Gomez M, del Rio L, Sandalio LM (2004) Cadmium-induced subcellular accumulation of $O_2^{\bullet -}$ and H_2O_2 in pea leaves. Plant Cell Environ 27:1122–1134
Romero-Puertas MC, Corpas F, Rodríguez-Serrano M, Gómez M, del Rio L, Sandalio LM (2007) Differential expression and regulation of antioxidative enzymes by Cd in pea plants. J Plant Physiol 164:1346–1357
Rosa M, Prado C, Podazza G, Interdonato R, González JA, Hilal M, Prado FE (2009) Soluble sugars—metabolism, sensing and abiotic stress: a complex network in the life of plants. Plant Signal Behav 4:388–393
Sabatini SE, Juarez AB, Eppis MR, Bianchi L, Luquet CM, Rios de Molina MC (2009) Oxidative stress and antioxidant defenses in two green microalgae exposed to copper. Ecotoxicol Environ Saf 72:1200–1206
Sandalio LM, Dalurzo HC, Gómez M, Romero-Puertas MC, del Río LA (2001) Cadmium-induced changes in the growth and oxidative metabolism of pea plants. J Exp Bot 52:2115–2126
Sarma AD, Sreelakshimi Y, Sharma R (1997) Antioxidant ability of anthocyanins against ascorbic acid oxidation. Phytochemistry 45:671–674
Schat H, Sharma SS, Vooijs R (1997) Heavy metal-induced accumulation of free proline in a metal-tolerant and a nontolerant ecotype of *Silene vulgaris*. Physiol Planta 101:477–482
Schat H, Llugany M, Vooijs R, Hartley-Whitaker J, Bleeker PM (2002) The role of phytochelatins in constitutive and adaptive heavy metal tolerances in hyperaccumulator and non-hyperaccumulator metallophytes. J Exp Bot 53:2381–2392
Scheidegger C, Sigg L, Behra R (2011) Characterization of lead induced metal–phytochelatin complexes in *Chlamydomonas reinhardtii*. Environ Toxicol Chem 30:2546–2552
Schreck E, Foucault Y, Sarret G, Sobanska S, Cécillon L, Castrec-Rouelle M, Uzu G, Dumat C (2012) Metal and metalloid foliar uptake by various plant species exposed to atmospheric industrial fallout: mechanisms involved for lead. Sci Tot Environ 427–428:253–262
Schützendübel A, Polle A (2002) Plant responses to abiotic stresses: heavy metal-induced oxidative stress and protection by mycorrhization. J Exp Bot 53:1351–1365

Schützendübel A, Schwanz P, Teichmann T, Gross K, Langenfeld-Heyser R, Godbold DL, Polle A (2001) Cadmium induced changes in antioxidative systems, hydrogen peroxide content, and differentiation in Scots pine roots. Plant Physiol 127:887–898

Semane B, Cuypers A, Smeets K, Van Belleghem F, Horemans N, Schat H, Vangronsveld J (2007) Cadmium responses in *Arabidopsis thaliana*: glutathione metabolism and antioxidative defence system. Physiol Planta 129:519–528

Senden MHMN, van der Meer AJGM, Verburg TG, Wolterbeek HT (1995) Citric acid in tomato plant roots and its effect on cadmium uptake and distribution. Plant Soil 17:333–339

Shah KH, Ritambhara GK, Verma S, Dubey RS (2001) Effect of cadmium on lipid peroxidation, superoxide anion generation and activities of antioxidant enzymes in growing rice seedlings. Plant Sci 161:1135–1144

Shahid M, Pinelli E, Pourrut B, Silvestre J, Dumat C (2011) Lead-induced genotoxicity to *Vicia faba* L. roots in relation with metal cell uptake and initial speciation. Ecotoxicol Environ Saf 74:78–84

Sharma SS, Dietz KJ (2006) The significance of amino acids and amino acid-derived molecules in plant responses and adaptation to heavy metal stress. J Exp Bot 57:711–726

Sharma SS, Dietz KJ (2009) The relationship between metal toxicity and cellular redox imbalance. Trends Plant Sci 14:43–50

Sharma P, Dubey RS (2005) Lead toxicity in plants. Braz J Plant Physiol 17:35–52

Sharma SS, Schat H, Vooijs R (1998) In vitro alleviation of heavy metal-induced enzyme inhibition by proline. Phytochemistry 49:15311–15535

Sheldon AR, Menzies NW (2005) The effect of copper toxicity on the growth and root morphology of Rhodes grass (*Chloris gayana* Knuth.) in resin buffered solution culture. Plant Soil 278:341–349

Shu X, Yan Yin L, Fa Zhang Q, Bo Wang W (2011) Effect of Pb toxicity on leaf growth, antioxidant enzyme activities, and photosynthesis in cuttings and seedlings of *Jatropha curcas* L. Environ Sci Pollut Res 19:893–902

Shukla UC, Murthy RC, Kakkar P (2008) Combined effect of ultraviolet-B radiation and cadmium contamination on nutrient uptake and photosynthetic pigments in *Brassica campestris* L. seedlings. Environ Toxicol 23:712–719

Siedlecka A, Krupa Z (1999) Cd/Fe interactions in higher plants - Its consequences for the photosynthetic apparatus. Photosynth Res 36:321–331

Singh VP (2005) Metal toxicity and tolerance in plants and animals. Sarup, New Delhi

Skowroński T, De Knecht JA, Simons J, Verkleji JAC (1998) Phytochelatin synthesis in response to cadmium uptake in *Vaucheria* (Xanthophyceae). Eur J Phycol 33:87–91

Ślesak I, Libik M, Karpinska B, Karpinski S, Miszalski Z (2007) The role of hydrogen peroxide in regulation of plant metabolism and cellular signaling in response to environmental stresses. Acta Biochim Pol 54:39–50

Smeets K, Ruytinx J, Semane B, Van Belleghem F, Remans T, Van Sanden S, Vangronsveld J, Cuypers A (2008) Cadmium-induced transcriptional and enzymatic alterations related to oxidative stress. Environ Exp Bot 63:1–8

Smirnoff N, Wheeler GL (2000) Ascorbic acid in plants: biosynthesis and function. Crit Rev Biochem Mol Biol 35:291–314

Stroiński A, Kozłowska M (1997) Cadmium induced oxidative stress in potato tuber. Acta Soc Bot Pol 66:189–195

Su C, LiQin J, WenJun Z (2014) A review on heavy metal contamination in the soil worldwide: situation, impact and remediation techniques. Environ Skep Crit 3:24–38

Sun Q, Ye ZH, Wang XR, Wong MH (2007) Cadmium hyperaccumulation leads to an increase of glutathione rather than phytochelatins in the cadmium hyperaccumulator *Sedum alfredii*. J Plant Physiol 164:1489–1498

Surosz W, Palinska KA (2005) Effects of heavy-metal stress on cyanobacterium *Anabaena flos-aquae*. Arch Environ Contam Toxicol 48:40–48

Tahara S (2007) A journey of twenty-five years through the ecological biochemistry of flavonoids. Biosci Biotechnol Biochem 71:1387–1404

Taiz L, Zeiger E (2002) Plant physiology. Sinauer, Sunderland, MA

Tamás L, Dudíková J, Ďurčeková K, Halušková L, Huttová J, Mistrík I, Olle M (2008) Alterations of the gene expression, lipid peroxidation, proline and thiol content along the barley root exposed to cadmium. J Plant Physiol 165:1193–1203

Tang YL, Ren WW, Zhang L, Tang KX (2011) Molecular cloning and characterization of gene coding for γ-tocopherol methyltransferase from lettuce (*Lactuca sativa*). Genet Mol Res 10:320–412

Teklić T, Engler M, Cesar V, Lepeduš H, Parađiković N, Lončarić Z, Štolfa I, Marotti T, Mikac N, Žarković N (2008a) Influence of excess copper on lettuce (*Lactuca sativa* L.) grown in soil and nutrient solution. J Food Agric Environ 6:439–444

Teklić T, Hancock JT, Engler M, Parađiković N, Cesar V, Lepeduš H, Štolfa I, Bešlo D (2008b) Antioxidative responses in radish (*Raphanus sativus* L.) plants stressed by copper and lead in nutrient solution and soil. Acta Biol Cracov Bot 50:79–86

Tian S, Lu L, Yang X, Huang H, Wang K, Brown P (2011) Root adaptations to cadmium-induced oxidative stress contribute to Cd tolerance in the hyperaccumulator *Sedum alfredii*. Biol Plant 56:344–350

Tomašević M, Aničić M (2010) Trace element content in urban tree leaves and sem-edax characterization of deposited particles. FU Phys Chem Technol 8:1–13

Trebst A, Depka B, Holländer-Czytko H (2002) A specific role for tocopherol and of chemical singlet oxygen quenchers in the maintenance of photosystem II structure and function in *Chlamydomonas reinhardtii*. FEBS Lett 516:156–160

Tripathi BN, Gaur JP (2006) Physiological behavior of *Scenedesmus* sp. during exposure to elevated levels of Cu and Zn and after withdrawal of metal stress. Protoplasma 229:1–9

Tripathi BN, Mehta SK, Amar A, Gaur JP (2006) Oxidative stress in *Scenedesmus* sp. during short- and long-term exposure to Cu^{2+} and Zn^{2+}. Chemosphere 62:538–544

Tripathy BC, Oelmüller R (2012) Reactive oxygen species generation and signaling in plants. Plant Signal Behav 7:1621–1633

Tsuji N, Hirayanagi N, Okada M, Miyasaka H, Hirata K, Zenk MH, Miyamoto K (2002) Enhancement of tolerance to heavy metals and oxidative stress in *Dunaliella tertiolecta* by Zn-induced phytochelatin synthesis. Biochem Biophys Res Commun 293:653–659

Van Assche F, Clijsters H (1990) Effect of metal on enzyme activity on plants. Plant Cell Environ 13:195–206

Van Assche F, Put C, Clijsters HMM (1986) Heavy metals induce specific isozyme patterns of peroxidase in *Phaseolus vulgaris* L. Arch Int Physiol Biochim 94:60

Van den Ende W, Valluru R (2009) Sucrose, sucrosyl oligosaccharides, and oxidative stress: scavenging and salvaging? J Exp Bot 60:9–18

Verma S, Dubey RS (2003) Lead toxicity induces lipid peroxidation and alters the activities of antioxidant enzymes in growing rice plants. Plant Sci 164:645–655

Viehweger K (2014) How plants cope with heavy metals. Bot Stud 55:35

Volland S, Lütz C, Michalkec B, Lütz-Meindl U (2012) Intracellular chromium localization and cell physiological response in the unicellular alga *Micrasterias*. Aquat Toxicol 109:59–69

Wada N, Sakamoto T, Matsugo S (2013) Multiple roles of photosynthetic and sunscreen pigments in cyanobacteria focusing on the oxidative stress. Metabolites 3:463–483

Wang Y, Fang J, Leonard SS, Murali Krishna Rao K (2004) Cadmium inhibits the electron transfer chain and induces reactive oxygen species. Free Radic Biol Med 36:1434–1443

Wang R, Gao F, Guo BQ, Huang JC, Wang L, Zhou YJ (2013) Short term chromium stress induced alterations in the maize leaf proteome. Int J Mol Sci 14:11125–11144

Weckx EJ, Clijsters H (1996) Oxidative damage and defense mechanisms in primary leaves of *Phaseolus vulgaris* as a result of root assimilation of toxic amounts of copper. Physiol Plant 96:506–512

Willkenes H, Chamnongpol S, Davey M, Schraudner M, Langebartels C, Van Montagu M, Inzé D, Van Camp W (1997) Catalase is a sink for H_2O_2 and is indispensable for stress defence in C3 plants. EMBO J 16:4806–4816

Wu TM, Hsu YT, Lee TM (2009) Effects of cadmium on the regulation of antioxidant enzyme activity, gene expression, and antioxidant defences in the marine macroalga *Ulva fasciata*. Bot Stud 50:25–34

Xiang C, Oliver DJ (1998) Glutathione metabolic genes coordinately respond to heavy metals and jasmonic acid in *Arabidopsis*. Plant Cell 10:1539–1550

Xiong ZT (1998) Lead uptake and effects on seed germination and plant growth in a Pb hyperaccumulator *Brassica pekinensis* Rupr. Bull Environ Contam Toxicol 60:285–291

Xue T, Li X, Zhu W, Wu C, Yang G, Zheng C (2009) Cotton metallothionein GhMT3a, a reactive oxygen species scavenger, increased tolerance against abiotic stress in transgenic tobacco and yeast. J Exp Bot 60:339–349

Yamamoto HY, Bassi R (1996) Carotenoids: localization and function. In: Ort DR, Yocum CF (eds) Oxygenic photosynthesis: the light reactions. Kluwer, Dordrecht

Yan R, Gao S, Yang W, Cao M, Wang S, Chen F (2008) Nickel toxicity induced antioxidant enzyme and phenylalanine ammonia-lyase activities in *Jatropha curcas* L. cotyledons. Plant Soil Environ 54:294–300

Yang Y, Wei X, Lu J, You J, Wang W, Shi R (2010) Lead-induced phytotoxicity mechanism involved in seed germination and seedling growth of wheat (*Triticum aestivum* L.). Ecotoxicol Environ Saf 73:1982–1987

Yannarelli GG, Fernández-Alvarez AJ, Santa-Cruz DM, Tomaro ML (2007) Glutathione reductase activity and isoforms in leaves and roots of wheat plants subjected to cadmium stress. Phytochemistry 68:505–512

Yilmaz DD, Parlak KU (2011) Antioxidative parameters in the opposite-leaved pondweed (*Gronlendia densa*) in response to nickel stress. Chem Spec Bioavail 23:71–79

Yusuf MA, Sarin NB (2007) Antioxidant value addition in human diets: genetic transformation of *Brassica juncea* with γ-TMT gene for increased α-tocopherol content. Transgenic Res 16:109–113

Zengin F (2013) Physiological behaviour of bean (*Phaseolus vulgaris* L.) seedlings under metal stress. Biol Res 46:79–85

Zengin FK, Munzuroglu O (2005) Effects of some heavy metals on content of chlorophyll, proline and some antioxidant chemicals in bean (*Phaseolus vulgaris* L.) seedlings. Acta Biol Cracov Ser Bot 47:157–164

Zhou ZP, Liu LN, Chen XL, Wang JX, Chen M, Zheng YZ, Zhou BC (2005) Factors that affect antioxidant activity of c-phycocyanins from *Spirulina platensis*. J Food Biochem 29:313–322

Zou J, Yu K, Zhang Z, Jiang W, Liu D (2009) Antioxidant response system and chlorophyll fluorescence in chromium (VI)-treated *Zea mays* (L.) seedlings. Acta Biol Cracov Bot 51:23–33

Arsenic and Chromium-Induced Oxidative Stress in Metal Accumulator and Non-accumulator Plants and Detoxification Mechanisms

Sarita Tiwari and Bijaya Ketan Sarangi

Contents

Abstract Mitigation of arsenic (As) and chromium (Cr) pollution is a topical environmental issue of high R&D priority due to its toxicity on living organisms and deleterious effects on the environment. Following uptake by plants, As and Cr generate reactive oxygen species (ROS) and induce oxidative stress, which exerts negative effects on biochemical, molecular, and cellular levels that hinder plant growth and development. When the stressor level reaches the threshold level of plant tolerance, the stress response is manifested physiologically and beyond that level, the plant succumbs. However, some plants termed as hyperaccumulators, i.e., those accumulating metal ions inside their cellular milieu with $BF > 1$, have evolved detoxification mechanisms due to their physiological and genetic makeup which facilitates scavenging of indigenously generated ROS. Various enzymatic and non-enzymatic antioxidants such as superoxide dismutase (SOD), catalase (CAT), ascorbate peroxidase (APX), glutathione peroxidase (GPX), ascorbate, glutathione, and phenolic compounds have been reported to be involved in neutralising ROS. It seems that the antioxidant defence system plays a significant

S. Tiwari • B.K. Sarangi (✉)
Environmental Biotechnology Division, National Environmental Engineering Research Institute (CSIR-NEERI), Nehru Marg, Nagpur 440020, Maharashtra, India
e-mail: bk_sarangi@neeri.res.in

D.K. Gupta et al. (eds.), *Reactive Oxygen Species and Oxidative Damage in Plants Under Stress*, DOI 10.1007/978-3-319-20421-5_7

role in combating metal stress and confers metal tolerance to these plants. Understanding the biochemistry of plants exposed to As and Cr stress would be beneficial for selecting As and Cr tolerant plants that are better equipped with such defence mechanisms. This chapter reviews different aspects related to antioxidant defence mechanisms in As and Cr hyperaccumulator and non-hyperaccumulator plants. This chapter also highlights usefulness of these biomarkers for screening plants with competent biochemical mechanisms for metal stress tolerance. This information, in turn will help to design efficient phytoextraction treatment systems through deployment of such competent plants.

Keywords Arsenic • Chromium • Phytoremediation • Oxidative stress • Antioxidant enzyme

1 Introduction

Amongst all biological organisms, terrestrial plants are most vulnerable to environmental stress in their habitat. Being non-motile, they are exposed to the ecosystem dynamics at different stages of growth and times out of which abiotic stress is more acute. These plant species try to maintain homeostatic interrelationship between ecosystems dynamics for their existence and survival. However, sustained stresses for long duration become stressors which are more profound with abiotic stress. Industrialisation and urbanisation have become inevitable to sustain human needs resulting in a footprint in the ecosystem which is ever being magnified due to improper environmental management practices. Uncontrolled disposal of waste, natural catastrophes, reckless mining and smelting of metallic ores, agricultural use of waste sludge and water, accidental and process spillage, etc. have led to serious pollution resulting in conversion of natural resources and ecosystems to contaminated wastelands or waterbodies (Bartolomeo et al. 2004; Landajo et al. 2004; Lokeshwari and Chandrappa 2006; Ogbonna et al. 2009). Wastes emanating from different industrial processes are both inorganic and organic compounds comprising heavy metals, combustible and putrescible substances, hazardous wastes, explosives, and petroleum products which become recalcitrant for treatment due to their complex nature (Xia et al. 2011). Heavy metals and inorganic constituents are indestructible and persist in most environmental matrices (Alloway 1990; Ghosh and Singh 2005). Heavy metal toxicity such as arsenic (As) and chromium (Cr) is one of the major environmental concerns that warrant urgent attention for mitigation and sustainable management.

Several end-of-the-pipe treatment methods are available for mitigation of As and Cr pollutions, but they are invasive, *ex situ*, have secondary impacts on the environment, and are non-sustainable. Moreover, application of these methods in a large stretch of land and water ecosystem is techno-economically not feasible.

Phytoremediation is an in situ cleanup technology which entails use of green plants for environmental restoration (Cunningham and David 1996; USEPA 2000, 2001; Mwegoha 2008; Das and Maiti 2008; Purakayastha and Chhonkar 2010; Raymond and Felix 2011; Witters et al. 2012). This remediation process follows the principles of green technologies and is sustainable (Li et al. 2003). It is a solar-driven biological process and in average, tenfold cheaper than engineering-based remediation methods, such as soil excavation, soil washing or burning, or pump-and-treat systems (Glass 1999; USEPA 2001; Sheoran et al. 2011; Conesa et al. 2012; Pokhrel and Dubey 2013).

Metal accumulation in plant species growing in metal-polluted soils is due to uptake and eventual accumulation along with other inorganic nutrients (Baker 1981; Ross 1994; Verbruggen et al. 2009; Sarangi et al. 2009; Sheoran et al. 2011) in their biomass, although, heavy metals also exert toxic effects on their metabolism. Different plant species significantly vary in their ability to tolerate metal stress and metal accumulation in the biomass depending on the genetic and biochemical organisation of the plant system. The ideal plants for phytoremediation application should possess multiple traits, such as high contaminant uptake capacity, relatively fast absorption kinetics, high level of biomass production, and adaptability to different climates and soil environments (Pichai et al. 2001; Padmavathiamma and Li 2007; Sarangi et al. 2009; Verbruggen et al. 2009). Effective remediation of contaminated soil and water systems using a specific plant species is an immensely complex task whose success depends on a multitude of factors: first and foremost, the ability of the client plant to uptake, translocate, detoxify, and accumulate the target contaminant in its biomass. Heavy metals are toxic to plants, but some plants known as metallophytes have evolved mechanisms to sustain and cope with this stress, and hyperaccumulator plants accumulate high levels of heavy metals in their biomass with a higher bioaccumulation factor (Nedelkoska and Doran 2000; Sheoran et al. 2011; Sarma 2011; Raymond and Felix 2011). This chapter presents recent progress in the research on mechanisms of metal tolerance in non-hyperaccumulating and hyperaccumulating plants.

1.1 Prevalence of Arsenic and Chromium Stress

Arsenic (As) Arsenic exists in the earth's crust, averaging about 3 mg kg^{-1} (Mandal and Suzuki 2002; Zhao et al. 2010), but it locally gets concentrated from various anthropogenic activities (Buchet and Lison 1998; Mandal and Suzuki 2002; Benner 2010). In nature, As is distributed ubiquitously throughout earth crusts, soil, sediments, water, air, and living organisms in over 200 different mineral forms, of which approximately 60 % are arsenates, 20 % sulphides and sulphosalts, and the remaining 20 % includes arsenides, arsenites, oxides, silicates, and elemental As (Mukherjee et al. 2008; Kim et al. 2009). Arsenic exists in a variety of inorganic and organic compounds in the environment, and inorganic species are highly mobile and toxic (Meharg and Hartley-Whitaker 2002; Giacomino et al. 2010). So far, four As species: As(V), As(III), As(0) and As (-III) have been studied extensively in the water–soil–plant system (Onken and Hossner 1995; Smith et al. 1998; Kertulis et al. 2005). Under aerobic soil conditions, arsenate (AsO_4^{3-}) is the thermodynamically stable form, whereas arsenite (AsO_3^{3-}) is the predominant inorganic species under reducing conditions (Smith et al. 1998; Chen et al. 2009).

Chromium (Cr) Chromium is listed as the seventh most abundant metal in the earth's crust (Panda and Choudhury 2005) and exists in bound form in the range of 100–300 mg kg^{-1} soil (Sarangi et al. 2009). The prime sources of Cr contamination in the environment are utilisation of Cr-containing compounds, pesticides and insecticides, indiscriminate disposal of industrial, urban, and mine wastes, denudation of soil cover, over exploitation of aquifers resulting in geogenic formations, etc. (Han et al. 2002; Da Costa et al. 2012). Chromium-based compounds are widely used in different industries such as tannery, metal plating industries, metallurgical and processing, wood preservation, and alloy preparation (Dhal et al. 2013; Fan et al. 2012). The discharge limit of Cr from industries is less than 1 mg L^{-1}. In the environment, Cr occurs in different oxidation states but primarily exists as a soluble, highly toxic Cr(VI) anion and the less soluble, less toxic Cr(III) form (Mohanty et al. 2006) Usually it occurs in association with oxygen as chromate (CrO_4^{2-}) or dichromate ($Cr_2O_7^{2-}$) oxyanions. Although Cr is required as a trace element in humans and animals, it exerts toxicity above 0.5 mg L^{-1}, which is its permissible discharge limit. Cr(VI) is both toxic and mutagenic to biological organisms and induces cancer and teratogenic effects in animals and plants (Shanker et al. 2005).

2 Phytotoxic Effects of As and Cr Stress in Hyperaccumulator and Non-hyperaccumulator Plants

Arsenic and Cr are not essential nutrients for plants, and their entry inside plants has deleterious effects at multiple levels. As a chemical analogue of phosphate, As (V) interferes with oxidative phosphorylation, while As(III) exerts inhibitory effects on enzyme activity by binding to thiol groups (Dixon 1996; Zhao et al. 2010; Hughes 2002). The primary targets of As toxicity are unknown, but As(V) can substitute for phosphate in phosphorylation reactions, including ATP synthesis. However, in plant cells, As(V) is rapidly reduced to As(III), catalysed by ACR2 arsenate reductases (Bleeker et al. 2006; Dhankher et al. 2006). Toxicity of As(III) is probably due to its high sulphydryl reactivity. Arsenic, either supplied as As(V) or As(III), causes oxidative stress and can deplete reduced glutathione, an important cellular antioxidant, through the formation of As(III)–glutathione (As(III)–GSH) and As(III)-induced phytochelatin (PC) synthesis. As in animals, the mitochondrial electron transfer chain of plant cells is one of the major targets of

As toxicity and is the site of rapid As-induced ROS production (Heyno et al. 2008). Increased ROS production induces lipid peroxidation. Furthermore, As, either supplied as As(V) or As (III), is both mutagenic and an inactivator of DNA mismatch repair in yeast and human cells (Jin et al. 2003), and the same mechanisms may also be effective in plants.

2.1 *Alterations in Physiological and Biochemical Mechanisms of Stressed Plants*

A putative chromatin remodelling factor, named *OXS3*, was recently identified in a screen for metal tolerance of a *Brassica juncea* cDNA library in *Schizosaccharomyces pombe* (Blanvillain et al. 2008). An *OXS3* mutant was hypersensitive to metal stress and overexpression improved tolerance. It was postulated that *OXS3* might protect DNA or alter its transcriptional selectivity. Interestingly, *OXS3* overexpression also enhanced tolerance to As, Cu, and Zn or oxidising chemicals like diamide. As-induced increase in ROS (H_2O_2) production is likely mediated by glycolate oxidase and may act as a cellular signal triggering the stress response (Gupta et al. 2013). Stress-responsive mitogen-activated protein kinases (MAPK) seem to be involved in transcriptional responses possibly activated by ROS under excess As (Jonak et al. 2004). It is thought that MAPK activation is actually mediated by ROS production, a secondary effect of oxidative injury by heavy metals. Similar to other abiotic stresses, different plant hormones and growth regulators may also participate in the plant response to metal(loid) stress (Dalcorso et al. 2008). Exposure to As (Norton et al. 2008) or other metal ions has been reported (Weber et al. 2006; Craciun et al. 2006; Van de Mortel et al. 2008) to induce many different transcription factors. Some of them are constitutively upregulated in *Arabidopsis halleri* or *Thlaspi caerulescens*. Large arrays of genes are constitutively highly expressed in metal hyperaccumulators compared to a related non-hyperaccumulator species (Verbruggen et al. 2009). Gene duplication and modification of *cis*-regulatory elements are demonstrated mechanisms of enhanced expression of these stress-related genes (Hanikenne et al. 2008). The molecular study of metal hyperaccumulators has unravelled the role of genes involved in metal homeostasis and detoxification previously identified in *Arabidopsis thaliana*. The major metal(loid) detoxification mechanisms in plants are (1) transport to the major storage organs or tissues, (2) chelation, (3) subcellular compartmentalization, or (4) efflux from the plant body. The specific mechanisms of metal-induced stress response and mitigation measures evolved by plants are discussed in the following sections.

2.2 *Non-hyperaccumulators*

Induction of abiotic stress triggers production of several metabolites and proteins, some of which may be responsible for conferring a certain degree of protection to these stresses (Jaleel et al. 2009). Plant species found in metal-polluted/contaminated soils are prone to take up metals (Baker 1981; Ross 1994) in their biomass, which also includes edible parts. As(V) is easily incorporated into plant cells through the high-affinity Pi transport system. Non-hyperaccumulators generally rely on decreased As(V) uptake, because of suppression of the high-affinity Pi uptake system (Meharg and Hartley-Whitaker 2002). It has been shown in *Holcus lanatus* that As(V) hypertolerance is not associated with any hypertolerance to As (III) because of the different uptake mechanisms of As(V) and As(III) (Bleeker et al. 2006). Arsenite has a molecular size and structure similar to that of silicic acid, and the silicon transporter OsNIP2;1/Lsi1 was recently demonstrated to be responsible for most of the As(III) influx into rice roots (Ma et al. 2008). However, As(III) efflux from the roots into the rhizosphere has also been demonstrated in a number of plant species. In non-hyperaccumulators, very considerable fractions (50–85 %) of the arsenic taken up may be removed from the plant body via root As (III) efflux, unlike in the As hyperaccumulator, *Pteris vittata*, where hardly any As (III) efflux was observed (Zhao et al. 2009).

Arsenic and Cr-mediated disorders in plant physiology have been observed by many researchers (Panda and Choudhury 2005; Shanker et al. 2005; Dhankher et al. 2006; Kramer 2010). Chromium and As-induced phytotoxicity results in plant growth inhibition, decrease in protein content, nutrient imbalance, and inhibition on enzymatic activities and mutagenesis (Panda and Choudhury 2005; Shanker et al. 2005; Yadav et al. 2010; Sharma 2012). Requejo and Tena (2005) reported up or downregulation of about 10 and 15 %, respectively, of total detected maize root and shoot proteins under As stress. Various investigators have proved that As toxicity in non-resistant plants results in considerable stress upon exposure, ranging from inhibition of root growth to death (Meharg 2003; Barrachina et al. 1995). Length and biomass of root and shoot were significantly affected due to As stress. Reduced root length growth in response to As exposure has been observed by a number of investigators in different plants (Shri et al. 2009; Liu et al. 2005; Hartley-Whitaker et al. 2001).

Uptake and concentration of Cr in plants bring many disorders in physiological and biochemical processes that has been studied in different plants like mosses, rice, pea, wheat, etc. in relation to oxidative stress (Choudhury and Panda 2005; Panda 2007; Hayat et al. 2012). Both forms of Cr [Cr (III) and Cr (VI)] are found to have adverse impacts on plant growth, root wilting, and chlorophyll content of plants (Panda and Choudhury 2005). Chen et al. (2001) reported a decrease in root weight and root length in maize seedlings grown in soil containing 20 mg Cr (VI) kg^{-1}. The inhibition in root growth is due to shutdown of root cell division/elongation or extension of the cell cycle in roots. In the presence of both forms of Cr, the root gets affected due to direct contact between roots and metals which in turn inhibits the roots ability to absorb water from the medium (Barcelo et al. 1985). Effect on inhibition of seedling germination under different Cr concentration was reported by many investigators. Plants such as *Echinochloa colona*, barley, *Phaseolus vulgaris*, sugarcane bud, etc. were reported to display a reduction in seedling growth in response to Cr stress (Skeffington et al. 1976; Parr and Taylor 1982; Rout et al. 2000; Shanker et al. 2005). The reduction in seed germination under Cr stress has predicated the suppressive effect of Cr on the activity of amylases and on the subsequent transport of sugars to the embryo axes. Inhibition of chlorophyll biosynthesis by Cr has also been reported in terrestrial plants (Vajpayee et al. 2000). In fact, Cr also causes deleterious effects on plant's physiological processes such as photosynthesis, water relations, and mineral nutrition. Barcelo et al. (1985) described the inhibition of P, K, Zn, Cu, and Fe translocation within the plant parts when bean plants were exposed to Cr in nutrient solutions. Sujatha et al. (1996) reported that tannery effluent irrigation caused micronutrient deficiencies in several agricultural crops. Metabolic alterations by Cr exposure have also been described in plants either by a direct effect on enzymes and metabolites or by the ability of Cr to generate ROS (Shanker et al. 2005).

2.3 Reactive Oxygen Species Generation Under Metal Stress

Under standard growth conditions, uneven harmony of a plant scavenging system with ROS buildup leads to uncontrolled oxidation and free radical chain reactions, which result in oxidative stress to the plant (Srivastava et al. 2005). During stress, electrons that have a high-energy state are transferred to molecular oxygen (O_2) to form ROS. ROS that most often occur under metal toxicity are singlet oxygen ($^1O_2$), superoxide anion ($O_2^{\bullet -}$), peroxides, hydroxyl radicals ($OH^{\bullet}$), and the widely distributed hydrogen peroxide (H_2O_2) (Ahmad et al. 2010). Heavy metals including As and Cr have been reported to stimulate the formation of ROS, mostly free radicals, leading to oxidative stress (Singh et al. 2007). Excess ROS formed within cells can provoke oxidation and modification of cellular amino acids, proteins, membrane lipids, and DNA. The functionality of protein can be affected by ROS either by oxidation of amino acid side chains or by secondary reactions with aldehydic products of lipid peroxidation (Reinheckel et al. 1998). Plant cell membranes are generally considered to be primary sites of metal injury. Membrane destabilisation is frequently attributed to lipid peroxidation due to an enhanced production of toxic amounts of oxygen free radicals after exposure to metal. The polyunsaturated fatty acids (PUFAs) are the major fatty acids in the plant membrane, and ROS attacks unsaturated fatty acids inducing peroxidation which gives rise to complex mixtures of lipid hydroperoxide products like malondialdehyde (MDA) (Mueller 2004; Singh et al. 2007; Mishra et al. 2008). Malondialdehyde (MDA) is one of the final decomposition products of lipid peroxidation and has been used as an index for lipid peroxidation status (Zhou 2001; Yamauchi et al. 2008; Singh et al. 2010). In rice seedlings, the levels of MDA were used as an indicator of lipid peroxidation under As stress (Shri et al. 2009). Extensive PUFA peroxidation decreases the fluidity of the membrane, increases leakiness, and causes secondary damage to membrane proteins (Ahmad et al. 2010). In high concentrations, ROS are predominantly employed in causing cell damage but in lower concentrations, they play a major physiological role in several aspects of intracellular signalling and regulation (Palmer and Paulson 1997). It has been clearly demonstrated that ROS interfere with the expression of a number of genes and signal transduction pathways (Ahmad et al. 2008; Jaleel et al. 2009) playing a significant role as secondary messengers.

In case of As, although As is a non-redox active element, there are significant evidences that both inorganic forms of As i.e., As(III) and As(V) generate ROS which cause toxicity and damage to cell systems (Meharg and Hartley-Whitaker 2002). There is significant evidence that exposure to inorganic As species results in the generation of ROS. Using a proteomics approach, it has been reported that As exposure induces proteins related to oxidative stress in plants (Requejo and Tena 2005). *P. vittata* is a well-known established hyperaccumulator which has a capacity of accumulating high concentrations of As inside plants without any toxicity symptoms (Ma et al. 2001). Whereas in most of the plants, As accumulation causes adverse effects by enhancement of ROS (Kertulis et al. 2005). After entry into the plant's cellular milieu, arsenic causes injury to the major biomolecules such as proteins, DNA, and lipids of the cell system (Miller et al. 2008; Mittler et al. 2004). However, the complete mechanism of As toxicity in plants is not clear so far, but inactivation of essential enzymes by SH-As binding, displacement of essential ions from enzymes, and interference of As with phosphate metabolism are some predictable mechanisms of As toxicity in plants (Sharma 2012). A comparative study by Srivastava et al. (2005) between *P. vittata* (As hyperaccumulator) and *P. ensiformis* and *N. exaltata* (As sensitive) observed elevated levels of TBARS (thiobarbituric acid reactive substances), which are lipid peroxidation products, in the fronds of *P. ensiformis* and *N. exaltata* in comparison to *P. vittata* after As exposure. The increase in levels of TBARS in the fronds with increasing concentrations of arsenate indicates that As induces oxidative stress in fern plants. Lower production of TBARS in the fronds of *P. vittata* corresponds to its higher As accumulation.

Chromium(III), on the other hand, apart from generating ROS, if presents in high concentrations, coordinates various organic compounds resulting in inhibition of some metalloenzyme systems. Significant increase in lipid peroxidation and H_2O_2 generation was seen in Cr-treated green gram (*Vigna radiata*) plants (Shanker et al. 2004). H_2O_2 levels increased in both roots and leaves of sorghum treated with either 50 AM Cr(VI) or 100 AM Cr(III) (Shanker et al. 2005). Under Cr stress, the plants induce certain metabolic modifications such as alterations in pigment production, e.g., chlorophyll, anthocyanin (Boonyapookana et al. 2002), increased production of antioxidant metabolites (e.g., glutathione, ascorbic acid) as a direct response to Cr stress which may cause damage to the plants (Shanker 2003), and production of biochemically related metabolites which confer resistance or tolerance to Cr stress (Schmfger 2001).

3 Mechanism of As and Cr Detoxification in Hyperaccumulator Plants

3.1 ROS Scavenging Mechanisms

Plants under metal stress induce different detoxification mechanisms, of which compartmentalization of metals inside the plant biomass is a primary action. However, failure to concentrate toxic metals inside plants turns on different biochemical processes through maximising the metal thiol binding, resulting in lower levels of free ionic species involved in stress-related detoxification reactions (Srivastava et al. 2005; Singh et al. 2006).To neutralise effects of generated ROS, endogenous defence mechanisms are activated which lead to the synthesis of adaptive biochemical responses such as (1) synthesis of phytochelatins, (2) increased production and activity of enzymatic and non-enzymatic antioxidants, and (3) prevention of the degradation of biomolecules (Shri et al. 2009; Schmoger et al. 2000). Antioxidant enzyme activity assays have been commonly used as indicators for evaluation of oxidative stress in some plants (Foyer and Noctor 2005; Zhang et al. 2007; Gusman et al. 2013) since the induction of antioxidants is often the only evidence that oxidative stress has occurred in vivo (Halliwell and Gutteridge 1999). Metal-tolerant plants have a well-organised defence system to combat metal stress. Antioxidative systems play a key role in maintaining ROS levels below its toxic threshold limits which get imbalanced due to abiotic stresses. Literature review reveals that heavy metal tolerance by plants is the result of enhanced production of antioxidant enzymes (Gratao et al. 2005). Stress-tolerant plants had evolved intricate biochemical mechanisms to sustain metal toxicity through reduced influx, complexation of ions, and enhanced production of antioxidants that detoxify ROS generated in response to the exposure to toxic metals (Gunes et al. 2009). The stimulation of enzymatic activities of, e.g., superoxide dismutase (SOD), catalase (CAT), ascorbate peroxidases (APX), glutathione reductase (GR) and non-enzymatic antioxidants like glutathione, ascorbate and carotenoids have been considered as basic defence mechanisms against As-induced oxidative stress in ferns and some higher plants (Sharma et al. 2007). Many reports demonstrate that As and Cr hyperaccumulators show an increase in enzymatic antioxidants in response to As and Cr stress (Singh et al. 2010; Srivastava et al. 2005; Diwan et al. 2010). Hyperaccumulator plants have developed efficient mechanisms to combat such adverse environmental stresses like heavy metal toxicity. The Chinese brake fern (*P. vittata*), a versatile and fast growing As hyperaccumulator, showed no toxicity symptoms, low levels of lipid peroxidation, and high activities of SOD, CAT, and APX in response to elevated As concentrations when compared with two other As-sensitive ferns, namely *P. ensiformis* and *N. exaltata*, implicating the antioxidative defence in As tolerance (Srivastava et al. 2005; Singh et al. 2010).

3.1.1 Enzymatic Antioxidants

Enzymatic antioxidants act as shields against metal stress by eliminating ROS and preventing degradation of biomolecules in plants (Schmoger et al. 2000; Shri et al. 2009). The major antioxidant enzymes which are involved in ROS scavenging mechanisms of plants include SOD, CAT, APX, and GPX (Ashraf 2009; Wang et al. 2010). In plant cells, each cellular compartment has been reported to contain more than one type of enzyme to neutralise ROS (Table 1). That is, normally, each cellular compartment contains more than one enzymatic activity that detoxifies a particular ROS. For example, the cytosol contains at least three different enzymatic activities that scavenge H_2O_2: APX, GPX, and PrxR (Suzuki and Mittler 2006). Specific roles for antioxidant enzymes have been explored via transgenic approaches. Reactive oxygen species initially produced during metal stress are eliminated, e.g. SOD, by virtue of converting $O_2^{\bullet-}$ to H_2O_2 which can further be removed by CAT, APX, and GPX enzymes as shown in Fig. 1 (Sharma and Dietz 2008). Enhanced activity of these enzymes had been reported in plants exposed to metal stress (Sun et al. 2007; Shri et al. 2009; Gusman et al. 2013). A comparative study of enzyme activity between As hyperaccumulator and non-accumulator plants showed an increased enzymatic activity in stress-resistant (i.e., As hyperaccumulator) plants (Srivastava et al. 2005).

3.1.1.1 Superoxide Dismutase, EC 1.15.1.1

SOD is one of the early induced enzymes known to scavenge ROS in response to metal stress (Gechev et al. 2006; Bowler et al. 1992). Several isomeric forms of SOD exist in plants which convert superoxide radicals ($O_2^{\bullet-}$) to hydrogen peroxide (H_2O_2). The upregulation of SODs is implicated in combating oxidative stress caused due to abiotic stress and have a critical role in the survival of plants. SOD activity as a function of treated As concentration was evident in *P. vittata,* and higher activity was observed in aerial parts due to a higher concentration of As in fronds. SOD activity in treated fronds was 46–61 % higher as compared to controls (Srivastava et al. 2005; Singh et al. 2010). An increase in SOD activity in response to As(V) and/or As(III) toxicity was also reported in *H. lanatus* (Hartley Whitaker et al. 2001) and *Zea mays* (Requejo and Tena 2005). Comparative analysis of SOD activity vs. As tolerance between different species of *Pteris* revealed that SOD

Table 1 Localization of enzymatic antioxidants in plant cells

Organelles	Enzyme
Chloroplast	APX/SOD/GR/glutathione
Mitochondria	APX/GR/Mn-SOD/glutathione
Cytosol	APX/GR/Mn-SOD/glutathione
Vacuole	Glutathione/peroxidase
Peroxisomes	CAT/SOD

Cited from Ahmad et al. (2010)

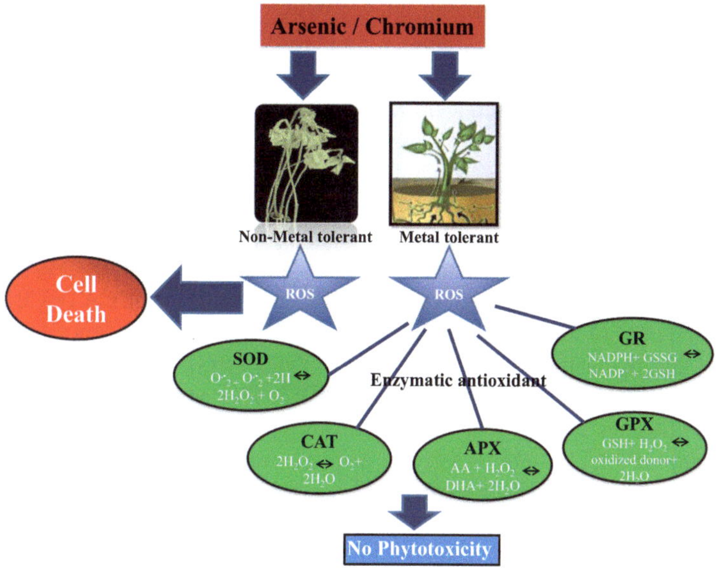

Fig. 1 Enzymatic antioxidant defence mechanism in hyperaccumulator and non-accumulator plants

activity was the highest in *P. vittata* followed by *P. ensiformis* and *N. exaltata* showing highest and lowest As tolerance, respectively (Srivastava et al. 2005). Similarly, decreased SOD activity in pea and other plants is related to their inability to fight against oxidative stress caused by As (Gong et al. 2005; Gunes et al. 2009). The study by Hartley-Whitaker et al. (2001) on *H. lanatus* concluded that SOD levels are depressed at higher As concentrations but enhanced at lower As concentrations. In maize, it has been shown that expression of *sod* genes in the leaves was increased on exposure to low levels of arsenate and arsenite but was decreased at higher levels of As treatment (Mylona et al. 1998). Altered expression of SOD may in turn lead to expression of other enzymes associated with the stress resistance (McKersie et al. 1999). Similarly, the highest level of SOD activity was observed in the fronds of the As hyperaccumulator *P. vittata*. Similar to As, exposure of pea to low Cr concentrations (20 μM) increased SOD activity, and this activity was inhibited at much higher (200 μM) concentrations which was also lethal to the plant. In another study, it has been shown that SOD was active in scavenging the superoxide produced by both Cr species in green gram, whereas catalase did not participate in active H_2O_2 reduction irrespective of Cr speciation (Shanker et al. 2004).

3.1.1.2 Catalase, EC1.11.1.6

A higher SOD activity during detoxification of ROS results in an increased generation of H_2O_2, and its higher concentrations may promote further oxidative damage. Therefore, a rapid removal of H_2O_2 produced as a result of SOD activity is another important requirement. The generated H_2O_2 is more permeable across membranes than other ROS (Mittler et al. 2004) and also toxic. Therefore, scavenging of the generated H_2O_2 is followed by an increase in the activity of other enzymes responsible for scavenging such as CAT, APX, GPX, and POX (Asada 1992; Srivastava et al. 2005; Gill and Tuteja 2010). The function of CAT is the direct degradation of H_2O_2 (i.e., without the involvement of an electron donor), a potential source of highly reactive hydroxyl radical and singlet oxygen and the conversion of H_2O_2 into water and oxygen (Ahmad et al. 2010). This enzyme is mainly localised in the peroxisomes, and the fact that peroxisomes proliferate during stress optimises the scavenging of H_2O_2 by CAT that diffuses from the cytosol. However, it is important to point out that under extreme conditions of stress, a plant may be too weak to produce enough antioxidant enzymes to protect it. Increased CAT activity in response to metal stress is observed in many plants (Fayiga et al. 2004), and an increase in catalase activity is supposed to be an adaptive trait (Sekmen et al. 2007; Vital et al. 2008). CAT has been shown to be upregulated by As at the transcriptional level. Srivastava et al. (2005) found that *P. vittata* (As hyperaccumulator) and *P. ensiformis* (an As-sensitive fern) had similar CAT activities in the fronds in the absence of As. After exposure to As, CAT activity significantly increased in the fronds of both plants (21–28 %). This may imply the role of CAT in As tolerance by both plants. CAT was not only induced in the fronds of *P. vittata*, but it was also activated, up to 300 %, upon As exposure. Although a similar activation pattern was observed in the non-hyperaccumulator *P. ensiformis*, the magnitude was much smaller. The activation of CAT at two different concentrations of arsenate may be the result of different CAT isozymes present in the fern. It was also observed that in the As-treated *P. vittata*, CAT activity was found primarily in fronds rather than in roots. This response is corroborated by the study of Duquesnoy et al. (2010), where high As(V) and As(III) concentrations increased CAT activity in leaves and roots of *Zea maize*. Activity of CAT in response to Cr has also been studied in many crops like rice, wheat, green gram, and mosses (Panda and Patra 2000; Choudhury and panda 2005; Panda and Choudhury 2005). In most of the studies, a gradual decrease in CAT activity was observed in the treated plants with increasing Cr concentrations. Chatterjee and Chatterjee (2000) had reported restricted CAT activity in cauliflower leaves at 0.5 mM Cr.

3.1.1.3 Ascorbate Peroxidase, EC 1.11.1.1

APX, which is produced mainly in the chloroplasts and other cell organelles, is required to scavenge H_2O_2 and to maintain the redox state of the cell (Asada 1992). APX has a higher affinity for H_2O_2 than CAT and POD, and it may have a more crucial role in the management of ROS stress or may be responsible for the fine modulation of ROS signalling. In contrary to SOD levels, APX was remarkably high in roots of As-treated *P. vittata* and *P. ensiformis* but not in *N. exaltata*. The As hyperaccumulator, *P. vittata*, demonstrated a significant increase in APX activity in its frond, rhizome, and root tissues under As treatment (Srivastava et al. 2005). Similar results were obtained in *P. ensiformis*. However, in the case of the sensitive species, *N. exaltata*, there was no significant increase in the activity of APX in the frond tissues (Singh et al. 2006).

3.1.1.4 Glutathione Peroxidase, EC 1.11.1.9

Glutathione peroxidases are the family of multiple isozymes which catalyse the reduction of H_2O_2 and cytotoxic hydroperoxides to alcohols (Dixon 1996). Thus, besides scavenging of H_2O_2, GPX also serves to detoxify products of lipid peroxidation formed due to the activity of ROS. The activity of GPX under As stress in frond tissues of *N. exaltata* and rhizome tissues of *P. ensiformis* and *N. exaltata* was significantly high. This indicates that an enhancement of GPX activity upon exposure to As serves as an intrinsic defence tool to resist As-induced oxidative damage in these plants. Similar increases in GPX activity in leaves of *Zea Mays* and *Oryza sativa* in response to As stress was also observed (Shri et al. 2009). In certain plants, like lettuce roots, APX activity in treated plants was lower than in controls. It was presumed that this could be due to non-equilibrium between the activation of CAT, SOD, POX, and APX (Gusman et al. 2013).

3.1.1.5 Glutathione Reductase, EC 1.6.4.2

GR is responsible for scavenging H_2O_2, catalysing disulphide bonds of glutathione, and maintaining the levels of glutathione (Yannarelli et al. 2007). GR is mainly localised in the chloroplast stroma but is also found in mitochondria, cytosol, and peroxisomes. GR catalyses the rate limiting last step of the ascorbate–glutathione pathway. A study by Kertulis-Tartar et al. (2009) after observation of combined results of induction, kinetics, and inhibition studies interpreted that GR is not affected by As in *P. vittata* and *P. ensiformis*. This non-induction of GR activity upon As exposure in *P. vittata* and *P. ensiformis* assumes no significant role of GR in *Pteris* in As tolerance (Cao et al. 2004; Srivastava et al. 2005; Kertulis-Tartar et al. 2009).

3.1.2 Non-enzymatic Antioxidants

3.1.2.1 Phenolic Compounds

Phenolics are diverse secondary metabolites (flavonoids, tannins, hydroxycinnamate esters, and lignin) abundant in plant tissues, having at least one aromatic ring (C_6) and bearing one or more hydroxyl groups. The phenolic compounds are potential non-enzymatic antioxidant acting as ROS scavenging compounds (Blokhina et al. 2003). Polyphenols possess an ideal structural chemistry for free radical scavenging activity, and they have been shown to be more effective antioxidants in vitro than tocopherols and ascorbate (Rice-Evans et al. 1997; Balasundram et al. 2006). Antioxidative properties of polyphenols arise from their high reactivity as hydrogen or electron donors, ability of the polyphenol-derived radical to stabilize and delocalize the unpaired electron (chain-breaking function), and the ability to chelate transition metal ions (termination of the ROS-generating Fenton reaction). Phenolic antioxidants (PhOH) interfere with the oxidation of lipids and other molecules by the rapid donation of hydrogen atom to radicals $ROO^{\bullet} + PhOH \rightarrow ROOH + PhO^{\bullet}$. The phenoxy radical intermediates are relatively stable, so they do not initiate further radical reactions. They even act as terminators of the chain reaction by interacting with other free radicals. Induction of polyhydroxy phenolic compounds in response to different heavy metals has been reported (Zhang et al. 2002). During heavy metal stress, phenolic compounds can act as metal chelators, on the other hand, phenolics can also directly scavenge molecular species of reactive oxygen. In an As non-accumulator, *A. filiculoides* Sánchez-Viveros et al. (2011) had observed a decrease in the total amount of phenolic compounds and presumed that this phenomenon is the reason for susceptibility of these plants to As-induced stress.

Flavonoids are one of the important phenolic compounds found abundantly in plant organs that act as ROS scavengers by identifying and neutralising radicals before onset of their toxic action (Lovdal et al. 2010). Their functionality depends on the number and arrangement of their hydroxyl groups attached to ring structures. Their ability to act as antioxidants depends on the reduction potentials of their radicals and accessibility of the radicals. Many flavonoid biosynthetic genes are induced under stress conditions, and there is significant enhancement in flavonoid level towards metal toxicity (Gill and Tuteja 2010). Another mechanism underlying the antioxidative properties of phenolics is the ability of flavonoids to alter peroxidation kinetics by modification of the lipid packing order and to decrease fluidity of the membranes. This could delay diffusion of free radicals and restrict peroxidative reactions. In the presence of H_2O_2 and phenolic substrates, peroxidases operate in the peroxidatic cycle and are engaged in the synthesis of lignin and other phenolic polymers. However, if the phenolic substrates are replaced by NADPH or related reduced compounds, a chain reaction starts that provides the basis for the H_2O_2-producing NADH-oxidase activity of peroxidises (Apel and Hirt 2004). Increased activity of peroxidise in root tips of pine were linearly related with elevated level of phenolics against cadmium stress (Schutzendubel and Polle 2002).

3.1.2.2 Ascorbate and Glutathione

These are probably the most extensively studied antioxidants and can be detected in the majority of plant cell types, organelles, and the apoplast (Potters et al. 2002;

Smirnoff 2000). Normally, ascorbate occurs in the reduced form (AsA), and its intracellular concentration ranges from 20 mM in the cytosol to 20–300 mM in the chloroplast stroma (Noctor and Foyer 1998). A fundamental role of AsA in the plant defence system is to protect metabolic processes against H_2O_2 and other potentially toxic derivatives of oxygen. AsA acts essentially as a reductant and scavenges many types of ROS. Two molecules of AsA are utilised by APX in the ascorbate–glutathione cycle to reduce H_2O_2 to water, with the concomitant generation of monodehydroascorbate (MDA) which is converted into dehydroascorbate (DHA) through regeneration of AsA. MDA can also be reduced directly to AsA. AsA can react directly and reduce superoxide, hydrogen peroxide, and hydroxyl radical or quench singlet oxygen. Ascorbate also functions as a co-substrate of plant oxidases, such as the ascorbate peroxidase system, which produces dehydroascorbate (Singh et al. 2006). Dehydroascorbate is reduced to ascorbic acid in a GSH (glutathione)-dependent reaction catalysed by dehydroascorbate reductase.

Glutathione (GSH) is a non-enzymatic low-molecular weight thiol-type antioxidant involved in cellular defence against toxic xenobiotics (Terry and Banuelos 2000; Gill and Tuteja 2010). Glutathione levels in plant tissues are known to increase under metal stress (Sun et al. 2007). Glutathione is involved in a range of metabolic processes and constitutes an important plant defence system against environmental stress, including heavy metals. Glutathione levels in plant tissues are known to increase under metal stress (Sun et al. 2007). Glutathione levels are constitutively higher in plants adapted to metal(loid) stress conditions (Mishra et al. 2008, 2009). Glutathione is the major source of nonprotein thiols and a precursor of the metal-binding phytochelatins (PC). The chemical reactivity of the thiol group of glutathione makes it particularly suitable to serve a broad range of biochemical functions in all organisms. Reaction of GSH by ROS forms oxidised glutathione (GSSG) and ascorbate is oxidised to monodehydroascorbate (MDA) and dehydroascorbate (DHA) through the ascorbate–glutathione cycle (Gill and Tuteja 2010). GSSG, MDA, and DHA can be reduced to reform GSH and ascorbate. A high ratio of reduced to oxidised ascorbate and GSH is essential for ROS scavenging in cells. GSH also prevented the toxic effects of As(V). Plants exposed to As substantially increase the synthesis of GSH and PC, which is a polymer of GSH. Depletion of GSH by oxidants may alter the redox status of the cell and present a stressful and toxic situation (Hughes 2002). However, the heritable metal tolerance in *P. vittata* (Kertulis-Tartar et al. 2009) and *Silene vulgaris* (De Vos et al. 1992) was not found to correlate with constitutively high GSH concentrations, implying the species specificity of GSH functions. GSH and Cys gave partial protection against As stress. Kertulis-Tartar et al. (2009) found a slight increase of ascorbate and glutathione in *P. vittata* and *P. ensiformis* under As exposure. The contents of non-enzymatic antioxidants (GSH, and –SH) were significantly increased at high levels of As exposure (>20 mg kg^{-1}), though no significant change at low As levels was observed (Cao et al. 2004). Mutants with decreased ascorbic acid levels or altered GSH contents are hypersensitive to As stress.

4 Importance of Detoxification Mechanisms for As and Cr Phytoremediation

There are a variety of existing physical and chemical technologies for remediation of As and Cr-contaminated sites (USEPA 1992; Choong et al. 2007). For As remediation treatment, conventional approaches like co-precipitation, adsorption, ion exchange, and membrane processes are effective in removing As from contaminated groundwater (Cheremisinoff 1998; Brenner and Lazarova 2012). On the other hand, in case of Cr, leaching of pollutants, vitrification, electro kinetic treatments, excavation, and off-site treatments are in use (Fu and Wang 2011). However, these remedial technologies are *ex situ*, expensive, and restricted for remediation of small areas due to techno-economic limitations (Paspaliaris et al. 2010). In recent times, R&D are more focused towards development of sustainable processes based on 'green chemistry' approaches taking into consideration human health, resources availability, energy input, and atom efficiency. Phytoremediation is a promising technology for pollution management in comparison to other physico-chemical methods with minimal maintenance requirement and public acceptance (Kramer 2010), and phytoextraction is suitable for metal extraction from contaminated matrixes. This plant-based biological remediation process can function with minimal maintenance after its establishment, as the costs of growing vegetation is minimal compared to those of soil removal, treatment, and replacement. Because biological processes are solar driven, phytoremediation is on average tenfold cheaper than engineering-based remediation methods, such as soil excavation, soil washing or burning, or pump-and-treat systems (Purakayastha and Chhonkar 2010; Macek et al. 2011). However, implementation of this technique for remediation of metal and other contaminants has met with limited success due to lack of enough evidence about competency of a potential plant for application based on phytoremediation principles (Zhao and McGrath 2009).

Plants ideal for phytoremediation should posses multiple traits viz. high contaminant uptake capacity, relatively fast absorption kinetics, high biomass, and adaptability to suit various climates and soil environments (Pichai et al. 2001; Sarangi et al. 2009). Employment of hyperaccumulator plants having higher ability for metal uptake and accumulation in their biomass has a promising scope to enhance efficacy of the phytoextraction process. A large number of hyperaccumulator plants specific for different metals have been experimentally discovered, and their extraordinary ability for phytoremediation has been proven. So far, many fern species within the *Pteridaceae* family have been reported to be As hyperaccumulators, some of these are listed in Table 2 (Zhao et al. 2009). *P. vittata* is an initially discovered and established As hyperaccumulator having an exceptional ability for As uptake and accumulate up to 2.3 % of its biomass (Ma et al. 2001; Zhao et al. 2009). In case of Cr, very few plants such as *Dicoma niccolifera*, *Sutera fodina,* and *Leersia hexandra* passed the criterion of being a Cr hyperaccumulator having >1000 mg kg^{-1} accumulation in their leaves (Reeves and Baker 2000; Zhang et al. 2007).

Table 2 Details of some As and Cr hyperaccumulator plants

Plant species	Metal concentration in frond (mg kg^{-1})	Treatment medium	Reference
Arsenic (As)			
Pteris vittata	4360	Soil	Ma et al. (2001)
P. biaurita	1770	Soil	Srivastava et al. (2005)
P. quadriaurita	2800	Soil	Srivastava et al. (2005)
P. ryukyuensis	3700	Soil	Srivastava et al. (2005)
P. cretica	2046	Soil	Ma et al. (2001)
Pityrogramma calomelanos	5390	Soil	Francesconi et al. (2001)
P. longifolia	2361	Soil	Meharg (2003)
P. umbrosa	5300	Soil	Zhao et al. (2002)
Chromium (Cr)			
Leersia hexandra	2978	Water	Zhang et al. (2007)
Dicoma niccolifera	1500	Water	Wild (1974)
Sutera fodina	2400	Water	Baker and Brooks (1989)

Identification of plant species suitable for phytoremediation is crucial for a successful deployment of phytoremediation technologies. The conventional approach for identification of plant species based on dose-dependent responses could be erroneous, which has been found to be true in some plants such as *H. lanatus* and other fern species from the genus *Pteris* which were considered to be As tolerant/accumulators (Meharg and Hartley-Whitaker 2002). Biochemical mechanisms in plants for scavenging stress-induced reactive molecules are efficient strategies evolved in those species that confer metal accumulation in the plant biomass with a high bioaccumulation factor. It is essential to be sure about such a capability of the plant by using molecular tools. Molecular markers using proteomics and genomics with a state-of-the-art knowledge base are potential tools to generate science-based evidence and overcome the bias in selection. The biochemical mechanisms discussed in this chapter could be used as molecular markers for identification of suitable candidate plants with hyperaccumulation ability for As and Cr phytoextraction.

5 Conclusions and Prospective

Metals are indestructible in the environment, and their increased concentrations act as stressors to plants at multiple levels. Remediation of As and Cr pollution through phytoextraction by hyperaccumulator plants is being increasingly investigated to offer a better alternative over other existing conventional *ex situ* methods. Plants respond differently under As and Cr stress due to differential endogenous mechanisms. Oxidative damages are caused by ROS, and excess amounts of ROS are harmful to many cellular components, including membrane lipids. The difference in antioxidant functions between metal tolerant and non tolerant plants can be further utilised for the selection of appropriate plants for phytoremediation. However, the plants encompassing systemic stress-response mechanisms are able to tolerate the oxidative stress generated by metals. For optimising and strengthening phytoremediation technologies, there should be a better understanding of fundamental biochemical and molecular processes regulating metal uptake and detoxification in the potential plant species. The inherent ability of a plant for metal accumulation is the primary requisite to make phytoextraction a practicable alternative to the existing conventional processes. Effective remediation of contaminated soil and water systems using a specific plant species is an immensely complex task whose success depends on a multitude of factors, first and foremost, the ability of the client plant to uptake, detoxify, translocate, and accumulate the contaminant in its biomass. The effectiveness of metal phytoremediation can be enhanced by two means, i.e., identification of plants with high metal hyperaccumulating potential and knowledge of factors to maximise metal accumulation. Antioxidant systems could be used as a defined set of markers to predict tolerance towards a particular type of stress. Enhancing antioxidant mechanisms of plants by manipulating biochemical mechanisms for more synthesis of antioxidant biomolecules and enzymes is a prospective research area to enhance As accumulation and simultaneously increase biomass generation in order to confer efficient phytoremediation.

References

Ahmad P, Sarwat M, Sharma S (2008) Reactive oxygen species, antioxidants and signaling in plants. J Plant Biol 51:167–173

Ahmad P, Jaleel CA, Salem MA, Nabi G, Sharma S (2010) Roles of enzymatic and nonenzymatic antioxidants in plants during abiotic stress. Crit Rev Biotechnol 30:161–175

Alloway BJ (1990) Soil processes and the behaviour of metals. In: Alloway BJ (ed) Heavy metals in soils. Chapman & Hall, London

Apel K, Hirt H (2004) Reactive oxygen species: metabolism, oxidative stress, and signal transduction. Annu Rev Plant Biol 55:373–399

Asada K (1992) Ascorbate peroxidase—a hydrogen peroxide-scavenging enzyme in plants. Physiol Plant 85:235–241

Ashraf M (2009) Biotechnological approach of improving plant salt tolerance using antioxidants as markers. Biotechnol Adv 27:84–93

Baker AJM (1981) Accumulators and excluders: strategies in the response of plants to trace metals. J Plant Nutr 3:643–654

Baker AJM, Brooks R (1989) Terrestrial higher plants which hyperaccumulate metallic elements. A review of their distribution, ecology and phytochemistry. Biorecovery 1:81–126

Balasundram N, Sundram K, Samman S (2006) Phenolic compounds in plants and agri-industrial by-products: antioxidant activity, occurrence, and potential uses. Food Chem 99:191–203

Barcelo J, Poschenriender C, Ruano A, Gunse B (1985) Leaf water potential in Cr(VI) treated bean plants (*Phaseolus vulgaris* L). Plant Physiol Suppl 177:163–164

Barrachina AC, Carbonell FB, Beneyto JM (1995) Arsenic uptake, distribution, and accumulation in tomato plants-effect of arsenite on plant growth and yield. J Plant Nutr 18:1237–1250

Bartolomeo AD, Poletti L, Sanchini G, Sebastiani B, Morozzi G (2004) Relationship among parameters of lake polluted sediments evaluated by multivariate statistical analysis. Chemosphere 55:1323–1329

Benner S (2010) Hydrology: anthropogenic arsenic. Nat Geosci 3:5–6

Blanvillain R, Kim JH, Wu S, Lima A, Ow DW (2008) Oxidative stress 3 is a chromatin-associated factor involved in tolerance to heavy metals and oxidative stress. Plant J 57:654–665

Bleeker PM, Hakvoort HW, Bliek M, Souer E, Schat H (2006) Enhanced arsenate reduction by a CDC25-like tyrosine phosphatase explains increased phytochelatin accumulation in arsenate tolerant *Holcus lanatus*. Plant J 45:917–929

Blokhina O, Virolainen E, Fagerstedt KV (2003) Antioxidants, oxidative damage and oxygen deprivation stress: a review. Ann Bot 91:174–194

Boonyapookana B, Upatham ES, Kruatrachue M, Pokethitiyook P, Singhakaew S (2002) Phytoaccumulation and phytotoxicity of cadmium and chromium in duckweed *Wolffia globosa*. Int J Phytorem 4:87–100

Bowler C, Montagu MV, Inze D (1992) Superoxide dismutase and stress tolerance. Annu Rev Plant Biol 43:83–116

Brenner A, Lazarova Z (2012) Membrane processes. In: Mustafa E, Barrott L (eds) Best practice guide on metals removal from drinking water by treatment. IWA, London

Buchet JP, Lison D (1998) Mortality by cancer in groups of the Belgian population with a moderately increased intake of arsenic. Int Arch Occup Environ Health 71:125–130

Cao XD, Ma LQ, Tu C (2004) Antioxidative responses to arsenic in the arsenic-hyperaccumulator Chinese brake fern (*Pteris vittata* L.). Environ Pollut 128:317–325

Chatterjee J, Chatterjee C (2000) Phytotoxicity of cobalt, chromium and copper in cauliflower. Environ Pollut 109:69–74

Chen NC, Kanazawa S, Horiguchi T, Chen NC (2001) Effect of chromium on some enzyme activities in the wheat rhizosphere. Soil Microorgan 55:3–10

Chen Y, Parvez F, Gamble M, Islam T, Ahmed A, Argos M, Ahsan H (2009) Arsenic exposure at low-to-moderate levels and skin lesions, arsenic metabolism, neurological functions, and biomarkers for respiratory and cardiovascular diseases: review of recent findings from the health effects of arsenic longitudinal study (HEALS) in Bangladesh. Toxicol Appl Pharmacol 239:184–192

Cheremisinoff NP (1998) Groundwater remediation and treatment technologies. Noyes, Westwood, NJ

Choong TS, Chuah TG, Robiah Y, Koay FG, Azni I (2007) Arsenic toxicity, health hazards and removal techniques from water: an overview. Desalination 217:139–166

Choudhury S, Panda SK (2005) Toxic effects, oxidative stress and ultrastructural changes in moss *Taxithelium nepalense* (Schwaegr) broth under chromium and lead phytotoxicity. Water Air Soil Pollut 167:73–90

Conesa HM, Evangelou MW, Robinson BH, Schulin R (2012) A critical view of current state of phytotechnologies to remediate soils: still a promising tool? Sci World J 2012: 168–214

Craciun AR, Courbot M, Bourgis F, Salis P, Saumitou-Laprade P, Verbruggen N (2006) Comparative cDNA–AFLP analysis of Cd tolerant and sensitive genotypes derived from crosses between the Cd hyperaccumulator *Arabidopsis halleri* and *Arabidopsis lyrata* ssp. petraea. J Exp Bot 57:2967–2983

Cunningham SD, David W (1996) Promises and prospects of phytoremediation. Plant Physiol 110:715–719

Da Costa TC, De Brito KCT, Rocha JAV, Leal KA, Rodrigues MLK, Minella JPG, Matsumoto ST, Vargas VMF (2012) Runoff of genotoxic compounds in river basin sediment under the influence of contaminated soils. Ecotoxicol Environ Saf 75:63–72

DalCorso G, Farinati S, Maistri S, Furini A (2008) How plants cope with cadmium: staking all on metabolism and gene expression. J Integr Plant Biol 10:1268–1280

Das M, Maiti SK (2008) Comparison between availability of heavy metals in dry and wetland tailing of an abandoned copper-tailing pond. Environ Monit Assess 137:343–350

De Vos CR, Vonk MJ, Vooijs R, Schat H (1992) Glutathione depletion due to copper-induced phytochelatin synthesis causes oxidative stress in *Silene cucubalus*. Plant Physiol 98:853–858

Dhal B, Thatoi HN, Das NN, Pandey BD (2013) Chemical and microbial remediation of hexavalent chromium from contaminated soil and mining/metallurgical solid waste: a review. J Hazard Mater 250:272–291

Dhankher OP, Rosen BP, McKinney EC, Meagher RB (2006) Hyperaccumulation of arsenic in the shoots of *Arabidopsis* silenced for arsenate reductase (ACR2). Proc Natl Acad Sci USA 103:5413–5418

Diwan H, Khan I, Ahmad A, Iqbal M (2010) Induction of phytochelatins and antioxidant defence system in *Brassica juncea* and *Vigna radiata* in response to chromium treatments. Plant Growth Regul 61:97–107

Dixon HB (1996) The biochemical action of arsonic acids especially as phosphate analogues. Adv Inorg Chem 44:191–227

Fan Y, Ovesen JL, Puga A (2012) Long-term exposure to hexavalent chromium inhibits expression of tumor suppressor genes in cultured cells and in mice. J Trace Elem Med Biol 26:188–191

Fayiga AO, Ma LQ, Cao X, Rathinasabapathi B (2004) Effects of heavy metals on growth and arsenic accumulation in the arsenic hyperaccumulator *Pteris vittata* L. Environ Pollut 132:289–296

Foyer CH, Noctor G (2005) Oxidant and antioxidant signalling in plants: a re-evaluation of the concept of oxidative stress in a physiological context. Plant Cell Environ 28:1056–1071

Francesconi K, Visoottiviseth P, Sridokchan W, Goessler W (2001) Arsenic species in an arsenic hyperaccumulating fern *Pityrogramma calomelanos*: a potential phytoremediator of arsenic-contaminated soils. Sci Total Environ 284:27–35

Fu F, Wang Q (2011) Removal of heavy metal ions from wastewaters: a review. J Environ Manag 92:407–418

Gechev TS, Van Breusegem F, Stone JM, Denev I, Laloi C (2006) Reactive oxygen species as signals that modulate plant stress responses and programmed cell death. Bioessays 28:1091–1101

Ghosh M, Singh SP (2005) A review on phytoremediation of heavy metals and utilization of its byproducts. Appl Ecol Environ Res 3:1–18

Giacomino A, Malandrino M, Abollino O, Velayutham M, Chinnathangavel T, Mentasti E (2010) An approach for arsenic in a contaminated soil: speciation, fractionation, extraction and effluent decontamination. Environ Pollut 158:416–423

Gill SS, Tuteja N (2010) Reactive oxygen species and antioxidant machinery in abiotic stress tolerance in crop plants. Plant Physiol Biochem 48:909–930

Glass DJ (1999) United States and international markets for phytoremediation, Needham, Mass.: D. Glass Associates Inc. hyperaccumulating plant. Sci Total Environ 300:167–177

Gong H, Zhu X, Chen K, Wang S, Zhang C (2005) Silicon alleviates oxidative damage of wheat plants in pots under drought. Plant Sci 169:313–321

Gratao PL, Polle A, Lea PJ, Azevedo RA (2005) Making the life of heavy metal-stressed plants a little easier. Funct Plant Biol 32:481–494

Gunes A, Pilbeam DJ, Inal A (2009) Effect of arsenic–phosphorus interaction on arsenic-induced oxidative stress in chickpea plants. Plant Soil 314:211–220

Gupta DK, Inouhe M, Rodríguez-Serrano M, Romero-Puertas MC, Sandalio LM (2013) Oxidative stress and arsenic toxicity: role of NADPH oxidases. Chemosphere 90:1987–1996

Gusman GS, Oliveira JA, Farnese FS, Cambraia J (2013) Mineral nutrition and enzymatic adaptation induced by arsenate and arsenite exposure in lettuce plants. Plant Physiol Biochem 71:307–314

Halliwell B, Gutteridge JMC (1999) Free radicals in biology and medicine. Oxford University Press, New York

Han FX, Banin A, Su Y, Monts DL, Plodinee MJ, Kingery WL, Triplet GE (2002) Industrial age anthropogenic inputs of heavy metals into the pedosphere. Naturwissenschaften 89:497–504

Hanikenne M, Talke IN, Haydon MJ, Lanz C, Nolte A, Motte P, Kroymann J, Weigel D, Kramer U (2008) Evolution of metal hyperaccumulation required cis-regulatory changes and triplication of HMA4. Nature 453:391–395

Hartley-Whitaker J, Ainsworth G, Meharg A (2001) Copper and- arsenic induced oxidative stress in *Holcus lanatus* L. cloned with differential sensitivity. Plant Cell Environ 24:713–722

Hayat S, Khalique G, Irfan M, Wani AS, Tripathi BN, Ahmad A (2012) Physiological changes induced by chromium stress in plants: an overview. Protoplasma 249:599–611

Heyno E, Klose C, Krieger-Liszkay A (2008) Origin of cadmium-induced reactive oxygen species production: mitochondrial electron transfer versus plasma membrane NADPH oxidase. New Phytol 179:687–699

Hughes MF (2002) Arsenic toxicity and potential mechanisms of action. Toxicol Lett 133:1–16

Jaleel CA, Gopi R, Manivannan P, Gomathinayagam M, Riadh K, Inès J, Chang-Xing Z, Hong-Bo S, Panneerselvam R (2009) Antioxidant defense responses: physiological plasticity in higher plants under abiotic constraints. Acta Physiol Plant 31:427–436

Jin YH, Clark AB, Slebos RJ, Al-Refai H, Taylor JA, Kunkel TA, Resnick MA, Gordenin DA (2003) Cadmium is a mutagen that acts by inhibiting mismatch repair. Nat Genet 34:326–329

Jonak C, Nakagami H, Hirt H (2004) Heavy metal stress. Activation of distinct mitogen-activated protein kinase pathways by copper and cadmium. Plant Physiol 136:3276–3283

Kertulis GM, Ma LQ, MacDonald GE, Chen R, Winefordner JD, Cai Y (2005) Arsenic speciation and transport in *Pteris vittata* L. and the effects on phosphorus in the xylem sap. Environ Exp Bot 54:239–247

Kertulis-Tartar GM, Rathinasabapathi B, Ma LQ (2009) Characterization of glutathione reductase and catalase in the fronds of two Pteris ferns upon arsenic exposure. Plant Physiol Biochem 47:960–965

Kim KW, Bang S, Zhu Y, Meharg AA, Bhattacharya P (2009) Arsenic geochemistry, transport mechanism in the soil–plant system, human and animal health issues. Environ Int 35:453–454

Kramer U (2010) Metal hyperaccumulation in plants. Annu Rev Plant Biol 61:517–534

Landajo A, Arana G, de Diego A, Etxebarria N, Zuloaga O, Amouroux D (2004) Analysis of heavy metal distribution in superficial estuarine sediments (estuary of Bilbao, Basque Country) by open-focused microwave-assisted extraction and ICP-OES. Chemosphere 56:1033–1041

Li YM, Chaney R, Brewer E, Roseberg R, Angle JS, Baker A, Reeves R, Nelkin J (2003) Development of a technology for commercial phytoextraction of nickel: economic and technical considerations. Plant Soil 249:107–115

Liu X, Zhang S, Shan X, Zhu YG (2005) Toxicity of arsenate and arsenite on germination, seedling growth and amylolytic activity of wheat. Chemosphere 61:293–301

Lokeshwari H, Chandrappa GT (2006) Impact of heavy metal contamination of Bellandur Lake on soil and cultivated vegetation. Curr Sci 91:622–627

Lovdal T, Olsen KM, Slimestad R, Verheul M, Lillo C (2010) Synergetic effects of nitrogen depletion, temperature, and light on the content of phenolic compounds and gene expression in leaves of tomato. Phytochemistry 71:605–613

Ma LQ, Komar KM, Tu C, Zhang W, Cai Y, Kennelley ED (2001) A fern that hyperaccumulates arsenic: a hardy, versatile, fast-growing plant helps to remove arsenic from contaminated soils. Nature 409:579

Ma JF, Yamaji N, Mitani N, Xu XY, Su YH, McGrath SP, Zhao FJ (2008) Transporters of arsenite in rice and their role in arsenic accumulation in rice grain. Proc Natl Acad Sci USA 105:9931–9935

Macek T, Francova K, Kochankova L, Lovecka P, Ryslava E, Rezek J, Mackova M (2011) Phytoremediation-biological cleaning of a polluted environment. Rev Environ Health 19:63–82

Mandal BK, Suzuki KT (2002) Arsenic round the world: a review. Talanta 58:201–235

McKersie BD, Bowley SR, Jones KS (1999) Winter survival of transgenic alfalfa overexpressing superoxide dismutase. Plant Physiol 119:839–848

Meharg AA (2003) Variation in arsenic accumulation: hyperaccumulation in ferns and their allies. New Phytol 157:25–31

Meharg AA, Hartley-Whitaker J (2002) Arsenic uptake and metabolism in arsenic resistant and non resistant plant species. New Phytol 154:29–43

Miller G, Vladimir S, Mittler R (2008) Reactive oxygen signaling and abiotic stress. Physiol Plant 133:481–489

Mishra S, Srivastava S, Tripathi RD, Trivedi PK (2008) Thiol metabolism and antioxidant systems complement each other during arsenate detoxification in *Ceratophyllum demersum* L. Aquat Toxicol 86:205–215

Mishra S, Tripathi RD, Srivastava S, Dwivedi S, Trivedi PK, Dhankher OP, Khare A (2009) Thiol metabolism play significant role during cadmium detoxification by *Ceratophyllum demersum* L. Bioresour Technol 100:2155–2161

Mittler R, Vanderauwera S, Gollery M, Van Breusegem F (2004) Reactive oxygen gene network of plants. Trends Plant Sci 9:490–498

Mohanty K, Jha M, Meikap BC, Biswas MN (2006) Biosorption of Cr (VI) from aqueous solutions by *Eichhornia crassipes*. Chem Eng J 117:71–77

Mueller MJ (2004) Archetype signals in plants: the phytoprostanes. Curr Opin Plant Biol 7:441–448

Mukherjee A, Bhattacharya P, Savage K, Foster A, Bundschuh J (2008) Distribution of geogenic arsenic in hydrologic systems: controls and challenges. J Contam Hydrol 99:1–7

Mwegoha JSW (2008) The use of phytoremediation technology for abatement soil and groundwater pollution in Tanzania: opportunities and challenges. JSDA 10:140–156

Mylona PV, Polidoros AN, Scandalios JG (1998) Modulation of antioxidant responses by arsenic in maize. Free Radic Biol Med 25:576–585

Nedelkoska TV, Doran PM (2000) Characteristics of heavy metal uptake by plant species with potential for phytoremediation and phytomining. Miner Eng 13:549–561

Noctor G, Foyer CH (1998) Ascorbate and glutathione: keeping active oxygen under control. Annu Rev Plant Biol 49:249–279

Norton GJ, Lou-Hing DE, Meharg AA, Price AH (2008) Rice–arsenate interactions in hydroponics: whole genome transcriptional analysis. J Exp Bot 59:2267–2276

Ogbonna DN, Kii BL, Youdeowei PO (2009) Some physico-chemical and heavy metal levels in soils of waste dumpsites in Port Harcourt municipality and environs. J Appl Sci Environ Manag 13:65–70

Onken BM, Hossner LR (1995) Plant uptake and determination of arsenic species in soil solution under flooded conditions. J Environ Qual 24:373–381

Padmavathiamma PK, Li LY (2007) Phytoremediation technology: hyper-accumulation metals in plants. Water Air Soil Pollut 184:105–126

Palmer HJ, Paulson KE (1997) Reactive oxygen species and antioxidants in signal transduction and gene expression. Nutr Rev 55:353–361

Panda SK (2007) Chromium-mediated oxidative stress and ultrastructural changes in root cells of developing rice seedlings. J Plant Physiol 164:1419–1428

Panda SK, Choudhury S (2005) Chromium stress in plants. Braz J Plant Physiol 17:95–102

Panda SK, Patra HK (2000) Nitrate and ammonium ions effect on the chromium toxicity in developing wheat seedlings. Pro Natl Acad Sci India B 70:75–80

Parr PD, Taylor FG Jr (1982) Germination and growth effects of hexavalent chromium in Orocol TL (a corrosion inhibitor) on *Phaseolus vulgaris*. Environ Int 7:197–202

Paspaliaris I, Papassiopi N, Xenidis A, Hung YT (2010) Soil remediation. In: Wang LK, Hung YT, Shammas NK (eds) Handbook of advanced industrial and hazardous wastes treatment. CRC, Boca Raton, FL

Pichai NMR, Samjamjaras R, Thammanoon H (2001) The wonders of a grass, Vetiver and its multifold applications. Asian Infrastruct Res Rev 3:1–4

Pokhrel LR, Dubey B (2013) Global scenarios of metal mining, environmental repercussions, public policies, and sustainability: a review. Crit Rev Environ Sci Technol 43:2352–2388

Potters G, De Gara L, Asard H, Horemans N (2002) Ascorbate and glutathione: guardians of the cell cycle, partners in crime? Plant Physiol Biochem 40:537–548

Purakayastha TJ, Chhonkar PK (2010) Phytoremediation of heavy metal contaminated soils. In: Sherameti I, Varma A (eds) Soil heavy metals. Springer, Berlin

Raymond AW, Felix EO (2011) Heavy metals in contaminated soils: a review of sources, chemistry, risks and best available strategies for remediation. ISRN Ecol 2011:402647

Reeves RD, Baker AJM (2000) Metal-accumulating plants. In: Raskin I, Ensley BD (eds) Phytoremediation of toxic metals: using plants to clean up the environment. Wiley, New York

Reinheckel T, Noack H, Lorenz S, Wiswedel I, Augustin W (1998) Comparison of protein oxidation and aldehyde formation during oxidative stress in isolated mitochondria. Free Radic Res 29:297–305

Requejo R, Tena M (2005) Proteome analysis of maize roots reveals that oxidative stress is a main contributing factor to plant arsenic toxicity. Phytochemistry 66:1519–1528

Rice-Evans C, Miller N, Paganga G (1997) Antioxidant properties of phenolic compounds. Trends Plant Sci 2:152–159

Ross SM (1994) Toxic metals in soil and plant systems. Wiley, Chichester, UK

Rout GR, Sanghamitra S, Das P (2000) Effects of chromium and nickel on germination and growth in tolerant and non-tolerant populations of *Echinochloa colona* (L). Chemosphere 40:855–859

Sánchez-Viveros G, Ferrera-Cerrato R, Alarcon A (2011) Short-term effects of arsenate-induced toxicity on growth, chlorophyll and carotenoid contents, and total content of phenolic compounds of *Azolla filiculoides*. Water Air Soil Pollut 217:455–462

Sarangi BK, Kalve SK, Pandey RA, Chakrabarti T (2009) Transgenic plants for phytoremediation of arsenic and chromium to enhance tolerance and hyperaccumulation. Transgen Plant J 3:57–86

Sarma H (2011) Metal hyperaccumulation in plants: a review focusing on phytoremediation technology. J Environ Sci Technol 4:118–138

Schmfger MEV (2001) Phytochelatins: complexation of metals and metalloids, studies on the phytochelatin synthase. Ph.D thesis, Munich University of Technology (TUM), Munich, Germany

Schmoger ME, Oven M, Grill E (2000) Detoxification of arsenic by phytochelatins in plants. Plant Physiol 122:793–801

Schutzendubel A, Polle A (2002) Plant response to abiotic stresses: heavy metal induced oxidative stress and protection by mycorrhization. J Exp Bot 53:1351–1365

Sekmen A, Turkan I, Takio S (2007) Differential responses of antioxidative enzymes and lipid peroxidation to salt stress in salt-tolerant *Plantago maritima* and salt-sensitive *Plantago media*. Physiol Plant 131:399–411

Shanker AK (2003) Physiological, biochemical and molecular aspects of chromium toxicity and tolerance in selected crops and tree species. Ph.D thesis, Tamil Nadu Agricultural University, Coimbatore, India

Shanker AK, Djanaguiraman M, Sudhagar R, Chandrashekar CN, Pathmanabhan G (2004) Differential antioxidative response of ascorbate glutathione pathway enzymes and metabolites to chromium speciation stress in green gram (*Vigna radiata* (L) R Wilczek, cv CO 4) roots. Plant Sci 166:1035–1043

Shanker AK, Cervantes C, Loza-Tavera H, Avudainayagam S (2005) Chromium toxicity in plants. Environ Int 31:739–753
Sharma I (2012) Arsenic induced oxidative stress in plants. Biologia 67:447–453
Sharma SS, Dietz KJ (2008) The relationship between metal toxicity and cellular redox imbalance. Trends Plant Sci 14:43–50
Sharma I, Singh R, Tripathi BN (2007) Biochemistry of arsenic toxicity and tolerance in plants. Biochem Cell Arch 7:165–170
Sheoran V, Sheoran AS, Poonia P (2011) Role of hyperaccumulators in phytoextraction of metals from contaminated mining sites: a review. Crit Rev Environ Sci Technol 41:168–214
Shri M, Kumar S, Chakrabarty D, Trivedi PK, Mallick S, Misra P, Shukla D, Mishra S, Srivastava S, Tripathi RD, Tuli R (2009) Effect of arsenic on growth, oxidative stress, and antioxidant system in rice seedlings. Ecotoxicol Environ Saf 72:1102–1110
Singh N, Ma LQ, Srivastava M, Rathinasabapathi B (2006) Metabolic adaptations to arsenic-induced oxidative stress in *Pteris vittata* L and *Pteris ensiformis* L. Plant Sci 170:274–282
Singh N, Raj A, Khare PB, Tripathi RD, Jamil S (2010) Arsenic accumulation pattern in 12 Indian ferns and assessing the potential of *Adiantum capillus*-veneris, in comparison to *Pteris vittata*, as arsenic hyperaccumulator. Bioresour Technol 101:8960–8968
Singh SK, Juwarkar AA, Kumar S, Meshram S, Fan M (2007) Effect of amendment on phytoextraction of arsenic by *Vetiveria zizanioides* from soil. Int J Environ Sci Technol 4:339–344
Skeffington RA, Shewry PR, Petersen PJ (1976) Chromium uptake and transport in barley seedlings *Hordeum vulgare*. Planta 132:209–214
Smirnoff N (2000) Ascorbate biosynthesis and function in photoprotection. Philos Trans R Soc Lond B Biol Sci 355:1455–1464
Smith E, Naidu R, Alston AM (1998) Arsenic in the soil environment: a review. Adv Agron 64:149–195
Srivastava M, Ma LQ, Singh N, Singh S (2005) Antioxidant responses of hyperaccumulator and sensitive fern species to arsenic. J Exp Bot 56:1335–1342
Sujatha P, Gupta A, Gupta A (1996) Tannery effluent characteristics and its effects on agriculture. J Ecotoxicol Environ Monit 6:45–48
Sun RL, Zhou QX, Sun FH, Jin CX (2007) Antioxidative defense and proline/phytochelatin accumulation in a newly discovered Cd-hyperaccumulator, *Solanum nigrum* L. Environ Exp Bot 60:468–476
Suzuki N, Mittler R (2006) Reactive oxygen species and temperature stresses: a delicate balance between signaling and destruction. Physiol Planta 126:45–51
Terry N, Banuelos GS (2000) Phytoremediation of contaminated soil and water. Lewis, Boca Raton, FL
USEPA (1992) Test methods for evaluating solid waste, physical/chemical methods. USEPA, SW-846, 3rd edn. United States Environmental Protection Agency
USEPA (2000) Introduction to phytoremediation. EPA/600/R-99/107. Washington, DC
USEPA (2001) Remediation technology cost compendium-year 2000, EPA-542-R-01 009, US Environmental Protection Agency, Office of Solid Waste and Emergency Response Technology Innovation Office, Washington, DC. Available at http://www.epa.gov
Vajpayee P, Tripathi RD, Rai UN, Ali MB, Singh SN (2000) Chromium (VI) accumulation reduces chlorophyll biosynthesis, nitrate reductase activity and protein content in *Nymphaea alba* L. Chemosphere 41:1075–1082
Van de Mortel JE, Schat H, Moerland PD, Loren V, van Themaat E, Van Der Ent S, Blankestijn H, Ghandilyan A, Tsiatsiani S, Aarts MG (2008) Expression differences for genes involved in lignin, glutathione and sulphate metabolism in response to cadmium in *Arabidopsis thaliana* and the related Zn/Cd hyperaccumulator *Thlaspi caerulescens*. Plant Cell Environ 31:301–324
Verbruggen N, Hermans C, Schat H (2009) Molecular mechanisms of metal hyperaccumulation in plants. New Phytol 181:759–776

Vital SA, Fowler RW, Virgen A, Gossett DR, Banks SW, Rodriguez J (2008) Opposing roles for superoxide and nitric oxide in the NaCl stress-induced upregulation of antioxidant enzyme activity in cotton callus tissue. Environ Exp Bot 62:60–68

Wang X, Ma LQ, Rathinasabapathi B, Liu YG, Zeng GM (2010) Uptake and translocation of arsenite and arsenate by *Pteris vittata* L.: effects of silicon, boron and mercury. Environ Exp Bot 68:222–229

Weber M, Trampczynska A, Clemens S (2006) Comparative transcriptome analysis of toxic metal responses in Arabidopsis thaliana and the Cd(2+)-hypertolerant facultative metallophyte *Arabidopsis halleri*. Plant Cell Environ 29:950–963

Wild H (1974) Arsenic tolerant plant species established on arsenical mine dumps in Rhodesia. Kirkia 9:265–278

Witters N, Mendelsohn RO, Van Slycken S, Weyens N, Schreurs E, Meers E, Tack F, Carleer R, Vangronsveld J (2012) Phytoremediation, a sustainable remediation technology? Conclusions from a case study in: energy production and carbon dioxide abatement. Biomass Bioenergy 39:454–469

Xia X, Chen X, Liu R, Liu H (2011) Heavy metals in urban soils with various types of land use in Beijing, China. J Hazard Mater 186:2043–2050

Yadav SK, Dhote M, Kumar P, Sharma J, Chakrabarti T, Juwarkar AA (2010) Differential antioxidative enzyme responses of *Jatropha curcas* L. to chromium stress. J Hazard Mater 180:609–615

Yamauchi Y, Furutera A, Seki K, Toyoda Y, Tanaka K, Sugimoto Y (2008) Malondialdehyde generated from peroxidized linolenic acid causes protein modification in heat-stressed plants. Plant Physiol Biochem 46:786–793

Yannarelli GG, Fernandez-Alvarez AJ, Santa-Cruz DM, Tomaro ML (2007) Glutathione reductase activity and isoforms in leaves and roots of wheat plants subjected to cadmium stress. Phytochemistry 68:505–512

Zhang W, Cai Y, Tu C, Ma LQ (2002) Arsenic speciation and distribution in an arsenic hyperaccumulating plant. Sci Total Environ 300:167–177

Zhang XH, Jie L, Huang HT, Chen J, Zhu YN, Wang DQ (2007) Chromium accumulation by the hyperaccumulator plant *Leersia hexandra* Swartz. Chemosphere 67:1138–1143

Zhao FJ, Dunham SJ, McGrath SP (2002) Arsenic hyperaccumulation by different fern species. New Phytol 156:27–31

Zhao FJ, Ma JF, Meharg AA, McGrath SP (2009) Arsenic uptake and metabolism in plants. New Phytol 181:777–794

Zhao FJ, McGrath SP (2009) Biofortification and phytoremediation. Curr Opin Plant Biol 12:373–380

Zhao FJ, McGrath SP, Meharg AA (2010) Arsenic as a food chain contaminant: mechanisms of plant uptake and metabolism and mitigation strategies. Annu Rev Plant Biol 6:535–559

Zhou Q (2001) The measurement of malondialdehyde in plants. In: Zhou Q (ed) Methods in plant physiology. China Agricultural Press, Beijing

Phytochelatin and Oxidative Stress Under Heavy Metal Stress Tolerance in Plants

Weitao Liu, Xue Zhang, Lichen Liang, Chen Chen, Shuhe Wei, and Qixing Zhou

Contents

Abstract With the rapid developing of industry and agriculture, heavy metal pollution in environment has been both serious and widespread worldwide. To cope with adverse environmental heavy metal toxicity, plants have evolved a variety of adaptive responses, which include immobilization, exclusion, chelation, and compartmentalization of metal ions and often involve metal-binding ligands. Particularly, phytochelatins (PCs), a family of peptides, have been regarded as the

W. Liu • X. Zhang • L. Liang • C. Chen • Q. Zhou
Key Laboratory of Pollution Processes and Environmental Criteria, Ministry of Education/ Tianjin Key Laboratory of Urban Ecology Environmental Remediation and Pollution Control, College of Environmental Science and Engineering, Nankai University, 94 Weijin Road, Tianjin 300071, China
e-mail: lwt@nankai.edu.cn

S. Wei (✉)
Key Laboratory of Pollution Ecology and Environmental Engineering, Institute of Applied Ecology, Chinese Academy of Sciences, Wenhua Road 72, Shenyang, Liaoning, Shenhe 110016, China
e-mail: Shuhewei@iae.ac.cn

D.K. Gupta et al. (eds.), *Reactive Oxygen Species and Oxidative Damage in Plants Under Stress*, DOI 10.1007/978-3-319-20421-5_8

best-characterized heavy metal chelators especially in detoxication of heavy metals such as cadmium (Cd) in plants and some microorganisms. Generally, PCs have the general structure (γ-Glu-Cys)$_n$-Gly ($n = 2$–11) and are produced by the enzyme phytochelatin synthases, which can bind to various metals including Cd, As Cu, or Zn. In this chapter, we focused on the biosynthesis and function of PCs and the role of PCs in metal detoxification and tolerance. Finally, the molecular biology of PCs has been briefly reviewed.

Keywords Heavy metals • Plants • Oxidative damage • Tolerance • Phytochelatins

1 Introduction

The term heavy metal generally refers to a series of metals and metalloids with atomic mass over 20 and specific gravity above 5, which can be toxic to both plants and animals even at much low concentrations (Liu et al. 2013; Rascio and Navari-Izzo 2011). Heavy metal contamination in soils has dramatically increased during the last century due to mining, smelting, manufacturing, agricultural soils amended with municipal sewage sludge, and waste disposal practices (Liu et al. 2015). Unlike organic pollutants, heavy metals could not be degraded naturally or by microbial communities. Therefore, heavy metals could persist in soils for a long period of time after its introduction (Zhou and Song 2004).

Many heavy metals are essential for plant physiological processes e.g., copper (Cu), zinc (Zn), manganese (Mn), and nickel (Ni), while other metals especially cadmium (Cd), lead (Pb), and mercury (Hg) are nonessential (Pal and Rai 2010). Both excessive essential metals and a small quantity of nonessential metals could be toxic to plant species. Heavy metal toxicity to plant is mainly induced oxidative stress. Reactive oxygen species (ROS), e.g., $^{\bullet}O_2^{-}$, H_2O_2, and $^{\bullet}OH$, was the primary response of plants exposure to high levels of heavy metals (Mithöfer et al. 2004). Malondialdehyde (MDA), one of the decomposition products of polyunsaturated fatty acids of membrane, is regarded as a reliable indicator of oxidative stress (Demiral and Türkan 2005).

Plants have developed a variety of mechanisms to prevent excessive accumulation of nonessential metals within cells and/or transform these metals into less toxic forms (Cobbett 2000b). Some plants produce metabolites that bind to heavy metals in the cytosol, such as glutathione (GSH), polypeptides, and proteins (e.g., metallothioneins (MTs) and phytochelatins (PCs)) (Zenk 1996). Particularly, the principal mechanism of intracellular detoxification of heavy metals in plants is complexation by phytochelatins (PCs). PCs are a family of metal-binding peptides with the general structure (γ-Glu-Cys)$_n$-Gly ($n = 2$–11), which have been regarded as the best-characterized heavy metal chelators in plants, especially in the context of Cd tolerance (Cobbett and Goldsbrough 2002; Cobbett 2000a). The enzyme catalyzing the biosynthesis of PCs from GSH, phytochelatin synthase, was first characterized by Grill et al. (Grill et al. 1989). After that, the presence of phytochelatins has been reported in a wide variety of plant species, including algae, mosses, ferns, gymnosperms, monocots, dicots, and some fungi species (Wójcik and Tukiendorf 2004). PCs are synthesized in the cytosol and then transported as complexes to the vacuole. Moreover, the synthesis of PCs is rapidly activated in the presence of heavy metals such as Cd, Cu, Zn, Ag, Au, Hg, and Pb (Manara 2012).

During the last decades, multiple excellent research articles and reviews have been published describing our knowledge on PCs (Pawlik-Skowrońska et al. 2002; Clemens 2006; Clemens and Peršoh 2009; Cobbett 2000b; Hirata et al. 2005; Lindberg et al. 2007; Liu et al. 2011a; Machado-Estrada et al. 2013; Pal and Rai 2010; Rauser 2003; Rea 2012; Stolt et al. 2003; Zhang et al. 2010, 2012; Schulz et al. 2008). This chapter reviews the recent advances in understanding of the biosynthesis and functions of PCs, involving metal detoxification especially in plants to provide updated information about these unique peptides applicable to heavy metal detoxification in plant. Furthermore, the genetic manipulation of plants with various genes involving directly or indirectly in phytochelatins metabolism and their role in heavy metal stress tolerance have also been discussed.

2 Metal Toxicity in Plants

Heavy metals contamination of soils has become a critical environmental issue (Wei et al. 2005, 2008a, b). Metal concentrations in soil typically range from less than one to as high as 100,000 mg kg^{-1} (Pal and Rai 2010). Many heavy metals are essential micronutrients for plants (i.e., Co, Cu, Fe, Mn, Mo, Ni, and Zn) because they are involved in numerous metabolic processes as constituents of enzymes and other proteins (Hirata et al. 2005). However, they can become toxic if their concentration is higher than a specific critical point, as they can lead to a range of interactions at the cellular and molecular level (Baldisserotto et al. 2007; Prasad and de Oliveira Freitas 2003). Soil and plant characteristics, e.g., mineralogical and organic matter properties of the soil and plant metal susceptibility and prediction of heavy metal uptake by plants by the common soil, have influence on bioavailability of heavy metals in soils (Liu et al. 2010a; Wei et al. 2011). Therefore, heavy metal toxicity in plants usually varied, and plant chemical analysis techniques are often unreliable (Zhou and Song 2004).

Soils contain high levels of Cd because of natural source and human activity (Liu et al. 2010a, 2011b). However, Cd is nonessential to plant nutrition, and high levels of Cd can cause great damage to the growth and development of plant (Manara 2012). Growth and seed germination inhibition, chlorosis, leaf roll, dwarfing, phenological phase delay, and biological production reduction that finally even lead to plant death are obvious visible symptoms (Haghiri 1974; Mohanpuria et al. 2007; Wójcik and Tukiendorf 2004). Cd can lead to Fe(III) reductase inhibition of root, reduce the generation of Fe(II), and cause the lack of Fe(II), which finally affect the photosynthesis of plants (Alcántara et al. 1994). Photosynthesis is one of the most important factors in plant growth and the lower of photosynthesis can lead to reduction of plant production directly. Previous studies have highlighted that, even at low level, soil Cd contamination could pose a significant risk to human health through the soil–crop–human exposure pathway (Liu et al. 2010a, 2011b).

Lead (Pb) is one of the major heavy metals of soils and has gained considerable importance as a potent environmental pollutant (Sharma and Dubey 2005). Pb contamination in soil has been serious and is not likely to decrease in the near future due to mining and smelting activities, Pb-containing paints, gasoline, and the disposal of municipal sewage sludge enriched with Pb (Yang et al. 2000). Moreover, Pb is easily taken up by plants from soils and accumulated in different organs (Liu et al. 2010b). When Pb uptake in plants exceeds a certain value, it will produce certain toxicity to plants, disorder the metabolism of plants, inhibit growth of plants, and even cause plant death (Pourrut et al. 2011; Sharma and Dubey 2005). The visual symptoms of Pb toxicity are rapid inhibition of root growth, stunted growth of plant, and chlorosis (Burton et al. 1984; Pourrut et al. 2011). Pb toxicity also causes inhibition of enzyme activities, water imbalance, alterations in membrane permeability, and disturbs mineral nutrition (Sharma and Dubey 2005). Plants under heavy metal stress can induce the generation of reactive oxygen and show the effect of plant poisoning (Liu et al. 2010b). Lead stress caused significant changes in the activity of antioxidative enzymes. Plants exposed to Pb can increase malondialdehyde (MDA) content coupled with the increase in the activities of superoxide dismutase (SOD), catalase (CAT), peroxidase (POD), glutathione reductase (GR), and glutathione S-transferase (GST) compared to control (untreated) plants (Reddy et al. 2005).

Copper (Cu) is an essential micronutrient to plants growth and development, which widely participates in various life activities (Thomas et al. 1998). Cu plays an important role in the synthesis of cytochrome oxidase, ascorbic acid oxidase, polyphenol oxidase, and polyamine oxidase. However, excessive Cu causes injury to plants, including plant growth retardation and leaf chlorosis (Lewis et al. 2001). The dynamic balance of copper in cellular is controlled mainly from three aspects: (1) absorption, (2) intracellular storage, and (3) exosmosis of copper (Nelson 1999). Phytotoxic effect of Cu on plant is higher than most of heavy metal contaminants (Xu et al. 2006). It was reported that toxic effect of Cu and other heavy metals on wheat growth was in the following order: Cd > Cu > Ni > Zn > Pb > Cr (Athar and Ahmad 2002). Most of the excess Cu in soil was transferred to and accumulated in plant leaves in which the storage rate in vacuoles and chloroplasts was 48 % and 7 %, respectively (Boojar and Tavakoli 2010). In response to copper exposure, plants showed significant induction of proteins and enzymes like superoxide dismutase (SOD), ascorbate peroxidase (APX), guaiacol peroxidase (GPX), catalase (CAT) and glutathione reductase (GR), however, only up to moderate exposures (Srivastava et al. 2006). Stress due to copper toxicity resulted in an increase in total catalase (CAT) and superoxide dismutase (SOD) activity and a simultaneous induction of SOD and CAT gene expression (Lombardi and Sebastiani 2005). Moreover, Cu toxicity removed the correlation between sulphur metabolism-related gene expression and the suggested regulatory metabolites (Shahbaz et al. 2014).

Arsenic (As) has been known as a poison for years (Sun et al. 2009). Arsenic is used in several industries like paints, dyes, metals, soaps, insecticides, and semiconductors and is also released into the environment through burning fossil fuels, paper production, cement manufacturing, and mining activities (Singh et al. 2015). Arsenic exists mainly in four oxidation states—arsenate (As^{V}), arsenite (As^{III}), arsenic(As^{0}), and arsine (As^{-III}), but the two biologically important species are As^{V} and As^{III}, which are interconvertible depending on the redox status of the environment (Elangovan and Chalakh 2006; Singh et al. 2015). Plants take up As mainly as As^{V}, which causes considerable stress in plants, including inhibition of growth, physiological disorders, and finally death (Garg and Singla 2011). Plants with As toxicity have obvious symptoms, such as leaf roll, chlorosis, evidently inhibition of root growth, and root and leaf tissue destruction (Rascio and Navari-Izzo 2011). Plants exposed to As substantially increase the synthesis of glutathione (GSH) and phytochelatins (PCs), the polymers of GSH (Tripathi et al. 2007).

Natural source, mining and smelting activities, sewage water irrigation, and fertilizers and pesticides cause mercury (Hg) contamination in soils (Xu et al. 2015). Hg is not an essential nutrition element to plant growth, which can cause plant damage. Hg has various existence forms, such as HgS, Hg^{2+}, Hg^{0}, and methyl-Hg (Yadav 2010). Many studies have shown that Hg is preferentially accumulated in plant roots (Beauford et al. 1977; Chen et al. 2009b; Iglesia-Turino et al. 2006). Cellular integrity and biological activity might be compromised due to its strong affinity for sulfhydryl residues of proteins and other biomolecules (Hall 2002). Mercury has also been found to be a potent inducer of oxidative stress (Rellan-Alvarez et al. 2006).

Manganese (Mn) is an essential trace element for plant tissues, but it can cause phytotoxicity at supraoptimal level (Lei et al. 2007). High concentrations of Mn in soil are a widespread environmental problem in the world (Yang et al. 2013). Natural processes and anthropogenic activities, e.g., acid rain, acidic fertilizer, mines, and industrial wastes, cause serious soil contamination of Mn (Yang et al. 2013). Brown spots on mature leaves, interveinal chlorosis and necrosis, deformation of young leaves, and growth retardation are obvious symptoms of Mn toxicity (Baldisserotto et al. 2007; Khabaz-Saberi et al. 2010).

Nickel (Ni) is an essential micronutrient for higher plants (Brown et al. 1987). However, Ni may be toxic to plants at high levels of contamination (Bingham et al. 1986). Nickel phytotoxicity varies with concentration of Ni in soil solution as well as with the plant species (Kramer et al. 1997; Sreekanth et al. 2013). Excess Ni can affect physiological/biochemical process like decreasing leaf chlorophyll contents and leaf photosynthetic and transpiration activities and impairing membrane permeability associated with enhanced extracellular peroxidase activity (Yang et al. 1996). Toxic levels of Ni may depress yield and disturb the pattern of nutrient uptake, resulting in reduced uptake of some nutrient elements and increased supply of the others (Chen et al. 2009a; Kramer et al. 1997).

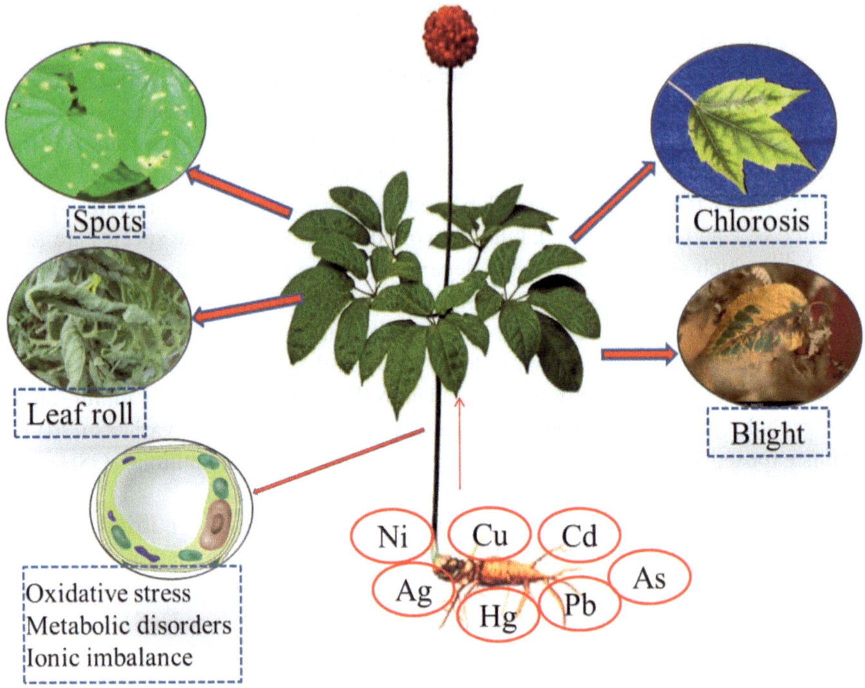

Fig. 1 Metal toxicity to plants

Further, heavy metals also can induce oxidative stress. Plants respond to oxidative stress by increasing the production of antioxidant enzymes such as superoxide dismutase (SOD), ascorbate peroxidase (APX), or peroxidase (POD) (Shri et al. 2009). The excess of Cd causes a decline in the antioxidant protection of the organ in *Triticum aestivum* L., with the consequent generation of considerable amounts of H_2O_2, a direct agent of oxidative stress (Ranieri et al. 2005). Requejo and Tena (2005) studied the effect of arsenic exposure on maize (*Zea mays* L.) root proteome and found that the induction of oxidative stress is the main process underlying arsenic toxicity in plants. Pb and Ni can induce the overproduction of reactive oxygen species (ROS) such as superoxide radicals ($^{\bullet}O_2^-$), singlet oxygen ($^1O_2$), and hydrogen peroxide (H_2O_2) in plant cells (Reddy et al. 2005). Indian mustard (*Brassica juncea* L.) effectively generated an enzymatic antioxidant defense system (especially CAT) to scavenge H_2O_2, resulting in lower H_2O_2 in shoots with higher mercury concentrations and generated adaptation system to mercury-induced oxidative stress (Shiyab et al. 2009).

As mentioned above, high levels of heavy metals can cause great damage to plants, and the main visible symptoms are shown in Fig. 1. To cope with excessive heavy metals uptake by plants, plants have evolved a variety of adaptive responses, which include immobilization, exclusion, chelation, and compartmentalization of metal ions and often involve metal-binding ligands. For example, plant cells developed a mechanism by which the metal ion, entering the cytosol of the cell, is immediately complexed and inactivated, thus preventing the metal from inactivating catalytically active or structural proteins (Zenk 1996). Shimwell and Laurie (1972) firstly pointed out that plants can excrete excessive heavy metals to ensure their normal growth. Similarly, the toxic metal ions in Lichens were immobilized in biologically non-active forms and then were secreted by plants glands (Tyler 1989). Heavy metals uptake by plants can be compartmentalized and sequestrated in less-active organelles (e.g., cell wall and vacuole) and/or be chelated with metal-binding ligands to improve the resistance to heavy metals (Wierzbicka 1986). It was reported that Cd can deposit in the vacuole in the form of oxalic acid salt (Van Balen et al. 1980). Up to now, it has been acknowledged that heavy metals can induce phytochelatins in plants and that the binding of heavy metals with phytochelatins is one of the main detoxification mechanisms (Chen et al. 2008, 2009b; Clemens 2006, 2009; Cobbett and Goldsbrough 2002).

3 Phytochelatin Biosynthesis

The peptide was first discovered by Hayashi and his groups as the Cd-binding complexes in fission yeast, *Schizosaccharomyces pombe*, exposed to Cd^{2+} and was named as "cadystins" (Murasugi et al. 1981). In 1984, cadystins was second discovered in wine yeast (*S. pombe*), which can chelate with metals (Kondo et al. 1984). A set of novel heavy-metal complexing peptides were isolated from plant cell suspension cultures. These peptides, named as phytochelatins (PCs), appear upon induction of plant cells with heavy metals and represent the principal metal-binding activities in the cells (Grill et al. 1985). PCs are synthesized inductively by exposure to not only Cd but also by other heavy metals such as Hg, Cu, Zn, Pb, and Ni (Yadav 2010).

3.1 *Structure of PCs*

PCs are a class of heavy-metal-binding peptides previously isolated from cell suspension cultures of several dicotyledonous and monocotyledonous plants. These peptides consist of repetitive γ-glutamylcysteine units with a carboxyl-terminal glycine and range from 5 to 17 amino acids in length and with the general structure of $(\gamma\text{-Glu-Cys})_n$-Gly ($n = 2$–11) (Fig. 2) (Grill et al. 1987). They are structurally related to glutathione (GSH; γ-Glu-Cys-Gly) and were presumed to be the products of a biosynthetic pathway. However, in some plants, the C-terminal glycine can be replaced by serine, glutamine, glutamate, or alanine, such as (-γ-Glu-Cys)$_n$-β-Ala, $(\gamma\text{-Glu-Cys})_n$-Ser, and $(\gamma\text{-Glu-Cys})_n$-Glu (Rauser 1995, 1999; Zenk 1996) (Table 1). These peptides were unequivocally identified as homologues of the phytochelatins by chemical analysis, and the chain length pattern of these compounds is identical to the PCs series. These compounds were called "homo-phytochelatins" (h-PCs) or iso-phytochelatins (iso-PCs) (Ducruix et al. 2006; Gekeler et al. 1989) (Fig. 3). Synthesis of iso-PCs mostly depends upon the

Fig. 2 Structure of PCs (after Pal and Rai 2010)

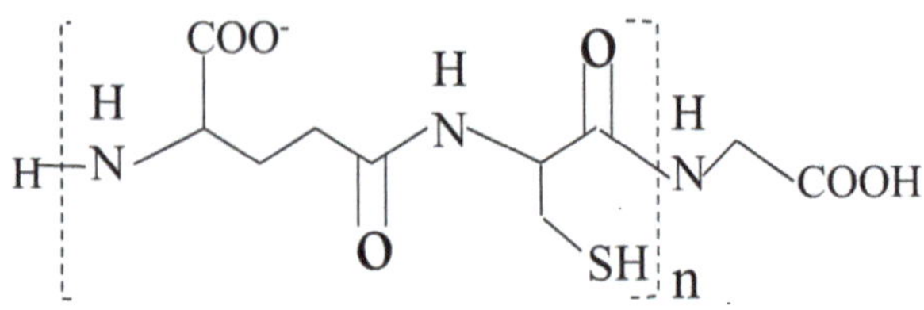

Table 1 Species of PCs in plants

PCs and h-PCs	Main plant species	References
(γ-Glu-Cys)$_n$-Gly	*Rauvolfia serpentina*, *Lycopersicon esculentum*	Grill et al. (1985); Rauser (1990)
(γ-Glu-Cys)$_n$-β-Ala	*Glycine max*, *Pisum sativum* L.	Klapheck et al. (1995); Rauser (1990)
(γ-Glu-Cys)$_n$-Ser	*Oryza sativa* L. cv Strella	Klapheck et al. (1994)
(γ-Glu-Cys)$_n$-Glu	*Zea mays* L.	Meuwly et al. (1993)
(γ-Glu-Cys)$_n$	*Armoracia rusticana*	Kubota et al. (2000)

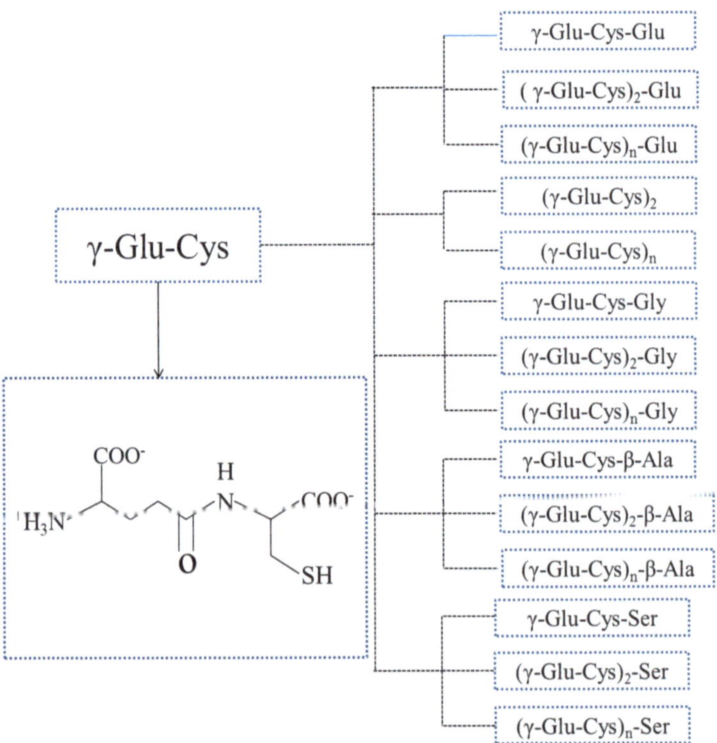

Fig. 3 Structure and classification of γ-Glu-Cys

availability of Gly or glutathione synthetase in the cells and may switch over to synthesize iso-PCs when the plant comes under stress (Rea 2012).

3.2 *PCs Biosynthesis*

3.2.1 Pathway of PCs Biosynthesis

Plants can develop a very potential mechanism to combat with heavy metal toxicity. Therefore, plants produce low-molecular weight thiols that show high affinity for toxic metals such as GSH—a sulfur containing tri-peptide thiol (Rauser 1995). The formula of GSH is γ-glutamate-cysteine-glycine. Steffens (1990) first suggested GSH is the precursor for PCs, heavy-metal-binding peptides involved in heavy metal tolerance and sequestration. The steady-state glutathione concentration in *Arabidopsis* plants was modified by expressing the cDNA for γ-glutamylcysteine synthetase (*GSH1*) in both the sense and antisense orientation, and the resulting plants had glutathione levels that ranged between 3 and 200 % in wild-type plants. *Arabidopsis* plants with low glutathione levels were hypersensitive to Cd due to the limited capacity of these plants to make phytochelatins (Xiang et al. 2001). Therefore the pathway of PCs biosynthesis is mostly related with that of GSH biosynthesis, which is generally divided into two stages (Fig. 4) (Noctor and Foyer 1998). The first step is that γ-glutamylcysteine synthetase (γ-ECS, GSH1, EC 6.3.2.2) catalyzes the formation of a peptide bond between the carboxyl group of glutamate and the amino group of cysteine, to yield γ-glutamylcysteine (γ-Glu-Gys, γ-EC), and glutathione synthetase (GS, GSH2, EC 6.3.2.3) catalyzes ligation between the carboxyl group of γ-Glu-Gys and glycine (Gly), to form γ-Glu-Gys-Gly (GSH) (Noctor and Foyer 1998). In this step, γ-ECS is one of the most important/critical synthetase to form GSH, and ATP is essential to complete this process. The second step is that PCs were synthesized, which derived from GSH by the function of phytochelatin synthases (PC synthases, EC 2.3.2.15) (Rea 2012). The biosynthesis of PCs is catalyzed by phytochelatin synthase, an enzyme that requires heavy metals as activating factors (Grill et al. 1989). It has been shown that Cd is the most effective activator of this enzyme, whereas other heavy metals activate it to a much lesser extent than Cd (Hirata et al. 2005). The γ-ECS is a rate-limiting

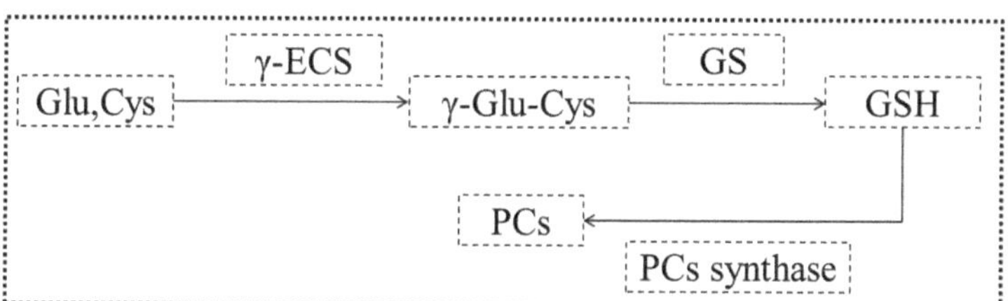

Fig. 4 Pathway of PCs biosynthesis

enzyme for GSH synthesis, and its activity is enhanced by metal ions like Cd^{2+} and suppressed by treatment with buthionine sulfoximine (BSO) (Inouhe 2005). In particular, genetic studies have confirmed that GSH-deficient mutants of *S. pombe* as well as *Arabidopsis* are PC deficient and hypersensitive to Cd (Cobbett and Goldsbrough 2002). Cells treated with BSO fail to produce phytochelatin and become extremely sensitive to the metals used to induce phytochelatin synthesis (Rauser 1990).

PCs synthases play a key role in the process of PCs synthesis, mainly for catalysis. Catalytic process is divided into two periods. The first period is activation. The signal detection zone of the carboxyl group of PC synthases detects signal of Cd^{2+} in the medium and immediately combines Cd^{2+} and Cys to form special space structure. Therefore, the amino group of PC synthases has catalytic activity (Yadav 2010). The second period is catalytic. The transfer of the γ-Glu-Cys moiety of a molecule of GSH onto a second molecule of GSH (or an existing PC molecule) to form a PC product by the function of the amino group of PC synthases activation. The reaction process can be repeated until the synthetic of $(\gamma\text{-Glu-Cys})_n$-Gly ($n=2$–11) forms (Cobbett 2000a), which was firstly synthesized by the use of GSH as the substrate in *Silene cucubalus* cell suspension cultures (Grill et al. 1989).

3.2.2 Regulation of PCs Biosynthesis

Lots of researchers have dedicated to the studies on the regulation of PCs biosynthesis since the success of PCs synthesis. At present, regulation of GSH biosynthesis and regulation of PC synthase activity are the two key strategies for regulating PCs biosynthesis (Zhou and Song 2004).

The intracellular level of GSH has been found to regulate PC synthesis (Chen et al. 2008; Collin-Hansen et al. 2007), and oxidative stress can regulate GSH synthesis (Liang et al. 2009). PC biosynthesis and Cd tolerance are increased in transgenic *Brassica juncea* in which either γ-glutamylcysteine synthetase (γ-ECS) or glutathione synthetase (GS) was overexpressed, which proved the importance of the regulation of GSH biosynthesis in modulating PC expression (Zhu et al. 1999a, b). The GS enzyme is rate limiting for the biosynthesis of glutathione and phytochelatins in the presence of Cd (Zhu et al. 1999b). Similarly, it was also found that γ-ECS and GS were both increased in *Arabidopsis* plants when exposed to Cd and Cu (Xiang and Oliver 1998). Indian mustard (*Brassica juncea*) was genetically engineered to overexpress the *Escherichia coli gshI* gene encoding γ-ECS, targeted to the plastids, and the results show that the γ-ECS transgenic seedlings showed increased tolerance to Cd and had higher concentrations of PCs, γ-Glu-Cys, glutathione, and total non-protein thiols compared with wild-type seedlings (Zhu et al. 1999a). Similarly, it was indicated that in roots, Cd-induced PC synthesis correlates with a moderate increase of expression of genes involved in GSH synthesis, the change for γ-ECS being most pronounced (Schäfer et al. 1998). Moreover, the activity of γ-ECS was approximately twofold higher in *CdR6-0* cells than in *CdS* cells of *Lycopersicon esculentum*, which thus increases the activity of the first enzyme of GSH biosynthesis in *CdR6-0* cells and can cause higher cadmium tolerance and capacity to synthesize GSH and PCs (Chen and Goldsbrough 1994). Therefore, regulation of PCs biosynthesis can be achieved by regulating the activity of γ-ECS.

Regulation of PC synthase is the other important way to regulate PCs biosynthesis. PC synthase is isoelectric near pH 4.8 and has a temperature optimum at 35 °C and pH 7.9 (Zenk 1996). It has been demonstrated that PC biosynthesis occurs within minutes and is independent of de novo protein synthesis using plant cell cultures exposed to Cd, consistent with the observation of enzyme activation in vitro (Cobbett 2001). In *Arabidopsis*, the *CAD1* gene, identified by using Cd-sensitive, PC-deficient *cad1* mutants, has been proposed to encode PC synthase (Ha et al. 1999). It was reported the suppression cloning of a cDNA (*AtPCS1*) from *A. thaliana* encoding a 55-kDa soluble protein that enhances heavy-metal tolerance and elicits Cd^{2+}-activated phytochelatin accumulation when expressed in *Saccharomyces cerevisiae* (Vatamaniuk et al. 1999). *Cad1* mutants of *A. thaliana* are sensitive to cadmium to different extents, deficient in their ability to form cadmium–peptide complexes, and deficient in its ability to accumulate phytochelatins (PCs) (Howden et al. 1995b). As mentioned above, PC synthase is primarily regulated by activation of the enzyme in the presence of heavy metals.

3.3 Factors Affecting PCs Biosynthesis

3.3.1 Types of Heavy Metals

The biosynthesis of PCs can be affected by the types of heavy metal (Cobbett 2001). Although a range of heavy metals can induce PCs in vivo, only a few heavy metals can form stable complex compound with PCs (Cobbett 2000a). It was reported that exogenous lead can induce a stronger stimulation of the synthesis of PCs by the *Chlorella vulgaris* cell (Bajguz 2002). Transgenic *Arabidopsis* plants exposed to high levels of Cd^{2+} could form PCs quickly, but the number of PCs reduces and eventually disappears with the decrease of metal concentration (Lee et al. 2003). PCs are rapidly induced in vivo by a wide range of metal ions, which follow the order $Cd^{2+} > Pb^{2+} > Zn^{2+} > Sb^{3+} > Ag^{+} > Hg^{2+} > As^{5-} > Cu^{+} > Sn^{2+} > Au^{3+} > Bi^{3+}$ (Grill et al. 1987). Similarly, it was found that the order of PCs induction by heavy metals in *Stigeoclonium* sp. was Cd > Pb > Zn (Pawlik-Skowronska 2001). PCs were the most strongly induced peptides under Cd and Hg stress, whereas As only tended to synthesize small thiols such as glutathione and γ-glutamylcysteine, which indicates that PCs are induced at different rates depending on the metal stressor used. (Dago et al. 2014).

3.3.2 Concentrations of Heavy Metals

The concentrations of heavy metals can also influence the induction of PCs. PCs induced significantly at 10 μM Cd in roots and decreased in roots at 50 μM Cd, which may be correlated with reduced level of GSH (Mishra et al. 2006). However, PCs binding efficiency in *Nitzschia palea* was enhanced with the increase of Cd exposure concentration (Figueira et al. 2014). It has been highlighted that PC_n ranged from PC_1 to PC_3 in Vetiver (*Vetiveria zizanioides*) roots exposed to 1200 mg Pb L^{-1}. The most abundant PC_n in both root and shoot tissues of Vetiver was PC_1, while PC_3 was observed only in the root tissues owing to the high levels of Pb (Andra et al. 2010). PC formation could be induced in the leaf, stem, and root tissues of *Sedum alfredii* upon exposure to 400 μM Cd and only in the stem and root when exposed to 700 μM Pb. However, no PCs were found in any part of *S. alfredii* when exposed to 1600 μM Zn (Zhang et al. 2008). The maximum amount of As chelated by PCs was found to be about 39 % in *Hydrilla verticillata* (L.f.) Royle exposed to As^{III} (at 10 μM) and 35 % in As^{V} exposed plants (at 50 μM) (Srivastava et al. 2007). Therefore, the induction of PCs depends not only on the concentrations of heavy metals but also on the existing form of heavy metals.

3.3.3 Species and Growing Condition of Plant

In addition, the induction of PCs is correlated with the species and growing condition of plants. The total concentrations of PCs in plants under As stress confirm the order of *Agropyron repens* > *Lolium perenne* > *Leonurus marrubiastrum* > *Zea mays* > *Glecoma hederacea* > *Urtica dioica* (Schulz et al. 2008). During mercury stress, both free (*Hydrilla verticillata*) and rooted (*Vallisneria spiralis*) submerged plants synthesized different species of phytochelatins (PCs), which bind with the accumulated mercury and showed high levels of cysteine and non-protein thiols. However, plants of *V. spiralis* showed more amount of PCs than *H. verticillata* despite relatively more Hg tolerance of *H. verticillata* (Gupta et al. 1998). Application of heavy metals to cell suspension cultures and whole plants of *Silene vulgaris* and tomato induces the formation of heavy metal-phytochelatin complexes with Cu and Cd and the binding of Zn and Pb to lower molecular weight substances. However, *S. vulgaris* cells can tolerate 5–10 times higher concentrations of Cu^{2+}, Cd^{2+}, Zn^{2+}, and Pb^{2+} in comparison to tomato cell cultures (Leopold et al. 1999). Transgenic *A. thaliana* contained almost 2 and 1.6 times more cadmium in the form of Cd-PC_2 than did the corresponding wild-type plants in their roots and shoots, respectively (Sadi et al. 2008).

Moreover, the growing condition of plant (e.g., pH, light and temperature) can also influence the induction to PCs. Higher production of phytochelatins was observed in response to metals at the lower studied pH 6.8 than at the alkaline pH, which might be that zinc availability to algae was improved due to the changed chemical Zn speciation after water acidification than at alkaline pH (Pawlik-

Skowronska 2001). The production of PC in the light is increased as compared to the dark and also at high (16 °C) as compared to low (4 °C) temperatures in *Vaucheria compacta* and *V. debaryana* (Xanthophyceae). Increased production at high temperature was related with the temperature optimum (35 °C) of phytochelatin synthase responsible for phytochelatin biosynthesis. However, decreased production in the dark, as well as at low temperature (4 °C), was related with the decreased rates of other cellular reactions including the reduced transport of cadmium leading to its decreased availability for PC induction (Gaur and Rai 2001).

4 Function of PCs

The function of PCs mainly reflects in two aspects: improve resistance of plants to heavy metals and detoxify the toxicity of heavy metals and maintain intracellular metal ions homeostasis (Rauser 1995). Particularly, the main function of PCs is the ability of complexing metals (plant micronutrient and nonessential metal elements).

4.1 *Improve Resistance of Plants to Heavy Metals and Detoxify the Toxicity of Heavy Metals*

PCs are peptides that function in heavy-metal chelation and detoxification in plants and fungi (Hirata et al. 2005). Cadmium is a kind of important chelating peptide resultant. PC-metal complexes are one of the ways to improve resistance of plants to heavy metals, which proved that the roles of PCs are to protect plants against toxic metals (Cobbett and Goldsbrough 2002). There are enough and genetic evidences to prove the role of PCs in Cd detoxification. Two different forms of PC-metal complexes in fission yeast cells exposed to Cd were reported, i.e., the first, composed mostly of PCs and Cd, elutes from gel filtration columns with an apparent molecular weight of 3–4 kDa (LMW PC–Cd complex), and the second, more highly charged form has an apparent molecular weight of 6–9 kDa and contains sulfide in addition to PCs and Cd (Ortiz et al. 1992). In extracts of *S. pombe*, two PC–Cd complexes (referred to as HMW and LMW) can be clearly resolved using gel-filtration chromatography. Sulfide ions play an important role in the efficacy of Cd detoxification by PCs. HMW PC–Cd complexes contain both Cd and acid-labile sulfide. The incorporation of sulfide into the HMW complexes increases both the amount of Cd per molecule and the stability of the complex (Fig. 5) (Cobbett 2000b; Cobbett and Goldsbrough 2002). An allelic series of *cad1*, cadmium-sensitive mutants of *A. thaliana*, was isolated. These mutants were sensitive to cadmium to different extents and were deficient in their ability to form cadmium–peptide complexes. Moreover, the amount of PCs accumulated by

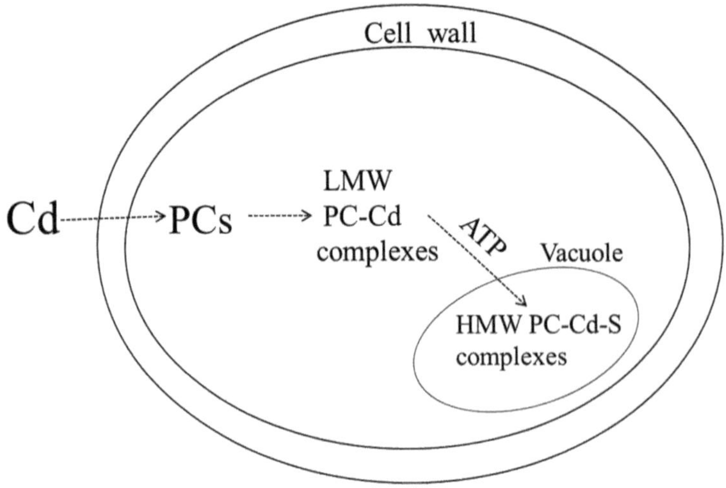

Fig. 5 Mechanisms of PCs detoxification (an example of cadmium)

each mutant correlated with its degree of sensitivity to cadmium (Howden et al. 1995b). In *B. juncea* L., Cd accumulation is accompanied by metabolic adaptation, in particular, the rapid induction of phytochelatin (PC) biosynthesis (Haag-Kerwer et al. 1999). It was highlighted that stringent control of Cd detoxification by PCs protects photosynthesis but does not prevent a decline in transpiration rate (Zhu et al. 1999a, b). Sequestration of Cd by PCs provides an essential cellular mechanism for Cd detoxification (Haag-Kerwer et al. 1999). Similarly, suspension-cultured cells of azuki bean (*Vigna angularis*) as well as the original root tissues were hypersensitive to Cd, and the azuki bean cells challenged to Cd did not contain phytochelatin (PC) peptides, unlike tomato (*L. esculentum*) cells that have a substantial tolerance to Cd (Inouhe et al. 2000). It was reported that enhanced Cd tolerance of Ce-PCS tobacco was accompanied by an increased cytosolic and vacuolar SH of PC/Cd ratio, suggesting the important role of mechanisms other than PC–Cd transport in Cd translocation to the vacuole. The key role of the PCs in Cd tolerance is temporary binding of Cd^{2+} in the cytosol (Sylwia et al. 2010). Exposure to Cd^{2+} leads to activation of phytochelatin synthase (PCs) and the formation of phytochelatins (PCs) in the cytosol. Binding of Cd by PCs and the subsequent transport of PC–Cd complexes to the vacuole are essential for Cd tolerance (Sylwia et al. 2010).

Induction of PCs in vivo and activation of PC synthase in vitro are conferred by a range of metal ions. However, there is little evidence supporting a role for PCs in the detoxification of such a wide range of metal ions (Cobbett 2000b). The ability of PCs detoxification is based on the ability of heavy metal ions complex. PCs were induced to various degrees by exposure to various metals (Ag^{+}, As^{3+}, As^{5+}, Cd^{2+}, Cu^{2+}, Ca^{3+}, Hg^{2+}, In^{3+}, Ni^{2+}, Pb^{2+}, Pd^{2+}, Se^{4+}, and Zn^{2+}) (Maitani et al. 1996). It was found that in a medium with low cation content, PC-deficient mutants show pronounced Zn^{2+} hypersensitivity. This phenotype is of comparable strength to the well-documented Cd^{2+} hypersensitivity of *cad1* mutants (Tennstedt et al. 2009). Andra et al. (2009) indicated that Vetiver grass (*Vetiveria zizanioides* L.) exposed to Pb induced the synthesis of phytochelatins and the formation of Pb-PC_n complexes to alleviate the phytotoxic effects of free Pb ions. Similarly, Estrella-Gómez et al. (2009) suggested that the accumulation of PC in *Salvinia minima* is a direct response to Pb^{2+} accumulation, and phytochelatins do participate as one of the mechanisms to cope with Pb^{2+} of this Pb-hyperaccumulator aquatic fern.

However, there are also some queries on the detoxifying of PCs. de Knecht et al. (1994) pointed out that in response to a range of Cd concentrations, the root tips of Cd-tolerant plants of *S. vulgaris* exhibit a lower rate of PC production accompanied by a lower rate of longer chain PC synthesis than those of Cd-sensitive plants. Similarly, phytochelatin synthesis and loss of total glutathione were observed in both the roots of copper-sensitive and tolerant *S. cucubalus* (L.) exposure to Cu (De Vos et al. 1992). A study on Cu-sensitive and Cu-tolerant populations of *S. vulgaris* found that both the populations produce PCs when exposed to Cu. However, when exposed to their own highest no-effect concentration or 50 %-effect concentration of Cu for root growth, the two plants exhibit equal phytochelatin contents in the root apex (Schat and Kalff 1992). It seems that Cu tolerance in *S. vulgaris* does not rely on phytochelatin production.

4.2 Maintain Intracellular Metal Ions Homeostasis

PCs are involved in metal ion homeostasis in plants. PCs accumulation is correlated with the initial concentration of zinc ions in the nutrient solution. PCs had almost disappeared from the cells after reaching stationary growth phase (Grill et al. 1988). PCs play a critical role in homeostasis of heavy metals in plants by regulating the availability of metal ions in the plant cell (Zenk 1996). Metal ions homeostasis requires intracellular complexation of metals when there is a cellular surplus and later release of metals to metal-requiring apoproteins and perhaps to final storage sites within cells (Rauser 1995). Kneer and Zenk found that heavy metal ions entering cells at sublethal concentrations are totally complexed by phytochelatins using $^{109}Cd^{2+}$ in *Rauvolfia serpentina* (and to a much lesser extent to some high molecular weight proteins) and a series of metal-sensitive plant enzymes such as alcohol dehydrogenase, glyceraldehyde-3-phosphate dehydrogenase, nitrate reductase, ribulose-1,5-diphosphate carboxylase, and urease tolerate Cd^{2+} in the form of a phytochelatin complex from 10- to 1000-fold the amount as compared with the free metal ion (Kneer and Zenk 1992). Therefore, the complexation of PCs and Cd protects the activity of the enzyme, so as to guarantee the normal metabolism of cells. It was reported that *AvPCS* (phytochelatin synthase genes isolated from *Avicennia germinans*) was the most active gene involved in the regulation of essential metals (e.g., Cu^{2+}) in *A. germinans* leaves (Gonzalez-Mendoza et al. 2007). PCs play a dual role in plants: the first is that PCs complex to the metal ion tightly to inactivate and store it in the vacuole, the second is that PCs transfer the essential metal to newly synthesized apoenzymes (Thumann et al. 1991).

Metal ions in multicellular organisms have to be moved to target cells, tissues, and organs distant from the site of uptake without being sequestered by other available sites along the way (Colangelo and Guerinot 2006). Key homeostatic processes include (1) tightly controlled transport of metal ions across membranes, (2) chelation by low-molecular weight molecules, (3) delivery to target proteins, e.g., via metallochaperones (Clemens and Peršoh 2009; Palmer and Guerinot 2009).

4.3 Other Functions

PCs can also function as long-distance carriers for heavy metal ions. Gong et al. (2003) found that transgenic expression of the *TaPCS1* gene increases long-distance root-to-shoot Cd^{2+} transport and reduces Cd^{2+} accumulation in roots. In the fungus *S. pombe*, complexes of heavy metals bound to PCs are transported across the tonoplast and sequestered in vacuoles by means of the ATP-binding cassette (ABC) transporter *HMT1* (Ortiz et al. 1995). The wheat (*Triticum aestivum*) PC synthase (*TaPCS1*) gene was expressed under the control of a shoot-specific promoter (*CAB2*) in an Arabidopsis (*A. thaliana*) PC-deficient mutant, *cad1-3* (CAB2: *TaPCS1*/*cad1-3*). The study shows that *TaPCS1* is expressed only in shoots and that CAB2: T*aPCS1*/*cad1-3* lines complement the cadmium (Cd) and arsenic metal sensitivity of *cad1-3* shoots, and PC_2 is transported over long distances in the shoot-to-root direction (Chen et al. 2006). These results imply that transported PCs play a role in the transport of Cd.

Vetiver grass (*V. zizanioides* L.) exposure to Pb induced the synthesis of phytochelatins (PC_n) and the formation of Pb–PC_n complexes, alleviating the phytotoxic effects of free Pb ions. PCs can transport Pb to vacuoles in vascular tissues of root and shoot (Andra et al. 2009). Li et al. (2006) showed the long-distance movement of thiol peptides from shoots down to roots by expressing a bacterial γ-glutamylcysteine synthetase (ECS) in the shoots of an *Arabidopsis* ECS-deficient mutant using a shoot-specific, light-induced regulatory cassette.

Based on the above evidences, the functions of PCs are summarized: (1) the phytochelatin pathway has an important role in essential metal homeostasis; (2) phytochelatin detoxifies the toxicity of heavy metals; and (3) Cd and As are essential elements, and the phytochelatin pathway is a part of the homeostatic network for these elements (Clemens 2006; Clemens and Peršoh 2009).

5 Role of PCs in Metal Detoxification and Tolerance

Response to heavy metal stress in plants involves immobilization, exclusion, chelation, and compartmentalization of metal ions and often involves metal-binding ligands (Cobbett 2000a, 2001). Since immobilized metals are less toxic than free ions, PCs are considered to be part of the mechanism detoxifying heavy metals in higher plants (Cobbett and Goldsbrough 2002). High concentrations of heavy metals such as Cd, Cu, Zn, Ag, Au, Co, Ni, and Bi can induce the synthesis of phytochelatin in plants, fungi, cyanobacteria, algae, and animals (Gaur and Rai 2001), and trace levels of essential metals can also induce PCs synthesis. Appearance of such metal-binding peptides (PCs, iso-PCs) in plants is one of the most important biochemical indications of heavy metal contamination in various environments (Li et al. 2006). The induction of PCs in *Phaeodactylum tricornutum* cells was tested, which exhibited a decrease of the PC pool and a concomitant increase of glutathione, suggesting a mechanism of degradation and release of metal-phytochelatin complexes (Morelli and Scarano 2001). More sensitivity was found toward Cd and arsenate for the *cad1-3* mutant of *A. thaliana* than wild-type plants. However, no considerable difference was found for metals like Zn, selenite, and Ni ions (Ha et al. 1999). Figueroa et al. (2008) revealed that metal bioavailability has effect on PCs production. Their results indicated that humic substance (HS) in soil restricted the availability of metal for creating stress and thereby resulted in reduced PC production in *Ricinus communis*. *Hydrilla verticillata* and *Vallisneria spiralis* can synthesize different species of phytochelatins (PCs), which bind with the accumulated mercury and showed high levels of cysteine and non-protein thiols during mercury stress and play a role in mercury detoxification (Gupta et al. 1998). Cysteine content and accumulation of acid soluble thiols inside *H. verticillata* tissue and synthesis of phytochelatins enhanced with concentrations of Pb, which indicated that PC synthesis was induced under lead stress utilizing glutathione as a substrate (Gupta et al. 1995). Pb-binding complex indicated involvement of phytochelatins in Pb detoxification. These evidences all demonstrated that PCs play a critical role in the detoxication of heavy metals.

Not all studies have supported the role of PCs. The phytochelatin-inhibitor, BSO, can increase the Zn tolerance of roots, which argues against a key role of phytochelatins in the Zn-tolerance mechanism of roots of *Festuca rubra* L. (Davies et al. 1991). Further, this preliminary evidence suggests that BSO stimulates cell division activity in roots of *F. rubra* L. The role of PCs in metal tolerance in *T. caerulescens* and the related non-accumulator *T. caerulescens* was examined. Results show that total PCs levels in the hyperaccumulator were generally lower, and the lack of a role for PCs in the hyperaccumulator's response to metal stress suggests that other mechanisms are responsible for Cd tolerance, but PCs were generally present at lower levels than in the non-accumulator *T. caerulescens* (Ebbs et al. 2002). PCs may play a secondary role in Cd tolerance of *T. caerulescens*. PCs were most abundant in leaves followed by stems but hardly detected in roots of *S. alfredii*. There is a linear correlation between PC synthesis and Cd accumulation in leaves, suggesting that PCs do not detoxify Cd in roots of *S. alfredii* (Zhang et al. 2010). When Cd hyperaccumulating ecotype (HE) *S. alfredii* was exposed to Cd, larger proportion (about 70 % in leaves and 47 % in stems) of Cd was bound with malic acid (the major organic acid in the shoots of the plants) (Tian et al. 2011). Their results indicate that a majority of Cd accumulates in the parenchyma cells, especially in stems, and is likely associated with calcium pathways and bound with organic acid (malate). PCs might not act in the major intracellular heavy metals detoxification mechanism in plants.

6 Molecular Biology of PCs

In 1989, Grill had begun to study the purification of PCs in plants (Grill et al. 1989). However, it was not until 1999 that the phytochelatin synthase genes from *A. thaliana* and *S. pombe* wheat were isolated, cloned, and identified (Clemens et al. 1999; Ha et al. 1999; Vatamaniuk et al. 1999). They are *AtPCS1* (CAD1), *SpAtPCS*, and *TaPCS1*, respectively, encoding a 55-kDa soluble protein. These proteins have 40–50 % sequence similarity between each other.

Later, phytochelatin synthase genes were isolated and purified. Howden et al. (1995a) isolated *Cad1-1* and *Cad2-1* from cadmium-sensitive mutant of *A. thaliana*. Cazale and Clemens (2001) isolated *AtPCS2* from *Arabidopsis*, a highly homologous gene to *AtPCS1*, and they concluded that *AtPCS2* gene encodes a functional phytochelatin synthase as shown by expression in *S. cerevisiae* and the complementation of a *S. pombe* phytochelatin synthase knockout strain. When *AtPCS2* is expressed in yeast, it confers Cd-resistance, indicating that its gene product is active (Cobbett 2000a). And compared with *AtPCS1*, the expression of *AtPCS2* is at a relatively low level. Oven et al. (2002) isolated and functionally expressed a cDNA *GmhPCS1* encoding homo-phytochelatin synthase from *Glycine max* and proved that *GmhPCS1* can accept homo-glutathione as the sole substrate for the synthesis of homo-phytochelatins. Loscos et al. (2006) further isolated a cDNA clone encoding a protein (*LjPCS1*) from legume Lotus (*Lotus japonicus*), a highly homologous to a homo-phytochelatin synthase (hPCS) of *Glycine max* (*GmhPCS1*). Osaki et al. (2008) successfully isolated *CmPCS*, a product of a *PCS*-like gene from the genomic DNA of the red alga *Cyanidioschyzon merolae*. Dong et al. (2005) reported the isolation of a full-length cDNA sequence encoding a phytochelatin synthase (PCS) from *P. vittata* L., which designated *PvPCS1*, predicts a protein of 512 amino acids with a molecular weight of 56.9 kDa. And they also demonstrated that *PvPCS1* expressed in *S. cerevisiae*, and PvPCS1 mediated increased Cd tolerance. Ramos et al. (2007) identified *LjPCS1*, *LjPCS2*, and *LjPCS3* genes in *L. japonicus*, through screening of transformation-competent artificial chromosome libraries, encoding protein products viz. *LjPCS2-7N* and *LjPCS3-7N* which conferred Cd tolerance when expressed in *S. cerevisiae*. It was reported that *CdPCS1* (from *Ceratophyllum demersum*, an aquatic As-accumulator plant) expressing rice transgenic lines showed marked increase in PCs activity and enhanced synthesis of PCs in comparison to non-transgenic plant (Shri et al. 2014).

The *Escherichia coli gshII* gene encoding glutathione synthetase (GS) was overexpressed in the cytosol of Indian mustard (*B. juncea*). The results indicate that the transgenic GS plants accumulated significantly more Cd than the wild type, and the GS plants showed enhanced tolerance to Cd at both the seedling and mature-plant stages. Cd accumulation and tolerance are correlated with the *gshII* expression level, and transgenic plants have more glutathione, phytochelatin, thiol, S, and Ca than wild-type plants (Zhu et al. 1999b).

7 Conclusive Remarks

PCs play a key role in heavy metal detoxification and accumulation in plants. The most significant recent advances of PC biosynthesis and function have come from molecular genetic studies using a variety of model systems. The biosynthesis of PC should be regulated to make the balance between PCs and GSH (the PCs synthesis substrate). In addition, further study should be made in the function of antioxidant molecular genetic regulation mechanism of the PCs synthesis. Especially, the pathway of biosynthesis and regulation mechanism of PCs in transgenic plants and the detoxification mechanism of transgenic plants to heavy metal stress need deep explorations for the breeding of suitable transgenic plants used for phytoremediation of heavy metal-contaminated soil.

The studies on the processes of different heavy-metal detoxification mechanisms will benefit to select plant species capable of hyperaccumulating heavy metal(s). Since it is an established fact that PCs are involved in metal detoxification and/or accumulation in plants, the above problem can be overcome by engineering PC synthase genes in common plants capable of growing fast and producing large biomass (Pal and Rai 2010). Therefore, understanding well of the function of PCs will potentially improve remediation efficiency of heavy metal pollution in future. Moreover, much more attention should also be paid on the influence of PCs on crops cultivated in contaminated agricultural soil. The surprising evidence for long-distance transport of PC-metal complexes raises the question of a PC involvement in intercellular metal ion transport, which should be intensively studied in future (Gong et al. 2003).

Acknowledgments This work was supported by the National Natural Science Foundation of China (41471411, 21107052, 31270540, 31070455, and 40971184). The authors would also like to thank the Open Fund of Key Laboratory of Contaminated Environment Control and Regional Ecology Safety (grant No. SYU-KF-L-03) & Open Fund of Key Laboratory of Regional Environment Eco-remediation, Education Ministry (grant No. SYU-KF-E-03) for a partial support.

References

Alcántara E, Romera FJ, Cañete M, Manuel D (1994) Effects of heavy metals on both induction and function of root Fe (III) reductase in Fe-deficient cucumber (*Cucumis sativus* L.) plants. J Exp Bot 45:1893–1898

Andra SS, Datta R, Sarkar D, Makris KC, Mullens CP, Sahi SV, Bach SB (2009) Induction of lead-binding phytochelatins in *Vetiver Grass* [(L.)]. J Environ Qual 38:868–877

Andra SS, Datta R, Sarkar D, Makris KC, Mullens CP, Sahi SV, Bach SBH (2010) Synthesis of phytochelatins in vetiver grass upon lead exposure in the presence of phosphorus. Plant Soil 326:171–185

Athar R, Ahmad M (2002) Heavy metal toxicity: effect on plant growth and metal uptake by wheat, and on free living azotobacter. Water Air Soil Pollut 138:165–180

Bajguz A (2002) Brassinosteroids and lead as stimulators of phytochelatins synthesis in *Chlorella vulgaris*. J Plant Physiol 159:321–324

Baldisserotto C, Ferroni L, Anfuso E, Pagnoni A, Fasulo MP, Pancaldi S (2007) Responses of *Trapa natans* L. floating laminae to high concentrations of manganese. Protoplasma 231:65–82

Beauford W, Barber J, Barringer AR (1977) Uptake and distribution of mercury within higher plants. Physiol Plant 39:261–265

Bingham FT, Pereyea F, Jarrell WM (1986) Metal toxicity to agricultural crops. Met Ions Biol Syst 20:119–156

Boojar MMA, Tavakoli Z (2010) Role of antioxidant enzyme responses and phytochelatins in tolerance strategies of *Alhagi camelorum* Fisch growing on copper mine. Acta Bot Croat 69:107–121

Brown PH, Welch RM, Cary EE (1987) Nickel: a micronutrient essential for higher plants. Plant Physiol 85:801–803

Burton K, Morgan E, Roig A (1984) The influence of heavy metals upon the growth of sitka-spruce in South Wales forests. Plant Soil 78:271–282

Cazale AC, Clemens S (2001) *Arabidopsis thaliana* expresses a second functional phytochelatin synthase. FEBS Lett 507:215–219

Chen J, Goldsbrough PB (1994) Increased activity of γ-glutamylcysteine synthetase in tomato cells selected for cadmium tolerance. Plant Physiol 106:233–239

Chen A, Komives EA, Schroeder JI (2006) An improved grafting technique for mature *Arabidopsis* plants demonstrates long-distance shoot-to-root transport of phytochelatins in *Arabidopsis*. Plant Physiol 141:108–120

Chen L, Guo Y, Yang L, Wang Q (2008) Synergistic defensive mechanism of phytochelatins and antioxidative enzymes in *Brassica chinensis* L. against Cd stress. Chin Sci Bull 53:1503–1511

Chen CY, Huang DJ, Liu JQ (2009a) Functions and toxicity of nickel in plants: recent advances and future prospects. Clean 37:304–313

Chen L, Yang L, Wang Q (2009b) In vivo phytochelatins and Hg-phytochelatin complexes in Hg-stressed *Brassica chinensis* L. Metallomics 1:101–106

Clemens S (2006) Evolution and function of phytochelatin synthases. J Plant Physiol 163:319–332

Clemens S, Peršoh D (2009) Multi-tasking phytochelatin synthases. Plant Sci 177:266–271

Clemens S, Kim EJ, Neumann D, Schroeder JI (1999) Tolerance to toxic metals by a gene family of phytochelatin synthases from plants and yeast. EMBO J 18:3325–3333

Cobbett CS (2000a) Phytochelatin biosynthesis and function in heavy-metal detoxification. Curr Opin Plant Biol 3:211–216

Cobbett CS (2000b) Phytochelatins and their roles in heavy metal detoxification. Plant Physiol 123:825–832

Cobbett CS (2001) Heavy metal detoxification in plants: phytochelatin biosynthesis and function. IUBMB Life 51:183–188

Cobbett C, Goldsbrough P (2002) Phytochelatins and metallothioneins: roles in heavy metal detoxification and homeostasis. Annu Rev Plant Biol 53:159–182

Colangelo EP, Guerinot ML (2006) Put the metal to the petal: metal uptake and transport throughout plants. Curr Opin Plant Biol 9:322–330

Collin-Hansen C, Pedersen SA, Andersen RA, Steinnes E (2007) First report of phytochelatins in a mushroom: induction of phytochelatins by metal exposure in *Boletus edulis*. Mycologia 99:161–174

Dago À, González I, Ariño C, Manuel Díaz-Cruz J, Esteban M (2014) Chemometrics applied to the analysis of induced phytochelatins in *Hordeum vulgare* plants stressed with various toxic non-essential metals and metalloids. Talanta 118:201–209

Davies K, Davies M, Francis D (1991) The influence of an inhibitor of phytochelatin synthesis on root growth and root meristematic activity in *Festuca rubra* L. in response to zinc. New Phytol 118:565–570

de Knecht JA, van Dillen M, Koevoets PL, Schat H, Verkleij JA, Ernst WH (1994) Phytochelatins in cadmium-sensitive and cadmium-tolerant *Silene vulgaris* (chain length distribution and sulfide incorporation). Plant Physiol 104:255–261

De Vos CH, Vonk MJ, Vooijs R, Schat H (1992) Glutathione depletion due to copper-induced phytochelatin synthesis causes oxidative stress in *Silene cucubalus*. Plant Physiol 98:853–858

Demiral T, Türkan I (2005) Comparative lipid peroxidation, antioxidant defense systems and proline content in roots of two rice cultivars differing in salt tolerance. Environ Exp Bot 53:247–257

Dong R, Formentin E, Losseso C, Carimi F, Benedetti P, Terzi M, Schiavo F (2005) Molecular cloning and characterization of a phytochelatin synthase gene, *PvPCS1*, from *Pteris vittata* L. J Ind Microbiol Biotechnol 32:527–533

Ducruix C, Junot C, Fievet JB, Villiers F, Ezan E, Bourguignon J (2006) New insights into the regulation of phytochelatin biosynthesis in *A. thaliana* cells from metabolite profiling analyses. Biochimie 88:1733–1742

Ebbs S, Lau I, Ahner B, Kochian L (2002) Phytochelatin synthesis is not responsible for Cd tolerance in the Zn/Cd hyperaccumulator *Thlaspi caerulescens* (J. and C. Presl). Planta 214:635–640

Elangovan D, Chalakh M (2006) Arsenic pollution in west Bengal. Tech Digest 9:31–35

Estrella-Gómez N, Mendoza-Cozatl D, Moreno-Sanchez R, Gonzalez-Mendoza D, Zapata-Perez-O, Martinez-Hernandez A, Santamaria JM (2009) The Pb-hyperaccumulator aquatic fern *Salvinia minima* Baker, responds to Pb^{2+} by increasing phytochelatins via changes in *SmPCS* expression and in phytochelatin synthase activity. Aquat Toxicol 91:320–328

Figueira E, Freitas R, Guasch H, Almeida SP (2014) Efficiency of cadmium chelation by phytochelatins in *Nitzschia palea* (Kützing) W. Smith. Ecotoxicology 23:285–292

Figueroa JAL, Wrobel K, Afton S, Caruso JA, Felix Gutierrez Corona J, Wrobel K (2008) Effect of some heavy metals and soil humic substances on the phytochelatin production in wild plants from silver mine areas of Guanajuato, Mexico. Chemosphere 70:2084–2091

Garg N, Singla P (2011) Arsenic toxicity in crop plants: physiological effects and tolerance mechanisms. Environ Chem Lett 9:303–321

Gaur J, Rai L (2001) Heavy metal tolerance in algae. In: Rai LC, Gaur JP (eds) Algal adaptation to environmental stresses. Springer, Berlin

Gekeler W, Grill E, Winnacker EL, Zenk MH (1989) Survey of the plant kingdom for the ability to bind heavy metals through phytochelatins. Z Naturforsch 44:361–369

Gong JM, Lee DA, Schroeder JI (2003) Long-distance root-to-shoot transport of phytochelatins and cadmium in *Arabidopsis*. Proc Natl Acad Sci USA 100:10118–10123

Gonzalez-Mendoza D, Moreno AQ, Zapata-Perez O (2007) Coordinated responses of phytochelatin synthase and metallothionein genes in black mangrove, *Avicennia germinans*, exposed to cadmium and copper. Aquat Toxicol 83:306–314

Grill E, Winnacker EL, Zenk MH (1985) Phytochelatins: the principal heavy-metal complexing peptides of higher plants. Science 230:674–676

Grill E, Winnacker EL, Zenk MH (1987) Phytochelatins, a class of heavy-metal-binding peptides from plants, are functionally analogous to metallothioneins. Proc Natl Acad Sci USA 84:439–443

Grill E, Thumann J, Winnacker EL, Zenk MH (1988) Induction of heavy-metal binding phytochelatins by inoculation of cell cultures in standard media. Plant Cell Rep 7:375–378

Grill E, Loffler S, Winnacker EL, Zenk MH (1989) Phytochelatins, the heavy-metal-binding peptides of plants, are synthesized from glutathione by a specific gamma-glutamylcysteine dipeptidyl transpeptidase (phytochelatin synthase). Proc Natl Acad Sci USA 86:6838–6842

Gupta M, Rai U, Tripathi R, Chandra P (1995) Lead induced changes in glutathione and phytochelatin in *Hydrilla verticillata* (lf) Royle. Chemosphere 30:2011–2020

Gupta M, Tripathi RD, Rai UN, Chandra P (1998) Role of glutathione and phytochelatin in *Hydrilla verticillata* (lf) Royle and *Vallisneria spiralis* L. under mercury stress. Chemosphere 37:785–800

Ha SB, Smith AP, Howden R, Dietrich WM, Bugg S, O'Connell MJ, Goldsbrough PB, Cobbett CS (1999) Phytochelatin synthase genes from arabidopsis and the yeast *Schizosaccharomyces pombe*. Plant Cell 11:1153–1163

Haag-Kerwer A, Schafer HJ, Heiss S, Walter C, Rausch T (1999) Cadmium exposure in *Brassica juncea* causes a decline in transpiration rate and leaf expansion without effect on photosynthesis. J Exp Bot 50:1827–1835

Haghiri F (1974) Plant uptake of cadmium as influenced by cation exchange capacity, organic matter, zinc, and soil temperature. J Environ Qual 3:180–183

Hall JL (2002) Cellular mechanisms for heavy metal detoxification and tolerance. J Exp Bot 53:1–11

Hirata K, Tsuji N, Miyamoto K (2005) Biosynthetic regulation of phytochelatins, heavy metal-binding peptides. J Biosci Bioeng 100:593–599

Howden R, Andersen CR, Goldsbrough PB, Cobbett CS (1995a) A cadmium-sensitive, glutathione-deficient mutant of *Arabidopsis thaliana*. Plant Physiol 107:1067–1073

Howden R, Goldsbrough PB, Andersen CR, Cobbett CS (1995b) Cadmium-sensitive, cad1 mutants of *Arabidopsis thaliana* are phytochelatin deficient. Plant Physiol 107:1059–1066

Iglesia-Turino S, Febrero A, Jauregui O, Caldelas C, Araus JL, Bort J (2006) Detection and quantification of unbound phytochelatin 2 in plant extracts of *Brassica napus* grown with different levels of mercury. Plant Physiol 142:742–749

Inouhe M (2005) Phytochelatins. Braz J Plant Physiol 17:65–78

Inouhe M, Ito R, Ito S, Sasada N, Tohoyama H, Joho M (2000) Azuki bean cells are hypersensitive to cadmium and do not synthesize phytochelatins. Plant Physiol 123:1029–1036

Khabaz-Saberi H, Rengel Z, Wilson R, Setter TL (2010) Variation of tolerance to manganese toxicity in *Australian hexaploid* wheat. J Plant Nutr Soil Sci 173:103–112

Klapheck S, Fliegner W, Zimmer I (1994) Hydroxymethyl-phytochelatins [(γ-glutamylcysteine) n-serine] are metal-induced peptides of the poaceae. Plant Physiol 104:1325–1332

Klapheck S, Schlunz S, Bergmann L (1995) Synthesis of phytochelatins and homo-phytochelatins in *Pisum sativum* L. Plant Physiol 107:515–521

Kneer R, Zenk MH (1992) Phytochelatins protect plant enzymes from heavy metal poisoning. Phytochemistry 31:2663–2667

Kondo N, Imai K, Isobe M, Goto T, Murasugi A, Wada-Nakagawa C, Hayashi Y (1984) Cadystin a and b, major unit peptides comprising cadmium binding peptides induced in a fission yeast—separation, revision of structures and synthesis. Tetrahedron Lett 25:3869–3872

Kramer U, Smith RD, Wenzel WW, Raskin I, Salt DE (1997) The role of metal transport and tolerance in nickel hyperaccumulation by *Thlaspi goesingense* Halacsy. Plant Physiol 115:1641–1650

Kubota H, Sato K, Yamada T, Maitani T (2000) Phytochelatin homologs induced in hairy roots of horseradish. Phytochemistry 53:239–245

Lee S, Moon JS, Ko T-S, Petros D, Goldsbrough PB, Korban SS (2003) Overexpression of Arabidopsis phytochelatin synthase paradoxically leads to hypersensitivity to cadmium stress. Plant Physiol 131:656–663

Lei Y, Chen K, Tian X, Korpelainen H, Li C (2007) Effect of Mn toxicity on morphological and physiological changes in two *Populus cathayana* populations originating from different habitats. Trees 21:569–580

Leopold I, Gunther D, Schmidt J, Neumann D (1999) Phytochelatins and heavy metal tolerance. Phytochemistry 50:1323–1328

Lewis S, Donkin M, Depledge M (2001) Hsp70 expression in *Enteromorpha intestinalis* (Chlorophyta) exposed to environmental stressors. Aquat Toxicol 51:277–291

Li Y, Dankher OP, Carreira L, Smith AP, Meagher RB (2006) The shoot-specific expression of γ-glutamylcysteine synthetase directs the long-distance transport of thiol-peptides to roots conferring tolerance to mercury and arsenic. Plant Physiol 141:288–298

Liang G, Liao X, Du G, Chen J (2009) A new strategy to enhance glutathione production by multiple H2O2-induced oxidative stresses in *Candida utilis*. Bioresour Technol 100:350–355

Lindberg S, Landberg T, Greger M (2007) Cadmium uptake and interaction with phytochelatins in wheat protoplasts. Plant Physiol Biochem 45:47–53

Liu WT, Zhou QX, An J, Sun YB, Liu R (2010a) Variations in cadmium accumulation among Chinese cabbage cultivars and screening for Cd-safe cultivars. J Hazard Mater 173:737–743

Liu WT, Zhou QX, Zhang YL, Wei SH (2010b) Lead accumulation in different Chinese cabbage cultivars and screening for pollution-safe cultivars. J Environ Manage 91:781–788

Liu GY, Zhang YX, Chai TY (2011a) Phytochelatin synthase of *Thlaspi caerulescens* enhanced tolerance and accumulation of heavy metals when expressed in yeast and tobacco. Plant Cell Rep 30:1067–1076

Liu WT, Zhou QX, Zhang ZN, Hua T, Cai Z (2011b) Evaluation of cadmium phytoremediation potential in Chinese cabbage cultivars. J Agric Food Chem 59:8324–8330

Liu WT, Ni JC, Zhou QX (2013) Uptake of heavy metals by trees: prospects for phytoremediation. Mater Sci Forum 743–744:768–781

Liu W, Liang L, Zhang X, Zhou Q (2015) Cultivar variations in cadmium and lead accumulation and distribution among 30 wheat (*Triticum aestivum* L.) cultivars. Environ Sci Pollut Res 22:8432–8441

Lombardi L, Sebastiani L (2005) Copper toxicity in *Prunus cerasifera*: growth and antioxidant enzymes responses of in vitro grown plants. Plant Sci 168:797–802

Loscos J, Naya L, Ramos J, Clemente MR, Matamoros MA, Becana M (2006) A reassessment of substrate specificity and activation of phytochelatin synthases from model plants by physiologically relevant metals. Plant Physiol 140:1213–1221

Machado-Estrada B, Calderón J, Moreno-Sánchez R, Rodríguez-Zavala J (2013) Accumulation of arsenic, lead, copper, and zinc, synthesis of phytochelatins by indigenous plants of a mining impacted area. Environ Sci Pollut Res 20:3946–3955

Maitani T, Kubota H, Sato K, Yamada T (1996) The composition of metals bound to class III metallothionein (phytochelatin and its desglycyl peptide) induced by various metals in root cultures of *Rubia tinctorum*. Plant Physiol (Rockville) 110:1145–1150

Manara A (2012) Plant responses to heavy metal toxicity. In: Furini A (ed) Plants and heavy metals. Springer, Dordrecht

Meuwly P, Thibault P, Rauser WE (1993) γ-Glutamylcysteinylglutamic acid—a new homologue of glutathione in maize seedlings exposed to cadmium. FEBS Lett 336:472–476

Mishra S, Srivastava S, Tripathi RD, Govindarajan R, Kuriakose SV, Prasad MNV (2006) Phytochelatin synthesis and response of antioxidants during cadmium stress in *Bacopa monnieri* L. Plant Physiol Biochem 44:25–37

Mithöfer A, Schulze B, Boland W (2004) Biotic and heavy metal stress response in plants: evidence for common signals. FEBS Lett 566:1–5

Mohanpuria P, Rana NK, Yadav SK (2007) Cadmium induced oxidative stress influence on glutathione metabolic genes of *Camellia sinensis* (L.) O. Kuntze. Environ Toxicol 22:368–374

Morelli E, Scarano G (2001) Synthesis and stability of phytochelatins induced by cadmium and lead in the marine diatom *Phaeodactylum tricornutum*. Mar Environ Res 52:383–395
Murasugi A, Chiaki W, Hayashi Y (1981) Cadmium-binding peptide induced in fission yeast, *Schizosaccharomyces pombe*. J Biochem 90:1561–1565
Nelson N (1999) Metal ion transporters and homeostasis. EMBO J 18:4361–4371
Noctor G, Foyer CH (1998) Ascorbate and glutathione: keeping active oxygen under control. Annu Rev Plant Biol 49:249–279
Ortiz DF, Kreppel L, Speiser DM, Scheel G, McDonald G, Ow DW (1992) Heavy metal tolerance in the fission yeast requires an ATP-binding cassette-type vacuolar membrane transporter. EMBO J 11:3491–3499
Ortiz DF, Ruscitti T, McCue KF, Ow DW (1995) Transport of metal-binding peptides by HMT1, a fission yeast ABC-type vacuolar membrane protein. J Biol Chem 270:4721–4728
Osaki Y, Shirabe T, Tamura S, Yoshimura E (2008) A functional putative phytochelatin synthase from the primitive red alga *Cyanidioschyzon merolae*. Biosci Biotechnol Biochem 72:3306–3309
Oven M, Page JE, Zenk MH, Kutchan TM (2002) Molecular characterization of the homo-phytochelatin synthase of soybean *Glycine max*—relation to phytochelatin synthase. J Biol Chem 277:4747–4754
Pal R, Rai J (2010) Phytochelatins: peptides involved in heavy metal detoxification. Appl Biochem Biotechnol 160:945–963
Palmer CM, Guerinot ML (2009) Facing the challenges of Cu, Fe and Zn homeostasis in plants. Nat Chem Biol 5:333–340
Pawlik-Skowronska B (2001) Phytochelatin production in freshwater algae Stigeoclonium in response to heavy metals contained in mining water; effects of some environmental factors. Aquat Toxicol 52:241–249
Pawlik-Skowrońska B, Sanità di Toppi L, Favali MA, Fossati F, Pirszel J, Skowroński T (2002) Lichens respond to heavy metals by phytochelatin synthesis. New Phytol 156:95–102
Pourrut B, Shahid M, Dumat C, Winterton P, Pinelli E (2011) Lead uptake, toxicity, and detoxification in plants. Rev Environ Contam Toxicol 213:113–136
Prasad MNV, de Oliveira Freitas HM (2003) Metal hyperaccumulation in plants: biodiversity prospecting for phytoremediation technology. Electron J Biotechnol 6:285–321
Ramos J, Clemente MR, Naya L, Loscos J, Perez-Rontome C, Sato S, Tabata S, Becana M (2007) Phytochelatin synthases of the model legume *Lotus japonicus*: a small multigene family with differential response to cadmium and alternatively spliced variants. Plant Physiol 143:1110–1118
Ranieri A, Castagna A, Scebba F, Careri M, Zagnoni I, Predieri G, Pagliari M, di Toppi LS (2005) Oxidative stress and phytochelatin characterisation in bread wheat exposed to cadmium excess. Plant Physiol Biochem 43:45–54
Rascio N, Navari-Izzo F (2011) Heavy metal hyperaccumulating plants: how and why do they do it? And what makes them so interesting? Plant Sci 180:169–181
Rauser WE (1990) Phytochelatins. Annu Rev Biochem 59:61–86
Rauser WE (1995) Phytochelatins and related peptides; Structure, biosynthesis, and function. Plant Physiol 109:1141–1149
Rauser WE (1999) Structure and function of metal chelators produced by plants—the case for organic acids, amino acids, phytin, and metallothioneins. Cell Biochem Biophys 31:19–48
Rauser WE (2003) Phytochelatin-based complexes bind various amounts of cadmium in maize seedlings depending on the time of exposure, the concentration of cadmium and the tissue. New Phytol 158:269–278
Rea PA (2012) Phytochelatin synthase: of a protease a peptide polymerase made. Physiol Plant 145:154–164
Reddy AM, Kumar SG, Jyothsnakumari G, Thimmanaik S, Sudhakar C (2005) Lead induced changes in antioxidant metabolism of horsegram (*Macrotyloma uniflorum* (Lam.) Verdc.) and bengalgram (*Cicer arietinum* L.). Chemosphere 60:97–104

Rellan-Alvarez R, Ortega-Villasante C, Alvarez-Fernandez A, del Campo FF, Hernandez LE (2006) Stress responses of *Zea mays* to cadmium and mercury. Plant Soil 279:41–50

Requejo R, Tena M (2005) Proteome analysis of maize roots reveals that oxidative stress is a main contributing factor to plant arsenic toxicity. Phytochemistry 66:1519–1528

Sadi BBM, Vonderheide AP, Gong JM, Schroeder JI, Shann JR, Caruso JA (2008) An HPLC-ICP-MS technique for determination of cadmium-phytochelatins in genetically modified *Arabidopsis thaliana*. J Chromatogr B 861:123–129

Schäfer HJ, Haag-Kerwer A, Rausch T (1998) cDNA cloning and expression analysis of genes encoding GSH synthesis in roots of the heavy-metal accumulator *Brassica juncea* L.: evidence for Cd-induction of a putative mitochondrial γ-glutamylcysteine synthetase isoform. Plant Mol Biol 37:87–97

Schat H, Kalff MM (1992) Are phytochelatins involved in differential metal tolerance or do they merely reflect metal-imposed strain? Plant Physiol 99:1475–1480

Schulz H, Haertling S, Tanneberg H (2008) The identification and quantification of arsenic-induced phytochelatins-comparison between plants with varying As sensitivities. Plant Soil 303:275–287

Shahbaz M, Stuiver CEE, Posthumus FS, Parmar S, Hawkesford MJ, De Kok LJ (2014) Copper toxicity in Chinese cabbage is not influenced by plant sulphur status, but affects sulphur metabolism-related gene expression and the suggested regulatory metabolites. Plant Biol 16:68–78

Sharma P, Dubey RS (2005) Lead toxicity in plants. Braz J Plant Physiol 17:35–52

Shimwell D, Laurie A (1972) Lead and zinc contamination of vegetation in the Southern Pennines. Environ Pollut 3:291–301

Shiyab S, Chen J, Han FXX, Monts DL, Matta FB, Gu MM, Su Y, Masad MA (2009) Mercury-induced oxidative stress in Indian mustard (*Brassica juncea* L.). Environ Toxicol 24:462–471

Shri M, Kumar S, Chakrabarty D, Trivedi PK, Mallick S, Misra P, Shukla D, Mishra S, Srivastava S, Tripathi RD, Tuli R (2009) Effect of arsenic on growth, oxidative stress, and antioxidant system in rice seedlings. Ecotoxicol Environ Saf 72:1102–1110

Shri M, Dave R, Diwedi S, Shukla D, Kesari R, Tripathi RD, Trivedi PK, Chakrabarty D (2014) Heterologous expression of *Ceratophyllum demersum* phytochelatin synthase, *CdPCS1*, in rice leads to lower arsenic accumulation in grain. Sci Rep 4:5784. doi:10.1038/srep05784

Singh R, Singh S, Parihar P, Singh VP, Prasad SM (2015) Arsenic contamination, consequences and remediation techniques: a review. Ecotoxicol Environ Saf 112:247–270

Sreekanth TVM, Nagajyothi PC, Lee KD, Prasad TNVKV (2013) Occurrence, physiological responses and toxicity of nickel in plants. Int J Environ Sci Technol 10:1129–1140

Srivastava S, Mishra S, Tripathi RD, Dwivedi S, Gupta DK (2006) Copper-induced oxidative stress and responses of antioxidants and phytochelatins in *Hydrilla verticillata* (L.f.) Royle. Aquat Toxicol 80:405–415

Srivastava S, Mishra S, Tripathi RD, Dwivedi S, Trivedi PK, Tandon PK (2007) Phytochelatins and antioxidant systems respond differentially during arsenite and arsenate stress in *Hydrilla verticillata* (L.f.) Royle. Environ Sci Technol 41:2930–2936

Steffens J (1990) The heavy metal-binding peptides of plants. Annu Rev Plant Biol 41:553–575

Stolt JP, Sneller FEC, Bryngelsson T, Lundborg T, Schat H (2003) Phytochelatin and cadmium accumulation in wheat. Environ Exp Bot 49:21–28

Sun Y, Zhou Q, Liu WT, An J, Xu ZQ, Wang L (2009) Joint effects of arsenic and cadmium on plant growth and metal bioaccumulation: a potential Cd-hyperaccumulator and As-excluder *Bidens pilosa* L. J Hazard Mater 165:1023–1028

Sylwia W, Anna R, Ewa B, Stephan C, Danuta Maria A (2010) The role of subcellular distribution of cadmium and phytochelatins in the generation of distinct phenotypes of *AtPCS1*- and *CePCS3*-expressing tobacco. J Plant Physiol 167:981–988

Tennstedt P, Peisker D, Bottcher C, Trampczynska A, Clemens S (2009) Phytochelatin synthesis is essential for the detoxification of excess zinc and contributes significantly to the accumulation of zinc. Plant Physiol 149:938–948

Thomas JC, Malick FK, Endreszl C, Davies EC, Murray KS (1998) Distinct responses to copper stress in the halophyte *Mesembryanthemum crystallinum*. Physiol Plant 102:360–368

Thumann J, Grill E, Winnacker EL, Zenk MH (1991) Reactivation of metal-requiring apoenzymes by phytochelatin—metal complexes. FEBS Lett 284:66–69

Tian S, Lu L, Labavitch J, Yang X, He Z, Hu H, Sarangi R, Newville M, Commisso J, Brown P (2011) Cellular sequestration of cadmium in the hyperaccumulator plant species *Sedum alfredii*. Plant Physiol 157:1914–1925

Tripathi RD, Srivastava S, Mishra S, Singh N, Tuli R, Gupta DK, Maathuis FJM (2007) Arsenic hazards: strategies for tolerance and remediation by plants. Trends Biotechnol 25:158–165

Tyler G (1989) Uptake, retention and toxicity of heavy metals in lichens. Water Air Soil Pollut 47:321–333

Van Balen E, Van de Geijn S, Desmet G (1980) Autoradiographic evidence for the incorporation of cadmium into calcium oxalate crystals. Z Pflanzenphysiol 97:123–133

Vatamaniuk OK, Mari S, Lu YP, Rea PA (1999) AtPCS1, a phytochelatin synthase from *Arabidopsis*: isolation and *in vitro* reconstitution. Proc Natl Acad Sci USA 96:7110–7115

Wei S, Zhou Q, Wang X (2005) Identification of weed plants excluding the uptake of heavy metals. Environ Int 31:829–834

Wei S, Teixeira da Silva JA, Zhou Q (2008a) Agro-improving method of phytoextracting heavy metal contaminated soil. J Hazard Mater 150:662–668

Wei SH, Zhou QX, Saha UK (2008b) Hyperaccumulative characteristics of weed species to heavy metals. Water Air Soil Pollut 192:173–181

Wei SH, Zhu JG, Zhou QX, Zhan J (2011) Fertilizer amendment for improving the phytoextraction of cadmium by a hyperaccumulator *Rorippa globosa* (Turcz.) Thell. J Soils Sediments 11:915–922

Wierzbicka M (1986) The effect of lead on the ultrastructure changes in the root-tip of Onion-*Allium cepa* L. Folia Histochem Cytobiol 24:340–341, Vesaluis Medical Publishing, Wislisko, Krakow, Poland

Wójcik M, Tukiendorf A (2004) Phytochelatin synthesis and cadmium localization in wild type of *Arabidopsis thaliana*. Plant Growth Regul 44:71–80

Xiang C, Oliver DJ (1998) Glutathione metabolic genes coordinately respond to heavy metals and jasmonic acid in *Arabidopsis*. Plant Cell Online 10:1539–1550

Xiang CB, Werner BL, Christensen EM, Oliver DJ (2001) The biological functions of glutathione revisited in *Arabidopsis* transgenic plants with altered glutathione levels. Plant Physiol 126:564–574

Xu J, Yang L, Wang Z, Dong G, Huang J, Wang Y (2006) Toxicity of copper on rice growth and accumulation of copper in rice grain in copper contaminated soil. Chemosphere 62:602–607

Xu J, Bravo AG, Lagerkvist A, Bertilsson S, Sjöblom R, Kumpiene J (2015) Sources and remediation techniques for mercury contaminated soil. Environ Int 74:42–53

Yadav SK (2010) Heavy metals toxicity in plants: an overview on the role of glutathione and phytochelatins in heavy metal stress tolerance of plants. S Afr J Bot 76:167–179

Yang X, Baligar V, Martens D, Clark R (1996) Plant tolerance to nickel toxicity: II Nickel effects on influx and transport of mineral nutrients in four plant species. J Plant Nutr 19:265–279

Yang YY, Jung JY, Song WY, Suh HS, Lee Y (2000) Identification of rice varieties with high tolerance or sensitivity to lead and characterization of the mechanism of tolerance. Plant Physiol 124:1019–1026

Yang Q, Zeng Q, Xiao F, Liu X, Pan J, He J, Li Z (2013) Investigation of manganese tolerance and accumulation of two Mn hyperaccumulators *Phytolacca americana* L. and *Polygonum hydropiper* L. in the real Mn-contaminated soils near a manganese mine. Environ Earth Sci 68:1127–1134

Zenk MH (1996) Heavy metal detoxification in higher plants—a review. Gene 179:21–30

Zhang Z, Gao X, Qiu B (2008) Detection of phytochelatins in the hyperaccumulator *Sedum alfredii* exposed to cadmium and lead. Phytochemistry 69:911–918

Zhang ZC, Chen BX, Qiu BS (2010) Phytochelatin synthesis plays a similar role in shoots of the cadmium hyperaccumulator *Sedum alfredii* as in non-resistant plants. Plant Cell Environ 33:1248–1255

Zhang X, Uroic MK, Xie WY, Zhu YG, Chen BD, McGrath SP, Feldmann J, Zhao FJ (2012) Phytochelatins play a key role in arsenic accumulation and tolerance in the aquatic macrophyte *Wolffia globosa*. Environ Pollut 165:18–24

Zhou Q, Song Y (2004) Principles and methods of contaminated soil remediation. Science, Beijing (in Chinese)

Zhu YL, Pilon-Smits EA, Tarun AS, Weber SU, Jouanin L, Terry N (1999a) Cadmium tolerance and accumulation in Indian mustard is enhanced by overexpressing gamma-glutamylcysteine synthetase. Plant Physiol 121:1169–1178

Zhu YL, Pilon-Smits EA, Jouanin L, Terry N (1999b) Overexpression of glutathione synthetase in Indian mustard enhances cadmium accumulation and tolerance. Plant Physiol 119:73–79

General Roles of Phytochelatins and Other Peptides in Plant Defense Mechanisms Against Oxidative Stress/Primary and Secondary Damages Induced by Heavy Metals

M. Inouhe, Y. Sakuma, S. Chatterjee, S. Datta, B.L. Jagetiya, A.V. Voronina, C. Walther, and Dharmendra K. Gupta

Contents

Abstract Phytochelatins (PCs) are nonprotein cysteine-rich oligopeptides having the general structure of (γ-glutamyl-cysteinyl)n-glycine ($n = 2$–11). They are synthesized from the precursor glutathione (a reduced form, GSH) by the activity of

M. Inouhe (✉) • Y. Sakuma
Plant Physiology Laboratory, Department of Biology, Faculty of Science, Graduate School of Science and Engineering, Ehime University, Matsuyama 790-8577, Japan
e-mail: inouhe@sci.ehime-u.ac.jp

S. Chatterjee • S. Datta
Defence Research Laboratory, DRDO, Post Bag 2, Tezpur 784001, Assam, India

B.L. Jagetiya
Laboratory of Plant Physiology and Biotechnology, Department of Botany, M.L.V. Government College, Bhilwara 311001, Rajasthan, India

A.V. Voronina
Physical Technology Institute, Ural Federal University, Mira str., 19, Ekaterinburg, Russia

C. Walther • D.K. Gupta
Institut für Radioökologie und Strahlenschutz (IRS), Gottfried Wilhelm Leibniz Universität Hannover, Herrenhäuser Str. 2, Gebäude 4113, 30419 Hannover, Germany
e-mail: guptadk1971@gmail.com

D.K. Gupta et al. (eds.), *Reactive Oxygen Species and Oxidative Damage in Plants Under Stress*, DOI 10.1007/978-3-319-20421-5_9

phytochelatin synthase (PCS). The biosynthesis is stimulated by several heavy metals (HMs), especially Cd and metalloid As. PCs can bind to various HMs like Cd, As, Cu, Pb, Zn, and Ag, via their sulfhydryl (–SH) and carboxyl (–COOH) groups. The complexations become more stable and massive in vacuole where acid-labile sulfides (S^{2-}) are incorporated to make the PCs–S–HMs conjugates. Both the thiols and S^{2-} are originated from sulfate through a partially common energy-dependent metabolism (sulfur assimilation), which is again enhanced by Cd, besides essential metals (Co, Mg). To date, fundamental roles of PCs and also related iso-peptides such as hPCs in intracellular detoxification and/or transport of HMs are well demonstrated in various plants, especially in experiments targeting genes and enzymes for PC and GSH biosynthesis. However, how they function as a defense molecule in the oxidative stresses or other biological processes are still unknown or conceiving subtle problems. Some of the possible functions are highlighted in this chapter as tentative examples for further discussion: (1) PCs–S–HMs complex as a potent pool/stock of thiols or reducing powers to be reusable for further robustious responses by the tolerant plants against various abiotic and biotic stresses including oxidative stress and (2) PCs as a possible mediator for metal translocation or redistribution via phloem rather than xylem, regardless of a trait of "hyperaccumulator" for HMs in land plants. Apart from the positive roles of PCs in HM-tolerant plants, arguments still hot arise an issue (3) the roles of PCs, GSH, and other thiols as delicate barometer or indicators in the mineral and redox balance and/or homeostasis, in addition to their well-known functions being substrates and antidotes. In the absence of HMs, the levels of PCs are too minute to account for their sufficient bindings to the essential metals. Although GSH is ubiquitous and abundant, it is a multifunctional peptide that rapidly consumed or oxidized for numerous enzymic or nonenzymic antioxidants/redox systems as well as direct substrate for PCS. Eventually, importance of preservation of thiols and sulfide (S^{2-}) as resource for reducing powers in sensitive sessile plants against various oxidative stresses is again emphasized in return for PCs in the HM-tolerant plants in metalliferous habitats.

Keywords Glutathione • Heavy metal • Phytochelatin • Reactive oxygen species • Thiol-sulfide pool

1 Introduction

Degradation of environmental quality due to metal and metalloid pollution has become a universal problem. Dispersions of untreated industrial and municipal wastes are widespread and create instability of natural equilibrium increasing the risk that the toxic pollutants would inflict their harmful effects upon the ecosystem and the individual organisms in the community. Constituting a diverse group of elements, heavy metals (HMs) having a density equal or greater than 4.0 or 5.0 g/cm^3 vary in their chemical characteristics, biological functions, and toxicity (Chatterjee et al. 2007). Most of them are microelements essential and vital for

Table 1 Some symptoms and remarks related to HM attacks in plant functions and metabolisms

HM	Essential	Redox active	ROS yield	Molecules targeted and damages (compositional molecules or enzymes)	HMs-binding ligands	Transporter: membrane (tonoplast)
Cd	n (no)	n (no)	y (yes)	Damages in Zn/Ca/Cu-enzymes, kinases, and DNA. Chorosis, lignification, PCD	PC, GSH, S^{2-}	ZIP, YSL, NRAMP (ABC, CAX, HMA, CDF)
Ni	n	n	y	Mg/Zn-protein, chlorosis, necrosis	GSH, PC, His	ZIP, YSL (CDF)
Pb	n	n	y	Damages in pigments, membrane permeability, enzyme activities	PC, S^{2-}	HMA, ABC? (HMA)
Hg	n	n	y	Membrane, aquaporin, exo-glucanase, other SH enzymes, phytotoxic, plastid	GSH, (PC), S^{2-}	? (ABC)
As[a]	n	n	y	Disturbance of P and Si, damages in DNA and SH enzymes	PC, GSH, Cys	PiT, NIP (ABC)
Zn	y	n	y	Lipid, protein, DNA damage if excess (SOD, RNA polymerase, ADH, CA, many Zn-enzymes)	PC, GSH, MA, CI, NA	ZIP, HMA, YSL (ZAT, HMA, FDR, CDF, CAX)
Co	y	y	y	Chlorosis, necrosis if excess (VitamineB_{12})	PC GSH	HMA (CDF)
Cr	n	y	y	Cell membrane, plastid	GSH, (PC?)	?
Cu	y	y	y	Chromosome or nucleus changes if excess (Cu-enzymes, SOD, cyt-c, cyanin)	GSH, (PC), S^{2-}, NA,	YSL (HMA)
Fe	y	y	y	Phytotoxic if excess (SOD, Cyt-c, Fd, other redox enzymes, heme, plastids)	MU, NA	ZIP, NRAMP, YSL (CDF)

PC phytochelatin (including hPC and iso-PC), *GSH* glutathione (+hGSH), *NA* nicotianamine, *MU* mugineic acids, *His* histidine, *MA* malate, *CI* citrate, *PiT* phosphate transporter, *CA* carbonic anhydrase, *ADH* alcohol dehydrogenase. See text for other abbreviations. After Shahid et al. (2014) and Socha and Guerinot (2014)
[a]Metalloid

plant growth and nutrition at optimum concentrations but toxic at excessive concentrations (i.e., supra-optimal concentration). The others are nonessential and just toxic even at low concentrations, simply called as "toxic HMs." Among the representatives as shown in Table 1, Cd, Pb, Hg, and Ni are very toxic. As is a metalloid but often put in to a member of HMs for convenience mainly because of the highly poisonous chemical property. Those elements are actually listed up in the high rankings applied for the various toxic and hazardous chemicals, i.e., As, Pb, Hg are no.1–3, and Cd is no.7 of the list (CERCLA 2007).

The toxicity of HMs lead to interference with metabolism and other biological activities through the generation of reactive oxygen species (ROS) such as superoxide radical ($O_2^{\bullet-}$), hydroxyl radical ($^{\bullet}OH$), and hydrogen peroxide molecule (H_2O_2), in addition to their direct disruptive functions to the essential enzymes and other molecules (Prasad and Freitas 2003). HMs are divided into two groups (redox-active or -inactive) for the characteristics whether or not having a redox activity by itself to produce free radicals (Viehweger 2014). According to this category, Fe, Cu, Cr, Co, Mn, and V are redox-active, and most of the rests (including Cd, Zn, Ni, Pb, and As, as noted above) are redox-inactive like Al and Na in light metals (Hossain et al. 2012a; Shahid et al. 2014). The redox-active metals are directly involved in the redox reaction in cells resulting in the formation of $O_2^{\bullet-}$ and $^{\bullet}OH$ radicals, while the redox-inactive metals indirectly increase the levels of $O_2^{\bullet-}$, $^{\bullet}OH$, and H_2O_2 by inhibiting various enzyme activities directing towards ROS sequestrations (Hossain et al. 2012b; Shahid et al. 2014). After all HMs are responsible in activating ROS formation and causing strong oxidative stress in plant cells. In conjunction to this commonality, anti-oxidation systems and processes that resume the HMs-induced stresses have attracted attention of scientists and experts (Shahid et al. 2014). Various types of anti-oxidative systems operate using many types of molecules. These are named as the redox (reduction and oxidation) cycling molecules, quenching antioxidant molecules (low molecular substances with anti-oxidation powers), detoxification enzymes (high molecular proteins involved in sequestration of ROS using low molecular antioxidants or substrates), etc. In these systematic strategies, quantitatively abundant molecules used are small antioxidants such as GSH and ascorbic acid (AsA), which reach or exceed 1 mM order at the intracellular concentrations. These play a central role in the maintenance of redox status and the nutritional homeostasis by buffering or pooling of the reducing powers within the soluble organic matters that are usable for the respective cases and places in plants under HMs and/or oxidative stresses. Concurrently, biological roles of these and other antioxidation systems especially in relation to plant's HM stress have been reviewed by several researchers (Shahid et al. 2014).

As measures to deduce the toxic HM ions within plant cells, vacuolar compartmentalization has been also suggested. However, traffic movements of HMs as an inert binding form from the outside to vacuole are necessary through the cytoplasm, the vital site of cells. Here, plant cells produce quite unique HMs-chelating thiol peptides named phytochelatins (PCs) (Grill et al. 1987; Rauser 1999; Inouhe 2005). These peptides were first recognized to be present in Cd-binding complexes in plants and yeast cells exposed to Cd as peptides having a function homologous to metallothioneins (MTs) in animals and other living organisms (Rauser 1999). It has been demonstrated that PCs are derivatives from glutathione (GSH), after all, not a protein unlike MTs (Rauser 1999; Inouhe 2005). The roles of PCs in HMs binding and detoxification have been demonstrated in various plants. However, their roles in the other biological processes and functions are still unknown in plants and other living organisms and now being very important questions to be addressed.

Further, mechanisms that appear independent from PCs and PC-related peptides have been reported from different aspects, i.e., plant species, HMs, and various molecular species involved in HM sequestration and/or transportation. Especially, regulation at transportation level can be useful as HMs defense mechanism by blocking the entrance via traffic channel or eliminating the toxic matter to outside of cells or plants. These defense mechanisms as well as intracellular PC- and antioxidant-mediated mechanisms are all required for normal life cycles of sessile plants that depend on the mineral uptakes and photosynthetic activities under different environmental conditions.

In this chapter, we summarize some information's of HMs-induced oxidative stress in plants, especially focusing on the functions of PCs and other thiols. Thereafter, some pros and cons for their biological functions of them and later are taken for discussion about a possible benefit for the pooling of PCs, GSH, and S metabolites. All these topics are highlighted on what factors are best involved in the decrease in ROS evolution in cells.

2 Input and Impact of HMs

2.1 Route into Plant Cells from Environment

Natural or anthropogenic routes are the major source of Cd contamination in soil. Natural or edaphic stress factors may influence plants development, growth, or productivity due to alteration of concentrations of different bio-reactive metals (Schützendübel and Polle 2002; Chatterjee et al. 2011). Natural phenomenon like Cd-rich rock weathering can enhance natural mineral outcrops which in turn pollute the environment. While, burning of fossil fuels such as coal or oil and the incineration of municipal wastes, cement factories, and as a by-product of phosphate fertilizers are the major anthropogenic sources of the Cd in environment (Mengel et al. 2001; Chen 2005; Kirkby and Johnson 2008; Lux et al. 2011). The concentrations of Cd may be up to 40–300 nM in natural non-polluted soils; however, the concentration may increase with clay concentration up to 1 μg/g dry soil (Wagner 1993; Mengel et al. 2001; Inaba et al. 2005). Availability of Cd to plants is greater in acid soils and its solubility increases with exudates of roots (Mengel et al. 2001; Lux et al. 2011). Delivery of Cd to plant roots is dominated by a transpiration-driven mass-flow process of the soil solution (Sterckeman et al. 2004; Ingwersen and Streck 2005).

Accumulation of HMs such as Ni, Pb, and Hg is also the result of several anthropogenic activities (Gupta et al. 2013a). Nonessential metals like Cd, As, Pb, and Hg present along with the essential one may also enter into the plant systems. Varied tactics are followed by plants in response to HMs toxicity, which include immobilization, exclusion, chelation, and compartmentalization of the metal ions, and expression of the general stress responses (Cobbett 2000). Several plants have been identified that possess the unique capability to live on toxic conditions at HMs contaminated sites and also been found to accumulate a considerable amount of such metals within their biomass (hyperaccumulators). Various studies have shown that natural hyperaccumulators like As hyperaccumulating fern species *Pteris vittata* (Gumaelius et al. 2004) and Ni hyperaccumulating species *Thlaspi caerulescens* (Freeman et al. 2004) can withstand higher amount of metal accumulation without having significant damage within cell system. Further studies on the conspicuous properties and functions of the hyperaccumulators will disclose the different and diverse mechanisms for HMs detoxification by plants (Inouhe et al. 2012).

2.2 *Toxicity to Plant Cells*

Biological impacts of HMs are different by the metal species, as well as plant species, their origins, growth stage, and condition. For example, Cd is very toxic for plants even at low concentrations and often interferes with other essential metals containing enzymes (enzymes of Zn, Fe, Cu, Mn, Mg, and Ca) by displacing these elements (Wagner 1993). Cd primarily damages photosystems and some other enzyme systems in plants. As noted earlier, Cd is a redox-inactive metal, like Zn or Ni, but usually accompanies an oxidative stress by causing a transient depletion of GSH and inhibition of antioxidant enzymes (Romero-Puertas et al. 1999). Thus, the strong and versatile phytotoxicity of Cd in growth, cell death, photosynthesis, and induced lignification, etc. are due to these direct and indirect effects (Hossain et al. 2012b). Ni causes inhibition of growth, chlorosis, necrosis, and flaccidity in plants, and this toxicity is also due to the generation of oxidative stress. Pb affects many processes of plants, causing a decrease in photosynthetic pigments, an increase in membrane permeability, and a disturbance of the mineral nutrition and affecting many enzyme activities. Hg is known to provoke oxidative stress in many plants accompanying overall increases in the antioxidant enzyme systems. Arsenic is not strictly a heavy metal since it is classified as a metalloid. However, it is an important poison, which induces toxicity in plants. Usually, As is present in two toxic inorganic forms, arsenate (AsO_4^{2-}) and arsenite. Arsenate disrupts energy flows in cells and is taken up by plants through high-affinity phosphate transporters. Arsenite provokes toxicity by reacting with sulfhydryl groups of enzymes and tissue proteins and consequently resulted in inhibition of cellular function. Both forms of arsenic induce the formation of ROS leading to oxidative stress.

Apart from these toxic HMs, some others (microelements) are indispensable for living organisms at low doses, but exposure of plants above certain metal threshold concentrations, specific for each one, develops damaging effects linked to disturbances of the oxidative balance. Thus, contrary to other HMs reported above, Cu at an adequate concentration is strictly necessary for plants, since it serves as a cofactor of enzymes required for normal growth and development such as Cu/Zn-superoxide dismutase (Cu/Zn-SOD), cytochrome *c*, or plastocyanin. However, Cu at high concentrations causes multiple toxic effects in plants. Fe is also a key element in a large number of plant metabolic routes requiring a redox exchange. Although Fe is abundant in soils, its availability is depressed in alkaline soils provoking to plant Fe deficiency, which is a common nutritional disorder for many dicotyledonous species. However, an excess of Fe have also phytotoxic effects. Conclusive remarks with examples for the hazardous effects of HMs in plants are also shown in Table 1.

Almost all HMs are potently to be toxic if present in excess as free ions (e.g., Hg^{2+}) or some organo-metallic forms (e.g., methyl-Hg) that are hydrophilic and hydrophobic, respectively. As mentioned above, free HM ions penetrated via root systems are the most probable xenobiotics for plants. Further considerations about the organic forms of HM contaminants are put aside in this chapter, while this will be a serious problem if the environmental pollution and contamination proceed and become more complex in the ecosystems during artificial activities.

2.3 ROS Production

Since pathways in which different ROS evolve are quite complex even in common plants under the influence of HMs, candidate pathways are shown here briefly (Fig. 1). ROS are produced during normal aerobic metabolisms, especially in

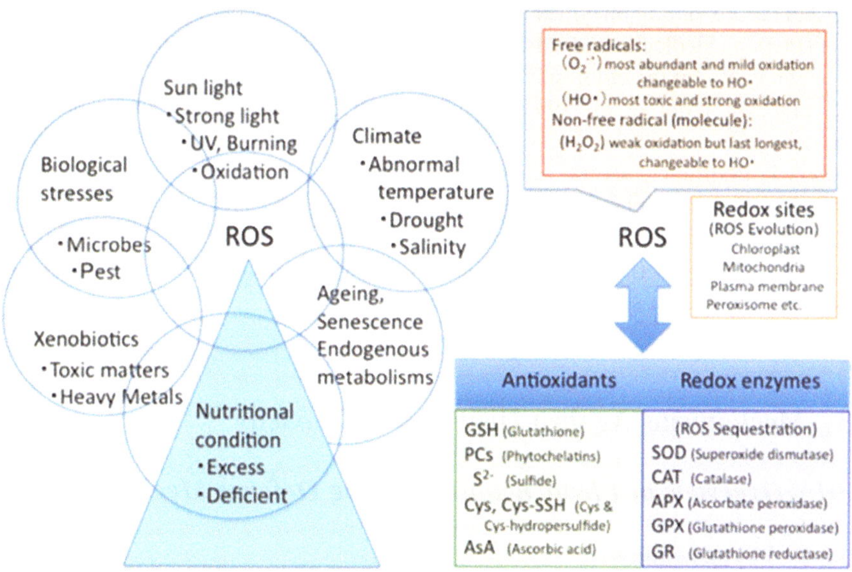

Fig. 1 Reactive oxygen species (ROS) induced by stresses and the possible antioxidants and detoxifications enzymes involved in the ROS sequestration in plants. Different abiotic and biotic stresses including HMs and oxidative stresses induce ROS in different sites in cells of plant tissues. The antioxidants (mainly in water-soluble forms) and enzymes are collaborating to the ROS sequestration in cases and places under stresses. GSH is a multifunctional thiol peptide which is also an important precursor for PCs and other HM-binding substances

chloroplast, mitochondria, peroxisome, apoplast (cell wall), and plasma membrane. HMs enhance the formations of ROS such as superoxide anion ($O_2^{\bullet-}$), hydroxyl radical ($^{\bullet}OH$), and H_2O_2 in those sites by inhibiting the enzymatic or nonenzymatic antioxidant systems (Shahid et al. 2014). The $O_2^{\bullet-}$ is generated when oxygen molecule (O_2) is reduced via electron transfer or energy transfer reactions. This radical is toxic but short-lived and readily converted to (H_2O_2) by the enzyme superoxide dismutase (SOD). The H_2O_2 molecule is weaker in toxicity than $O_2^{\bullet-}$ but stable, long lasting, and permeable across membrane and hence also serves as inter/intracellular second signals for various oxidative stresses. These ROS are able to generate much toxic $^{\bullet}OH$ radical, in the presence of redox-active HMs such as Fe and Cu. This radical is extremely reactive and causes strong oxidation damages in bio-membranes (usually known as lipid peroxidation reaction trigging a self-propagating chain reaction in membranes resulting in serious problems of the cell viability) and in other macromolecules including proteins, DNA, conjugated lipids, and photosynthetic pigments (Gechev et al. 2006; Hossain et al. 2012a, b). After all, ROS interact with HMs resulting in various damages in several cell sites and components (Shahid et al. 2014). ROS production is common in all living cells but the rate of production of ROS in chloroplasts is increased by influence of excessive light energy, and HM contamination is quite unique to plants. Typically, HM stress reduces photosynthesis rate and hence lead to increased production of ROS such as $O_2^{\bullet-}$ and H_2O_2 (Takahashi and Murata 2008). These adverse effects of metal stress can be observed in several places of photosynthesis, including PS I, PS II, and carboxylating enzymes like RuBisCO and phosphoenol pyruvate carboxylase (PEPC) (Siedlecka and Baszynaski 1993; Hossain et al. 2012b). Usually PS II reaction center in the chlorophyll is mostly affected by metals like Cd that replaces Ca and Mn (Atal et al. 1991). Similarly, ROS generation is also evident in mitochondria at complex I and the ubiquinone (Q) zone (Blokhina and Fagerstedt 2006). Furthermore, several reports suggest that NADPH oxidase-dependent ROS induction can take place in response to Cd stress in *Pisum sativum* (Rodrıguez-Serrano et al. 2006), As stress in *Arabidopsis thaliana* (Gupta et al. 2013a), Pb stress in *Vicia faba* (Pourrut et al. 2008), Cd and Cu stress in *A. thaliana* (Remans et al. 2010), and Ni stress in wheat (Hao et al. 2006).

3 Mechanisms Against Heavy Metal Toxicity

3.1 *Overview of Phytochelatin-Binding Defense Mechanism*

3.1.1 Phytochelatins

Because details for research history and topics of findings for PCs are available in reviews (Grill et al. 1987; Cobbett 2000; Inouhe 2005), the framework is shown briefly. PCs are nonprotein cysteine-rich oligopeptides having the general structure of (γ-glutamyl-cysteinyl)*n*-glycine ($n=2$–11). They are synthesized from the precursor GSH by the activity of phytochelatin synthase (PCS), encoded in genes (PCS1, CAD1) isolated in 1999 at three laboratories (Ha et al. 1999; Vatamaniuk et al. 1999; Clemens et al. 1999). The PCs biosynthesis is stimulated by several HMs, especially Cd and metalloid As. PCs can bind to various HMs like Cd, As, Cu, Pb, Zn, and Ag, via their sulfhydryl (–SH) and carboxyl (–COOH) groups, and these complexations become more stable and massive in vacuole where acid-labile S^{2-} are incorporated to make the PCs–S–HMs conjugates. Both the thiols and S^{2-} are originated from sulfate through a partially common energy-dependent metabolism (sulfur assimilation), which is again enhanced by Cd, besides essential metals. Coordinative roles and functional linkages in these sulfur-containing compounds are expectable as discussed in detail later.

PCs and structurally PC-related peptides have been described in various plants and other organisms. Initially, PCs with different degrees of polymerization were reported from more than 300 species of plants and other organisms (Grill et al. 1989; Gekeler et al. 1989). In angiosperms, more than 23 species of monocotyledonous plants and 90 species of dicotyledonous plants were tested, and all of the plants were shown to produce PCs after Cd treatments (Gekeler et al. 1989). The PC synthesis was also confirmed either in suspension cultures or differentiated plant seedlings of various higher plants as well as in the lower plants that belong to groups of mosses or ferns (Gekeler et al. 1989). Such ubiquitous occurrences of PC peptides with the same structures among plant kingdom strongly suggest significant role of PCs as common metabolites in plants, while their physiological roles in the absence of heavy metals are still in open question at present, as described later.

Fundamental roles of PCs in intracellular detoxification are well demonstrated in various plants but especially for Cd (and As). This metal might be nonessential for most living organisms as so far known; nevertheless, why most plants have the most favorite response with PCs to this metal remains as an interesting question. Recent molecular phylogenetics approaches have started disclosing the ubiquitous roles of widespread PCS enzymes and genes in various organisms, which would have been diverged, specified, or converged during more than hundred billions of years of evolutionally time-span. However, evidences still show very low levels of PCs produced in the ancient type of plants like bryophyte as compared with their substantial levels in the some group of fungi (yeast), green algae, and various species of vascular plants (Hayashi et al. 1991; Mehra and Winge 1990; Inouhe et al. 1996; Murasugi 2008). Besides Cd, in response to other HMs stress in plants, PCs may play a significant role in detoxification in higher plants (Cobbett and Goldsbrough 2002) and make a complex, as immobilized metals are less toxic than the free ions. Synthesis and emergence of these metal-binding peptides in plants indicate HMs contamination under various environments (Gupta et al. 2002, 2004). It has been reported that in plants, PCs–HMs complexes form during detoxification process against a wide range metal ions, like Cd, Pb, As, Ag, Hg or Zn, Cu, Ni (Maitani et al. 1996; Mehra et al. 1996; Rauser 1999; Ha et al. 1999; Manara 2012). More direct evidence for the role of PCs against HMs contamination was presented through study on isolated PC-deficient cad1 mutants in heavy metal stress condition. Indeed, such a mutant of *A. thaliana* was more sensitive towards Cd and arsenate (AsO_4^{2-}) than wild-type plants; however, no considerable difference was found for the others like Zn, selenite (SeO_3^{2-}), and Ni (Ha et al. 1999). A PCS-deficient mutant of *Schizosaccharomyces pombe* showed moderate sensitivity to Cu and Hg and modestly to Ag (Maitani et al. 1996; Ha et al. 1999; Manara 2012). Further evidence for Cu-induced triggering of PCs biosynthesis in Cu tolerance has been shown in Cu-tolerant species *Mimulus guttatus*. In contrast, a differential tolerance was reported in *Silene vulgaris*, on exposing root tips to Cu; both the Cu-tolerant and Cu-sensitive ecotypes produced comparable quantity of PCs. It is also manifested that PC–Cu complexes are comparatively transient and relatively poorly sequestered to the vacuole (Schat and Kalff 1992; De Knecht et al. 1994; Cobbett and Goldsbrough 2002). However, in plant *Rubia tinctorum*, exposure to different heavy metals leads to the formation of PC-metal complexes in the roots. Heavy metals ions like Ag, As, Cd, Cu, Hg, and Pb were appeared most effective in stimulation of PCs, though, PC complexes known in vivo were with Cd, Ag, and Cu ions (Maitani et al. 1996; Cobbett and Goldsbrough 2002). Moreover, the key role of PCs against heavy metal stress in plants and detoxification of different heavy metals has been corroborated in many studies. However, why such a change appears in the contribution of PCs to tolerance and detoxification against different HMs is still not fully understood.

3.1.2 Variation in Phytochelatins: Homo- and Iso-phytochelatins

PCs and structurally PC-related peptides have been described in various plants and other organisms. Such a ubiquitous occurrence of the PC peptides with the similar structures among plant kingdom strongly suggests their significant roles as primary metabolites common in the plants. However, here are some exceptional cases for the ubiquity of PCs in some restricted plants and yeast, i.e., some diversity is known for the molecules. PC peptides have Gly in the C-terminal end in general. The presence of some des-Gly variants of PCs in Cd-binding complexes was reported in *S. pombe* (Hayashi et al. 1991) and *Candida glabrata* (Mehra and Winge 1990). They have a structure of (γ-Glu-Cys)n. Similar peptides were not abundant in many higher plants but its substantial level can be found in *Zea mays* roots treated with Cd ions. Furthermore, four other PC-related peptides were discovered from plant sources. They have different amino acid residues at the C-terminal end of (γ-Glu-Cys)n peptides, Ala, Ser, Glu, or Gln. The (γ-Glu-Cys)n-Ala peptides first isolated from plants belonging to Fabaceae (Phaseoleae) are called homo-phytochelatins (hPCs) because they are synthesized from homo-glutathione (hGSH) with the structure of γ-Glu-Cys-Ala. Some other variants of those peptides have been also detected in maize and other plants and named as iso-phytochelatins (iso-PCs), which have the structures of (γ-Glu-Cys)n–Ser, (γ-Glu-Cys)n–Glu, or (γ-Glu-Cys)n–Gln (Cobbett 2000; Rea 2012). Biological roles of these variant peptides have not been well understood for a long time; however, their biochemical functions as thiol peptides are assumed to be basically equivalent to that of PCs. They also have common pathways in metabolisms, at least, some enzymes and precursors such as (γ-Glu-Cys) dipeptidyl units, or glutamate (Glu, E) and cysteine (Cys, C), except glycine (Gly, E), for biosynthesis. Amount and distribution of these PC-related peptides may differ in different plant species; as for example, cells of *A. thaliana* are capable of synthesizing different PCs and iso-PCs (Ducruix et al. 2006). Synthesis of iso-PCs typically depends upon the availability of Gly or GSH synthetase in the cells that helps to switch over to synthesize the peptide (as, e.g., synthesis of dipeptide γ-glutamyl cysteine (γ-EC) when plant comes under stress (Ducruix et al. 2006; Rea 2012). The appearance of the mixture of PCs and iso- PCs such as (γ-Glu-Cys)n–Ser, (γ-Glu-Cys)n–Glu, or (γ-Glu-Cys)n–Gln, and (γ-Glu-Cys)n has been conceivably demonstrated as characteristic of Poaceae family (grasses) under Cd stress by several workers (Klapheck et al. 1994; Cobbett and Goldsbrough 2002) and also under As stress (Zhang et al. 2010; Duan et al. 2011; Batista et al. 2014). As induced hPCs and other variants PCs in *Lotus japonicus* (Ramos et al. 2008). However, it is suggested that for the PCS1 and hPCS enzymes, hGSH is a good acceptor, but a poor donor, of γ-EC units. Purified AtPCS1 and LjPCS1 were activated (in decreasing order) by Cd, Zn, Cu, and Fe, but not by Co or Ni, in the presence of 5 mM GSH and 50 mM metal ions. Activation of both enzymes by Fe was proven by the complete inhibition of PC synthesis by the Fe-specific chelator, desferrioxamine. *Arabidopsis* and Lotus plants accumulated hPCs only in response to a large excess of Cu and Zn, but to a much lower extent than did with Cd, indicating that hPC synthesis may not significantly contribute in vivo to Cu, Zn, and Fe detoxification.

3.1.3 Glutathione and Homo-glutathione

GSH is a direct precursor for PC synthesis but itself a very multifunctional metabolite and antioxidant important for metal tolerance and many other biological processes. GSH synthesis consists of two steps of energy-dependent processes that can occur in the cytosol or in the cell organelle like chloroplasts and mitochondria (Zechmann and Müller 2010). First step is an ATP-dependent rate-limiting reaction catalyzed by γ-EC synthetase (EC 6.3.2.2) producing γ-EC from glutamate and cysteine. Second step is an addition of glycine to γ-EC by GSH synthetase (EC 6.3.2.3) activity (Noctor et al. 2012). Both enzymes (named as GSH1 and GSH2, respectively) are encoded by single genes with alternate transcription initiation sites, and GSH1 is exclusively localized in the plastids, whereas GSH2, albeit also present in the chloroplasts, is to a large extent, a cytosolic protein. Thus, the compartmentalization of GSH synthesis functionally links the different cellular compartments and may provide a platform for intracellular redox signaling (Wachter et al. 2005). These and other data lead to the view that the synthesis of γ-EC is restricted to the plastid but that GSH synthesis can also occur in the cytosol using γ-EC, transported from plastids (Wachter et al. 2005; Noctor et al. 2011). In general, GSH synthetase expression and activity increased concurrently with that of γ-EC synthetase, both of which are otherwise indispensable for early developmental stages in plants. GSH-deficit *Arabidopsis* resulted in increased sensitivity to Cd

(Sengupta et al. 2012). The γ-EC synthetase is a rate-limiting enzyme for GSH synthesis (Noctor and Foyer 1998) whose activity is elevated by the presence of metal ions like Cd^{2+} and repressed by the treatment with buthionine sulfoximine (BSO), a specific inhibitor of this enzyme (Grill et al. 1987; Scheller et al. 1987). GSH plays also an important role in Pb detoxification in *Sedum alfredii*, under stress conditions, where PCs are absent, and chelated Pb, in conjunction with PCs synthesis and complexation, reduces stress in Pb-tolerant plants (Gupta et al. 2010, 2013b). Likewise, other reports suggested that *Vigna radiata* under a long-term stress with water deficit condition showed a decrease in both γ-ECS activity and its transcript levels in roots but with higher mRNA levels during the recovery period (Zagorchev et al. 2013). Homo-glutathione (hGSH) has antioxidant activity and serves functions in the transport of reduced sulfur and as direct substrate for hPC synthesis in legumes, as GSH does for PCs in these and other plants (Sobrino-Plata et al. 2009; Zagorchev et al. 2013). If the PCS or hPCS activities are same or samely reduced, these tripeptide levels become important factors that control the major antioxidative reactions for HMs and ROS sequestration. Here, the γ-ECS or similar enzyme has been shown to be involved in hGSH synthesis, while still unknown for the other iso-peptides. Whereas PC synthases (PCSs) are categorized as the γ-EC dipeptidyltranspeptidase (EC 2.3.2.15) that adds a γ-EC-unit of GSH to PCs or another GSH in vitro (Grill et al. 1989; Loeffler et al. 1989), which has been reported to be effective for the formations of hPCs and other iso-PCs (Ramos et al. 2008). Biochemical functions of hGSH and GSH are therefore basically similar if the metabolic or enzymic backgrounds are fulfilled in plants. It was shown that hGSH is an important regulator of root nodule formation, symbiotic interactions and nitrogen fixation in legumes (Zagorchev et al. 2013). Furthermore, their levels, distributions, and redox balances change differently in specific plants in response to different stress or hormone treatments and the developmental stages of the plants (Becana et al. 2010; Clemente et al. 2012; Zagorchev et al. 2013). The biological and evolutionary importance of hGSH and those for substitutions or deletion of the C-terminal amino acid in GSH to form other isotypes in different plants or organ sites await further investigations.

3.2 Other Mechanisms

3.2.1 Transport

Metal ions are vital for life and therefore maintenance of homeostasis of those ions is tremendously important (Fig. 2). Loss of homeostatic balance of elements may create severe metabolic and physiological dysfunction leading to death or severe illness of the plants. The homeostatic maintenance is a highly regulated process integrating uptake, storage, and secretion, where a number of transporters and antiporters proteins are involved. Inhibition in essential nutrient will decrease the plant vitality and its ability to cope with (metal) stress (Huang et al. 2008). Precise

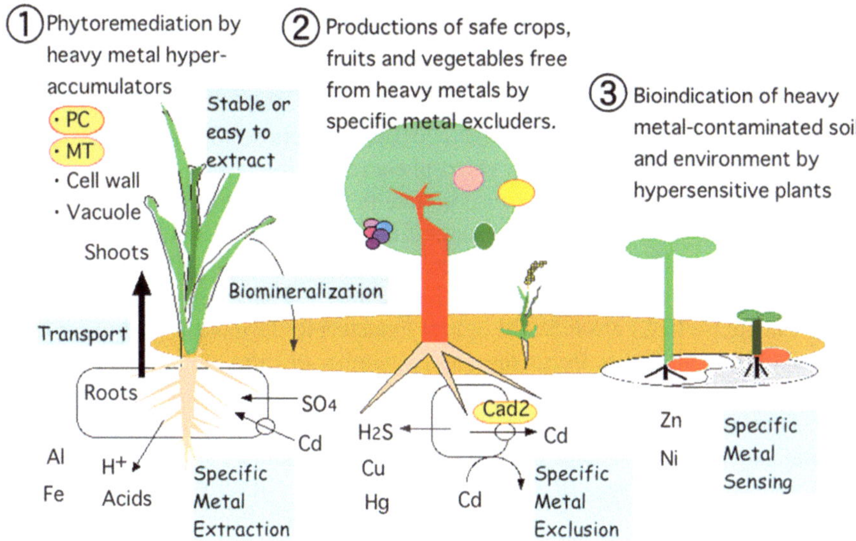

Fig. 2 Impact and route of HM in three ideal types of plants. (1) Hyperaccumulators absorb HMs from roots and transport them to shoots via xylem transport, where various kinds of HM-transports have critical roles. In shoots (and roots), special detoxification/sequestration mechanisms operate. (2) Plants that developed the HM exclusion mechanisms at roots can be useful in agricultural purpose as safety products for other organisms. (3) HM-sensitive plant will be good biological/biochemical index or markers against HMs contaminations

activation of metal-responsive genes to counteract the stress through the synthesis of proteins and signaling molecules related to stress takes place (Maksymiec 2007). Physiological transport of nutrients like Ca, Fe, Mg, Mn, Co, and Zn is unique in plants and some of these are inhibited by HMs. For example, Cd competes with these essential nutrients during transportation through transmembrane energy-dependent nutrient transporters and ion channels (Clemens et al. 1998; Curie et al. 2000; Thomine et al. 2000; Papoyan and Kochian 2004). Cortical tissues of the root help entering metal ions and usually become accumulated in the roots. It gets into the xylem by apoplastic and/or symplastic pathway and further transported to shoots. During the journey, the metal may be complexed by a number of ligands such as organic acids and/or PCs. Here, plant roots have the ability either to exclude and/or chelate or sequester Cd and other HMs from the plant tissues (Cataldo and Wildung 1983; Salt et al. 1995).

HMs-hyperaccumulator plant ecotypes include the Cd-hyperaccumulators such as *Noccaea caerulescens* (J&C Presl.) FK Mey, *Phytolacca americana* L., and *A. halleri* (L.) O'Kane and Al-Shehbazsetc (Lux et al. 2011). These plants also have the defensive mechanism through the production of Cd-chelators (such as organic acids, etc., as described below) other than PCs at the root zone that confining the entry of Cd to the xylem to prevent the metal accumulation in shoot tissues (Liu et al. 2010; Lux et al. 2011). Cd accumulation in shoot of species of the

Caryophyllales and Lamiales was much higher than other species (Broadley et al. 2001). In general, Cd concentrations are mostly (but not always) higher in roots than in shoots, indicating that transportation of Cd to the xylem and phloem is limited in most plants and lowest in seeds, fruits, and tubers (Seregin and Kozhevnikova 2008; Conn and Gilliham 2010). Absorption of HMs in higher plants is a critical issue, where especially rhizosphere region interacts with HMs (Wenzel et al. 2003). They are usually cotransported in the form of cation across the plasma membrane (Manara 2012). Reports suggest that plant roots primarily secrete exudates in its surrounding soil matrix that helps in the chelation of unwanted metals to prevent transportation inside the cell (Marschner 1995). For example, histidine (His) and citrate (CA) are secreted as root exudates to prevent the Ni uptake from the soil (Salt et al. 2000). Pectic sites and a number of extra cellular carbohydrate molecules present on the cell wall play an important role for immobilization of toxic heavy metal ions (Manara 2012). However, HM homeostasis is mainly maintained by transporters present on the plasma membrane. Typical examples of these transporters are the ZIP, the HMA, the YSL, the NRAMP, the CDF, and the CAX families (Williams et al. 2000; Guerinot 2000; Hossain et al. 2012a, b; Sochia and Guerinot 2014), as shown below briefly.

1. The ZIP (zinc-regulated transporter/iron-regulated transporter [ZRT/IRT1]-related protein) family transporters are well characterized for divalent metal uptake, which consists of eight transmembrane domains with similar topology at N- and C-termini exposed to apoplast also containing a histidine-rich domain supposed to involve in specific metal binding (Guerinot 2000; Nishida et al. 2008). ZIP protein gets activated in response to Fe or Zn loading. In *A. thaliana*, IRT1, the founding member of ZIP family, was the first reported transporter in root cells and has an important role in Fe uptake from the soil (Vert et al. 2002). IRT1 can also transport Mn, Zn, and Cd (Korshunova et al. 1999). AtIRT1 in yeast enhanced the Ni-uptake activity (Nishida et al. 2011). Furthermore, AtZIP4 proteins, expressed in roots and shoots, are involved in Zn transport and also helps in Cd uptake from soil into the root cells and Cd transport from root to shoot (Krämer et al. 2007).
2. The HMAs family transporters (P_{1B}-type ATPases that belong to P-type ATPase superfamily) efflux various metal cations across biological membranes (Axelsen and Palmgren 2001). They are basically internal transporters to load Cd and Zn metals into the xylem from the surrounding tissues and act as an efflux pump. The HMAs were categorized as both monovalent and divalent cation transporters (Baxter et al. 2003; Krämer et al. 2007). In *A. thaliana*, AtHMA3 transporter helps in sequestration of a wider range of HMs, and its overexpression increases the tolerance to HMs like Cd, Pb, Co, and Zn (Morel et al. 2009; Manara 2012). In ABC transporter family, AtPDR8 was discovered in the plasma membrane of *A. thaliana* root hairs and epidermal cells that helps in effluxing of Cd and Pb from plasma membrane (Kobae et al. 2006; Kim et al. 2007).
3. Oligopeptide transporters (OPTs) are a group of membrane-localized proteins. The OPT proteins belong to a small gene family in plants, named as the YSL

(yellow stripe-like) subfamily, taken its name from the maize Yellow stripe 1 protein (ZmYS1), and are involved in uptake of Fe by transporting Fe(III)-phytosiderophore complexes (Curie et al. 2000). Heavy metal ions like Fe, Zn, Cu, Ni, and to a lesser extent Mn and Cd are transported by ZmYS1 transporter (Schaaf et al. 2004). Based on sequence similarity with maize gene, eight presumed YSL transporters have been identified in *A. thaliana* (Colangelo and Guerinot 2006). AtYSL1 is expressed in the leaf xylem parenchyma, in pollen, and in young siliques, whereas AtYSL2 is expressed in shoot and root vascular tissues and is present in the lateral plasma membrane, steady with a role in the lateral movement of metals into the veins (DiDonato et al. 2004; Schaaf et al. 2004).

4. Metal Tolerance Proteins (MTPs) are metal efflux transporters in plants that belongs to CDF (cation diffusion facilitator) transporter family involved in the pumping divalent metal cations like Zn, Cd, Co, Fe, Ni, and Mn and transportation from the cytoplasm to the vacuole (Nies 1992; Krämer et al. 2007; Peiter et al. 2007; Montanini et al. 2007; Manara 2012). CDF transporters consist of six transmembrane domains, a C-terminal cation efflux domain, and a histidine-rich region between transmembrane domains IV and V (Mäser et al. 2001) which probably act as a sensor for heavy metal concentration (Kawachi et al. 2008).
5. Natural resistance-associated macrophage protein (NRAMP) transporters such as AtNRAMP3 or AtNRAMP4 are localized in the tonoplast and help in the transport of Fe from the vacuole (Thomine et al. 2003; Lanquar et al. 2005). Overexpression of AtNRAMP3 increases Cd sensitivity and prevents the accumulation of Mn, indicating a possible role in the homeostasis of metals other than Fe (Thomine et al. 2003).
6. The CAX (cation exchanger) proteins are one of five transporter families that constitute the Ca/cation antiporters (CaCA) superfamily (Shigaki et al. 2006; Emery et al. 2012). The CAX family members were first identified as Ca transporters but later it was revealed that they are capable of transporting more kinds of HMs. Typical CAX proteins contain 11 transmembrane domains. They facilitate the redistribution of cations across a membrane using electrochemical energy generated by a proton pump in order to maintain optimal ionic concentrations in the cell (Socha and Guerinot 2014).

Among these, the HM transporters which are involved in transport to vacuole or in exclusion at plasma membrane are effective in reducing HMs levels in the cytological active compartments including cytoplasm and plasmids in the cells, which reduce the toxicity exerted by HMs as free radicals or indirect inducers of ROS in the sites. There are strongly convincing proofs that many hyperaccumulator plants for various HMs are prevailed for these transportation mechanisms via xylem transport systems rather than their special detoxification mechanisms in the cells. For details about the respective functions of the transports in hyperaccumulators or HM-tolerant plants, see recent reviews cited above and others (Hossain et al. 2012b; Socha and Guerinot 2014).

3.2.2 Redox Enzymes

Antioxidant system in plants is an intrinsic defense mechanism that regulates ROS levels according to the cellular requirements at a certain period. This system is actually governed under the catalytic activities by the several cooperative enzymes, named redox enzymes or detoxification enzymes (Fig. 1), which involve superoxide dismutase (SOD; EC 1.15.1.1), monodehydroascorbate reductase (MDHAR; EC 1.6.5.4), ascorbate peroxidase (APX; EC 1.11.1.11), catalase (CAT; EC 1.11.1.6), glutathione peroxidase (GPX; EC 1.11.1.9), dehydroascorbate reductase (DHAR; EC 1.8.5.1), and glutathione reductase (GR; EC 1.6.4.2). These enzymes are regulated under HM stresses and hence consequently participate in the mechanism of protection against oxidative stress mediated by HMs. Glutathione S-transferase (GST; EC 2.5.1.18) that catalyzes the GSH-dependent conjunction with various types of substrate molecules to form thioether bond between them also contribute to detoxification of xenobiotics by conjugation reactions (Sherratt and Hayes 2001). There are many reports that support the positive effects of HM on the enzymic defense mechanisms expressed prior to or simultaneously with the enhanced tolerance characteristics to the stress. The water-soluble compounds such as AsA and GSH are used central substrates (Hossain et al. 2012b), but little is known for the role of PCs and h-PCs in the enzymic antioxidant systems at present. Gene expressions related with stress responses to quench directly ROS and develop further tolerance appear to be mediated by GSH and its oxidates (GSSG). In addition, abiotic stress tolerance through the glyoxalase pathway is widely been reported which consists of glyoxalase I (Gly I; lactoylglutathionelyase; EC 4.4.1.5) and glyoxalase II (Gly II; hydroxy-acylglutathione hydrolase; EC 3.1.2.6) (Hossain et al. 2009).

As mentioned earlier, ROS generation is evident in chloroplast, mitochondria, peroxisome, and apoplast adjacent to membrane. Here, several reports suggest that NADPH oxidase-dependent ROS induction can take place in response to Cd stress in *P. sativum* (Rodrıguez-Serrano et al. 2006), As stress in *A. thaliana* (Gupta et al. 2013a), Pb stress in *V. faba* (Pourrut et al. 2008), Cd and Cu stress in *A. thaliana* (Remans et al. 2010), and Ni stress in wheat (Hao et al. 2006). There is no evidence that cytoplasmic PCs have a role preventing the ROS induction at plasma membrane-associated ROS formation or at apoplast. However, it is acceptable that cell-wall-associated peroxidase catalyzes formation of membrane-permeable H_2O_2 in apoplast and then makes it possible to interact with cytosolic PCs and other thiol peptides. It is interesting that living with the appropriate concentration level of ROS like H_2O_2 can promote plant development and support resistance to environmental stressors by controlling the expression of genes and redox signaling (Neill et al. 2002). However, direct interactions between PCs and ROS that result in induction of any HM-tolerant mechanisms are not yet demonstrated; therefore, their actions might be independent but be synergistic.

3.2.3 Sulfur Assimilation

Sulfur (S) is an essential and ubiquitous element involved in a large number of vital biochemical and physiological processes. It is also responsible for developing hypersensitivity to HMs. Earlier studies on transgenic plants revealed that excessive PC levels helps the plant to accumulate more amounts of HMs without enhancing tolerance conferring HM-hypersensitivity (Lee et al. 2003; Pomponi et al. 2006; Manara 2012). Sulfur is known for its catalytic or electrochemical properties and having a capacity to react with a broad spectrum of agents, like cytotoxic electrophilic organic xenobiotics, HMs, and free radicals. GSH and PCs biosynthesis is highly regulated and coordinates to meet the demand for Cys consuming activities, which indirectly explain the overall S demand by plants. Sulfur requirement by plants vary under the diverse environmental conditions, biotic and abiotic stresses including HMs (Rausch and Wachter 2005). It has been previously observed that the withdrawal of S from the growing medium dramatically decreases the levels of S, Cys, and GSH in plant tissues (Lappartient and Touraine 1996; Lappartient et al. 1999; Saito 2004; Nocito et al. 2007). In addition, importance of PCs in homeostasis of metals, antioxidant property, and also in S metabolism was suggested (Dietz et al. 1999; Cobbett 2000). Furthermore, when more stable and massive Cd-PCs conjugates are formed in vacuole under Cd stress, large quantity of S^{2-} is incorporated to the complexes. All of these thiols and S^{2-} are originated through the common energy-dependent S assimilation using sulfate taken by roots. These processes are summarized in Fig. 3 (top), as a series of five process termed as A to E for convenience, where several transporters and/or enzymes with some intermediates as the key factors interconnecting functions among S-containing substances are also shown. Simultaneous or cooperative stimulation of these processes may result in total increase in the level of S-containing compounds as well as the total reducing power in the plants. The diversity of the components accumulated in the tissues also increases. Therefore, gross activation of the processes by HMs or other stressors may contribute to the stress tolerance mechanisms in many plants. However, it is important to note that such influences on the process or component are quite different in case and place, as shown in Fig. 3 (bottom) as tentative examples. Furthermore, it is already evident that diverse species of S-containing substances have diverse functions in plants. Briefly, Cys is the first organic product of S assimilation in plants and is notably used for synthesis of proteins (amino acids) directly or after converting to methionine (Leustek et al. 2000; Saito 2004). These amino acids are thought to be an important sink for reduced S, as well known as for GSH (Noctor et al. 2011). Such a supply or increased sink for reduced S in plants has been positively correlated with resistance to some pathogens, a phenomenon termed S-induced resistance (SIR) by some researchers (Bloem et al. 2007). Here, tissue contents of GSH or precursors are thought to be the factors linking S nutrition to the responses of plants to fungal and viral infection (Noctor et al. 2011).

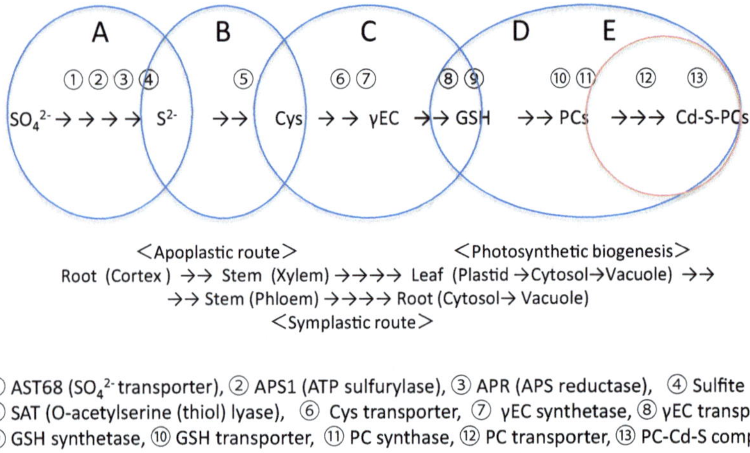

Case	Activation	Ideal Pool-Stock Reservation
1(Cd)	A,B,C,D,E	**Abundant S** (**S^{2-}**, Cys, GSH, **PCs**, **PC-S**)
2(As)	A,B,C,D	**Moderate S** (S^{2-}, Cys, **GSH**, **PCs**, PC-S)
3(Zn)	A,B,C	Shortage S (**S^{2-}**, Cys, **GSH**, PCs, PC-S)
4(Cu)	A (?)	Deficient S (S^{2-}, Cys, GSH, PCs, PC-S)

Fig. 3 Diagram of path for PCs, GSH, Cys, and S^{2-}. *Top*, examples for via root viz shoot circulate paths. (1) AST68 (SO_4^{2-} transporter), (2) APS1 (ATP sulfurylase), (3) APR (APS reductase), (4) Sulfite reductase, (5) SAT (O-acetylserine (thiol) lyase), (6) Cys transporter, (7) γEC synthetase, (8) γEC transporter, (9) GSH synthetase, (10) GSH transporter, (11) PC synthase, (12) PC transporter, (13) PC-Cd-S complexation. *Bottom*, ideal changes in the S and thiol pools as affected by representative HMs. The data are tentative ones and never cover all cases or previously reported respective evidence

Besides their roles as the major storage and transport forms of reduced S, the importance of GSH and Cys have been implicated in the regulation of S metabolism (Kopriva and Rennenberg 2004). GSH inhibits sulfate uptake and S assimilation by repressing the activities and expressions of several functional proteins mediating the earlier steps of the process, such as sulfate transporter (AST), ATP sulfurylase (APS1), and adenosine 5′-phosphosulfate reductase (APR) in plants (Lappartient et al. 1999; Leustek 2002; Noctor et al. 2011). In most studies, the increased concentrations of Cys in response to oxidative stresses have been reported together with increased GSH concentrations, leading to the conclusion that Cys is mainly needed for the biosynthesis of S-rich compounds with antistress activity, such as GSH and stress-related proteins. However, it is generally thought that at concentrations above 50 μM, Cys is toxic for plants (Meyer and Hell 2005). Cys is a potent chelator of heavy metals ions, but the formed Cys-metal complexes can trigger the Fenton reaction, thereby producing the highly toxic $^{\bullet}OH$ radical, and free Cys is often irreversibly oxidized to different by-products. Sulfide is more toxic than sulfate but stabilized in the complex such as Cd-S-PCs. GSH is quickly turning over but the balance of it and its oxidized form (GSSG) are strictly regulated under enzymic redox systems. Therefore, like a general rule for stock chemicals, more stable S-containing substances (or ones under more strict control) may play a central role as a stable state S-donors in plants. Cd-S-PC complex may be a potential stable S donor as well as other S-containing storage proteins or polysaccharides.

3.2.4 Other Mechanisms: Hypothetical View

The detoxification and sequestration mechanisms driven by PCs may be restricted to the HMs and ROS invaded or occurring in the cell compartments in hyperaccumulator plants or tolerant plants grown under naturally metalliferous or artificially contaminated environments. Here, it is important to note that the direct precursor of PCs is GSH, which is multifunctional and thus required for the many other biochemical processes in response to or not to various HMs. In these plants, by blocking or inefficiency of or inefficiency of HMs and other oxidants to the synthesis of PC peptides and other S-containing oligo-/poly- peptides, can increase the small organic and inorganic S-substances such as S^{2-} or Cys, and thus GSH. This can be mimicked by the case if the final S-conjugated products are stored or immobilized in a certain compartment apart from cytoplasm. These increased S-substances are alone or in combination, capable of increasing the binding and quenching of the free radicals of HMs and ROS more. Furthermore, S can make more stable complex with PCs-Cd, and the core Cd-S formed inside of the PC-Cd-S complexes can act as the semi-conductance micro-devise to control electron flow or exchange between two or more different molecules in cytosol or protein complexes on the bio-membrane, while a direct association of PC peptides to other biopolymer/oligomer has not yet been proven. PC has been associated with many important functions in plants against HM ions, including intracellular binding, detoxification, transport, following compartmentalization in vacuole, and also as a substantial coordinator for the distinguished characteristics of several plant-hyperaccumulators of different HM species. However, relatively less attention has been paid on their potential roles in defense mechanism against ROS induced by various abiotic or biotic oxidative stresses, other than by HMs. This is mainly due to the smaller contents of PCs (less than 0.01 mM) than the other antioxidants under normal condition with little contaminated HMs. Intracellular levels of total PCs rise drastically up to 1–10 mM and 0.1–0.5 mM orders as maximum so far reported, when exposed to 0.2–0.8 mM of metal Cd and 0.02–0.5 mM of metalloid As respectively; hence, the detoxification and sequestration mechanisms driven by PCs may be restricted to the HMs and ROS invaded or occurring in the cell compartments in hyperaccumulator plants or tolerant plants grown under naturally metalliferous or artificially contaminated environments. However, more interestingly, PC-HM complex has been detected in phloem sap rather than xylem sap, indicating that PC-functioning site is again intracellular well-reduced state or symplastic place but not oxidative places to produce ROS like apoplastic sites with high oxygen pressure and organelle having active electron transport system. Further assessments are needed but phloem loading and transport will be the attractive performance stage as nutritional PCs.

4 Conclusion and Future Prospective

PCs and related thiol peptides have been associated with many important functions in plants against HM ions, including intracellular binding, detoxification, transport, following compartmentalization in vacuole, and also as a substantial coordinator for the distinguished characteristics of several plant-hyperaccumulators of different HM species. However, relatively less attention has been paid on their potential roles in defense mechanism against ROS induced by HMs and/or other various abiotic or biotic stresses. This is mainly due to the trace levels of PCs (generally less than 10 μM) constitutively maintained in control plants and their trivial changes after either the treatment with HMs besides Cd or the influence of the other abiotic/biotic stresses, where instead evoked are drastic change and activation of the other antioxidants and redox systems. However, we need further consideration for the potential roles of PCs in the plants growing or habituated on the grounds with artificially or naturally HM-dense conditions, as core or polymeric absorbent for nutritional elements and especially S-coordinated substances, resulting also in a dominant storage bank for thio-mediated reduction power sources. As linking to a phenomenon SIR or a concept GSH/thiol pools, it can be urged for our reconsideration that these stocks will play an important role in the maintenance or increase of the robustious characteristics of the plants against various oxidative stresses under biotic and abiotic occasions with no further excess biochemical or metabolic costs. Potential roles of PCs as the key peptidic thiol and reducing agents in the interconnection to other antioxidant compounds and components, such as LWM soluble thiols, AsA, sugars and sugar alcohols, proline and other amino acids and compatible solutes, as well as membrane and cell-wall-associated redox cycle systems will be highlighted in future.

Acknowledgments The authors apologize for the many colleagues who are not referenced in this work due to space limitations.

References

Atal N, Saradhi PP, Mohanty P (1991) Inhibition of the chloroplast photochemical reactions by treatment of wheat seedlings with low concentrations of cadmium: analysis of electron transport activities and changes in fluorescence yield. Plant Cell Physiol 32:943–951

Axelsen KB, Palmgren MG (2001) Inventory of the superfamily of P-Type ion pumps in Arabidopsis. Plant Physiol 126:696–706

Batista BL, Nigar M, Mestrot A, Rocha BA, Júnior FB, Price AH, Raab A, Feldmann J (2014) Identification and quantification of phytochelatins in roots of rice to long-term exposure: evidence of individual role on arsenic accumulation and translocation. J Exp Bot 65:1467–1479

Baxter I, Tchieu J, Sussman MR, Boutry M, Palmgren MG, Gribskov M, Harper JF, Axelsen KB (2003) Genomic comparison of P-Type ATPase ion pumps in Arabidopsis and rice. Plant Physiol 132:618–628

Becana M, Matamoros MA, Udvardi M, Dalton DA (2010) Recent insights into antioxidant defenses of legume root nodules. New Phytol 188:960–976

Bloem E, Haneklaus S, Salac I, Wichkenhauser P, Schnug E (2007) Facts and fiction about sulfur metabolism in relation to plant-pathogen interaction. Plant Biol 9:596–607

Blokhina O, Fagerstedt K (2006) Oxidative stress and antioxidant defenses in plants. In: Singh KK (ed) Oxidative stress, disease and cancer. Imperial College Press, London

Broadley MR, Willey NJ, Wilkins JC, Baker AJM, Mead A, White PJ (2001) Phylogenetic variation in heavy metal accumulation in angiosperms. New Phytol 152:9–27

Cataldo DA, Wildung RE (1983) The role of soil and plant metabolic processes in controlling trace element behavior and bioavailability to animals. Sci Total Environ 28:159–168

CERCLA (2007) Priority list of hazardous substances. http://www.atsdr.cdc.gov/spl/supportdocs/appendix-d.pdf

Chatterjee S, Chattopadhyay B, Mukhopadhyay SK (2007) Sequestration and localization of metals in two common wetland plants of contaminated east Calcutta wetlands: a Ramsar Site in India. Land Contam Reclamat 15:437–452

Chatterjee S, Chetia M, Singh L, Chattopadhyay B, Datta S, Mukhopadhyay SK (2011) A study on the phytoaccumulation of waste elements in wetland plants of a Ramsar site in India. Environ Monit Assess 178:361–371

Chen A (2005) Long distance transport of phytochelatins in Arabidopsis and the isolation and characterization of cadmium tolerant mutants in Arabidopsis. PhD thesis, UC San Diego Electronic Theses and Dissertations. http://escholarship.org/uc/item/0fm285zm

Clemens S, Antosiewicz DM, Ward JM, Schachtman DP, Schroeder JI (1998) The plant cDNA LCT1 mediates the uptake of calcium and cadmium in yeast. Proc Natl Acad Sci USA 95:12043–12048

Clemens S, Kim EJ, Neumann D, Schroeder JI (1999) Tolerance to toxic metals by a gene family of phytochelatin synthases from plants and yeast. EMBO J 18:3325–3333

Clemente MR, Bustos-Sanmamed P, Loscos J, James EK, Pérez-Rontomé C, Navascués J, Gay M, Becana M (2012) Thiol synthetases of legumes: immunogold localization and differential gene regulation by phytohormones. J Exp Bot 63:3923–3934

Cobbett CS (2000) Phytochelatins and their roles in heavy metal detoxification. Plant Physiol 123:825–832

Cobbett C, Goldsbrough P (2002) Phytochelatin and metallothioneins: roles in heavy metal detoxification and homeostasis. Annu Rev Plant Biol 53:159–182

Colangelo EP, Guerinot ML (2006) Put the metal to the petal: metal uptake and transport throughout plants. Curr Opin Plant Biol 9:322–330

Conn S, Gilliham M (2010) Comparative physiology of elemental distributions in plants. Ann Bot 105:1081–1102

Curie C, Alonso JM, Le Jean M, Ecker JR, Briat JF (2000) Involvement of NRAMP1 from *Arabidopsis thaliana* in iron transport. Biochem J 347:749–755

De Knecht JA, van Dillen M, Koevoets PLM, Schat H, Verkleij JAC, Ernst WHO (1994) Phytochelatins in cadmium-sensitive and cadmium-tolerant *Silene vulgaris*: chain length distribution and sulfide incorporation. Plant Physiol 104:255–261

DiDonato RJ Jr, Roberts LA, Sanderson T, Eisley RB, Walker EL (2004) Arabidopsis Yellow Stripe-Like2 (YSL2): a metal-regulated gene encoding a plasma membrane transporter of nicotianamine–metal complexes. Plant J 39:403–414

Dietz KJ, Baier M, Krämer U (1999) Free radicals and reactive oxygen species as mediators of heavy metal toxicity in plants. In: Prasad MNV, Hagemeyer J (eds) Heavy metal stress in plants: from molecules to ecosystems. Springer, Heidelberg

Duan GL, Hu Y, Liu WJ, Kneer R, Zhao FJ, Zhu YG (2011) Evidence for a role of phytochelatins in regulating arsenic accumulation in rice grain. Environ Exp Bot 71:416–421

Ducruix C, Junot C, Fiévet JB, Villiers F, Ezan E, Bourguignon J (2006) New insights into the regulation of phytochelatin biosynthesis in *A. thaliana* cells from metabolite profiling analyses. Biochimie 88:1733–1742

Emery L, Whelan S, Hirschi KD, Pittman JK (2012) Protein phylogenetic analysis of $Ca^{(2+)}$/cation Antiporters and insights into their evolution in plants. Front Plant Sci 3:1

Freeman JL, Persans MW, Nieman K, Albrecht C, Peer W, Pickering IJ, Salt DE (2004) Increased glutathione biosynthesis plays a role in nickel tolerance in Thlaspi nickel hyperaccumulators. Plant Cell 16:2176–2191

Gechev TS, Van Breusegem F, Stone JM, Denev I, Laloi C (2006) Reactive oxygen species as signals that modulate plant stress responses and programmed cell death. Bioessays 11:1091–1101

Gekeler W, Grill E, Winnacker EL, Zenk MH (1989) Survey of the plant kingdom for the ability to bind heavy metals through phytochelatins. Z Naturforsh 44:361–369

Grill E, Winnacker EL, Zenk MH (1987) Phytochelatins, a class of heavy-metal-binding peptides from plants, are functionally analogous to metallothioneins. Proc Natl Acad Sci USA 84:439–443

Grill E, Loffler S, Winnacker EL, Zenk MH (1989) Phytochelatins, the heavy-metal-binding peptides of plants, are synthesized from glutathione by a specific γ-glutamylcysteine dipeptidyl transpeptidase (phytochelatin synthase). Proc Natl Acad Sci USA 86:6838–6842

Guerinot ML (2000) The ZIP family of metal transporters. Biochim Biophys Acta 1465:190–198

Gumaelius L, Lahner B, Salt DE, Banks JA (2004) Arsenic hyperaccumulation in gametophytes of *Pteris vittata*: a new model system for analysis of arsenic hyperaccumulation. Plant Physiol 136:3198–3208

Gupta DK, Tohoyama H, Joho M, Inouhe M (2002) Possible roles of phytochelatins and glutathione metabolism in cadmium tolerance in chickpea roots. J Plant Res 115:429–437

Gupta DK, Tohoyama H, Joho M, Inouhe M (2004) Changes in the levels of phytochelatins and related metal binding peptides in chickpea seedlings exposed to arsenic and different heavy metal ions. J Plant Res 117:253–256

Gupta DK, Huang HG, Yang XE, Razafindrabe BHN, Inouhe M (2010) The detoxification of lead in *Sedum alfredii* H. is not related with phytochelatins but the glutathione. J Hazard Mater 177:437–444

Gupta DK, Inouhe M, Rodríguez-Serrano M, Romero-Puerta MC, Sandalio LM (2013a) Oxidative stress and arsenic toxicity: role of NADPH oxidases. Chemosphere 90:1987–1996

Gupta DK, Huang HG, Corpas FJ (2013b) Lead tolerance in plants: strategies for phytoremediation. Environ Sci Pollut Res 20:2150–2161

Ha SB, Smith AP, Howden R, Dietrich WM, Bugg S, O'Connell MJ, Goldsbrough PB, Cobbett CS (1999) Phytochelatin synthase genes from Arabidopsis and the yeast, *Schizosaccharomyces pombe*. Plant Cell 11:1153–1164

Hao F, Wang X, Chen J (2006) Involvement of plasma membrane NADPH oxidase in nickel-induced oxidative stress in roots of wheat seedling. Plant Sci 170:151–158

Hayashi Y, Nakagawa CW, Mutoh N, Isobe M, Goto T (1991) Two pathways in the biosynthesis of cadystins (gEC)nG in the cell-free system of the fission yeast. Biochem Cell Biol 69:115–121

Hossain MA, Hossain MZ, Fujita M (2009) Stress-induced changes of methylglyoxal level and glyoxalase I activity in pumpkin seedlings and cDNA cloning of glyoxalase I gene. Aust J Crop Sci 3:53–64

Hossain MA, Hossain MD, Rohman MM, da Silva JAT, Fujita M (2012a) Onion major compounds (flavonoids, organosulfurs) and highly expressed glutathione-related enzymes: possible physiological interaction, gene cloning and abiotic stress response. In: Aguirre CB, Jaramillo LM (eds) Onion consumption and health. Nova, New York

Hossain MA, Piyatida P, Teixeira da Silva JA, Fujita M (2012b) Molecular mechanism of heavy metal toxicity and tolerance in plants: central role of glutathione in detoxification of reactive oxygen species and Methylglyoxal and in heavy metal chelation. J Bot 872875:37

Huang HG, Li TX, Tian S, Gupta DK, Zhang X, Yang XE (2008) Role of EDTA in alleviating lead toxicity in accumulator species of *Sedum alfredii* H. Bioresour Technol 99:6088–6096

Inaba T, Kobayashi E, Suwazono Y, Uetani M, Oishi M, Nakagawa H, Nogawa K (2005) Estimation of cumulative cadmium intake causing Itai-itai disease. Toxicol Lett 15:192–201

Ingwersen J, Streck T (2005) A regional-scale study on the crop uptake of cadmium from sandy soils: measurement and modeling. J Environ Qual 34:1026–1035

Inouhe M (2005) Phytochelatins. Braz J Plant Physiol 17:65–78

Inouhe M, Sumiyoshi M, Tohoyama H, Joho M (1996) Resistance to cadmium ions and formation of a cadmium-binding complex in various wild-type yeasts. Plant Cell Physiol 37:341–346

Inouhe M, Huang HG, Chaudhary SK, Gupta DK (2012) Heavy metal bindings and its interactions with thiol peptides and other biological ligands in plant cells. In: Gupta DK, Sandalio LM (eds) Metal toxicity in plants: perception, signaling and remediation. Springer, Heidelberg

Kawachi M, Kobae Y, Mimura T, Maeshima M (2008) Deletion of a histidine-rich loop of AtMTP1, a vacuolar $Zn^{2+}/H+$ antiporter of *Arabidopsis thaliana*, stimulates the transport activity. J Biol Chem 283:8374–8383

Kim DY, Bovet L, Maeshima M, Martinoia E, Lee Y (2007) The ABC transporter AtPDR8 is a cadmium extrusion pump conferring heavy metal resistance. Plant J 50:207–218

Kirkby EA, Johnson AE (2008) Soil and fertilizer phosphorus in relation to crop nutrition. In: White PJ, Hammond JP (eds) The ecophysiology of plant-phosphorus interactions. Springer, Dordrecht

Klapheck S, Fliegner W, Zimmer I (1994) Hydroxymethyl-phytochelatins [(gamma-glutamylcysteine)n-serine] are metal-induced peptides of the Poaceae. Plant Physiol 4:1325–1332

Kobae Y, Sekino T, Yoshioka H, Nakagawa T, Martinoia E, Maeshima M (2006) Loss of AtPDR8, a plasma membrane ABC transporter of *Arabidopsis thaliana*, causes hypersensitive cell death upon pathogen infection. Plant Cell Physiol 47:309–318

Kopriva S, Rennenberg H (2004) Control of sulphate assimilation and glutathione synthesis: interaction with N and C metabolisms. J Exp Bot 55:1831–1842

Korshunova YO, Eide D, Clark WG, Guerinot ML, Pakrasi HB (1999) The IRT1 protein from *Arabidopsis thaliana* is a metal transporter with a broad substrate range. Plant Mol Biol 40:37–44

Krämer U, Talke IN, Hanikenne M (2007) Transition metal transport. FEBS Lett 581:2263–2272

Lanquar V, Lelièvre F, Bolte S, Hamès C, Alcon C, Neumann D, Vansuyt G, Curie C, Schröder A, Krämer U, Barbier-Brygoo H, Thomine S (2005) Mobilization of vacuolar iron by AtNRAMP3 and AtNRAMP4 is essential for seed germination on low iron. EMBO J 24:4041–4051

Lappartient AG, Touraine B (1996) Demand-driven control of root ATP sulfurylase activity and SO42- uptake in intact canola. Plant Physiol 111:147–157

Lappartient AG, Vidmar JJ, Leustek T, Glass AMD, Touraine B (1999) Inter-organ signalling in plants: regulation of ATP sulfurylase and sulfate transporter genes expression in roots mediated by phloem-translocated compound. Plant J 18:89–95

Lee S, Moon JS, Ko TS, Petros D, Goldsbrough PB, Korban SS (2003) Overexpression of Arabidopsis phytochelatin synthase paradoxically leads to hypersensitivity to cadmium stress. Plant Physiol 131:656–663

Leustek T, Martin MN, Bich JN, Davies JP (2000) Pathways and regulation of sulfur metabolism revealed through molecular and genetic studies. Annu Rev Plant Physiol Plant Mol Biol 51:141–165

Leustek T (2002) Sulfate metabolism. In: Arabidopsis book. doi:10.1199/tab.0017

Liu F, Tang Y, Du R, Yang H, Wu Q, Qiu R (2010) Root for aging for zinc and cadmium requirement in the Zn/Cd hyperaccumulator plant *Sedum alfredii*. Plant Soil 327:365–375

Loeffler S, Hochberger A, Grill E, Winnacker EL, Zenk MH (1989) Termination of the phytochelatin synthase reaction through sequestration of heavy metals by the reaction product. FEBS Lett 258:42–46

Lux A, Martinka M, Vaculık M, White PJ (2011) Root responses to cadmium in the rhizosphere: a review. J Exp Bot 62:21–37

Maitani T, Kubota H, Sato K, Yamada T (1996) The composition of metals bound to class III metallothionein (phytochelatins and its desglycyl peptide) induced by various metals in root cultures of *Rubia tinctorum*. Plant Physiol 110:1145–1150

Maksymiec W (2007) Signaling responses in plants to heavy metal stress. Acta Physiol Plant 29:177–187

Manara A (2012) Plant responses to heavy metal toxicity. In: Furini A (ed) Plants and heavy metals, Springer briefs in biometals. Springer, New York. doi:10.1007/978-94-007-4441-7_2

Marschner H (1995) Mineral nutrition of higher plants, 2nd edn. Academic, London

Mäser P, Thomine S, Schroeder JI, Ward JM, Hirschi K, Sze H, Talke IN, Amtmann A, Maathuis FJM, Sanders D, Harper JF, Tchieu J, Gribskov M, Persans MW, Salt DE, Kim SA, Guerinot ML (2001) Phylogenetic relationships within cation transporter families of Arabidopsis. Plant Physiol 126:1646–1667

Mehra RK, Winge DR (1990) Metal ion resistance in fungi: molecular mechanisms and their regulated expression. J Cell Biochem 45:1–11

Mehra RK, Tran K, Scott GW, Mulchandani P, Sani SS (1996) Ag(I)-binding to phytochelatins. J Inorg Biochem 61:125–142

Mengel K, Kirkby EA, Kosegarten H, Appel T (2001) Principles of plant nutrition. Kluwer, Dordrecht

Meyer AJ, Hell R (2005) Glutathione homeostasis and redox regulation by sulfhydryl groups. Photosynth Res 86:435–457

Montanini B, Blaudez D, Jeandroz S, Sanders D, Chalot M (2007) Phylogenetic and functional analysis of the cation diffusion facilitator (CDF) family: improved signature and prediction of substrate specificity. BMC Genomics 8:107

Morel M, Crouzet J, Gravot A, Auroy P, Leonhardt N, Vavasseur A, Richaud P (2009) AtHMA3, a P1B-ATPase allowing Cd/Zn/Co/Pb vacuolar storage in Arabidopsis. Plant Physiol 149:894–904

Murasugi A (2008) Cadystin (phytochelatin) synthesis induced by cadmium and resulted formation of cadmium sulphide nanoparticles in *Schizosaccharomyces pombe*. Curr Top Biotechnol 4:65–73

Neill S, Desikan R, Hancock J (2002) Hydrogen peroxide signalling. Curr Opin Plant Biol 5:388–395

Nies DH (1992) Resistance to cadmium, cobalt, zinc, and nickel in microbes. Plasmid 27:17–28

Nishida S, Mizuno T, Obata H (2008) Involvement of histidine-rich domain of ZIP family transporter TjZNT1 in metal ion specificity. Plant Physiol Biochem 46:601–606

Nishida S, Tsuzuki C, Kato A, Aisu A, Yoshida J, Mizuno T (2011) AtIRT1, the primary iron uptake transporter in the root, mediates excess nickel accumulation in *Arabidopsis thaliana*. Plant Cell Physiol 52:1433–1442

Nocito FF, Lancilli C, Giacomini B, Sacchi GA (2007) Sulfur metabolism and cadmium stress in higher plants. Plant Stress 1:142–156, © Global Science Books (downloaded on 21-2-2013 from: http://globalsciencebooks.info/JournalsSup/images/Sample/PS_1(2)142-156.pdf)

Noctor G, Foyer CH (1998) Ascorbate and glutathione: keeping active oxygen under control. Annu Rev Plant Physiol Plant Mol Biol 49:249–279

Noctor G, Queval G, Mhamdi A, Chaouch S, Foyer CH (2011) Glutathione. The Arabidopsis book. American Society of Plant Biologists, Rockville, MD. doi:10.1199/tab.0142, e0142

Noctor G, Mhamdi A, Chaouch S, Han Y, Neukermans J, Marquez-Garcia B, Queval G, Foyer CH (2012) Glutathione in plants: an integrated overview. Plant Cell Environ 35:454–484

Papoyan A, Kochian LV (2004) Identification of *Thlaspi caerulescens* genes that may be involved in heavy metal hyperaccumulation and tolerance: characterization of a novel heavy metal transporting ATPase. Plant Physiol 136:3814–3823

Peiter E, Montanini B, Gobert A, Pedas P, Husted S, Maathuis FJM, Blaudez D, Chalot M, Sanders D (2007) A secretory pathway-localized cation diffusion facilitator confers plant manganese tolerance. Proc Natl Acad Sci USA 104:8532–8537

Pomponi M, Censi V, Di Girolamo V, De Paolis A, Di Toppi LS, Aromolo R, Costantino P, Cardarelli M (2006) Overexpression of Arabidopsis phytochelatin synthase in tobacco plants enhances Cd^{2+} tolerance and accumulation but not translocation to the shoot. Planta 223:180–190

Pourrut B, Perchet G, Silvestre J, Cecchi M, Guiresse M, Pinelli E (2008) Potential role of NADPH-oxidase in early steps of lead-induced oxidative burst in *Vicia faba* roots. J Plant Physiol 165:571–579

Prasad MNV, Freitas HMO (2003) Metal hyperaccumulation in plants-biodiversity prospecting for phytoremediation technology. Electron J Biotechol 6:285–321

Ramos J, Naya L, Gay M, Abián J, Becana M (2008) Functional characterization of an unusual phytochelatin synthase, LjPCS3, of *Lotus japonicus*. Plant Physiol 148:536–545

Rausch T, Wachter A (2005) Sulfur metabolism: a versatile platform for launching defence operations. Trends Plant Sci 10:503–509

Rauser WE (1999) Structure and function of metal chelators produced by plants: the case for organic acids, amino acids, phytin and metallothioneins. Cell Biochem Biophys 31:19–48

Rea PA (2012) Phytochelatin synthase: of a protease a peptide polymerase made. Physiol Plant 145:154–164

Remans T, Opdenakker K, Smeets K, Mathijsen D, Vangronsveld J, Cuypers A (2010) Metal-specific and NADPH oxidase dependent changes in lipoxygenase and NADPH oxidase gene expression in *Arabidopsis thaliana* exposed to cadmium or excess copper. Funct Plant Biol 37:532–544

Rodrıguez-Serrano M, Romero-Puertas MC, Zabalza A, Corpas FJ, Gomez M, Del Rio LA, Sandalio LM (2006) Cadmium effect on oxidative metabolism of pea (*Pisum sativum* L.) roots. Imaging of reactive oxygen species and nitric oxide accumulation in vivo. Plant Cell Environ 29:1532–1544

Romero-Puertas MC, McCarthy I, Sandalio LM, Palma JM, Corpas FJ, Gómez M, del Río LA (1999) Cadmium toxicity and oxidative metabolism of pea leaf peroxisomes. Free Radic Res 31(Suppl 1):25–31

Saito K (2004) Sulfur assimilatory metabolism: the long and smelling road. Plant Physiol 136:2443–2450

Salt DE, Prince RC, Pickering IJ, Raskin I (1995) Mechanisms of cadmium mobility and accumulation in Indian mustard. Plant Physiol 109:1427–1433

Salt DE, Kato N, Krämer U, Smith RD, Raskin I (2000) The role of root exudates in nickel hyperaccumulation and tolerance in accumulator and nonaccumulator species of Thlaspi. In: Terry N, Bañuelos G (eds) Phytoremediation of contaminated soils and waters. CRC, Boca Raton, FL

Schaaf G, Ludewig U, Erenoglu BE, Mori S, Kitahara T, von Wirén N (2004) ZmYS1 functions as a proton-coupled symporter for phytosiderophore- and nicotianamine-chelated metals. J Biol Chem 279:9091–9096

Schat H, Kalff MMA (1992) Are phytochelatins involved in differential metal tolerance or do they merely reflect metal-imposed strain? Plant Physiol 99:1475–1480

Scheller HV, Huang B, Hatch E, Goldsbrough PB (1987) Phytochelatin synthesis and glutathione levels in response to heavy metals in tomato cells. Plant Physiol 85:1031–1035

Schützendübel A, Polle A (2002) Plant responses to abiotic stresses: heavy metal-induced oxidative stress and protection by mycorrhization. J Exp Bot 53:1351–1365

Sengupta D, Ramesh G, Mudalkar S, Kumar KRR, Kirti PB, Reddy AR (2012) Molecular cloning and characterization of γ-glutamyl cysteine synthetase (VrγECS) from roots of *Vigna radiate* (L.) Wilczek under progressive drought stress and recovery. Plant Mol Biol Rep 30:894–903

Seregin IV, Kozhevnikova AD (2008) Roles of root and shoot tissues in transport and accumulation of cadmium, lead, nickel, and strontium. Russ J Plant Physiol 55:1–22

Shahid M, Pourrut B, Dumat C, Nadeem M, Aslam M, Pinel E (2014) Heavy-metal-induced reactive oxygen species: phytotoxicity and physicochemical changes in plants. Rev Environ Contam Toxicol 232:1–44

Sherratt PJ, Hayes JD (2001) Glutathione *S*-transferases. In: Ioannides C (ed) Enzyme systems that metabolise drugs and other xenobiotics. Willey, New York

Shigaki T, Rees I, Nakhleh L, Hirschi KD (2006) Identification of three distinct phylogenetic groups of CAX cation/proton antiporters. J Mol Evol 63:815–825

Siedlecka A, Baszynaski T (1993) Inhibition of electron flow around photosystem I in chloroplasts of Cd-treated maize plants is due to Cd-induced iron deficiency. Physiol Plant 87:199–202

Sobrino-Plata J, Ortega-Villasante C, Flores-Cáceres ML, Escobar C, Del Campo FF, Hernández LE (2009) Differential alterations of antioxidant defenses as bioindicators of mercury and cadmium toxicity in alfalfa. Chemosphere 77:946–954

Socha AL, Guerinot ML (2014) Mn-euvering manganese: the role of transporter gene family members in manganese uptake and mobilization in plants. Front Plant Sci 5:106

Sterckeman T, Perriguey J, Caël M, Schwartz C, Morel JL (2004) Applying a mechanistic model to cadmium uptake by Zea mays and *Thlaspi caerulescens*: consequences for the assessment of the soil quantity and quality factors. Plant Soil 262:289–302

Takahashi S, Murata N (2008) How do environmental stresses accelerate photo inhibition? Trends Plant Sci 13:178–182

Thomine S, Wang R, Ward JM, Crawford NM, Schroeder JI (2000) Cadmium and iron transport by members of a plant metal transporters family in Arabidopsis with homology to Nramp genes. Proc Natl Acad Sci USA 97:4991–4996

Thomine S, Lelievre F, Debarbieux E, Schroeder JI, Barbier-Brygoo H (2003) AtNRAMP3, a multi specific vacuolar metal transporter involved in plant responses to iron deficiency. Plant J 34:685–695

Vatamaniuk OK, Mari S, Lu YP, Rea PA (1999) AtPCS1, a phytochelatins synthase from Arabidopsis: isolation and in vitro reconstitution. Proc Natl Acad Sci USA 96:7110–7115

Vert G, Grotz N, Dédaldéchamp F, Gaymard F, Guerinot ML, Briat JF, Curie C (2002) IRT1, an Arabidopsis transporter essential for iron uptake from the soil and for plant growth. Plant Cell 14:1223–1233

Viehweger K (2014) How plants cope with heavy metals. Bot Stud 55:35

Wachter A, Wolf S, Steininger H, Bogs J, Rausch T (2005) Differential targeting of GSH1 and GSH2 is achieved by multiple transcription initiation: implications for the compartmentation of glutathione biosynthesis in the Brassicaceae. Plant J 41:15–30

Wagner GJ (1993) Accumulation of cadmium in crop plants and its consequences to human health. Adv Agron 51:173–212

Wenzel WW, Bunkowski M, Puschenreiter M, Horak O (2003) Rhizosphere characteristics of indigenously growing nickel hyperaccumulator and excluder plants on serpentine soil. Environ Pollut 123:131–138

Williams LE, Pittman JK, Hall JL (2000) Emerging mechanisms for heavy metal transport in plants. Biochim Biophys Acta 77803:1–23

Zagorchev L, Seal CE, Kranner I, Odjakova M (2013) A central role for thiols in plant tolerance to abiotic stress. Int J Mol Sci 14:7405–7432

Zechmann B, Müller M (2010) Subcellular compartmentation of glutathione in dicotyledonous plants. Protoplasma 246:15–24

Zhang J, Zhao QZ, Duan GL, Huan YC (2010) Influence of sulphur on arsenic accumulation and metabolism in rice seedlings. Environ Exp Bot 72:34–40

Role of Polyphenols as Antioxidants in Native Species from Argentina Under Drought and Salinization

Mariana Reginato, Celeste Varela, Ana M. Cenzano, and Virginia Luna

Contents

Abstract Plants inhabiting arid environments are exposed to drought and, in several cases, also to salt stress as usually occurs in different regions in Argentina. These plants have developed different strategies to avoid or tolerate the lack of water and/or the excess of toxic ions during their development. As drought and salt stress lead to increased production of reactive oxygen species (ROS) in plant cells, halophytic and xerophytic species have the ability to reduce these toxic ROS by means of a powerful antioxidant system that includes enzymatic and nonenzymatic components. Production of phenolic compounds is one of the strategies used by some native species of these adverse environments, principally to protect their cells from the oxidative damage caused by drought and salinity. This chapter provides an

*Mariana Reginato and Celeste Varela contribute equally to this chapter

M. Reginato • C. Varela • V. Luna (✉)
Laboratorio de Fisiología Vegetal, Departamento de Ciencias Naturales, Universidad Nacional de Río Cuarto, 5800 Río Cuarto, Argentina
e-mail: vluna@exa.unrc.edu.ar

A.M. Cenzano
Laboratorio de Ecofisiología y Bioquímica Vegetal, Centro Nacional Patagónico-Consejo Nacional de Investigaciones Científicas y Técnicas, Boulevard Brown 2915, 9120 Puerto Madryn, Chubut, Argentina

D.K. Gupta et al. (eds.), *Reactive Oxygen Species and Oxidative Damage in Plants Under Stress*, DOI 10.1007/978-3-319-20421-5_10

overview of the oxidative response and the polyphenols involvement as part of a tolerance mechanism in five native species from Argentina: the shrubs *Prosopis strombulifera*, *Larrea divaricata*, and *Lycium chilense* and the grasses *Pappostipa speciosa* and *Poa ligularis* which have shown to have a high polyphenol production correlated with a high antioxidant capacity. Also, two of these native species may be considered as important sources of antioxidants and biomolecules for biotechnological purposes.

Keywords Drought • Salinity • Native species • Oxidative stress • Polyphenols

1 Introduction

Oxidative stress is a central factor in abiotic and biotic stress phenomena and is defined as the chemical toxic effect of reactive oxygen species (ROS) on different plant structures. Oxidative stress occurs when there is a serious imbalance in any cell compartment between ROS production and antioxidant defense, leading to dramatic physiological challenges (Foyer and Noctor 2003). It was considered that ROS concentration needs to be maintained as low as possible, although this concept is changing because of the multiple functions that are currently being discovered for these molecules (Mittler and Blumwald 2010).

ROS production is known to be increased dramatically under different stressing conditions. A rapid increase in ROS production (known as "oxidative burst") has been reported in response to a variety of abiotic stresses including drought, salinity, flooding, heat, cold, ozone, heavy metals, air pollution, nutrient deprivation, excess of light, and UV radiation (Mittler 2002; Miller et al. 2008). Among the above mentioned stresses, drought and salinity are two of the most substantial factors affecting and limiting crop production worldwide. Oxidative stress induced by salinity and drought has been reported by numerous authors (Mittler 2002; Wang et al. 2003; Pang and Wang 2008; Tounekti et al. 2010).

To prevent damage caused by ROS, plants possess a complex antioxidant defense system that is generally able to balance these ROS and keep them at nontoxic levels. This system includes enzymatic and nonenzymatic components. Nonenzymatic antioxidants are vital because they can scavenge ROS that cannot be detoxified by enzymatic systems; polyphenols are an example of a group of compounds that exhibit a high antioxidant activity (Chanwitheesuk et al. 2005). These natural antioxidants are the result of the plant secondary metabolism; they occur in all plant organs and their enhanced synthesis under stressful conditions is believed to protect the cellular structures from oxidative effects (Jaleel et al. 2007). Among these compounds, mainly flavonoids play an important role in the defense against ROS.

Production of phenolic compounds is one of the strategies used by some native species that grow in adverse environments, principally to protect their cells from the oxidative damage caused by drought and salinity.

This chapter provides an overview of the oxidative response and the polyphenols involvement as part of a tolerance mechanism to drought and salinity in five native species from Argentina (*Prosopis strombulifera*, *Larrea divaricata*, and *Lycium chilense* as shrub and *Pappostipa speciosa* and *Poa ligularis* as grasses) that have shown a high polyphenols production correlated with high antioxidant capacity. Also, two of these native species may be considered as important sources of natural products.

2 ROS Production and Oxidative Damage in Plants

ROS is used as a collective term for several O_2-derived radicals like superoxide anion ($O_2^{\bullet -}$), peroxyl radicals (RO_2^-), alkoxyl radicals ($RO^\bullet$), hydroxyl radical ($HO^\bullet$), and also for non-radical singlet oxygen ($^1O_2$) and hydrogen peroxide (H_2O_2). H_2O_2 is mostly produced as a result of dismutation of $O_2^{\bullet -}$, which can occur spontaneously or enzymatically in the cell or apoplast.

ROS integrate signaling pathways involved in plant growth, development, gravitropism, hormonal action, and many other physiological phenomena (Mittler 2002; Apel and Hirt 2004; Foyer and Noctor 2005; Miller et al. 2008). In many of these cases, ROS production is genetically programmed, and ROS are used as second messengers (Foyer and Noctor 2005). Thus, it appears that ROS have pleiotropic effects in plants as they do in animals (Storey 1996).

When ROS are produced in a controlled manner within specific cell compartments, they have key roles in plant growth and development. When ROS are produced in excess, an uncontrolled oxidation occurs, leading to cellular damage and eventual cell death. Thus, it is very important for cells to keep a tight control of ROS concentration (Schützendübel and Polle 2002). Under salinity and drought stresses, stomata are closed as a rapid response to reduce water loss. However, this response limits gas exchange, which is vital for effective photosynthesis. Limitation of gas exchange causes a variety of redox changes in the cell due to variations in the effectiveness of photosynthesis. Decreased availability of CO_2 can be dramatic in terms of ROS production when photosynthetic active radiation (PAR) is high. Uncoupling of light reactions and the Calvin–Benson cycle is the main cause of ROS production in chloroplasts. When electron transport chains are overloaded, $^1O_2$ can be produced in PSII (Asada 2006). Also, photorespiration is another ROS-producing process caused by limited gas exchange under salinity.

Besides photosynthesis related ROS, the production of these species can also occur during respiration in mitochondria; over reduction of the mitochondrial electron transport chain can cause $O_2^{\bullet -}$ production. Electrons can escape from complexes I and III to molecular oxygen if the other electron acceptors in the line are over reduced (Noctor et al. 2007). For more detailed information on the ROS production in chloroplasts and mitochondria, see chapter "Production Sites of Reactive Oxygen Species in Organelles from Plant Cells" by Corpas and colleagues.

Lipids and proteins are the main targets of damage caused by ROS in the cell. ROS can cause the oxidative decomposition of polyunsaturated lipids in membranes, known as lipid peroxidation, which starts a reaction chain that can also create other reactive products such as aldehydes, ketones, and hydroxyl acids and can modify proteins, by oxidation of some amino acid residues.

To prevent those oxidative processes caused by ROS, plants have developed a complex antioxidant defense system that includes enzymatic antioxidants such as superoxide dismutase (SOD), catalase (CAT), peroxidase (POX), and ascorbate peroxidase (APX) and nonenzymatic antioxidants such as ascorbic acid, glutathione, a-tocopherols, and b-carotenes besides the polyphenols mentioned above. Also, other redox regulatory enzymes such as the ascorbate-glutathione cycle enzymes, monodehydroascorbate reductase (MDAR), dehydroascorbate reductase (DAR), glutathione reductase (GR), glutathione peroxidase (GPX), and glutathione-S-transferases (GST), are also key components of the antioxidant defense.

3 Polyphenol Accumulation Under Stress Conditions

Plants may vary widely in their phenolic content and composition, with both genetics and environment affecting the type and level of these compounds (Awika and Rooney 2004; De Abreu and Mazzafera 2005). Generally, their accumulation is stimulated in response to ROS increases under biotic/abiotic stresses (Dixon and Paiva 1995; Roberts and Paul 2006; Julkunen-Tiito et al. 2015).

Phenolic compounds exhibit antioxidant activity in tissues exposed to a wide range of environmental stressors, by inactivating lipid free radicals or preventing decomposition of hydroperoxides into free radicals as mentioned above (Krishnaiah et al. 2011; Agati et al. 2012; Brunetti et al. 2013). In recent years, phenolic compounds research has focused the interest in their antioxidant role as nonenzymatic mechanism (Isabelle et al. 2010; Pollastri and Tattini 2011). Thus, in several species around the world, a role for polyphenols under different stresses has been proposed. To mention some examples, *Larrea tridentata*, a native plant from North America deserts, was found to have a dense resin in the leaves with high amount of phenolic compounds, providing drought tolerance through avoiding transpiration, protecting the photosynthetic system against UV radiation, preventing herbivores, and it can be also involved in allelopathy (Hyder et al. 2002; Martins et al. 2010). *Hypericum brasiliense*, a medicinal herb that produces several phenolic compounds with important pharmacological activity, increased the level of phenolic compounds under water stress (De Abreu and Mazzafera 2005). *Echinacea purpurea*, the purple-coneflower native from the dry hills of North America, enhanced total phenols content under brief drought stress periods (Gray et al. 2003). *Labisia pumila*, a small woody plant from the tropical forest of Malaysia, increases total phenols, anthocyanins, and flavonoids production under severe water stress (Jaafar et al. 2012). In tea plants, water stress increases the production of two important flavan-3-ols (epicatechin and epigallocatechin gallate) with potent antioxidant properties (Hernández et al. 2006). Hawthorns native from North America showed an increase in polyphenols, majority flavonoids, under water and cold stress while under flooding and herbivores it showed a variable response, increasing some polyphenols and decreasing others (Kirakosyan et al. 2004). *Ligustrum vulgare*, a woody plant from the rain forest of Europe and Asia, increased flavonoids and hydroxycinnamates under drought and light stress (Tattini et al. 2004). Moreover, increases in total polyphenol content in different tissues under rising salinity has been reported in a number of plants (Agastian et al. 2000; Muthukumarasamy et al. 2000; Navarro et al. 2006). *Cakile maritima*, a native plant from the Atlantic coast, the Mediterranean Sea, and the Black Sea, enhanced phenolic compounds with high antioxidant capacity under salt stress (Ksouri et al. 2007). Also, different maize varieties showed an increase in phenolic compounds under water and salt stress (Hajlaoui et al. 2009).

4 The Importance of Polyphenols as Antioxidants

From the above reports, it can be deduced that plants produce thousands of phenolic and polyphenolic compounds. Polyphenols constitute one of the most numerous and widely distributed groups of natural products in the plant kingdom. It includes an ample variety of molecules with phenol ring structure. At present, there is not any universal classification for polyphenols, but the most current classification divides this type of molecules in different groups according to their chemical structure (Motilva et al. 2013). There are three main groups: non-flavonoids, flavonoids, and tannins. The non-flavonoids are molecules that have at least one phenolic ring with different reactive groups (hydroxyl, nitrosyl, SH, etc.); this group includes simple phenolic acids, phenyl alcohols, stilbenes, and chalcones (Fig. 1). The flavonoids are characterized by a phenyl chromane structure of C_{15} formed by two aromatic rings bound with a carbon chain (C_6–C_3–C_6) and include antocyanidins, flavonols, flavones, flavonons, isoflavons, flavan-n-ols, etc. The third group, called tannins, is subdivided into two groups: condensed tannins and hydrolysable tannins. Condensed tannins are flavonoid polymers type A (C_7–C_2 and an ether bond) and type B (C_4–C_8 or C_4–C_6) and hydrolysable tannins are phenolic acids polymers bound to a five or six carbon ring (Khanbabaee and Van Ree 2001).

Polyphenols are essential for the physiology of plants. These compounds have been implicated in diverse functional roles, including structure, pigmentation, lignification, plant resistance against microbial pathogens, and animal herbivores such as insects (antibiotic and anti-feeding actions), protection against solar radiation (screen against DNA-damaging UV-B light), and they were probably important during early terrestrial plant evolution (Hatier and Gould 2008; Lattanzio et al. 2008). The majority of polyphenols are synthesized by the highly branched phenylpropanoid pathway, which is responsible for the biosynthesis of a large

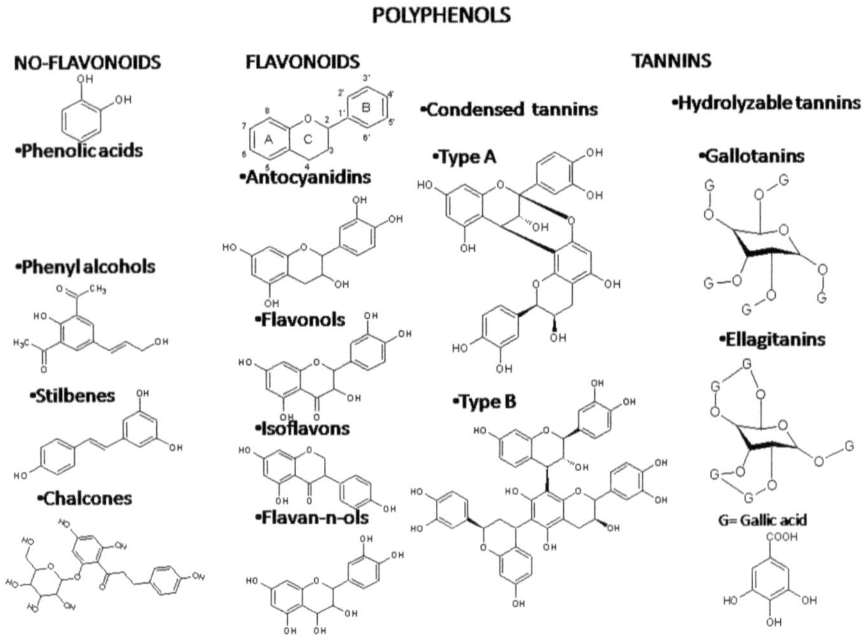

Fig. 1 Main groups of polyphenols according to their chemical structure. *G* gallic acid

number of chemical compounds with considerable structural diversity (Duthie et al. 2003).

Plant polyphenols exhibit a wide range of bio-physicochemical properties bundled within the phenol functional group, which makes them remarkably versatile metabolites and unique and intriguing natural products. The major characteristic of polyphenols is their capability to scavenge ROS, as well as oxidatively generate free radicals RO$^{\bullet}$ and ROO$^{\bullet}$ such as those derived from biomolecules like the low-density lipoproteins proteins (LDLs) (Neudörffer et al. 2004) and oligonucleic acids (DNA and RNA) (Li et al. 2000).

In its most elementary structural form, namely a benzene ring bearing a hydroxy group (PhOH), a phenol function constitutes an amphiphilic moiety that combines the hydrophobic character of its planar aromatic nucleus with the hydrophilic character of its polar hydroxy substituent, which can act either as a hydrogen-bond donor or as an acceptor. Hydrophobic p-stacking (van der Waals) interactions and the formation of hydrogen bonds are seemingly dichotomic, yet are often complementary effects that plant phenolics can use to interact physically with other biomolecules, among which proteins are often first in line (Dangles and Dufour 2008).

Polyphenols are also able to chelate transition metals through their multiple hydroxy groups on a phenyl ring open (Andjelkovic et al. 2008). By chelating metal ions, such as iron(II)/copper(I) and iron(III)/copper(II) that are involved in the conversion of $O_2^{\bullet-}$ and H_2O_2 into highly reactive hydroxyl radicals (HOl), polyphenols can act as DNA protective agents from free radicals damages (Heim et al. 2002; Perron and Brumaghin 2009; Leopoldini et al. 2011). This capacity for metal chelating also contributes to plant pigmentation (Lattanzio et al. 2008) as well as cationic nutrient (e.g., Ca, Mg, Mn, Fe, Cu) cycling through plant–litter–soil interactions.

Polyphenols act as scavengers of free radicals and ROS, when these are unable to be subdued by the regular action of other antioxidants such as glutathione (GSH), glutathione peroxidase, or superoxide dismutase or by vitamins (e.g., vitamins E and C, carotenoids). Two main intimate antioxidative mechanisms have been proposed for polyphenols (Wright et al. 2001).The first is based on the capacity of the phenol functional group to donate a hydrogen atom to a free radical $R^{\bullet}$ such as peroxy radicals $LOO^{\bullet}$ generated during lipid (LH) autoxidation (peroxidation; LH $\rightarrow L^{\bullet}$, then $L^{\bullet} + {}^3O_2 \rightarrow LOO^{\bullet}$) such as peroxy radicals $LOO^{\bullet}$ generated during lipid peroxidation, acting as chain-breaking antioxidants. The second mechanism is the single-electron transfer from the phenolic antioxidant (ArOH) to a free radical $R^{\bullet}$ with formation of a stable radical cation $ArOH^{\bullet+}$.

5 Oxidative Stress and Phenolic Compounds in Native Species from Argentina

5.1 Xerophytic Species from the Patagonian Monte

Argentina has a large territorial extension with a variety of climates and plant species. The Phytogeographic Province of Monte is the most arid rangeland of Argentina located from 24°15′ S to 44°20′ S and 64° to 68° W (Fernandez and Busso 1999). The Southern part of Monte is characterized by strong water deficits in spring and summer, high evaporation enhanced by westerly winds, mean annual temperature of 13.9 °C, annual rainfalls between 200 and 260 mM, and unpredictable rainfall events (Fernandez and Busso 1999; Campanella and Bertiller 2008). Thus, in this region there are many native species exposed to severe drought periods, which are mostly evergreen or deciduous shrubs and perennial grasses. Within the shrubs we can find *Larrea divaricata, Chuquiraga hystrix, Lycium chilense, Junellia alatocarpa, Condalia microphylla, Prosopidastrum globosum, Schinus johnstonii, Monttea aphylla, Atriplex lampa, Chuquiraga avellanedae, Prosopis denudans, and Prosopis alpataco* and within the grasses *Stipa tenuis, Poa ligularis, Pappostipa speciosa* and *Stipa humilis* are the most common species (Beeskow et al. 1987; Ares et al. 1990). All these species have developed different strategies to survive and to avoid oxidative damage in their tissues through the production of secondary metabolites with antioxidant capacity.

Four of these native species were selected to carry out an ecophysiological study on the relationship between phenolic compound production, antioxidant capacity, and lipid peroxidation throughout the annual seasons in year 2013. During this year, the rainfalls were concentrated in spring rendering the highest soil water content. The lowest soil water content was found in autumn. The highest temperatures were measured in summer and the lowest in winter. Taking into account the water availability in the soil and the evapotranspiration rate, it was assumed that plants were exposed to drought stress throughout the whole year being more severe during autumn and/or summer.

Among the dominating shrubs, *Larrea divaricata* (chaparral) and *Lycium chilense* (coralillo) are common species, and among grasses *Poa ligularis* (blue-grass) and *Pappostipa speciosa* (desert needlegrass) are common species (Ares et al. 1990). *Larrea divaricata* and *Pappostipa speciosa* are evergreen species grouped as xerophytic and resource-conservative species, whilst *Lycium chilense* and *Poa ligularis* are deciduous species grouped as mesophytic and resource-acquisitive species (Campanella and Bertiller 2008). From an ecophysiological point of view, species from the first group show morpho-functional traits typical of drought tolerance mechanisms and species from the second group show drought avoidance mechanisms (Campanella and Bertiller 2008; Cenzano et al. 2013, 2014).

In our study, the evergreen shrub *L. divaricata* increased total phenols and flavonoids production in autumn and consequently a major antioxidant capacity along with an increase in lipid peroxidation. *L. chilense*, a deciduous shrub which loses their leaves during autumn and winter, did not show differences in polyphenols production and antioxidant capacity among seasons, but the lipid peroxidation was higher in summer. In comparison with *L. divaricata*, the total phenols content by this species was two times smaller and flavonoids content was ten times smaller. These results were interpreted in the light of the drought avoidance mechanism displayed by the deciduous *L. chilense* which would have no need to waste carbon resources in polyphenols production. On the contrary, *L. divaricata* produces polyphenols as part of its tolerant mechanism to cope against drought.

P. speciosa, an evergreen grass, increased total phenols and flavonoids during autumn and winter in concordance with an enhanced antioxidant capacity and increased lipid peroxidation. *P. ligularis*, a deciduous grass, increased total phenols and flavonoids with high antioxidant capacity in the roots during autumn, accompanied by high lipid peroxidation. These results suggest that phenolic compounds, principally flavonoids, have an important antioxidant role during the dry seasons, especially in the species that maintain their leaves (*L. divaricata* and *P. speciosa*). Additionally, condensed tannins production was higher in roots than in leaves in the four species analyzed during the four seasons suggesting that this kind of molecules may have a structural role combined with an antioxidant role as reported by Ayres et al. (1997) and/or have an allelopathic role limiting growth of others plants, insects attacks, and/or microorganism invasion from the environment (Thelen et al. 2005; Li et al. 2010).

5.2 *Prosopis strombulifera*, a Native Halophyte

Several studies have established that most of the salt tolerance mechanisms used by halophytes are similar to those displayed by glycophytes (Sengupta and Majumder 2010; Bartels and Dinakar 2013). Nonetheless, although the salt tolerance mechanisms are similar, halophytes may either constitutively turn on salt tolerance mechanisms or exhibit changes in their transcriptional and posttranscriptional regulation, which causes large variation in the salt tolerance levels between glycophytes and halophytes (Oh et al. 2010; Kosová et al. 2013).

Halophytes could be defined as plants that are well adapted to live in soils containing a high concentration of salts and benefiting from it; thus, this group of plants represents a very interesting model to understand complex physiological and genetic mechanisms of salt tolerance.

P. strombulifera (Burkart 1976) is a halophytic spiny shrub widely distributed in America and particularly abundant in high-salinity areas of central Argentina. In these soils, proportions of two salts, NaCl and Na_2SO_4, are generally similar, although in previous studies we found that Na_2SO_4 was more abundant than NaCl (up to three times in several soil samples) (Sosa et al. 2005). Similarly, this situation occurs in many countries such as Pakistan, India, Egypt, China, Tunisia, and some regions of United States, where Na_2SO_4 is highly abundant, even more than NaCl in the soils and groundwater of some areas (Bie et al. 2004; Shi and Sheng 2005; Manivannan et al. 2008; Tarchoune et al. 2010; Peterson and Murphy 2015). However, most studies concerning salt tolerance of plants have been focused only in NaCl, and injury symptoms are often ascribed to the toxicity of Na^+ and Cl^- ions. For that reason, it is important to analyze the effects of different salts on plant growth, in order to have a complete view and to understand better the physiological responses of plants in natural environments.

In previous studies using NaCl, Na_2SO_4, and their iso-osmotic mixture, a great variability in the response of *P. strombulifera* plants depending on the type of salt (s) used and the osmotic potential (Ψo) generated in the culture medium was observed. *P. strombulifera* responded to NaCl treatment by increasing their growth up to 500 mM (−1.9 MPa) which is an interesting halophytic response, distinct from findings in other woody *Prosopis* species (Felker 2007). Moreover, our previous studies indicate that the NaCl tolerance of *P. strombulifera* exceeds the limits described for most halophytic plants (Catalán et al. 1994).

However, *P. strombulifera* has different response to Na_2SO_4, being much less tolerant and showing a strong growth inhibition from the beginning of salinization. These plants grown in the presence of Na_2SO_4 showed immediate and sustained reduction of shoot height and leaf number per plant that was accompanied by senescence symptoms such as chlorosis, necrosis, and leaf abscission (Reinoso et al. 2005; Reginato et al. 2012). Furthermore, treatment with Na_2SO_4 induced several structural alterations in cells and tissues and modification of growth patterns in roots, stems, and leaves that were different to that induced by NaCl (Reinoso et al. 2004, 2005). *P. strombulifera* responds to progressive salt stress by changing leaf development, particularly Na_2SO_4-treated plants, leading to modifications in leaf size and micro-morphology of leaf cells (increase in stomata density and epidermal cell density) (Reginato et al. 2013). These anatomical modifications are consistent with our previous physiological studies which demonstrated that the adaptive success of *P. strombulifera* grown under high NaCl salinity seems to involve: (1) a delicate balance among Na^+ accumulation (and its use for osmotic adjustment) and efficient compartmentation in vacuoles (Reginato et al. 2014a); (2) the ability of the whole plant to maintain a sufficient K^+ supply by a high degree of K^+/Na^+ discrimination; (3) maintenance of normal Ca^{2+} levels in leaves (Reginato et al. 2014a); and (4) osmotic balance and protection by compatible solutes like proline, polyols (Llanes et al. 2013), and polycations such as polyamines under salt stress (Reginato et al. 2012).

Salt-induced damage to cellular membranes due to lipid peroxidation was also studied in *P. strombulifera*. This halophyte showed an important oxidative damage induced in tissues when the SO_4^{-2} anion was present in the medium, showing once again a greater sensitivity to Na_2SO_4 than to NaCl. The significant increase in malondialdehyde (MDA) that resulted from lipid peroxidation under Na_2SO_4 treatment (Reginato et al. 2014b) was correlated with the growth inhibition and several metabolic disorders mentioned above. Also, H_2O_2 concentration was significantly increased in roots of Na_2SO_4 and Na_2SO_4 + NaCl-treated plants at moderate salinity (−1.9 MPa) in correlation with increased lipid peroxidation in the former. Recently, several authors have demonstrated that H_2O_2 accumulation under highly saline concentrations acts as a signal for adaptive responses to stress (Miller et al. 2010). Therefore, a tight control of H_2O_2 concentration is critical for cell homoeostasis.

An interesting observation in our studies is that when SO_4^{-2} and Cl^- anions are both present in the growth medium (in the salt mixture treatment), ionic interactions occur possibly at the membrane level between both anions, causing a partial reversion of the oxidative damage caused by SO_4^{-2} in roots. This response is in agreement with the observation that NaCl + Na_2SO_4-treated plants showed intermediate values between those obtained with mono-saline treatments in growth parameters, compatible solute synthesis, and ion content, as previously demonstrated (Llanes et al. 2013; Reginato et al. 2014a). In regards to photosynthetic pigments, *P. strombulifera* showed a good ability to tolerate elevated NaCl concentrations without changes in chlorophyll levels, while Na_2SO_4 stress significantly reduced chlorophylls with respect to controls.

Carotenoids and vitamins are other typical compounds that exhibit high antioxidant activities besides polyphenols (Chanwitheesuk et al. 2005). Similar to the results reported by Ramani et al. (2006) in the halophytes *Sesuvium portulacastrum* and *Aster tripholium*, carotenoid levels remained unchanged in *P. strombulifera* plants (Reginato et al. 2014b). Carotenoids (β-carotene in particular), in addition to their role as secondary light-absorbing pigments, are able to reduce the Chl triplet state and to prevent the formation of the harmful singlet oxygen or to scavenge it after its production by the interaction of triplet chlorophyll with O_2 (Ramani et al. 2006). In our experiments, Na_2SO_4-grown plants showed unchanged carotenoid level despite reduction in chlorophyll concentration, resulting in an increased carotenoid/chlorophyll ratio which may represent a strategy to protect photosystems against photooxidation. One of the most effective mechanisms of excess energy dissipation is de-epoxidation of violaxanthin to antheraxanthin and zeaxanthin through the xanthophyll cycle (VAZ) (Demmig-Adams and Adams 1992). Such a protective mechanism seems to be carried out by salinized plants of *P. strombulifera*, principally those grown in the presence of Na_2SO_4 with the maximal de-epoxidation index (DEPS index), indicating the need to alleviate excessive excitation pressure (Reginato et al. 2014b). Concomitantly, these plants showed a remarkable decrease in the maximal photochemical efficiency (Fv/Fm) and electron transport rate at the end of the experiment (Devinar, PhD Thesis). If there were inactive units in photosystem II, there would be great potential for ROS formation.

Several studies showed that halophytes have higher constitutive antioxidant defense activity as compared with glycophytes. Some authors proposed that antioxidant enzymes in halophytes enhance their activity with increasing salt concentrations at which plants are exposed (Bose et al. 2014). In relation to antioxidant enzymes, catalase (CAT) and superoxide dismutase (SOD) activities in *P. strombulifera* exhibited differential responses in different organs to different salts. NaCl-treated plants showed an increase in CAT activity in roots under moderate and high salinity (−1.9 MPa, −2.6 MPa). Nevertheless, leaves showed an increase in CAT activity at moderate salinity followed by a marked decrease at high salinity. In Na_2SO_4-treated plants, a gradual decrease in CAT activity in roots was observed, while leaves showed a marked increase at high salinity. In contrast, a strong SOD activity was observed both in leaves and roots of these plants, with activity levels much higher than in NaCl-treated plants. In Na_2SO_4 + NaCl-treated plants, while CAT activity decreased with salinity, SOD activity decreased in roots but increased in leaves. From these results, it seems that CAT would play an important role in early detoxification of H_2O_2 in roots, while SOD would have a predominant role in leaves for anion superoxide detoxification. According to our previous results, we propose that at higher salt concentrations the antioxidant enzymatic activity allows H_2O_2 to act as a stress signal to trigger some adaptive physiological responses such as early suberification and lignification in roots (Reinoso et al. 2004, 2005). The intense activity of SOD in Na_2SO_4-treated plants indicates an effort of plants to counteract the severe oxidative stress caused by this salt (Fig. 2).

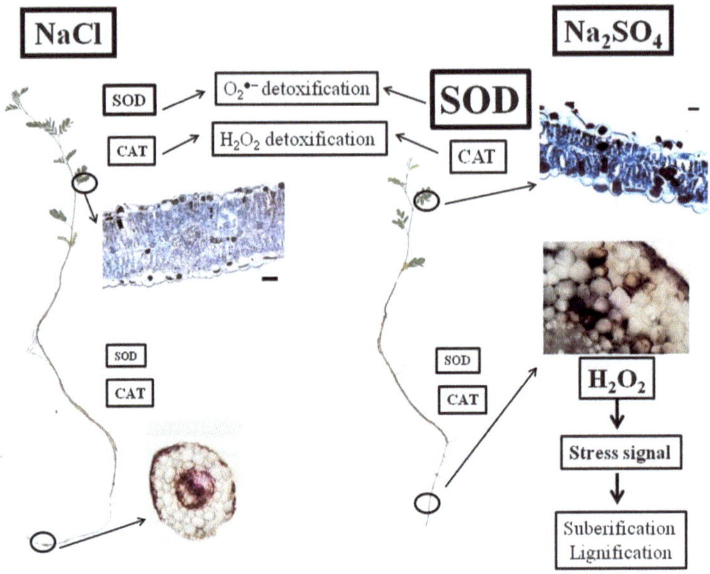

Fig. 2 Effects of NaCl and Na_2SO_4 on H_2O_2 accumulation, SOD and CAT activity, and polyphenol production in *P. strombulifera* plants ($\Psi o = -2.6$ MPa, 48 days). Areas of squares for each enzyme are approx. proportional to their activity. In leaves, cross section of leaflets of 48-days-old plants showing reduction in mesophyll thickness and polyphenols accumulation in Na_2SO_4-treated plants (scale = 50 μm). In roots, cross section indicating the H_2O_2 localization in cells (scale = 50 μm). In Na_2SO_4-treated plants, H_2O_2 can act as a stress signal to trigger some adaptive physiological responses such as early suberification and lignification in roots. In leaves of these plants, the intense accumulation of polyphenols may indicate a role for these compounds in counteracting the strong oxidative damage induced by severe salt stress when other ROS-detoxifying systems are not effective enough

5.2.1 Synthesis of Polyphenols: An Expensive Cost to Survive

As mentioned above, to counteract oxidative stress induced by salinity, plants have developed different strategies including the stimulation of the synthesis of secondary metabolites.

In the halophyte *P. strombulifera*, NaCl treatment did not increase the synthesis of polyphenols, differently from Na_2SO_4 treatment which induced a sharp increase in total polyphenols and consequently, in the antioxidant activity in both leaves and roots. Na_2SO_4 treatment sharply induced an accumulation of flavonoids and flavan-3-ols in leaves (40 % approx.) that accumulated mainly in mesophyll and epidermal cells, as evidenced by microscopical analysis (Reginato et al. 2014b) in coincidence with Tattini et al. (2005) who proposed that the main sites of flavonoid accumulation in plants (including glycosylated forms) are the mesophyll, epidermis, and subepidermis of photosynthetic tissues. A similar increase in flavonoids and flavan-3-ols was observed (30 % approx) in roots. HPLC analysis of extracts from leaves of these plants showed high levels of different polyphenols as rutin, catechin, epicatechin, and proanthocyanidins.

The increase in total flavonoids and flavan-3-ols when SO_4^{2-} anion is present in the growth solution may indicate a role for these compounds in counteracting the strong oxidative damage induced by severe salt stress. Thus, when other ROS-detoxifying systems such as the xanthophyll cycle are not effective enough as it seems to be the case of Na_2SO_4-treated plants (Reginato et al. 2014b), polyphenols production is increased as an alternative detoxifying system. Accordingly, Agati et al. (2012) reported that antioxidant mesophyll flavonoids, at micromolar range, may effectively avoid reactive oxygen forms generation (e.g., by chelating transition metal ions).

Nevertheless, despite their protective functions, stress-induced increase of secondary metabolites was often counteracted by a corresponding decrease in biomass in several species (Selmar and Kleinwächter 2013). In coincidence with these observations, in *P. strombulifera*, the large increase in total polyphenols found in Na_2SO_4-treated plants was accompanied by a strong growth inhibition. In contrast, in NaCl-treated plants, polyphenols did not increase with salinization and growth was not affected; plants were healthy and without toxicity symptoms. It might be thought that under NaCl treatment, more efficient energy dissipation mechanisms such as the xanthophyll cycle or detoxification of the oxygen radicals by SOD would render unnecessary to invest resources in flavonoid synthesis (Reginato et al. 2014b). These observations lead to the proposal of a fundamental role of polyphenols in the protection of the photosynthetic apparatus under severe oxidative stress when other ROS-detoxifying systems are not sufficient, at the expense of growth.

6 *P. strombulifera* and *Larrea divaricata*: Natural Sources of Antioxidants and Biomolecules

In recent years, polyphenols have had an increasing recognition by the scientific community and by the general public because of their abundance in many fruits, seeds, and vegetables. Numerous epidemiological studies support the evidence that health-promoting effects of certain polyphenols are beneficial to human (World Health Organization 2003; Boeing et al. 2012). For this reason, regular consumption of vegetables and fruits with high polyphenol content is known to be beneficial for human health. It is their capacity to scavenge oxidatively generated free radicals, such as those derived from lipids and nucleic acids, that has been highlighted as the fundamental chemical event that underlies their utility in reducing the risk of certain age related degenerations and diseases.

Plants growing in severe environments as halophytes and xerophytes are well adapted to survive under extreme conditions by means of multiple tolerance mechanisms including synthesis of different compounds which help them grow in conditions where other plants fail to survive (Khan and Qaiser 2006; Qasim et al. 2011). Several halophytic species are actually being used in folklore medicine, since their extracts have been proved to have activity against human, animal, and plant pathogens (Ksouri et al. 2012). A recent study of the ethno-botanical uses of medicinal plants from coastal areas of Pakistan showed that medicinal plants in this region were 54 % xerophytes and 40 % halophytes (including halophytes and xerohalophytes) and few of them (6 %) were glycophytes (Qasim et al. 2014). There is a growing interest in the use of plants as source of natural antioxidants (principally polyphenolic compounds) for many reasons, such as the replacement of synthetic antioxidants to enhance health and food preservation (Rice-Evans et al. 1996), search for natural healthy substances and additives, and the use of plant-derived therapeutics for cancer prevention and chemotherapy.

Plant polyphenols are one of the most important and extensively used classes of plant-derived therapeutics for cancer prevention and chemotherapy. Experimental evidence suggest that these protective effects could be in part explained by the capacity of plant polyphenols to act as antioxidants scavenging ROS which are involved in damaging mechanisms to DNA. Also, polyphenols can modulate proinflammatory and oncogenic signals acting as anti-invasive cancer agents (Pan et al. 2010; Vauzour et al. 2010).

Moreover, the cosmetic industry is currently exploiting polyphenols extracted from various plant parts to use them in the development of different products that aim to better protect the skin from damages caused by solar radiation and aging. For all these reasons, plant polyphenols can be considered as an important pool of bioactive natural products with potential benefits for human health.

In Argentina, the study of compounds obtained from regional plants is emerging. There are a lot of species for which "folkloric medicine" has described several uses to preserve and aid health (Roig 2002). Only a small number of them have recently been studied in detail to confirm their phytopharmaceutical properties. The halophyte *P. strombulifera* has been frequently used in folk medicine, mainly the fruits, which seem to have different properties as astringent, anti-inflammatory and odontalgic agent, and anti-diarrheic (Ariza Espinar et al. 2006; Ratera and Ratera 1980; Toursarkissian 1980). Recent scientific studies have confirmed part of these ethnopharmacological uses, describing the molecular mechanism involved in the analgesic effect of this plant (Saragusti et al. 2012) and its biological activity against several microorganisms such as *Escherichia coli*, *Staphylococcus aureus*, and *Salmonella typhi* (Anesini and Perez 1993; Pérez and Anesini 1994a, b).

More recently, Hapon et al. (2014) reported for the first time the cytotoxic activity of *P. strombulifera* extracts against human tumor cell lines. These authors demonstrated that the crude aqueous extract obtained from leaves of this halophyte can be used without risk, avoiding the unexpected effects observed as a consequence of DNA injuries. For all mentioned above, the halophyte *P. strombulifera* is a promising native plant to obtain natural products and biomolecules with different purposes, for cancer research and treatment and other pharmacological uses.

The xerophytic species *L. divaricata* and *L. tridentata* have also been used in folk medicine by native populations due to the amount of bioactive compounds that they contain, mostly in their leaves. Traditionally, leaves and twigs were used to prepare tea for treatment of stomach, kidney, or liver diseases and their extract as an ointment for skins or allergic problems (Arteaga et al. 2005). *L. tridentata* and L. *divaricata* are considerable similar species, the first one found in arid zones of North America and the second one in arid zones of South America, mostly in Argentina and Bolivia. Molecular studies have established that both are the same species with disjunctive distribution, even when taxonomically are classified as different (Lia et al. 2001).

L. divaricata is considered a source of natural products with approximately 50 % of the leaves (dry weight) as extractable matter. The resin that covers the leaves has large amount of flavonoids (principally kaemferol and quercetin), aglycones, lignans, sapogenins, essential oils, and nordihydroguaiaretic acid (NDGA) (Argueta 1994; Hyder et al. 2002). NDGA is considered a phenolic lignin with biological activity of interest in health area as antiviral, antifungal, antimicrobial, and antitumorgenic (Hwu et al. 2008). The therapeutic potential for tumors and cancer treatment of this compound has been demonstrated by the inhibition of cancer cell growth via an apoptotic mechanism (Zavodovskaya et al. 2008). Kaemferol and quercetin are the most abundant flavonoids in the resin of *L. divaricata* leaves (Palacio et al. 2012). These two flavonoids exist as a variety of glycosides or in aglycone form and are structurally similar, differing only by one hydroxyl group in the B-ring. Different researches have demonstrated important biochemical effects such as metal chelation and antioxidant properties (Brown et al. 1998; Boots et al. 2008). Interestingly, it has also been demonstrated that the interaction flavonoid–flavonoid enhances the antioxidant activity comparing with the individual activity (Hidalgo et al. 2010).

7 Conclusions and Perspectives

In recent years, a great deal of drought and salt stress research has been conducted, but knowledge about ecophysiology and stress tolerance mechanisms of native plants from Patagonian Monte is very scarce. Similarly, knowledge related with possible functions of secondary metabolites in drought and salinity responses is in its early phase.

Our studies show that native species adapted to semi-arid and saline environments display a variety of tolerance mechanisms to face these stressing conditions. The significant accumulation of phenolic compounds, mainly flavonoids, in tissues of these plants and their powerful antioxidant activity indicate an important role for these compounds in counteracting the oxidative damage induced by drought and severe salt stress. From a biotechnological point of view, the study of different wild species and their ability to produce different polyphenols, many of them of interest in the pharmacological and food industry, may help selecting plants species for the production of this bioactive compounds and knowing the better environmental conditions to enhance their production.

Halophytes and xerophytes are important sources of bioactive compounds, and more attention is needed to carry out detailed chemical and pharmacological evaluations. Such investigations may lead to the discovery of novel bioactive compounds that will help to assess the efficacy of herbal remedies. Particularly, the recent results obtained about biomolecules present in the leaves make *P. strombulifera* a promising plant as source of natural products for cancer research and treatment, as well as other pharmacological uses.

Acknowledgements We thank Mr. F. Sarasa who allowed us to harvest mother plants in his Estancia San Luis, Wildlife refuge "La Esperanza" of Natural Patagonia Foundation. This research was funded by the National Research Council of Argentina (CONICET), the Ministry of Science, Technology and Innovation of Chubut Province, and the National University of Río Cuarto, Córdoba, Argentina.

References

Agastian P, Kingsley SJ, Vivekanandan M (2000) Effect of salinity on photosynthesis and biochemical characteristics in mulberry genotypes. Photosynthetica 38:287–290

Agati G, Azzarelllo E, Pollastri S, Tattini M (2012) Flavonoids as antioxidants in plants: location and functional significance. Plant Sci 196:67–76

Andjelkovic M, van Camp J, de Meulenaer B, Depaemelaere G, Socaciu C, Verloo M, Verhe R (2008) Iron-chelation properties of phenolic acids bearing catechol and galloyl groups. Food Chem 98:23–31

Anesini C, Perez C (1993) Screening of plants used in Argentine folk medicine for antimicrobial activity. J Ethnopharmacol 39:119–128

Apel K, Hirt H (2004) Reactive oxygen species: metabolism, oxidative stress, and signal transduction. Annu Rev Plant Biol 55:373–399

Ares JO, Beeskow AM, Bertiller MB, Rostagno CM, Irisarri MP, Anchorena J, Defossé GE, Merino CA (1990) Structural and dynamic characteristics of vergrazed grasslands of northern Patagonia. In: Breymeyer A (ed) Managed grasslands. Regional studies. Elsevier, Amsterdam

Argueta V (1994) Atlas de las Plantas de la Medicina Tradicional Mexicana, vol II. Instituto Nacional Indigenista, Mexico

Ariza Espinar L, Barboza GE, Bonzani NE, Cantero JJ, Filippa EM (2006) Flora medicinal de la provincia de Córdoba. Pteridófitas y antófitas silvestres o naturalizadas Museo Botánico, Córdoba

Arteaga S, Andrade-Cetto A, Cardenas R (2005) *Larrea tridentata* (creosote bush), an abundant plant of Mexican and US-American deserts and its metabolite nordihydroguaiaretic acid. J Ethnopharmacol 98:231–239

Asada K (2006) Production and scavenging of reactive oxygen species in chloroplasts and their functions. Plant Physiol 141:391–396

Awika JM, Rooney LW (2004) Sorghum phytochemicals and their potential impact on human health. Phytochemistry 65:1199–1221

Ayres MP, Clausen TP, MacLean SF, Redman AM, Reichardt PB (1997) Diversity of structure and antiherbivore activity in condensed tannins. Ecology 78:1696–1712

Bartels D, Dinakar C (2013) Balancing salinity stress responses in halophytes and non-halophytes: a comparison between *Thellungiella* and *Arabidopsis thaliana*. Funct Plant Biol 40:819–831

Beeskow AM, Del Valle H, Rostagno CM (1987) Los Sistemas fisiográficos de la Región Árida y Semiárida de la Provincia de Chubut. SECYT-Delegación Patagonia, Bariloche

Bie Z, Tadashi I, Shinoara Y (2004) Effects of sodium sulfate and sodium bicarbonate on the growth, gas exchange and mineral composition of lettuce. Sci Hortic 99:215–224

Boeing H, Bechthold A, Bub A, Ellinger S, Haller D, Kroke A, Leschik-Bonne E, Muller MJ, Oberritter H, Schulze M, Stehle P, Watzl B (2012) Critical review: vegetables and fruit in the prevention of chronic diseases. Eur J Nutr 51:637–663

Boots AW, Haenen GRMM, Bast A (2008) Health effects of quercetin: from antioxidant to nutraceutical. Eur J Pharmacol 585:325–337

Bose J, Rodrigo-Moreno A, Shabala S (2014) ROS homeostasis in halophytes in the context of salinity stress tolerance. J Exp Bot 65:1241–1257

Brown JE, Khodr H, Hider RC, Rice-Evans CA (1998) Structural-dependence of flavonoid interactions with copper ions: implications for their antioxidant properties. Biochem J 330:1173–1178

Brunetti C, Di Ferdinando M, Fini A, Pollastri S, Tattini M (2013) Flavonoids as antioxidants and developmental regulators: relative significance in plants and humans. Int J Mol Sci 14:3540–3555

Burkart A (1976) A monograph of the genus *Prosopis* (Leguminosae subfam. Mimosoideae) Catalogue of the recognized species of *Prosopis*. J Arnold Arbor 57:450–525

Campanella MV, Bertiller MB (2008) Plant phenology, leaf traits and leaf litterfall of contrasting life forms in the arid Patagonian Monte, Argentina. J Veg Sci 19:75–85

Catalán L, Balzarini M, Taleisnik E, Sereno R, Karlin U (1994) Effects of salinity on germination and seedling growth of *Prosopis flexuosa* (D.C.). Forest Ecol Manag 63:347–357

Cenzano A, Varela MC, Bertiller M, Luna V (2013) Effect of drought on morphological and functional traits of *Poa ligularis* and *Pappostipa speciosa*, native perennial grasses with wide distribution in Patagonian rangelands, Argentina. Aust J Bot 61:383–393

Cenzano AM, Masciarelli O, Luna MV (2014) Abscisic acid metabolite profiling as indicators of plastic responses to drought in grasses from arid Patagonian Monte (Argentina). Plant Physiol Biochem 83:200–206

Chanwitheesuk A, Teerawutgulrag A, Rakariyatham N (2005) Screening of antioxidant activity and antioxidant compounds of some edible plants of Thailand. Food Chem 92:491–497

Dangles O, Dufour C (2008) Flavonoid-protein binding processes and their potential impact on human health. In: Daayf F, Lattanzio V (eds) Recent advances on polyphenol research, vol 1. Wiley-Blackwell, Oxford

De Abreu IN, Mazzafera P (2005) Effect of water and temperature stress on the content of active constituents of *Hypericum brasiliense* Choisy. Plant Physiol Biochem 43:241–248

Demmig-Adams B, Adams WW (1992) Photoprotection and other responses of plants to high light stress. Annu Rev Plant Physiol Plant Mol Biol 43:599–626

Dixon RA, Paiva N (1995) Stress-induced phenylpropanoid metabolism. Plant Cell 7:1085–1097

Duthie GG, Gardner PT, Kyle AM (2003) Plants polyphenols: are they the new magic bullet? Proc Nutr Sci 62:599–603

Felker P (2007) Unusual physiological properties of the arid adapted tree legume Prosopis and their applications in developing countries. In: De la Barrera E, Smith W (eds) Perspectives in biophysical plant ecophysiology: a tribute to Park Nobel. Mildred E. Mathias Botanical Garden, University of California Press, Los Angeles

Fernandez OA, Busso CA (1999) Arid and semi-arid rangelands: two thirds of Argentina. Rala Report 200, Agricultural Res Inst, Reykjavik, Iceland

Foyer CH, Noctor G (2003) Redox sensing and signaling associated with reactive oxygen in chloroplasts, peroxisomes and mitochondria. Physiol Plant 119:355–364

Foyer CH, Noctor G (2005) Redox homeostasis and antioxidant signalling: a metabolic interface between stress perception and physiological responses. Plant Cell 17:1866–1875

Gray DE, Pallardy SG, Garrett HE, Rottinghaus G (2003) Acute drought stress and plant age effects on alkamide and phenolic acid content in purple coneflower roots. Planta Med 69:50–55

Hajlaoui H, Denden M, El Ayeb N (2009) Differential responses of two maize (*Zea mays* L.) varieties to salt stress: changes on polyphenols composition of foliage and oxidative damages. Ind Crop Prod 30:144–151

Hapon MB, Hapon MV, Persia FA, Pochettino A, Lucero GS (2014) Aqueous extract of *Prosopis strombulifera* (LAM) Benth induces cytotoxic effects against tumor cell lines without systemic alterations in BALB/c mice. J Clin Toxicol 4:1–8

Hatier HB, Gould KS (2008) Anthocyanin function in vegetative organs. In: Gould K, Davies K, Winefield C (eds) Anthocyanins—biosynthesis, functions, and applications. Springer, New York

Heim KE, Tagliafeno AR, Bobilya DJ (2002) Flavonoid antioxidants: chemistry, metabolism and structure-activity relationships. J Nutr Biochem 13:572–584

Hernández I, Alegre L, Munné-Bosch S (2006) Enhanced oxidation of flavan-3-ols and proanthocyanidin accumulation in water- stressed tea plants. Phytochemistry 67:1120–1126

Hidalgo M, Sánchez-Moreno C, Pascual-Teresa S (2010) Flavonoid–flavonoid interaction and its effect on their antioxidant activity. Food Chem 121:691–696

Hwu JR, Hsu MH, Huang RC (2008) New nordihydroguaiaretic acid derivates as anti-HIV agents. Bioorg Med Chem Lett 18:1884–1888

Hyder PW, Fredrickson EL, Estell RE, Tellez M, Gibbens RP (2002) Distribution and concentration of total phenolics, condensed tannins, and nordihydroguaiaretic acid (NDGA) in creosotebush (*Larrea tridentata*). Biochem Syst Ecol 30:905–912

Isabelle M, Lee BL, Lim MT, Koh WP, Huang D, Ong Ch N (2010) Antioxidant activity and profiles of common vegetables in Singapore. Food Chem 120:993–1003

Jaafar HZ, Ibrahim MH, Fakri NFM (2012) Impact of soil field water capacity on secondary metabolites, phenylalanine ammonia-lyase (PAL), malondialdehyde (MDA) and photosynthetic responses of Malaysian kacipfatimah (*Labisia pumila* Benth). Molecules 17:7305–7322

Jaleel CA, Gopi R, Manivannan P, Panneerselvam R (2007) Antioxidative potentials as a protective mechanism in *Catharanthus roseus* (L.) G. Don. plants under salinity stress. Turk J Bot 3:245–251

Julkunen-Tiito R, Nenadis N, Neugart S, Robson M, Agati G, Vepsäläinen J, Zipoli G, Nybakken L, Winkler B, Jansen M (2015) Assessing the response of plant flavonoids to UV radiation: an overview of appropriate techniques. Phytochem Rev 14:273–297

Khan MA, Qaiser M (2006) Halophytes of Pakistan: characteristics, distribution and potential economic usages. In: Khan MA (ed) Sabkha ecosystems Volume II: West and Central Asia. Springer, Dordrecht

Khanbabaee K, Van Ree T (2001) Tannins: classification and definition. Nat Prod Rep 18:641–649

Kirakosyan A, Kaufman P, Warber S, Zick S, Aaronson K, Bolling S (2004) Applied environmental stresses to enhance the levels of polyphenolics in leaves of hawthorn plants. Physiol Planta 121:182–186

Kosová K, Vítámvás P, Urban MO, Prášil IT (2013) Plant proteome responses to salinity stress-comparison of glycophytes and halophytes. Funct Plant Biol 40:775–786

Krishnaiah D, Sarbatly R, Nithyanandam R (2011) A review on the antioxidant potential of medicinal plant species. Food Bioprod Process 89:217–233

Ksouri R, Megdiche W, Debez A, Falleh H, Grignon C, Chedly A (2007) Salinity effects on polyphenol content and antioxidant activities in leaves of the halophyte *Cakile maritima*. Plant Physiol Biochem 45:244–249

Ksouri R, Smaoui A, Isoda H, Abdelly C (2012) Utilization of halophyte species as new sources of bioactive substances. J Arid Land Stud 22:41–44

Lattanzio V, Kroon P, Quideau S, Treutter D (2008) Plant phenolics-secondary metabolites with diverse functions. In: Daayf F, Lattanzio V (eds) Recent advances in polyphenol research, vol 1. Wiley-Blackwell, Oxford

Leopoldini M, Russo N, Toscani M (2011) The molecular basis of working mechanism of natural polyphenolic antioxidants. Food Chem 125:288–306

Li AS, Bandy B, Tsang SS, Davison AJ (2000) DNA-breaking versus DNA-protecting activity of four phenolic compounds in vitro. Free Radic Res 33:551–566

Li Z, Qiang W, Xiao R, Cun-De P, De-An J (2010) Phenolics and plant allelopathy. Rev Mol 15:8933–8952

Lia V, Confalonieri C, Comas I, Hunziker J (2001) Molecular phylogeny of *Larrea* and its allies (Zygophyllaceae): reticulate evolution and the probable time of creosote bush arrival in North America. Mol Phylogenet Evol 21:309–320
Llanes A, Bertazza G, Palacio G, Luna V (2013) Different sodium salts cause different solute accumulation in the halophyte *Prosopis strombulifera*. Plant Biol 15:118–125
Manivannan P, Abdul Jaleel C, Sankar B, Kishorekumar A, Murali P, Somasundaram R, Panneerselvam R (2008) Mineral uptake and biochemical changes in *Helianthus annuus* under treatment with different sodium salts. Colloids Surf B Biointerfaces 62:58–63
Martins S, Aguilar CN, de la Garza-Rodriguez I, Mussatto SI, Teixeira SA (2010) Kinetic study of nordihydroguaiaretic acid recovery from *Larrea tridentata* by microwave-assisted extraction. J Chem Technol Biotechnol 85:1142–1147
Miller G, Shulaev V, Mittler R (2008) Reactive oxygen signaling and abiotic stress. Physiol Plant 133:481–489
Miller G, Suzuki N, Ciftci-Yilmaz S, Mittler R (2010) Reactive oxygen species homeostasis and signalling during drought and salinity stresses. Plant Cell Environ 33:453–467
Mittler R (2002) Oxidative stress, antioxidants and stress tolerance. Trends Plant Sci 9:405–410
Mittler R, Blumwald E (2010) Genetic engineering for modern agriculture: challenges and perspectives. Annu Rev Plant Biol 61:443–462
Motilva MJ, Serra A, Macià A (2013) Analysis of food polyphenols by ultra-high-performance liquid chromatography coupled to mass spectrometry: an overview. J Chromatogr A 1292:66–82
Muthukumarasamy M, Gupta SD, Pannerselvam R (2000) Enhancement of peroxidase, polyphenol oxidase and superoxide dismutase activities by triadimefon in NaCl stressed *Raphanus sativus* L. Biol Plant 43:317–320
Navarro JM, Flores P, Garrido C, Martinez V (2006) Changes in the contents of antioxidant compounds in pepper fruits at ripening stages, as affected by salinity. Food Chem 96:66–73
Neudörffer D, Bonnefont-Rousselot A, Legrand M, Fleury B, Largeron M (2004) 4-Hydroxycinnamic ethyl ester derivatives and related dehydrodimers: relationship between oxidation potential and protective effects against oxidation of low-density lipoproteins. J Agric Food Chem 52:2084–2091
Noctor G, De Paepe R, Foyer CH (2007) Mitochondrial redox biology and homeostasis in plants. Trends Plant Sci 12:125–134
Oh DH, Dassanayake M, Haas JS, Kropornika A, Wright C, D'Urzo MP, Hong H, Ali S, Hernandez A, Lambert GM, Inan G, Galbraith DW, Bressan RA, Yun DJ, Zhu JK, Cheeseman JM, Bohnert HJ (2010) Genome structures and halophyte-species gene expression of the extremophile *Thellungiella parvula* in comparison with *Thellungiella salsuginea* (*Thellungiella halophila*) and *Arabidopsis*. Plant Physiol 154:1040–1052
Palacio L, Cantero JJ, Cusidó RM, Goleniowski ME (2012) Phenolic compound production in relation to differentiation in cell and tissue cultures of *Larrea divaricata* (Cav.). Plant Sci 193–194:1–7
Pan MH, Lai CS, Ho CT (2010) Anti-inflammatory activity of natural dietary flavonoids. Food Funct 1:15–31
Pang CH, Wang BS (2008) Oxidative stress and salt tolerance in plants. In: Lütge U, Matyssek R, Ramos C, Miguel F (eds) Progress in botany. Springer, Berlin
Pérez C, Anesini C (1994a) Antibacterial activity of alimentary plants against *Staphylococcus aureus* growth. Am J Chin Med 22:169–174
Pérez C, Anesini C (1994b) In vitro antibacterial activity of Argentine folk medicinal plants against *Salmonella typhi*. J Ethnopharmacol 44:41–46
Perron NR, Brumaghin JL (2009) A review of the antioxidant mechanisms of polyphenol compounds related to iron binding. Cell Biochem Biophys 53:75–100
Peterson A, Murphy K (2015) Tolerance of lowland quinoa cultivars to sodium chloride and sodium sulfate salinity. Crop Sci 55:331–338
Pollastri S, Tattini M (2011) Flavonols: old compounds for old roles. Ann Bot 108:1225–1233

Qasim M, Gulzar S, Khan MA (2011) Halophytes as medicinal plants. In: Ozturk M, Mermut AR, Celik A (eds) Urbanisation, land use, land degradation and environment. Daya, New Delhi
Qasim M, Abideen Z, Adnan YM, Ansari R, Gul B, Ajmal Khan M (2014) Traditional ethnobotanical uses of medicinal plants from coastal areas of Pakistan. J Coast Life Med 2:22–30
Ramani B, Reeck T, Debez A, Stelzerd R, Huchzermeyer B, Schmidt A, Papenbrock J (2006) *Aster tripolium* L. and *Sesuvium portulacastrum* L.: two halophytes, two strategies to survive in saline habitats. Plant Physiol Biochem 44:395–408
Ratera EL, Ratera MO (1980) Plantas de la Flora Argentina empleadas en Medicina Popular. Hemisferio Sur S.A, Buenos Aires
Reginato M, Abdala G, Miersch O, Ruiz O, Moschetti E, Luna V (2012) Changes in the levels of jasmonates and free polyamines induced by Na_2SO_4 and NaCl in roots and leaves of the halophyte *Prosopis strombulifera*. Biologia 67:689–697
Reginato M, Reinoso H, Llanes A, Luna V (2013) Stomatal abundance and distribution in *Prosopis strombulifera* plants growing under different iso-osmotic salt treatments. Am J Plant Sci 4:80–90
Reginato M, Sosa L, Llanes A, Hampp E, Vettorazzi N, Reinoso H, Luna V (2014a) Na_2SO_4 and NaCl determine different growth responses and ion accumulation in the halophytic legume *Prosopis strombulifera*. Plant Biol 16:97–106
Reginato M, Castagna A, Furlán A, Castro S, Ranieri A, Luna V (2014b) Analysis of the oxidative damage in the halophyte *Prosopis strombulifera* salinized with NaCl and Na_2SO_4. Role of polyphenols as antioxidant protection. AoB Plants 6(SI: Physiology and ecology of halophytes—plants living in salt-rich environments):plu042. doi:10.1093/aobpla/plu042
Reinoso H, Sosa L, Ramirez L, Luna V (2004) Salt-induced changes in the vegetative anatomy of *Prosopis strombulifera* (Leguminosae). Can J Bot 82:618–628
Reinoso H, Sosa L, Reginato M, Luna V (2005) Histological alterations induced by sodium sulfate in the vegetative anatomy of *Prosopis strombulifera* (Lam.) Benth. World J Agric Sci 1:109–119
Rice-Evans CA, Miller NJ, Paganga G (1996) Structure-antioxidant activity relationships of flavonoids and phenolic acids. Free Radic Biol Med 20:933–956
Roberts MR, Paul ND (2006) Seduced by the dark side: integrating molecular and ecological perspectives on defence against pests and pathogens. New Phytol 170:677–699
Roig F (2002) Flora medicinal mendocina. Las plantas medicinales y aromáticas de la provincia de Mendoza (Argentina). EDIUNC Serie Manuales N° 33, Mendoza, Argentina
Saragusti AC, Bustos PS, Pierosan L, Cabrera JL, Chiabrando GA (2012) Involvement of the L-arginine-nitric oxide pathway in the antinociception caused by fruits of *Prosopis strombulifera* (Lam.) Benth. J Ethnopharmacol 140:117–122
Schützendübel A, Polle A (2002) Plant responses to abiotic stresses: heavy metal-induced oxidative stress and protection by mycorrhization. J Exp Bot 53:1351–1365
Selmar D, Kleinwächter M (2013) Influencing the product quality by deliberately applying drought stress during the cultivation of medicinal plants. Ind Crop Prod 42:558–566
Sengupta S, Majumder AL (2010) *Porteresia coarctata* (Roxb.) Tateoka, a wild rice: a potential model for studying salt-stress biology in rice. Plant Cell Environ 33:526–542
Shi D, Sheng Y (2005) Effect of various salt-alkaline mixed stress conditions on sunflower seedlings and analysis of their stress factors. Environ Exp Bot 54:8–21
Sosa L, Llanes A, Reinoso H, Reginato M, Luna V (2005) Osmotic and specific ion effects on the germination of *Prosopis strombulifera*. Ann Bot 96:261–267
Storey KB (1996) Oxidative stress: animal adaptations in nature. Braz J Med Biol Res 29:1715–1733
Tarchoune I, Sgherri C, Izzo R, Lachaal M, Ouerghi Z, Navari-Izzo F (2010) Antioxidative responses of *Ocimum basilicum* to sodium chloride or sodium sulphate salinization. Plant Physiol Biochem 9:772–777
Tattini M, Galardi C, Pinelli P, Massai R, Remorini D, Agati G (2004) Differential accumulation of flavonoids and hydroxycinnamates in leaves of *Ligustrum vulgare* under excess light and drought stress. New Phytol 163:547–561

Tattini M, Guidi L, Morassi-Bonzi L, Pinelli P, Remorini D, Degl'Innocenti E, Giordano C, Massai R, Agati G (2005) On the role of flavonoids in the integrated mechanisms of response of *Ligustrum vulgare* and *Phillyrea latifolia* to high solar radiation. New Phytol 167:457–470

Thelen GC, Vivanco JM, Newingham B, Good W, Bais HP, Landers P, Caesar A, Callaway RM (2005) Insect herbivory stimulates allelopathic exudation by an invasive plant and the suppression of natives. Ecol Lett 8:209–217

Tounekti T, Vadel A, Oñate M, Khemira H, Munné-Bosch S (2010) Salt-induced oxidative stress in rosemary plants: damage or protection? Environ Exp Bot 71:298–305

Toursarkissian M (1980) Plantas medicinales de la Argentina. Sus nombres botánicos, vulgares, usos y distribución geográfica. Hemisferio Sur SA, Buenos Aires

Vauzour D, Rodriguez-Mateos A, Corona G, Oruna-Concha MJ, Soencer JPE (2010) Polyphenols and human health: prevention of disease and mechanisms of action. Nutrients 2:1106–1131

Wang W, Vinocur B, Altman A (2003) Plant response to drought, salinity and extreme temperatures: toward genetic engineering for stress tolerance. Planta 218:1–14

World Health Organization (WHO) (2003) Diet, nutrition and the prevention of chronic diseases. WHO technical report series, no. 916, Geneva

Wright JS, Johnson ER, DiLabio GA (2001) Predicting the activity of phenolic antioxidants: theoretical method, analysis of substituent effects, and application to major families of antioxidants. J Am Chem Soc 123:1173–1183

Zavodovskaya M, Campbell MJ, Maddux BA, Shiry L, Allan G, Hodges L, Kushner P, Kerner JA, Youngren JF, Goldfine ID (2008) Nordihydroguaiaretic acid (NDGA), an inhibitor of the HER2 and IGF-1 receptor tyrosine kinases, blocks the growth of HER2-overexpressing human breast cancer cells. J Cell Biochem 103:624–635

Reactive Oxygen Species and Plant Disease Resistance

András Künstler, Renáta Bacsó, Yaser Mohamed Hafez, and Lóránt Király

Contents

A. Künstler • R. Bacsó • L. Király (✉)
Plant Protection Institute, Centre for Agricultural Research, Hungarian Academy of Sciences, PO Box 102, H-1525 Budapest, Hungary
e-mail: kiraly.lorant@agrar.mta.hu

Y.M. Hafez
Department of Agricultural Botany, Plant Pathology Branch, Faculty of Agriculture, Kafr-El-Sheikh University, 33516 Kafr-El-Sheikh, Egypt

D.K. Gupta et al. (eds.), *Reactive Oxygen Species and Oxidative Damage in Plants Under Stress*, DOI 10.1007/978-3-319-20421-5_11

Abstract Plants may successfully limit or even kill pathogens at least in part by eliciting spatial patterns of ROS production in different parts of invaded plant cells, e.g., the cell wall and plasma membrane. Recent research also suggests a significant contribution to plant disease resistance by ROS-mediated processes in the plant cuticle and intracellular organelles. The role of temporal patterns (i.e., proper timing) of ROS accumulation in eliciting an effective plant disease resistance is also discussed. Essentially, defense against pathogens could be very effective if it is a rapid, symptomless process, eliminating the pathogen in due time and not overusing resources of the plant, a process likely mediated by ROS. On the other hand, a delayed and failed attempt by the host to elicit resistance may result in massively stressed plant tissues and a partial or almost complete loss of control over pathogen invasion. Thus, it seems that when plants encounter pathogens they need to defend themselves simultaneously against biotic and abiotic stresses (i.e., pathogen accumulation and excessive cell/tissue death) by turning on two different types of—partially overlapping—signaling pathways that may function in parallel. Very recent interesting data suggest a pivotal role of autopropagating ROS waves in these signaling processes.

Keywords Plant disease resistance • Reactive oxygen species • Antioxidants • Hypersensitive response • Symptomless resistance • Plant cell death

1 Introduction

1.1 Early Research on the Role of ROS in Plant Disease Resistance

When plants are able to successfully resist pathogenic invaders, the ultimate result is inhibition or killing of these pathogens. In past years, several theories have tried to explain the possible mechanisms of disease resistance in plants, including accumulation of antimicrobial compounds, cell wall reinforcement, localized cell and tissue death (hypersensitive response, HR), reactive oxygen species (ROS), etc. (Király et al. 1972; Goodman et al. 1986; Jones and Dangl 2006; Spoel and Dong 2012). In the last three decades, the role of ROS, primarily superoxide ($O_2^{\cdot -}$) and hydrogen peroxide (H_2O_2), has become a pivotal research topic in studies of both animal (human) immunity and plant disease resistance mainly because of two reasons. First, a cause-and-effect relationship between animal phagocytosis and the accumulation of ROS has been demonstrated (Morel et al. 1991; Segal 2008). Second, ROS accumulation has been also associated and functionally linked to numerous plant disease resistance events (Doke 1983a, b; Doke and Ohashi 1988; Ádám et al. 1989; Levine et al. 1994; Baker and Orlandi 1995; Delledonne et al. 2001; Torres et al. 2006; Shang et al. 2010; Torres 2010; Dubiella et al. 2013).

In 1980s, the pioneering discovery of Doke (1983a, b) demonstrated that resistance of potato to an oomycete pathogen (*Phytophthora infestans*) is specifically associated with the generation of superoxide that is rapidly transformed to hydrogen peroxide. Doke and Ohashi (1988) also described a similar phenomenon occurring during resistance of tobacco to a viral pathogen (*Tobacco mosaic virus*). Ádám et al. (1989) were the first to report ROS generation, i.e., the accumulation of superoxide, in plants (tobacco) during resistance responses to bacterial infection. A similar, rapid production of hydrogen peroxide was also observed during an incompatible interaction between another bacterium (*Pseudomonas syringae* pv. *tomato* DC3000) and the plant *Arabidopsis thaliana* (Alvarez et al. 1998).

These early observations set forth a series of studies on ROS production in plants exposed to live pathogens or their elicitors. It has become apparent that ROS in plants are produced during both abiotic and biotic stresses and are key components of plant disease resistance responses.

1.2 The Two Main Lines of Plant Defense to Pathogens and the Oxidative (ROS) Burst

The essence of disease resistance is the recognition of and protection against the foreign (the non-self). During plant–pathogen interactions, the first line of plant defense consists primarily of the pathogen-associated molecular pattern (PAMP) recognition system that has been shown to confer a so-called basal resistance to a wide range of pathogens (Jones and Dangl 2006; Boller and Felix 2009). PAMPs regularly occur in bacteria, fungi, and in several other microbes. Examples of PAMPs include conserved bacterial cell surface structures like lipopolysaccharides, flagellin, peptidoglucans, or fungal cell wall components like glucan or chitin. In plants, PAMPs are recognized by distinct pattern recognition receptors which result in the activation of basal (nonhost) defense responses also termed as PAMP-triggered immunity (PTI). Pathogens use effector proteins to block this first line of plant defense (PTI) leading to virulence (i.e., a successful infection). In this case, the effector is termed as a virulence factor and the pathogen is virulent. When the PAMP system fails to recognize the pathogen as an invader, a second line of plant defense is induced. Plants may produce specific surveillance proteins encoded by resistance (R) genes to recognize effectors (called then avirulence—avr—factors and the pathogens being avirulent) and mount this second line of defense called effector-triggered immunity (ETI).

ROS production in plants (primarily $O_2^{\cdot -}$ and H_2O_2) can be detected both at the first and second line of defense to pathogenic infections (i.e., during PTI and ETI). Pathogens eliciting PTI or ETI (i.e., resistance) typically induce a biphasic oxidative burst (ROS production) in plants, consisting of a rapid but weak and transient first phase, followed by a massive and sustained ROS accumulation during the second phase (Lamb and Dixon 1997; Bolwell et al. 2002; Torres et al. 2006).

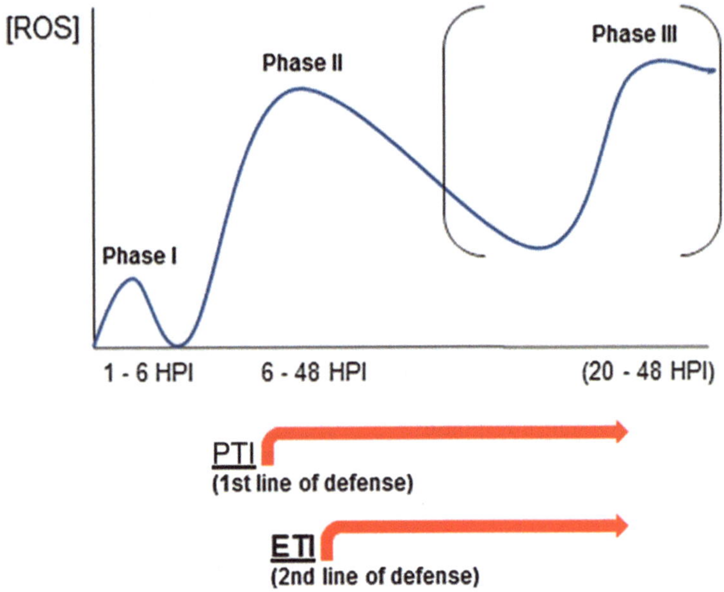

Fig. 1 The different phases of the oxidative (ROS) burst during elicitation of plant disease resistance (defense) by invading pathogens. Relationship of the ROS burst to the two lines of plant defense. *HPI* time (hours) after pathogen inoculation, *PTI* PAMP-triggered immunity, *ETI* effector-triggered immunity. Phase III of the ROS burst occurs only in certain host–pathogen combinations. For further explanations, see the text. Modified after Lamb and Dixon (1997), Hückelhoven and Kogel (2003)

In certain cases, a third phase of the oxidative burst has also been detected during plant disease resistance, e.g., to the powdery mildew pathogen (*Blumeria graminis* f. sp. *hordei*) attacking barley (Hückelhoven and Kogel 2003) and *Septoria tritici* infecting wheat (Shetty et al. 2003) (Fig. 1). These differences could be attributed to the complex development of these fungal pathogens and the influence of the host genotype, which might determine whether two or three phases of the oxidative burst may occur. Interestingly, in case of symbiotic plant–microbe interactions, ROS accumulation has also been observed but the second phase of the oxidative burst seems to be suppressed (Shaw and Long 2003; Lohar et al. 2007). Furthermore, usually only the transient first phase of ROS accumulation has been detected during plant interactions with virulent pathogens that avoid or suppress host recognition (Bolwell et al. 2002). These findings demonstrate that the first phase of the oxidative burst is a biologically nonspecific reaction. On the other hand, the second phase of ROS accumulation seems to play a pivotal role in the establishment of plant defense responses towards pathogens (Levine et al. 1994; Lamb and Dixon 1997).

The oxidative burst following pathogen recognition and initiation of resistance responses is indeed associated with a series of biochemical and pathophysiological events. ROS contribute to the establishment of physical reinforcements of plant cell walls (papillae) that are formed at the site of pathogen ingress by cross-linking of cell wall glycoproteins (Bradley et al. 1992) or via oxidative cross-linking of precursors during the localized biosynthesis of lignin and suberin polymers (Thordal-Christensen et al. 1997). ROS also have a direct antimicrobial effect in vitro and also in planta (Aver'yanov and Lapikova 1988; Peng and Kuć 1992; Király et al. 1993). Evidence suggests, however, that ROS also have diverse signaling functions that mediate defense gene activation and establishment of additional defenses, e.g., by redox control of transcription factors (Mou et al. 2003) or by interaction with other signaling components like phosphorylation cascades (Kovtun et al. 2000; Dubiella et al. 2013), nitric oxide (Delledonne et al. 1998, 2001; Mur et al. 2006), and the plant hormone salicylic acid (Chen et al. 1993; Torres et al. 2006). ROS also contribute to the production of phytoalexins and other secondary metabolites that may play a role in arresting pathogen growth (Lamb and Dixon 1997; Thoma et al. 2003). Last but not least, ROS development is closely linked to the hypersensitive response (HR), a localization of invading pathogens to their entry site accompanied by programmed plant cell and tissue death (necrosis) (Doke 1983a, b; Doke and Ohashi 1988; Ádám et al. 1989; Levine et al. 1994; Alvarez et al. 1998). The HR not only contributes to the direct limitation of pathogens but is also a source of signals for the establishment of further defenses, both locally and systemically (e.g., priming, local, and systemic acquired resistance) (Chester 1933; Ross 1961a, b; Sticher et al. 1997; Conrath et al. 2001; Fu and Dong 2013; Pastor et al. 2013).

1.3 Expression of ROS-Related Genes and Their Functions in Plant Disease Resistance

As mentioned above, pathogen-induced ROS themselves are considered as signaling molecules. Elevated levels of ROS may be perceived by different receptors, proteins, or enzymes. However, it is not clear exactly how plant cells sense ROS, as specific ROS receptors have not been identified so far. One possibility is that ROS directly inhibit/modify redox-sensitive proteins like phosphatases (Apel and Hirt 2004) and heat shock transcription factors (HSFs; Miller and Mittler 2006). Overexpression of the *Arabidopsis thaliana* gene encoding HSF-A1b confers enhanced resistance to both drought and bacterial infection, and this response is dependent on H_2O_2 signaling (Bechtold et al. 2013). Transcriptome analysis of rice revealed that HSF binding sites were significantly enriched in promoters of up-regulated genes in response to cold, heat, and oxidative stress suggesting a role of HSFs as central regulators of plant stress responses that involve ROS accumulation, e.g., disease resistance (Mittal et al. 2012).

Expression of ROS-related, plant defense-associated genes is likely regulated by ROS through the oxidation of cysteine residues of transcription factors. For example, the zinc finger transcription factor ZAT6 positively regulates resistance to abiotic stresses as well as resistance to bacterial infection by modulating gene expression related to ROS and salicylic acid (SA), a central component of plant defenses (Shi et al. 2014). Another family of transcription factors (WRKY) includes several members that play regulatory roles during plant disease resistance responses (see Pandey and Somssich 2009). For example, WRKY70 is a key regulator of plant disease resistance and a crossroads between two signaling pathways mediated by SA and jasmonic acid (JA), being induced by the former and inhibited by the latter (Li et al. 2004). Also, WRKY30 is rapidly induced by exposure to several pathogen-associated molecular patterns (PAMPs) and herbicide-induced oxidative stress, suggesting a ROS-dependent role in resistance to pathogens (Scarpeci et al. 2013).

One of the best characterized defense signaling pathways regulated by oxidation events (i.e., ROS accumulation) is the induction of SA-dependent resistance responses with the involvement of two key regulators, the SA receptor NPR1 and TGA transcription factors in *A. thaliana* (Liao et al. 2012; Fu and Dong 2013). In healthy plants, the NPR1 protein is S-nitrosylated at cysteine-156 by S-nitrosoglutathione (GSNO) and sequestered in the cytoplasm as an oligomer formed by disulfide bounds. SA accumulation upon pathogen attack alters the cellular redox state, causing a reduction of disulfide bonds in NPR1 by two thioredoxins, TRX-h3 and TRX-h5 (Tada et al. 2008). Furthermore, oxidation of cysteine residues of TGA transcription factors is also necessary to promote their interaction with NPR1 in the nucleus (Després et al. 2003). In fact, reduction of the NPR1 oligomer releases NPR1 monomers that translocate to the nucleus and interact with the oxidized TGA transcription factors, promoting the induction of defense response genes, primarily the so-called pathogenesis-related (PR) genes (Tada et al. 2008). Interestingly, however, ROS accumulation does not always result in the induction of these SA-dependent defense genes. During symbiotic infections in legume roots, a single phase oxidative burst (see above) suppresses the expression of PR genes/proteins (Peleg-Grossman et al. 2012), pointing to the complexity of ROS-dependent signal transduction in plant–microbe interactions.

PR proteins usually do not accumulate in healthy plants, but are induced by pathogen infection and abiotic stresses (Van Loon et al. 2006). In fact, an early, enhanced induction of PR proteins and their respective genes occurs during incompatible host–pathogen interactions (i.e., resistance), as compared to compatible interactions (Bell et al. 1986; Van Loon et al. 2006), suggesting a role during plant disease resistance. However, although some of these proteins exhibit potential in vitro antimicrobial activities and their accumulation is associated with plant resistance, a direct functional role in defense could not be demonstrated for all PR proteins (Van Loon and Van Strien 1999; Van Loon et al. 2006; Sels et al. 2008 and references within). Nevertheless, some PR proteins have activities that suggest a defense function via ROS generation. For example, PR-9 of tobacco has been shown to be a lignin-forming peroxidase with a likely role in producing H_2O_2 necessary for cell wall reinforcement during pathogen attack (Lagrimini et al. 1987). Also, it has been demonstrated that PR-15 and PR-16 in barley have oxalate oxidase and oxalate oxidase-like (i.e., superoxide dismutase) enzymatic activities, respectively, resulting in H_2O_2-accumulation during resistance responses to powdery mildew (*B. graminis* f. sp. *hordei*) (Zhang et al. 1995; Wei et al. 1998). Thus, PR genes/proteins are ROS-related not only in the sense of their induction (see above) but some of them directly produce ROS associated with disease resistance.

Probably the most obvious ROS-related genes in plants are those encoding antioxidants. Pathogenic infections and abiotic stresses are frequently associated with ROS accumulation, eventually resulting in oxidative stress. Up-regulation of antioxidants is often the only evidence that oxidative stress has occurred in vivo (Halliwell and Gutteridge 1999). On the other hand, repression of antioxidant genes/enzymes in early stages of HR-type resistance (i.e., during formation of localized necrotic lesions) implies that a programmed, transient decline in antioxidant capacity could also contribute to enhanced ROS accumulation functional in, e.g., defense responses. This is supported by several observations showing that expression of antioxidant genes/enzymes is transiently suppressed during relatively early stages of viral HR. The activity of four antioxidant enzymes (glutathione reductase, glutathione S-transferase, superoxide dismutase, and ascorbate peroxidase) was found to be transiently suppressed within the first 24 h of HR induced by *Tobacco mosaic virus* (TMV) in tobacco (Fodor et al. 1997). Also, Mittler et al. (1998) reported the posttranscriptional suppression of cytosolic ascorbate peroxidase during systemically induced, HR-type cell death in TMV-infected plants. Furthermore, mRNA levels of a tobacco catalase gene (*CAT1*) and total catalase activity are transiently suppressed in HR-type necrotic lesions elicited by TMV and *Tobacco necrosis virus* (TNV) (Dorey et al. 1998; Yi et al. 1999, 2003; Künstler et al. 2007). It was also shown that *CAT1* is repressed only in the vicinity of necrotic lesions, while its expression is high in adjacent, healthy tissues (Dorey et al. 1998; Yi et al. 1999). These findings suggest that temporal and spatial control of antioxidant gene expression at sites of pathogen entry could drive localized, transient ROS accumulation leading not only to programmed plant cell/tissue death but also a significant limitation of pathogen movement and/or replication.

2 Pathogen Limitation in Plant Cells: The Contribution of ROS

2.1 *Plant Cell Walls and Their ROS-Mediated Reinforcement: An Initial Barrier to Pathogen Ingress*

A first line of spatial defense against pathogens directly penetrating plant cells operates at the plant cell wall (apoplast) (Schulze-Lefert 2004). These can be preformed barriers, determined by a thick cuticle or epicuticular wax layer or the composition and physical properties of the cell wall. Induced formation of cell wall barriers in response to pathogen attack may also occur by reinforcement of the cell wall at attempted penetration sites by deposition of callose and lignification. These physical reinforcements of the plant cell wall are called papillae or cell wall appositions (Hückelhoven 2007).

ROS production has been shown to be associated with pathogen-induced cell wall reinforcements in several ways. *De novo* synthesized apoplastic class III plant peroxidases mediate ROS-dependent cross-linking of components of the cell wall including glycoproteins, lignin, and suberin (Almagro et al. 2009; O'Brien et al. 2012). Genetic evidence for this function was provided by *Arabidopsis* plants transformed with the antisense sequence of a class III peroxidase of french bean (*FBP1*). These plants showed a down-regulation of two related endogenous peroxidases and reduced levels of H_2O_2 production when challenged with a fungal phytotoxin (fumonisin B1). Furthermore, these plant lines displayed enhanced susceptibility against bacterial (*Pseudomonas syringae* pv. *maculicola* ES4326, *P. syringae* pv. *tomato* DC3000) and fungal (*Botrytis cinerea*, *Golovinomyces orontii*) pathogens (Bindschedler et al. 2006). It was suggested that down-regulation of apoplastic peroxidases might compromise resistance by preventing oxidative cross-linking, which could be effective in case of fungal pathogens that can penetrate plant cell walls. Interestingly, a similar apoplastic peroxidase in pepper ($CaPO_2$) plays a significant role in generating H_2O_2 and effective defense in response to the bacterium *Xanthomonas campestris* pv. *vesicatoria*, a further indication that apoplastic class III plant peroxidases may have additional roles in disease resistance besides contributing to cell wall reinforcement (Choi et al. 2007).

Indeed, it has been demonstrated that peroxidase-dependent H_2O_2 production (visualized by 3,3-diaminobenzidine staining) in barley infected with its powdery mildew pathogen (*Blumeria graminis* f. sp. *hordei*) leads to cross-linking (lignification) and protein immobilization in plant cell walls (Thordal-Christensen et al. 1997). The presence of H_2O_2 in papillae is in fact a biochemical marker of barley resistance to powdery mildew (Hückelhoven et al. 1999, 2000). A general role of H_2O_2 in resistance to pathogen penetration is supported by several observations. Overexpression of H_2O_2-generating enzymes (peroxidases, glucose oxidase) enhances resistance to penetration by wheat powdery mildew (*Blumeria graminis* f. sp. *tritici*) in transiently transformed wheat cells (Schweizer et al. 1999). Furthermore, enzymatic removal of H_2O_2 by catalase enhances fungal penetration of leaf epidermal cells in three different plant–fungus interactions (Mellersh et al. 2002). Interestingly, a catalase (CatB) secreted by *B. graminis* f. sp. *hordei* at penetration sites is involved in scavenging H_2O_2, demonstrating that a pathogen may actively interfere with penetration resistance (Zhang et al. 2004). In a similar experiment, it was shown that suppression of the oxidative burst in corn is facilitated by Pep1, a key virulence effector of the corn smut fungus *Ustilago maydis*. Pep1 inhibits the apoplastic oxidative burst by a direct interaction with, and inhibition of class III peroxidases (Hemetsberger et al. 2012). Another role of H_2O_2 in reinforcing the plant cell wall during pathogen attack besides lignification is the contribution to protein immobilization by cross-linking of proline-rich glycoproteins, thus conferring enhanced resistance to fungal cell wall-degrading enzymes (Bradley et al. 1992; Brisson et al. 1994). This process, together with lignification, may effectively entrap fungal penetration structures in a papilla.

Interestingly, research evidence suggests that H_2O_2 may also play a role in pathogen limitation upstream of papillae since compounds of papillae are oxidatively cross-linked on the way to the site of deposition. For example, Collins et al. (2003) showed that H_2O_2 was associated with vesicles containing cell wall components which were in transit to papillae. Similarly, An et al. (2006) demonstrated that multivesicular bodies, which might participate in the secretion of building blocks of papillae, are associated with H_2O_2 accumulation. Another interesting role of papillae besides arresting fungal penetration is the blockage of all plasmodesmata between intact cells and those undergoing localized necrosis during an HR, thereby containing hypersensitive cell death (Shetty et al. 2008).

A further remarkable role of ROS production in the first line of spatial defense against pathogens is the association of ROS with defects in the plant cuticle. A physical barrier to pathogen ingress that covers the outer surface of aerial plant parts, the cuticle consists mainly of the polyester cutin, polysaccharides, and waxes and is closely linked to the cell wall of underlying epidermal cells (Serrano et al. 2014). It is known that strong resistance to the infection of aggressive, virulent fungi like *Botrytis cinerea* can be observed both in wounded plants and in plants with cuticular defects (Chassot et al. 2008). L'Haridon et al. (2011) have observed that the production of ROS ($O_2^{\cdot-}$ and H_2O_2) and a permeable cuticle is common to all these situations. For example, the *A. thaliana* cuticular mutants, *bdg* and *lacs2*, constitutively produce ROS as shown by fluorescence assays and display enhanced resistance to *B. cinerea* (L'Haridon et al. 2011; Serrano et al. 2014). However, treatment of wild type leaf surfaces with fungal cutinase also results in ROS accumulation and resistance. It is possible that cutin monomers accumulating in mutants with a defective cuticle are perceived by the plant as a danger signal (danger-associated molecular pattern, DAMP) leading to the production of ROS and induction of further defense responses and resistance. In case of fungal pathogens that penetrate plant cell walls, this illustrates why plants that are in fact permanently exposed to external pathogen elicitors do not constantly induce resistance responses; this only takes place once the cuticle has been permeabilized, for example, after a successful pathogen attack.

From the above discussed data, it seems evident that ROS production associated with plant cell walls and the cuticle govern diverse pathophysiological processes that ultimately lead to the inhibition of microbial invaders in the initial stages of pathogenesis.

2.2 *Plant Stomatal Immunity: A Barrier to Pathogen Ingress Through Natural Openings is Mediated by ROS*

Openings in the epidermal layer of terrestrial plants are called stomata. These pores are surrounded by two guard cells regulating opening and closure in order to establish gas exchange between the leaf and the environment. This regulation system allows the control of transpiration. For example, during drought stress, the regulation of anion channels in guard cells is coordinated by abscisic acid (ABA). In fact, in *A. thaliana* the perception of ABA activates the guard cell-specific, ABA-related protein kinase OST1 (also called SNF1-RELATED KINASE 2.6) followed by ROS production and activation of Ca^{2+}-signaling (Mustilli et al. 2002). This results in activation of anion channels and K^+ efflux both through Ca^{2+}-independent and Ca^{2+}-dependent pathways, and a decrease in guard cell turgor leading to stomatal closure (Sawinski et al. 2013). Interestingly, stomatal closure is also tightly controlled during plant resistance responses, and it may have a crucial role in prevention of pathogen invasion. This phenomenon is referred to as stomatal immunity or stomatal defense response and functions as a physical barrier against the infection of pathogens preferring stomatal entry, e.g., bacteria (Melotto et al. 2008; Sawinski et al. 2013). An active role of guard cells in plant defense was first demonstrated when inoculation of leaf surfaces with *Pseudomonas syringae* pv. *tomato* DC3000 was found to decrease stomatal aperture within 1 to 2 h after inoculation (Melotto et al. 2006).

However, the guard cell response to biotic stress seems to differ from the response to abiotic stress in several aspects. First, pathogen-induced stomatal closure primarily depends on the elicitation of basal resistance, i.e., perception of pathogen-associated molecular patterns (PAMPs) by pattern recognition receptors (PRR) in guard cells (Sawinski et al. 2013). For instance, FLS2, the PRR detecting the PAMP flg22 (a conserved peptide of the bacterial flagellum), is indeed expressed in guard cells. In fact, mutant *fls2 Arabidopsis* plants are impaired in stomatal closure in response to flg22 and show increased susceptibility to *P. syringae* pv. *tomato* DC3000 when sprayed onto the leaf surface—as opposed to infiltration to the leaf intercellular space (Zipfel et al. 2004; Zeng and He 2010). Second, although ROS function in both ABA- and pathogen-induced stomatal closure, ROS production in response to PAMPs and pathogens seem to rely on RBOHD, a NADPH (reduced nicotinamide adenine dinucleotide phosphate) oxidase in plant cell membranes with a pivotal role in the generation of $O_2^{\cdot -}$ during plant disease resistance (discussed below), as opposed to the role of RBOHF during ABA-induced stomatal closure (Mersmann et al. 2010; Sawinski et al. 2013). Indeed, flg22- and elf18-dependent stomatal closure is abolished in the *rbohD* mutant of *Arabidopsis* (Mersmann et al. 2010; Macho et al. 2012). Third, mitogen-activated protein kinases like MPK3 and MPK6 are activated in *Arabidopsis* by flg22 downstream of ROS production, and these kinases are also highly responsive to drought stress (Kovtun et al. 2000; Liu et al. 2010b; Tsugama et al. 2012). Although this suggests an overlap in pathogen- and ABA-activated kinase signaling pathways, guard cell-specific *mpk3* antisense lines are affected in closing stomata when exposed to bacteria or the PAMP LPS but not ABA (Gudesblat et al. 2009), suggesting a specific requirement for MPK3 in pathogen-induced stomatal closure.

Based on the above data, it was suggested that stomatal closure during pathogen defense is controlled by an ABA-independent signaling pathway distinctly different from that activated during abiotic stress (Montillet et al. 2013). A stomatal defense response is induced upon perception of PAMPs like flg22 resulting in pathogen-induced oxidative stress (ROS production, oxidative burst) and induction of various defenses including activation of MPK3 and MPK6. These kinases would induce a guard cell-specific lipoxygenase, LOX1, that is required for signaling and the peroxidation of poly unsaturated fatty acids to oxylipins like *cis*-OPDA. Montillet and coworkers (2013) found that in guard cells of *Arabidopsis*, the accumulation of *cis*-OPDA in response to flg22-elicitation is followed by an increase in levels of salicylic acid (SA), a hormone playing a central role in regulating plant defenses. Finally, downstream of SA accumulation, the activation of the anion channel SLAC1 is required for stomatal closure to occur.

In summary, it seems that ROS play an instrumental role in mediating plant resistance at natural openings (stomata) to the infection of pathogens that prefer this mode of host entry, e.g., bacteria.

2.3 Pathogen Limitation by ROS at the Plasma Membrane: A Possible Role of NADPH Oxidases

Early research by Doke and coworkers (Doke 1995; Doke and Miura 1995) showed that an NADPH-dependent $O_2^{\cdot-}$-generating system is present in plasma membrane fractions of late blight-infected potato tubers. Incubation of tuber slices with an incompatible *Phytophthora infestans* race or treatment with pathogen-derived elicitors stimulated NADPH-dependent $O_2^{\cdot-}$-generating activity, while no stimulation of NADPH oxidase activity was detected in membrane fractions from tissues inoculated with a compatible race of the pathogen or from control tissues. The ROS burst could also be observed in elicitor-treated protoplasts (Doke 1983a), indicating that NADPH oxidase-mediated ROS generation does not necessarily require the presence of the cell wall or apoplastic enzymes. $O_2^{\cdot-}$ release was extracellular which is expected, since $O_2^{\cdot-}$ is hardly diffusible across membranes (Doke 1983a; Lamb and Dixon 1997). It has been shown in different experimental systems that this pathogen-elicited, plasma membrane-derived, NADPH-oxidase generated $O_2^{\cdot-}$ is rapidly dismutated to H_2O_2 via enzymatic catalysis by superoxide dismutases (Levine et al. 1994; Tenhaken et al. 1995).

Plasma membrane NADPH oxidases and their corresponding genes in plants have been discovered on the basis of their sequence similarity (homology) to the mammalian respiratory burst NADPH oxidase subunit gp91phox (Groom et al. 1996; Keller et al. 1998; Torres et al. 1998), hence the abbreviated name of these genes/proteins: RBOH. These NADPH oxidases are closely associated with ROS formation during plant–pathogen interactions (Suzuki et al. 2011; Marino et al. 2012). For example, two RBOH genes have been described in *Nicotiana benthamiana*: *NbRBOHA* being expressed constitutively at low levels and *NbRBOHB* induced by the elicitor INF1 from *Phytophthora infestans*. Transient silencing of these genes resulted in a reduction in ROS accumulation and

INF1-induced cell death, along with a loss of resistance to *P. infestans* (Yoshioka et al. 2003; Asai et al. 2008). In general, the absence of expression of NADPH oxidase (RBOH) genes confers enhanced susceptibility to biotrophic pathogens (preferring live host tissues, e.g., powdery mildews), as shown, e.g., in the pathosystems *Arabidopsis thaliana/Golovinomyces cichoracearum* (Berrocal-Lobo et al. 2010) and *Hordeum vulgare/Blumeria graminis* f. sp. *hordei* (Proels et al. 2010). This indicates that NADPH oxidase-derived ROS indeed has a role in limiting these pathogens, by inducing pathogen and host cell death (Király et al. 1993; El-Zahaby et al. 2004) and/or promoting cell wall reinforcements in attacked host cells (Proels et al. 2010). On the other hand, infection of NADPH oxidase mutant plants by necrotrophic pathogens often results in enhanced resistance, as demonstrated for *A. thaliana* infected by *Alternaria brassicicola* (Pogány et al. 2009) and *N. benthamiana* attacked by *Botrytis cinerea* (Asai and Yoshioka 2009). Apparently, RBOH-dependent ROS facilitate colonization of host tissues by necrotrophic pathogens by promoting death of attacked host cells (for a review see Barna et al. 2012).

However, the role of particular RBOH genes in pathogen limitation may be different. As mentioned above, the *NbRBOHA* gene of *N. benthamiana* is expressed constitutively, while another RBOH gene in the same host, *NbRBOHB*, is induced only upon elicitor treatment, although both genes are required for elevated ROS accumulation and disease resistance. Interestingly, a dual role during infection by *A. brassicicola* was shown for *AtRBOHD*, but not *AtRBOHF*: a functional RBOHD protein triggers death in cells that are damaged by fungal infection but simultaneously inhibits death in neighboring cells (Pogány et al. 2009).

Taken together, accumulated research evidence implies that, during plant–pathogen interactions, the regulatory mechanisms conferred by NADPH oxidases and ROS produced by these enzymes are highly sophisticated and may function not only in limitation of pathogens at the plasma membrane but also in maintenance of the normal physiological state of plant hosts.

2.4 Subcellular Localization of Intracellular ROS and Pathogen Limitation

Bacterial and fungal plant pathogens do not enter the intracellular space of attacked plant cells. In fact, fungi can only penetrate through plant cell walls, but not the plasma membrane, while bacteria multiply only in the intercellular space. However, plant viruses do enter plant cells and are intimately associated with the cytoplasm and cellular organelles during viral pathogenesis (i.e., they are obligate biotrophic pathogens). Nevertheless, limitation of all these pathogens is influenced by ROS-related biochemical processes that occur within plant cells (as opposed to the cell wall and plasma membrane).

2.4.1 Mitochondria

Both earlier and recent research suggests that plant mitochondria may play an important role in host defense responses to biotic stresses. Mitochondrial respiration is one of the most important metabolic processes of plant cells, and a pronounced increase in the respiratory rate of a host can be a marker of resistance during plant–pathogen interactions (Goodman and Novacky 1994; Hanquing et al. 2010 and references within). In mitochondria of higher plants, electrons produced by the respiratory oxidation of NADH (reduced form of nicotinamide adenine dinucleotide) can flow through the usual cytochrome respiratory pathway or the alternative respiratory pathway. It is well known that the alternative respiratory pathway is catalyzed by alternative oxidases (AOX), located in the mitochondrial inner membrane and acting as terminal oxidases in the mitochondrial electron transport chain (mtETC). The primary function of AOXs in mitochondria is to prevent overreduction of the cytochrome pathway and the resulting accumulation of ROS during abiotic stresses and pathogen attack (Chivasa and Carr 1998; Maxwell et al. 1999; Robson and Vanlerberghe 2002). In fact, mitochondria are a main source of ROS generation in plant cells contributing to 20–30 % of the cytoplasmic steady-state concentration of H_2O_2 (Hanquing et al. 2010). Therefore, AOX may potentially control ROS levels in the plant cytoplasm. This is supported by observations that inhibition of antioxidant (catalase) activity can enhance AOX mRNA expression. Furthermore, a lack of AOX is accompanied by an increase in cytoplasmic antioxidant defenses (Rizhsky et al. 2002; Mizuno et al. 2005; Amirsadeghi et al. 2006).

Pathogen elicitors and toxins can increase the production of mitochondrial (mt) ROS, suggesting that these organelles are a likely source of ROS in biotic stress, in particular, during the plant defense-related oxidative burst (Bolwell and Wojtaszek 1997; Krause and Durner 2004; Rhoads et al. 2006). Furthermore, defense-related compounds (e.g., salicylic acid and methyl jasmonate) associated with plant disease resistance responses are inhibitors of mtETC, and their application can cause an increase in mtROS (Norman et al. 2004; Zhang and Xing 2008). Indeed, it has been shown recently that salicylic acid may inhibit the cytochrome pathway of mtETC through binding to alpha-ketoglutarate dehydrogenase E2 (α-kGDH E2), a rate limiting enzyme controlling NADH supply for efficient mitochondrial electron transport. This results in limiting systemic infection and accumulation of *Tobacco mosaic virus* (TMV), a process that requires mRNA expression of AOX1a (Liao et al. 2015). The authors suggest that inhibition of α-kGDH E2 could result in elevated ROS generation that activates AOX and virus resistance. Accordingly, *Nicotiana sylvestris* mitochondrial mutants with elevated AOX protein levels display a higher degree of resistance to TMV (Dutilleul et al. 2003). Furthermore, AOX transcript and protein levels increase in tobacco that displays hypersensitive resistance (HR, local necrotic lesions) during TMV infection, but do not change during a compatible tobacco–TMV interaction (Lennon et al. 1997; Chivasa and Carr 1998). On the other hand, an *A. thaliana* mutant disrupted in stress responses (*dsr1*) exhibits lower mitochondrial ROS production and enhanced susceptibility to bacterial and fungal root pathogens, e.g., *Rhizoctonia solani* AG8 (Gleason et al. 2011). This indicates a potential role of plant mitochondria in resistance to pathogens that do not enter the cytoplasm, e.g., through ROS-mediated signaling. In general, increases in AOX gene and protein expression have been noted in plant resistance responses to viral and other pathogens but not during compatible infections (Chivasa et al. 1997; Lennon et al. 1997; Chivasa and Carr 1998; Lacomme and Roby 1999; Simons et al. 1999; Király et al. 2008). The above-discussed results indicate that increased production of mitochondrial ROS and its control by AOX could be a key factor of plant resistance to certain pathogens. This might be especially the case with viruses that could be in close contact with mitochondria and therefore might be directly inhibited by high concentrations of, e.g., mtROS during HR-type resistance. In fact, loss of HR-type resistance to TMV in tobacco at high temperatures (30 °C) is associated with a down-regulation of AOX mRNA expression and $O_2^{\cdot-}$ levels suggesting that AOX indeed contributes to the regulation of ROS levels during HR in order to prevent excessive plant cell death (Király et al. 2008). Importantly, AOX-mediated ROS regulation during HR-type resistance also involves a transient suppression of AOX gene expression with a likely role in transiently increasing ROS in, e.g., mitochondria to limit viral and bacterial pathogens. The transient nature of ROS increase would ensure the control of plant cell death during an HR-type resistance (Lacomme and Roby 1999; Künstler et al. 2007).

2.4.2 Chloroplasts

In photosynthetic organisms, chloroplasts transform light into reducing power resulting in CO_2 fixation. However, excess reducing power is inevitably produced during photosynthesis, since most plants use only ca. 50 % of the absorbed light energy for this process. Excess reducing power can increase the leakage of electrons from the photosynthetic electron transport chain of chloroplast thylakoid membranes (photosystems/PS/I and II) leading to ROS generation and oxidative damage to the photosynthetic apparatus (Dat et al. 2000). Environmental and pathogen stress may further decrease the CO_2 fixation of photosynthesis, therefore leading to enhanced ROS accumulation in chloroplasts. In fact, the plant defense related oxidative burst occurs in several cellular locations, including chloroplastic PSI and PSII (Sharma et al. 2012), implying that elevations of chloroplastic ROS may be exploited by plant hosts for successful defense against pathogens, especially viruses that are often associated with chloroplasts. Remarkably, it has been shown that compatible tobamovirus infections facilitate the inhibition of PSII electron transport by disturbing the oxygen-evolving complex (OEC), a potential site of ROS generation (Lehto et al. 2003). The levels of two PSII OEC proteins, PsbP and PsbQ, were lower in plants infected by *Pepper mild mottle virus* (PMMoV), as compared to healthy controls, and virus infection also affected oxygen evolution rates of thylakoid membranes (Rahoutei et al. 2000). Proteomic analyses have shown specific down-regulation of the PsbP protein not only by PMMoV (Perez-Bueno et al. 2004) but also by CMV (Di Carli et al. 2010) during compatible infections. Interestingly, Balasubramaniam and coworkers (2014) have recently demonstrated that PsbP interacts with the coat protein (CP) of *Alfalfa mosaic virus* (AMV) in the cytoplasm and in high concentrations PsbP may inhibit virus replication, suggesting a function in antiviral defense. The authors suggest that in compatible infections, AMV CP may act as an effector protein by interacting with PsbP to interfere with host defense signaling. AMV CP may sequester the chloroplast-targeted PsbP in the cytoplasm, thereby inhibiting the biogenesis of PSII and the consequent ROS generation and activation of other plant defense responses by chloroplasts. Such a strategy has been suggested for the interaction of chloroplast proteins with the TMV replicase protein and the *Plum pox virus* CI protein (Abbink et al. 2002; Jimenez et al. 2006) and, recently, for the interaction of *Rice stripe virus* SP protein with rice PsbP (Kong et al. 2014). However, modulation of chloroplast-controlled defense responses like ROS generation by pathogen effectors is not a virus-exclusive strategy for evading host defense responses. HopN1, an effector protein of the bacterium *P. syringae* pv. *tomato* DC3000, was shown to localize to the thylakoids of chloroplasts where it bounds the PSII OEC protein PsbQ and promotes its degradation (Rodriguez-Herva et al. 2012). The resulting interruption of PSII activity and suppression of ROS production compromised cell death during HR-type resistance and points to the role of chloroplastic PsbQ as a potential factor of antibacterial defense.

2.4.3 Peroxisomes

Excess reducing power generated in chloroplasts through photosynthesis can be diverted, e.g., to peroxisomes by the export of glycolate. In peroxisomes, glycolate is first metabolized to glyoxylate by glycolate oxidases, a reaction that generates H_2O_2 during a process called photorespiration. Recent evidence suggests that peroxisome-generated H_2O_2 indeed has a role in defense against virus infections. Mathioudakis et al. (2013) have demonstrated an in planta interaction between the triple gene block protein 1 (TGBp1) of *Pepino mosaic virus* (PepMV) and tomato catalase 1 (CAT1). The primary function of class 1 catalases is the removal of H_2O_2 produced during photorespiration in leaf peroxisomes (Willekens et al. 1995). Interaction of the viral TGBp1 and tomato CAT1 was observed in nuclei and the cytoplasm suggesting that during a compatible PepMV infection, virus limitation is evaded at least in part by scavenging peroxisomal H_2O_2 (Mathioudakis et al. 2013). Both PepMV infection and in planta overexpression of TGBp1 increased CAT activity and lowered H_2O_2 levels, while silencing of *CAT1* conferred a reduced accumulation of PepMV. The role of peroxisome-generated H_2O_2 in resistance to plant viruses is also supported by an earlier study (Talarczyk et al. 2002). Overexpression of a yeast peroxisomal catalase A1 gene (*CTA1*) in tobacco revealed that the CTA1 protein is indeed localized in peroxisomes of tobacco cells and likely confers a reduction in resistance to TMV. Interestingly, the size of HR-type (localized) necrotic lesions was significantly larger in the infected leaves of these transgenic plants coupled with lower H_2O_2 levels around lesions. This suggests that peroxisomal H_2O_2 is indeed functional in limiting virus replication and/or movement during HR-type virus resistance, which is also supported by observations that—similarly to AOX (see above)—plant *CAT1* genes and catalase activity are transiently suppressed in TMV-elicited HR-type necrotic lesions (Dorey et al. 1998; Yi et al. 1999, 2003; Künstler et al. 2007). Remarkably, peroxisome-generated H_2O_2 can also significantly limit plant pathogenic bacteria. Chamnongpol et al. (1998) showed that excess H_2O_2 produced in *CAT1* antisense tobacco dramatically suppresses levels of *P. syringae* pv. *syringae* within 24 h following inoculation.

The data discussed above point to a central role of ROS produced by various intracellular organelles (mitochondria, chloroplasts, peroxisomes) in the suppression of pathogens, even in case of bacteria and fungi that do not enter plant cells.

3 Temporal ROS Accumulation and the Efficiency of Pathogen Limitation in Plant Tissues: Timing is Everything?

3.1 ROS Accumulation may Result in Disease Resistance and Plant Cell/Tissue Death During the Hypersensitive Response

Recognition of a pathogen as an invader may result in the elicitation of localized programmed cell death associated with pathogen restriction at the infection site. This phenomenon often culminates in the formation of macroscopically visible localized necrotic lesions (hypersensitive response [HR], see, e.g., in Klement 1982; Goodman and Novacky 1994; Greenberg and Yao 2004). HR is known to occur during both the first and second line of plant defense (i.e., basal/nonhost resistance /PAMP-triggered immunity/ and R gene-mediated resistance /effector-triggered immunity/). Although macroscopic cell and tissue death often seems to be tightly linked to the onset of pathogen resistance, it is possible to uncouple cell death from resistance during certain fungal, bacterial, and viral infections (Király et al. 1972; Mittler et al. 1996; Schiffer et al. 1997; Yu et al. 1998; Bendahmane et al. 1999; Cole et al. 2001; Gassmann 2005; Coll et al. 2010; for a review see Király and Király 2006). However, the existence of independent resistance and cell death responses within an HR does not exclude the possibility that cell death has a role in reinforcing or stimulating the induction of defenses and pathogen localization (Lamb and Dixon 1997; Hatsugai et al. 2004; Greenberg and Yao 2004).

A substantial evidence indicates that the HR is associated with an oxidative burst, i.e., a rapid and localized accumulation of ROS like $O_2^{\cdot -}$, H_2O_2, and the hydroxyl radical ($OH^{\cdot}$) (Doke 1983a, b; Doke and Ohashi 1988; Ádám et al. 1989; Alvarez et al. 1998; Allan and Fluhr 1997; Lamb and Dixon 1997; Torres et al. 2006). Furthermore, it has been shown that artificial, external ROS treatments are indeed sufficient to induce cell death and plant resistance (an HR). For instance, it was possible to induce HR-type resistance to bacterial and fungal pathogens (powdery mildews and rusts) in originally susceptible plants by external application of 50 mM H_2O_2 or treatments with ROS-generating agents (riboflavin–methionine, xanthine–xanthine oxidase) (Hafez and Király 2003; El-Zahaby et al. 2004). These observations are supported by analysis of transgenic antisense tobacco plants with reduced expression of antioxidant genes (e.g., catalases) that display enhanced HR-type resistance during bacterial infections (Chamnongpol et al. 1998; Takahashi et al. 1997). All of the above cited reports demonstrated that ROS accumulate in resistant hosts that display an HR. However, certain data indicate that in some cases elicitors and pathogens may trigger a strong oxidative burst and associated defense mechanisms without causing an HR (Glazener et al. 1996; Jabs et al. 1997). This indicates that (1) cell and tissue death is not always a necessary component of ROS-induced resistance during an HR and (2) ROS could play other roles during plant disease resistance as direct antimicrobial agents.

3.2 *ROS as Antimicrobial Agents in Plants*

ROS were first regarded as antimicrobial agents during plant defense responses by demonstrating an in vitro sensitivity of bacterial and fungal pathogens to externally produced $O_2^{\cdot -}$ (Aver'yanov and Lapikova 1988; Tzeng et al. 1990; Jordan et al. 1992; Király et al. 1993; Ouf et al. 1993) and H_2O_2 (Peng and Kuć 1992; Király et al. 1993; Wu et al. 1995; Shetty et al. 2007). For example, micromolar concentrations (25 μM) of H_2O_2 may inhibit spore germination of a number of fungal pathogens in vitro (Peng and Kuć 1992). Also, 100 μM H_2O_2 completely hinders growth of cultured bacteria (*Erwinia carotovora* ssp. *carotovora*) and results in >95 % suppression of the growth of *Phytophthora infestans* (Wu et al. 1995). Interestingly, 5 mM H_2O_2 inhibited fungal development in young (4-day-old) *Septoria tritici* cultures while a much higher concentration (50 mM H_2O_2) was required to inhibit growth in older (16-day-old) cultures, reflecting the differing ability of a pathogen to tolerate H_2O_2 during different stages of its life cycle (Shetty et al. 2007).

The hypothesis that pathogens are also sensitive to ROS produced during infection of their plant hosts was first suggested and tested by Aver'yanov and Lapikova (1988) and Jordan et al. (1992). The former authors demonstrated that in rice, the toxicity of leaf diffusates towards the fungal pathogen *Pyricularia oryzae* is mediated by ROS, while the latter group showed that potato (*Solanum tuberosum* cv. Kennebec) pretreated with an $O_2^{\cdot -}$-generating riboflavin–methionine mixture develops fewer diseased leaves in response to *Phytophthora infestans*. Accordingly, high, sustained levels of ROS (H_2O_2) achieved by transgenic approaches in different crops also caused resistance to disease symptoms and pathogen multiplication at least in part by direct killing and/or inhibition of plant pathogens. Overexpression of a H_2O_2-producing fungal glucose oxidase gene in potato, tobacco, and cabbage conferred not only a reduction in necrotic disease symptoms in response to bacteria and the oomycete *P. infestans* but, in case of potato infection by *E. carotovora* ssp. *carotovora*, also a suppression of pathogen multiplication (Wu et al. 1995; Lee et al. 2002). Expression of the *BvGLP-1* gene of sugar beet encoding a germin-like protein in *A. thaliana* increased the H_2O_2 content in transgenic plants and conferred resistance of roots to the fungal pathogens *Verticillium longisporum* and *Rhizoctonia solani* (Knecht et al. 2010). Also, transgenic tomato expressing a wheat oxalate oxidase displayed oxalate oxidase activity (H_2O_2-generation) and reduced symptoms after inoculation with *Botrytis cinerea* (Walz et al. 2008). It must be noted here that oxalate (oxalic acid) is a major pathogenicity factor of necrotrophic fungi like *B. cinerea* and a suppressor of ROS (Cessna et al. 2000). Therefore, elevated oxalate oxidase expression in these tomato plants is not only an extra source of in planta ROS generation but ensures normal ROS production in the plant host by eliminating the ROS-suppressor activity of the attacking pathogen in due time. In fact, another necrotrophic fungus (*Sclerotinia sclerotiorum*) also confers an early, oxalic acid-mediated ROS-suppression in the host during initial stages of pathogenesis, as demonstrated by the elegant experiments (real-time GFP redox sensing, histological staining, and reverse fungal genetics) of Williams et al. (2011). Thus, when pathogens, including necrotrophic fungi, are exposed to an oxidative burst early, during the initial phase of infection, their invasion seems to be weakened.

3.3 Timing is Everything: Early ROS Accumulation Seems to Confer Efficient, Symptomless Disease Resistance in Plants

The fact that a constitutive overproduction of H_2O_2 in, e.g., transgenic plants confers resistance by killing/limiting pathogens (see above) indeed points to the role of the temporal patterns of plant ROS production in eliciting a successful defense. It was possible to induce resistance to powdery mildew (*B. graminis* f. sp. *hordei* A6) in susceptible barley plants (*Mlo* genotype) by external application of 50 mM H_2O_2 or treatment with ROS-producing agents (Hafez and Király 2003). Symptomless resistance occurred if ROS were administered early (ca. 1 day) after inoculation, while HR developed when ROS application was late (ca. 3 days after inoculation). Interestingly, a barley line expressing the resistance gene *mlo5* (resistant without HR) also developed HR upon late application of ROS, while the line expressing *Mla12* (resistant with HR symptoms) displayed a symptomless resistance upon early ROS application but an enhanced HR (i.e., an earlier development of a higher number of necrotic lesions) following a late ROS exposure. A similar phenomenon (induction of symptomless resistance following early ROS treatments and HR following later ROS exposure) was also demonstrated to occur during infection by fungal and bacterial pathogens in barley and other plants (El-Zahaby et al. 2004). Importantly, the antimicrobial and cell death-inducing effects of these ROS treatments were completely eliminated by an initial infiltration of antioxidant enzymes (superoxide dismutase and catalase) into leaves.

The above-mentioned studies imply that an early in planta ROS accumulation may efficiently kill/limit pathogens, while a later development of higher ROS levels may only partially hinder pathogen replication and spread, accompanied by death of attacked plant host cells (an HR). Therefore, defense against pathogens could be very effective if it is a rapid, symptomless process, eliminating the pathogen in due time and not overusing resources of the plant. In fact, our preliminary results suggest an important role of early ROS accumulation during symptomless (Type I) non-host resistance (i.e., resistance to pathogens adapted to other host plants, mechanistically the same process as basal resistance /PAMP-triggered immunity). For example, during symptomless non-host resistance of barley to wheat powdery mildew (*B. graminis* f. sp. *tritici*) a rapid, early accumulation of $O_2^{\cdot -}$ is already detectable 1 day after inoculation, much earlier than during HR-type host resistance of barley to its own powdery mildew (*B. graminis* f. sp. *hordei*), as assayed by nitro blue tetrazolium (NBT) tissue staining (Király et al. 2013).

The pivotal role of early ROS production in eliciting plant disease resistance responses also seems likely during plant–virus interactions. We have shown that not only HR-type resistance to TMV but NADPH oxidase-dependent $O_2^{\cdot -}$-generation is also suppressed in tobacco plants carrying the resistance gene *N* and exposed to higher (30 °C) temperatures (Király et al. 2008). However, the early, external application of ROS—2 h after inoculation—confers a symptomless resistance of tobacco otherwise susceptible to TMV, while ROS treatments 3 days after inoculation do not elicit resistance, only HR-like cell and tissue death in the host (Király et al. 2008; Bacsó et al. 2011). In fact, a rapid symptomless plant resistance to virus infections indeed occurs in nature and is termed extreme resistance. This is a very efficient resistance response of plant hosts conferring almost total immunity against viruses like, e.g., *Potato virus X* (PVX), *Soybean mosaic virus* (SMV), or *Turnip crinkle virus* (TCV) (Bendahmane et al. 1999; Cooley et al. 2000; Hajimorad and Hill 2001). Extreme resistance to PVX in potato is conditioned by the *Rx1* and *Rx2* resistance genes (Bendahmane et al. 1999, 2000). Interestingly, during transient expression of the PVX coat protein (the viral avirulence gene product) in *Rx1* plants, the coat protein elicits HR (Bendahmane et al. 1999). This implies that the resistance conditioned by *Rx* genes to PVX occurs so rapidly that the PVX coat protein cannot attain a concentration sufficient to elicit HR. A similar model has been proposed to explain the resistance conferred by the *HRT* gene in *Arabidopsis thaliana* to TCV. *A. thaliana* that carries *HRT* responds to TCV with HR (Cooley et al. 2000). Transgenic plants with moderately elevated *HRT*-expression exhibit a "micro-HR" (i.e., death of individual plant cells), while transgenic *A. thaliana* overexpressing *HRT* to high levels develops no HR upon TCV infection. In fact, the latter plants display extreme resistance to TCV because they respond rapidly to the virus (Cooley et al. 2000). These studies also suggest that the efficiency of disease resistance depends on the speed of host response: a rapid reaction of the plant host ensures early elimination of the pathogen without any disease symptoms. Although the biochemical mechanism of symptomless (extreme) resistance is unknown, Bendahmane and coworkers (1999) proposed the possible involvement of ROS. Thus, the extremely rapid induction of resistance governed by *Rx* genes would cause an early accumulation of ROS ($O_2^{\cdot -}$, H_2O_2, or perhaps $OH^{\cdot}$ that is considered to be one of the most reactive ROS) and an almost complete inhibition of virus replication in the initially attacked plant cells.

If early ROS accumulation is crucial for plant disease resistance, then the often very intensive in planta ROS production during the advanced stages of pathogenesis might be regarded, at least in part, as a late and failed attempt by the host to limit pathogens. For example, the correlation of high ROS levels with susceptibility to fungal pathogens that are hemibiotrophic (i.e., having an early biotrophic and late necrotrophic phase of pathogenesis) could also be explained by this phenomenon. This is the case with barley infected with *Bipolaris sorokiniana*, where $O_2^{\cdot -}$ can be detected in chloroplasts of cells surrounding necrotic spot blotch lesions (Schäfer et al. 2004). Also, an enhanced accumulation of H_2O_2 in the apoplastic space of systemically infected leaves is a characteristic of apricot cultivars susceptible to *Plum pox virus* (PPV) (Diaz-Vivancos et al. 2006; Hernández et al. 2006). The strong systemic burst of H_2O_2 in these plants that display severe chlorosis might also be attributed to a delayed and failed attempt by the host to elicit resistance in systemic tissues. It is likely that a similar phenomenon occurs in tobacco systemically infected by *Cucumber mosaic virus* (CMV), where Shang and coworkers (2010) have demonstrated that absence of the virus in so-called "dark green islands" of systemically infected leaf tissues correlates well with the presence of $O_2^{\cdot -}$.

From the above, it seems evident that the proper timing of ROS accumulation is a pivotal factor of plant disease resistance. Thus, even during advanced stages of pathogenesis, plant hosts may repeatedly try to limit pathogens by processes that involve ROS accumulation.

4 ROS-Mediated Signaling During Plant Disease Resistance: Regulating Abiotic Stress and Pathogen Levels in Concert

4.1 The Dual, Concentration-Dependent Role of ROS in Plant Disease Resistance

When plants encounter pathogens, a properly timed ROS accumulation seems to contribute to the development of disease resistance, either without any obvious side effects in the host (extreme resistance) or with the concomitant development of controlled and limited cell and tissue death (HR). Consequently, a delayed and failed attempt by the host to elicit resistance responses would result in massively stressed plant tissues (pathogen-elicited systemic chlorotic/necrotic symptoms) and a partial or almost complete loss of control over pathogen invasion.

Obviously, however, ROS produced by plants in response to pathogenic infections promote not only the direct killing/limitation of pathogens and cell/tissue death in the host but participate in various signaling pathways that elicit (1) further defense responses to pathogens (accumulation of antimicrobial compounds, induction of transcription factors and so-called pathogenesis-related genes, etc.) and (2) defenses to control unwanted cell/tissue death occurring during pathogen invasion (induction of plant antioxidants). Thus, it seems that when plants encounter pathogens they need to defend themselves simultaneously against biotic and abiotic stresses (pathogen accumulation and excessive cell/tissue death) by turning on two different types of—partially overlapping—signaling pathways that may function in parallel. These ROS-related plant signaling pathways have been thoroughly discussed in past years by several excellent reviews (Lamb and Dixon 1997; Dat et al. 2000; Mittler 2002; Torres and Dangl 2005; Hanquing et al. 2010; Barna et al. 2012; Barrios Perez and Brown 2014; Gilroy et al. 2014). Briefly, the main features that qualify ROS as important signaling molecules are (1) a dynamic control of ROS levels through scavenging by antioxidants in a cell autonomous manner, (2) ROS accumulation in different subcellular organelles, resulting in efficient intracellular signaling, (3) the chemical nature (reactivity and short half-life) of ROS confers targeted interaction/modification of proteins, (4) ROS-induced signaling is rapidly propagated from the origin of stimuli to other plant cells (Mittler et al. 2011). Indeed, plants exploit this versatility of ROS when mediating defenses to both abiotic stresses and pathogenic infections.

NADPH oxidase-dependent ROS ($O_2^{\cdot-}$ and H_2O_2) have been shown to have such a dual action, i.e., a participation in defense to both abiotic (excessive cell/tissue death) and biotic (pathogen accumulation) stresses. In fact, low concentrations of H_2O_2 act as a diffusible signal for the induction of antioxidant and pathogenesis-related genes in adjacent plant tissues. However, high concentrations of H_2O_2 are related to the induction of death (necrosis) of plant as well as invading pathogen cells. This was first demonstrated by inoculation of soybean cells with avirulent *P. syringae* pv. *glycinea;* NADPH oxidase-derived H_2O_2 mediated cell death and resistance in bacteria-infected plant cells, while accumulation of transcripts of, e.g., glutathione-S transferase and glutathione peroxidase occurred in adjacent healthy cells separated by dialysis membranes (Levine et al. 1994). These results were later confirmed by analysis of H_2O_2-producing transgenic plants (Wu et al. 1997; Chamnongpol et al. 1998; Vandenabeele et al. 2003). Furthermore, suppression of two NADPH oxidase genes (*AtRBOHD* and *AtRBOHF*) in *lsd1* (lesion simulating disease 1) mutants of *Arabidopsis* that form spontaneous, localized lesions gave unexpected results. Torres et al. (2005) have shown that suppression of these RBOH genes in the *lsd1* mutants leads to plants that develop enhanced, spreading necrotic lesions. The authors speculated that ROS produced by these RBOH genes could limit the spread of cell death around pathogen infection sites by, e.g., activating antioxidant enzymes. Indeed, Pogány et al. (2009) have convincingly demonstrated that during infection by the fungus *Alternaria brassicicola*, a functional *Arabidopsis* RBOHD protein triggers death in plant cells damaged by fungal infection but simultaneously inhibits death in neighboring plant cells.

The concentration-dependent role of ROS in mediating plant defenses to abiotic stresses and pathogens was further demonstrated by experiments analyzing the effects of exogenously applied H_2O_2. Indeed, an exogenous application of relatively low concentrations (5–20 mM) of H_2O_2 to tobacco leaves and pea seedlings stimulated the antioxidant capacity of plants, increasing thereby tolerance to abiotic stresses and enhancing seedling growth (Gechev et al. 2002; Barba-Espin et al. 2010). We have shown, however, that pretreatment of tobacco with 5 to 10 mM of H_2O_2 also induces a suppression of pathogen-induced necrotic symptoms associated with either HR-type resistance to viruses and bacteria or a compatible infection by a fungus (Hafez et al. 2012). This "immunization" of plant hosts was associated with the induction of transcription and activity of antioxidant enzymes and was shown to operate indeed by the enhancement of plant antioxidant capacity, since artificial, simultaneous application of superoxide dismutase and catalase suppressed necrosis caused by viral, bacterial, or fungal pathogens similarly as H_2O_2 pretreatment. Remarkably, however, pathogen multiplication did not change in "immunized" tobacco plants clearly differentiating this H_2O_2-mediated suppression of pathogen-induced necrosis from plant disease resistance responses (Hafez et al. 2012).

Besides mediating plant defenses to abiotic stresses and pathogens locally, at the site of plant stress, ROS are also involved in the induction of systemic defenses. This means acquired resistance responses to both abiotic stresses (systemic acquired acclimation, SAA) and pathogens (systemic acquired resistance, SAR, and induced systemic resistance, ISR). SAA is induced by usually mild abiotic stresses and confers enhanced resistance to a later exposure of distal tissues to the same or similar (stronger) stresses. SAA is accompanied by H_2O_2 accumulation, systemic redox changes, and the induction of antioxidant defenses (Karpinski et al. 1999; Miller et al. 2009). On the other hand, SAR is typically induced by pathogenic infections resulting in tissue necrosis (Ross 1961b; Sticher et al. 1997), although there are exceptions (see Liu et al. 2010a), while ISR is elicited by nonpathogenic plant growth-promoting rhizobacteria (PGPR) (Pieterse et al. 1996). In both cases, an initial microbial challenge will result in resistance against subsequent challenge by a pathogen in distal plant parts. ROS may be involved in ISR (Barna et al. 2012), while the contribution of ROS to SAR is much more evident. HR-type resistance is often associated with the activation of SAR (Chen et al. 1993; Gaffney et al. 1993; Grant and Lamb 2006). ROS, together with salicylic acid (SA), a central regulator of plant disease resistance, were suggested to play a pivotal role in the establishment of SAR (Durrant and Dong 2004). Specifically, SA potentiates the oxidative burst and facilitates programmed cell death. However, enzymatic, and nonenzymatic antioxidants are also activated during the development of SAR (Fodor et al. 1997) but down-regulated in a transgenic tobacco that fails to develop SAR following virus infection (Király et al. 2002). In fact, these results demonstrate that the dual role of ROS (concentration-dependent mediation of plant defenses to abiotic stresses and pathogens) is also functional during SAR.

4.2 ROS Waves in Plant Disease Resistance: An Integration of Signaling Pathways

Although ROS involvement in systemic defenses seems obvious, previously, the involvement of ROS molecules traveling as long distance signals during systemic acquired resistance responses (SAA, SAR/ISR) was questioned since most ROS are very sensitive to scavenging enzymes (antioxidants). Remarkably, a new mechanism for rapid, long-distance cell-to-cell signaling during SAA was described that utilizes ROS (Miller et al. 2009; Mittler et al. 2011; Gilroy et al. 2014). Exposure of a particular plant tissue to different abiotic stresses (high light, heat, salinity, cold, mechanical injury) initiates enhanced ROS production in the affected local tissue and triggers a systemic, autopropagating ROS producing wave traveling from the affected tissue to distal plant parts at a rate of up to ca. 8.4 cm min^{-1}. This ROS wave is dependent on the presence of the NADPH oxidase RBOHD and is accompanied by H_2O_2 accumulation in apoplast of cells along the systemic signal path. The autopropagating nature of the signal suggests that each cell along the way activates its own RBOHD and generates ROS that induce all adjacent cells to undergo this same process.

The existence of autopropagating systemic ROS waves seems also likely during elicitation of plant disease resistance responses. One of the first evidences was provided by Park et al. (1998) who demonstrated that the application of a fungal elicitor in potato tubers induced an oxidative burst in distal non-treated parts of the tissue. It was also shown that a primary inoculation of lower leaves with avirulent pathogens induces a secondary oxidative burst in distant upper leaves, leading to low-frequency systemic micro-HRs (single cell HRs) required for the establishment of SAR (Alvarez et al. 1998; Fodor et al. 2001). Operation of a ROS wave functional during plant disease resistance has been recently suggested by Dubiella et al. (2013). A Ca^{2+}-dependent protein kinase 5 (CPK5) in *Arabidopsis thaliana* was activated upon stimulation by pathogen-associated molecular patterns (PAMP) or ROS and phosphorylated (activated) RBOHD in vivo. This CPK5 signaling resulted in elevated SA-mediated resistance to the bacterial pathogen *P. syringae* pv. *tomato* DC3000, enhanced plant defense gene expression in distal plant parts, and ROS synthesis (Dubiella et al. 2013). These results suggest a model of ROS-mediated cell-to-cell communication, where a self-propagating activation circuit (the protein kinase CPK5 and the NADPH oxidase RBOHD) facilitates rapid signal propagation necessary for activation of defenses and pathogen resistance at distal plant sites.

As discussed above, autopropagating ROS waves may participate in eliciting plant defenses to both abiotic stresses and pathogens. In fact these two types of signaling pathways (conferring inhibition of abiotic and biotic stresses) could significantly overlap, as shown, e.g., by their requirement for NADPH oxidase (RBOHD)-dependent ROS production (Miller et al. 2009; Mittler et al. 2011) and the necessity of proper antioxidant responses during SAR (Fodor et al. 1997; Király et al. 2002). Nevertheless, the fact that plant pretreatments with low concentrations of exogenous ROS cause a reduction of necrotic disease symptoms but not of pathogen levels (Hafez et al. 2012) points to the existence of different types of ROS waves activated in response to abiotic and biotic stresses, respectively. In order to confer an efficient, simultaneous plant defense to pathogen infection and abiotic stress, these independently developed ROS waves should act in concert and be integrated.

5 Conclusions

ROS production associated with plant cell walls and the cuticle may lead to the inhibition of microbial invaders in the initial stages of pathogenesis. As a second line of spatial defense, ROS produced by various intracellular organelles (mitochondria, chloroplasts, peroxisomes) may also play a central role in the suppression of pathogens, even in case of bacteria and fungi that do not enter plant cells.

The proper timing of ROS accumulation is a pivotal factor of plant disease resistance. Thus, a very early and rapid ROS accumulation seems to contribute to the development of symptomless (extreme) resistance, while a moderately early ROS accumulation results in resistance with the concomitant development of controlled and limited cell and tissue death (HR). Accordingly, a delayed and failed attempt by the host to elicit resistance responses would result in massively stressed plant tissues (systemic necrosis) and a partial or almost complete loss of control over pathogen invasion. Thus, it seems that when plants encounter pathogens they need to defend themselves simultaneously against biotic and abiotic stresses (pathogen accumulation and excessive cell/tissue death) by turning on two different types of—partially overlapping—signaling pathways that may function in parallel.

From the above, it seems logical that autopropagating ROS waves are required for eliciting plant defenses to both abiotic stresses and pathogens. It is likely that these two signaling pathways are significantly overlapping, as shown by their requirement for NADPH oxidase (RBOHD)-dependent ROS production (Miller et al. 2009; Mittler et al. 2011) and the necessity of proper antioxidant responses during SAR (Fodor et al. 1997; Király et al. 2002). On the other hand, the fact that low concentrations of exogenous ROS confer a reduction of necrotic disease symptoms but not of pathogen levels (Hafez et al. 2012) suggests that different types of independent ROS waves could also be activated in response to abiotic and biotic stresses. Accordingly, these ROS waves should act in concert and be integrated to confer efficient defense to pathogen infection with the simultaneous control of abiotic stress.

Acknowledgments Research in the laboratory of the authors is supported by grants of the Hungarian Scientific Research Fund (OTKA K111995 and PD108455).

References

Abbink TE, Peart JR, Mos TN, Baulcombe DC, Bol JF, Linthorst HJ (2002) Silencing of a gene encoding a protein component of the oxygen-evolving complex of photosystem II enhances virus replication in plants. Virology 295:307–319

Almagro L, Gómez Ros LV, Belchi-Navarro S, Bru R, Ros Barceló A, Pedreño MA (2009) Class III peroxidases in plant defence reactions. J Exp Bot 60:377–390

Alvarez ME, Pennell RI, Meijer PJ, Ishikawa A, Dixon RA, Lamb C (1998) Reactive oxygen intermediates mediate a systemic signal network in the establishment of plant immunity. Cell 92:773–784

Allan AC, Fluhr R (1997) Two distinct sources of elicited reactive oxygen species in in tobacco epidermal cells. Plant Cell 9:1559–1572

Amirsadeghi S, Robson CA, McDonald AE, Vanlerberghe GC (2006) Changes in plant mitochondrial electron transport alter cellular levels of reactive oxygen species and susceptibility to cell death signaling molecules. Plant Cell Physiol 47:1509–1519

An Q, Ehlers K, Kogel KH, van Bel AJE, Hükelhoven R (2006) Multivesicular compartments proliferate in susceptible and resistant MLA12-barley leaves in response to infection by the biotrophic powdery mildew fungus. New Phytol 172:563–576

Apel K, Hirt H (2004) Reactive oxygen species: metabolism, oxidative stress, and signal transduction. Annu Rev Plant Biol 55:373–399

Asai S, Ohta K, Yoshioka H (2008) MAPK signaling regulates nitric oxide and NADPH oxidase-dependent oxidative bursts in *Nicotiana benthamiana*. Plant Cell 20:1390–1406

Asai S, Yoshioka H (2009) Nitric oxide as a partner of reactive oxygen species participates in disease resistance to necrotrophic pathogen *Botrytis cinerea* in *Nicotiana benthamiana*. Mol Plant Microbe Interact 22:619–629

Aver'yanov AA, Lapikova VP (1988) Fungi toxicity mediated by active oxygen species in rice leaf diffusates. Fiziologiya Rastenii 35:1142–1151

Ádám AL, Farkas T, Somlyai G, Hevesi M, Király Z (1989) Consequence of $O_2^{\cdot -}$ generation during a bacterially induced hypersensitive reaction in tobacco: deterioration of membrane lipids. Physiol Mol Plant Pathol 34:13–26

Bacsó R, Hafez YM, Király Z, Király L (2011) Inhibition of virus replication and symptom expression by reactive oxygen species in tobacco infected with *Tobacco mosaic virus*. Acta Phytopathol Entomol Hung 46:1–10

Baker CJ, Orlandi EW (1995) Active oxygen in plant pathogenesis. Annu Rev Phytopathol 33:299–321

Balasubramaniam M, Kim BS, Hutchens-Williams HM, Loesch-Fries LS (2014) The photosystem II oxygen-evolving complex protein PsbP interacts with the coat protein of *Alfalfa mosaic virus* and inhibits virus replication. Mol Plant Microbe Interact 27:1107–1118

Barba-Espin G, Diaz-Vivancos P, Clemente-Moreno MJ, Albacete A, Faize L, Faize M, Pérez-Alfocea F, Hernández JA (2010) Interaction between hydrogen peroxide and plant hormones during germination and the early growth of pea seedlings. Plant Cell Environ 33:981–994

Barna B, Fodor J, Harrach B, Pogány M, Király Z (2012) The Janus face of reactive oxygen species in resistance and susceptibility of plants to necrotrophic and biotrophic pathogens. Plant Physiol Biochem 59:37–43

Barrios Perez I, Brown PJ (2014) The role of ROS signaling in cross-tolerance: from model to crop. Front Plant Sci 5:754

Bechtold U, Albihlal WS, Lawson T, Fryer MJ, Sparrow PAC, Richard F, Mullineaux PM (2013) *Arabidopsis HEAT SHOCK TRANSCRIPTION FACTOR A1b* overexpression enhances water productivity, resistance to drought, and infection. J Exp Bot 64:3467–3481

Bell JN, Ryder TB, Wingate VPM, Bailey JA, Lamb CJ (1986) Differential accumulation of plant defense gene transcripts in a compatible and an incompatible plant-pathogen interaction. Mol Cell Biol 6:1615–1623

Bendahmane A, Kanyuka K, Baulcombe DC (1999) The *Rx* gene from potato controls separate virus resistance and cell death responses. Plant Cell 11:781–791

Bendahmane A, Querci M, Kanyuka K, Baulcombe DC (2000) *Agrobacterium* transient expression system as a tool for the isolation of disease resistance genes: application to the *Rx2* locus in potato. Plant J 21:73–81

Berrocal-Lobo M, Stone S, Yang X, Antico J, Callis J, Ramonell KM, Somerville S (2010) ATL9, a RING zinc finger protein with E3 ubiquitin ligase activity implicated in chitin- and NADPH oxidase-mediated defense responses. PLoS One 5:e14426

Bindschedler LV, Dewdney J, Blee KA, Stone JM, Asai T, Plotnikov J, Denoux C, Hayes T, Gerrish C, Davies DR, Ausubel FM, Bolwell GP (2006) Peroxidase-dependent apoplastic oxidative burst in Arabidopsis required for pathogen resistance. Plant J 47:851–863

Boller T, Felix GA (2009) A renaissance of elicitors: perception of microbe-associated molecular patterns and danger signals by pattern-recognition receptors. Annu Rev Plant Biol 60:379–406

Bolwell GP, Bindschedler LV, Blee KA, Butt VS, Davies DR, Gardner SL, Gerrish C, Minibayeva F (2002) The apoplastic oxidative burst in response to biotic stress in plants: a three-component system. J Exp Bot 53:1367–1376

Bolwell GP, Wojtaszek P (1997) Mechanisms for the generation of reactive oxygen species in plant defense: a broad perspective. Physiol Mol Plant Pathol 51:347–366

Bradley DJ, Kjellbom P, Lamb CJ (1992) Elicitor- and wound-induced oxidative cross-linking of a proline-rich plant cell wall protein – A novel, rapid defense response. Cell 70:21–30

Brisson LF, Tenhaken R, Lamb C (1994) Function of oxidative cross-linking of cell wall structural proteins in plant disease resistance. Plant Cell 6:1703–1712

Cessna SG, Sears VE, Dickman MB, Low PS (2000) Oxalic acid, a pathogenicity factor for *Sclerotinia sclerotiorum*, suppresses the oxidative burst of the host plant. Plant Cell 12:2191–2200

Chamnongpol S, Willekens H, Moeder W, Langebartels C, Sandermann H, Van Montagu M, Inzé D, Van Camp W (1998) Defense activation and enhanced pathogen tolerance induced by H_2O_2 in transgenic tobacco. Proc Natl Acad Sci USA 95:5818–5823

Chassot C, Nawrath C, Métraux JP (2008) The cuticle: not only a barrier for defence. Plant Signal Behav 3:142–144

Chen Z, Silva H, Klessig D (1993) Active oxygen species in the induction of plant acquired resistance by salicylic acid. Science 262:1883–1885

Chester KS (1933) The problem of acquired physiological immunity in plants. Q Rev Biol 8:275–324

Chivasa S, Carr JP (1998) Cyanide restores *N* gene-mediated resistance to tobacco mosaic virus in transgenic tobacco expressing salicylic acid hydroxylase. Plant Cell 10:1489–1498

Chivasa S, Murphy AM, Naylor M, Carr JP (1997) Salicylic acid interferes with tobacco mosaic virus replication via a novel salicylhydroxamic acid-sensitive mechanism. Plant Cell 9:547–557

Choi HW, Kim YJ, Lee SC, Hong JK, Hwang BK (2007) Hydrogen peroxide generation by the pepper extracellular peroxidase $CaPO_2$ activates local and systemic cell death and defense response to bacterial pathogens. Plant Physiol 145:890–904

Cole AB, Király L, Ross K, Schoelz JE (2001) Uncoupling resistance from cell death in the hypersensitive response of *Nicotiana* species to *Cauliflower mosaic virus* infection. Mol Plant Microbe Interact 14:31–41

Coll NS, Vercammen D, Smidler A, Clover C, van Breusegem F, Dangl JL, Epple P (2010) *Arabidopsis* Type I metacaspases control cell death. Science 330:1393–1397

Collins NC, Thordal-Christensen H, Lipka V, Bau S, Kombrink E, Qiu JL, Hückelhoven R, Stein M, Freialdenhoven A, Somerville SC, Schulze-Lefert P (2003) SNARE-protein-mediated disease resistance at the plant cell wall. Nature 425:973–977

Conrath U, Thulke O, Katz V, Schwindling S, Kohler A (2001) Priming as mechanism in induced systemic resistance of plants. Eur J Plant Pathol 107:113–119

Cooley MB, Pathirana S, Wu HJ, Kachroo P, Klessig DF (2000) Members of the *Arabidopsis HRT/RPP8* family of resistance genes confer resistance to both viral and oomycete pathogens. Plant Cell 12:663–676

Dat J, Vandenabeele S, Vranová E, van Montagu M, Inzé D, van Breusegem F (2000) Dual action of the active oxygen species during plant stress responses. Cell Mol Life Sci 57:779–795

Delledonne M, Xia Y, Dixon RA, Lamb C (1998) Nitric oxide functions as a signal in plant disease resistance. Nature 394:585–588

Delledonne M, Zeier J, Marocco A, Lamb C (2001) Signal interactions between nitric oxide and reactive oxygen intermediates in the plant hypersensitive disease resistance response. Proc Natl Acad Sci USA 98:13454–13459

Després C, Chubak C, Rochon A, Clark R, Bethune T, Desveaux D, Fobert PR (2003) The Arabidopsis NPR1 disease resistance protein is a novel cofactor that confers redox regulation of DNA binding activity to the basic domain/Leucine zipper transcription factor TGA1. Plant Cell 15:2181–2191

Diaz-Vivancos P, Rubio M, Mesonero V, Periago PM, Ros Barceló A, Martínez-Gómez P, Hernández JA (2006) The apoplastic antioxidant system in *Prunus*: response to long-term plum pox virus infection. J Exp Bot 57:3813–3824

Di Carli M, Villani ME, Bianco L, Lombardi R, Perrotta G, Benvenuto E, Donini M (2010) Proteomic analysis of the plant-virus interaction in *Cucumber mosaic virus* (CMV) resistant transgenic tomato. J Proteome Res 9:5684–5697

Doke N (1983a) Generation of superoxide anion by potato tuber protoplasts during the hypersensitive response to hyphal cell wall components of *Phytophthora infestans* and specific inhibition of the reaction by suppressors of hypersensitivity. Physiol Plant Pathol 23:359–367

Doke N (1983b) Involvement of superoxide anion generation in the hypersensitive response of potato tuber tissues to infection with an incompatible race of *Phytophthora infestans* and to the hyphal wall components. Physiol Plant Pathol 23:345–357

Doke N (1995) NADPH-dependent $O_2^{\cdot-}$ generation in membrane fractions isolated from wounded potato tubers inoculated with *Phytophthora infestans*. Physiol Plant Pathol 27:311–322

Doke N, Miura Y (1995) In-vitro activation of NADPH-dependent $O_2^{\cdot-}$ generating system in a plasma membrane-rich fraction of potato tuber tissues by treatment with an elicitor from *Phytophthora infestans* or with digitonin. Physiol Mol Plant Pathol 46:17–28

Doke N, Ohashi Y (1988) Involvement of an $O_2^{\cdot-}$ generating system in the induction of necrotic lesions on tobacco leaves infected with tobacco mosaic virus. Physiol Mol Plant Pathol 32:163–175

Dorey S, Baillieul F, Saindrenan P, Fritig B, Kaufmann S (1998) Tobacco class I and II catalases are differentially expressed during elicitor-induced hypersensitive cell death and localized acquired resistance. Mol Plant Microbe Interact 11:1102–1109

Dubiella U, Seybold H, Durian G, Komandera E, Lassiga R, Wittea CP, Schulzeb WX, Romeis T (2013) Calcium-dependent protein kinase/NADPH oxidase activation circuit is required for rapid defense signal propagation. Proc Natl Acad Sci USA 110:8744–8749

Durrant WE, Dong X (2004) Systemic acquired resistance. Annu Rev Phytopathol 42:185–209

Dutilleul C, Garmier M, Noctor G, Mathieu C, Chétrit P, Foyer CH, de Paepe R (2003) Leaf mitochondria modulate whole cell redox homeostasis, set antioxidant capacity, and determine stress resistance through altered signaling and diurnal regulation. Plant Cell 15:1212–1226

El-Zahaby HM, Hafez YM, Király Z (2004) Effect of reactive oxygen species on plant pathogens in planta and on disease symptoms. Acta Phytopathol Entomol Hung 39:325–345

Fodor J, Gullner G, Ádám AL, Barna B, Kőmíves T, Király Z (1997) Local and systemic responses of antioxidants to tobacco mosaic virus infection and to salicylic acid. Plant Physiol 114:1443–1451

Fodor J, Hideg É, Kecskés A, Király Z (2001) In vivo detection of tobacco mosaic virus-induced local and systemic oxidative burst by electron paramagnetic resonance spectroscopy. Plant Cell Physiol 42:775–779

Fu ZQ, Dong X (2013) Systemic acquired resistance: turning local infection into global defense. Annu Rev Plant Biol 64:839–863

Gaffney T, Friedrich L, Vernooij B, Negrotto D, Nye G, Uknes S, Ward E, Kessman H, Ryals J (1993) Requirement of salicylic acid for the induction of systemic acquired resistance. Science 261:754–756

Gassmann W (2005) Natural variation in the *Arabidopsis* response to the avirulence gene *hopPsyA* uncouples the hypersensitive response from disease resistance. Mol Plant Microbe Interact 18:1054–1060

Gechev T, Gadjev I, Van Breusegem F, Inzé D, Dukiandjiev S, Toneva V, Minkov I (2002) Hydrogen peroxide protects tobacco from oxidative stress by inducing a set of antioxidant enzymes. Cell Mol Life Sci 59:708–714

Gilroy S, Suzuki N, Miller G, Choi WG, Toyota M, Devireddy AR, Mittler R (2014) A tidal wave of signals: calcium and ROS at the forefront of rapid systemic signaling. Trends Plant Sci 19:623–630

Glazener JA, Orlandi EW, Baker CJ (1996) The active oxygen response of cell suspensions to incompatible bacteria is not sufficient to cause hypersensitive cell death. Plant Physiol 110:759–763

Gleason C, Huang SB, Thatcher LF, Foley RC, Anderson CR, Carroll AJ, Millar AH, Singh KB (2011) Mitochondrial complex II has a key role in mitochondrial-derived reactive oxygen species influence on plant stress gene regulation and defense. Proc Natl Acad Sci USA 108:10768–10773

Goodman RN, Király Z, Wood KR (1986) The biochemistry and physiology of plant disease. University Missouri Press, Columbia, MO

Goodman RN, Novacky AJ (1994) The hypersensitive reaction in plants to pathogens. APS Press, St. Paul, MN

Grant M, Lamb C (2006) Systemic immunity. Curr Opin Plant Biol 9:414–420

Greenberg JT, Yao N (2004) The role and regulation of programmed cell death in plant-pathogen interactions. Cell Microbiol 6:201–211

Groom QJ, Torres MA, Fordham-Skelton AP, Hammond-Kossack KE, Robinson NJ, Jones JDG (1996) *RbohA*, a rice homologue of the mammalian *gp91phox* respiratory burst oxidase gene. Plant J 10:515–522

Gudesblat GE, Torres PS, Vojnov AA (2009) *Xanthomonas campestris* overcomes Arabidopsis stomatal innate immunity through a DSF cell-to-cell signal-regulated virulence factor. Plant Physiol 149:1017–1027

Hafez YM, Bacsó R, Király Z, Künstler A, Király L (2012) Up-regulation of antioxidants in tobacco by low concentrations of H_2O_2 suppresses necrotic disease symptoms. Phytopathology 102:848–856

Hafez YM, Király Z (2003) Role of hydrogen peroxide in symptom expression of barley susceptible and resistant to powdery mildew. Acta Phytopathol Entomol Hung 38:227–236

Hajimorad MR, Hill JH (2001) Rsv1-mediated resistance against *Soybean mosaic virus*-N is hypersensitive response-independent at inoculation site, but has the potential to initiate a hypersensitive response-like mechanism. Mol Plant-Microbe Interact 14:587–598

Halliwell B, Gutteridge JMC (1999) Free radicals in biology and medicine. Oxford University Press, New York

Hanquing F, Kun S, Mingquan L, Hongyu L, Xin L, Yan L, Yifeng W (2010) The expression, function and regulation of mitochondrial alternative oxidase under biotic stresses. Mol Plant Pathol 11:429–440

Hatsugai N, Kuroyanagi M, Yamada K, Meshi T, Tsuda S, Kondo M, Nishimura M, Hara-Nishimura I (2004) A plant vacuolar protease, VPE, mediates virus-induced hypersensitive cell death. Science 305:855–858

Hemetsberger C, Herrberger C, Zechmann B, Hillmer M, Doehlemann G (2012) The *Ustilago maydis* effector Pep1 suppresses plant immunity by inhibition of host peroxidase activity. PLoS Pathog 8:e1002684

Hernández JA, Díaz-Vivancos P, Rubio M, Olmos E, Ros Barceló A, Martínez-Gómez P (2006) Long-term plum pox virus infection produces an oxidative stress in a susceptible apricot, *Prunus armeniaca*, cultivar but not in a resistant cultivar. Physiol Plant 126:140–152

Hückelhoven R (2007) Cell wall–associated mechanisms of disease resistance and susceptibility. Annu Rev Phytopathol 45:101–127

Hückelhoven R, Fodor J, Preis C, Kogel KH (1999) Hypersensitive cell death and papilla formation in barley attacked by the powdery mildew fungus are associated with H_2O_2 but not with salicylic acid accumulation. Plant Physiol 119:1251–1260

Hückelhoven R, Fodor J, Trujillo M, Kogel KH (2000) Barley *Mla* and *Rar* mutants compromised in the hypersensitive death response against *Blumeria graminis* fsp. *hordei* are modified in their ability to accumulate reactive oxygen intermediates at sites of fungal invasion. Planta 212:16–24

Hückelhoven R, Kogel KH (2003) Reactive oxygen intermediates in plant-microbe interactions: who is who in powdery mildew resistance? Planta 216:891–902

Jabs T, Tschöpe M, Colling C, Hahlbrock K, Scheel D (1997) Elicitor-stimulated ion fluxes and $O_2^{\cdot -}$ from the oxidative burst are essential components in triggering defense gene activation and phytoalexin biosynthesis in parsley. Proc Natl Acad Sci USA 94:4800–4805

Jimenez I, Lopez L, Alamillo JM, Valli A, Garcia JA (2006) Identification of a *Plum pox virus* CI-interacting protein from chloroplast that has a negative effect in virus infection. Mol Plant Microbe Interact 19:350–358

Jones AM, Dangl JL (2006) The plant immune system. Nature 444:323–329

Jordan CM, Wakeman RJ, DeVay JE (1992) Toxicity of free riboflavine and methionine riboflavin solutions to *Phytophthora infestans* and the reduction of potato late blight disease. Can J Microbiol 38:1108–1111

Karpinski S, Reynold H, Karpinska B, Wingsle G, Mullineaux PM (1999) Systemic signaling and acclimation in response to excess excitation energy in Arabidopsis. Science 284:654–657

Keller T, Damude HG, Werner D, Doerner P, Dixon RA, Lamb C (1998) A plant homologue of the neutrophyl NADPH oxidase gp91^{phox} subunit gene encodes a plasma membrane protein with Ca^{2+} binding motifs. Plant Cell 10:255–266

Király L, Hafez YM, Fodor J, Király Z (2008) Suppression of tobacco mosaic virus-induced hypersensitive-type necrotisation in tobacco at high temperature is associated with down-regulation of NADPH oxidase and superoxide and stimulation of dehydroascorbate reductase. J Gen Virol 89:799–808

Király L, Király Z (2006) To die or not to die – is cell death dispensable for resistance during the plant hypersensitive response? Acta Phytopathol Entomol Hung 41:11–21

Király L, Künstler A, Bacsó R, Hafez YM, Király Z (2013) Similarities and differences in plant and animal immune systems – What is inhibiting pathogens? Acta Phytopathol Entomol Hung 48:187–205

Király Z, Barna B, Érsek T (1972) Hypersensitivity as a consequence, not the cause of plant resistance to infection. Nature 239:456–458

Király Z, Barna B, Kecskés A, Fodor J (2002) Down-regulation of antioxidative capacity in a transgenic tobacco which fails to develop acquired resistance to necrotization caused by tobacco mosaic virus. Free Radic Res 36:981–991

Király Z, El-Zahaby H, Galal A, Abdou S, Ádám A, Barna B, Klement Z (1993) Effect of oxy free radicals on plant pathogenic bacteria and fungi and on some plant diseases. In: Mózsik GY, Emerit I, Fehér J, Matkovics B, Vincze Á (eds) Oxygen free radicals and scavengers in the natural sciences. Akad Kiadó Budapest, Hungary

Klement Z (1982) Hypersensitivity. In: Mount MS, Lacy GH (eds) Phytopathogenic prokaryotes II. Academic, New York

Knecht K, Seyffarth M, Desel C, Thurau T, Sherameti I, Lou B, Oelmuller R, Cai D (2010) Expression of *BvGLP-1* encoding a germin-like protein from sugar beet in *Arabidopsis thaliana* leads to resistance against phytopathogenic fungi. Mol Plant Microbe Interact 23:446–457

Kong L, Wu J, Lu L, Xu Y, Zhou X (2014) Interaction between Rice stripe virus disease-specific protein and host PsbP enhances virus symptoms. Mol Plant 7:691–708

Kovtun Y, Chiu WL, Tena G, Sheen J (2000) Functional analysis of oxidative stress-activated mitogen-activated protein kinase cascade in plants. Proc Natl Acad Sci USA 97:2940–2945

Krause M, Durner J (2004) Harpin inactivates mitochondria in *Arabidopsis* suspension cells. Mol Plant Microbe Interact 17:131–139

Künstler A, Hafez YM, Király L (2007) Transient suppression of a catalase and an alternative oxidase gene during virus-induced local lesion formation (hypersensitive response) is independent of the extent of leaf necrotization. Acta Phytopathol Entomol Hung 42:185–196

Lacomme C, Roby D (1999) Identification of new early markers of the hypersensitive response in *Arabidopsis thaliana*. FEBS Lett 459:149–153

Lagrimini LM, Burkhart W, Moyer M, Rothstein S (1987) Molecular cloning of complementary DNA encoding the lignin-forming peroxidase from tobacco: molecular analysis and tissue-specific expression. Proc Natl Acad Sci USA 84:7542–7546

Lehto K, Tikkanen M, Hiriart JB, Paakkarinen V, Aro EM (2003) Depletion of the photosystem II core complex in mature tobacco leaves infected by the *Flavum* strain of *Tobacco mosaic virus*. Mol Plant Microbe Interact 16:1135–1144

Lamb C, Dixon RA (1997) The oxidative burst in plant disease resistance. Annu Rev Plant Physiol Plant Mol Biol 48:251–275

Lee YH, Yoon IS, Suh SC, Kim HI (2002) Enhanced disease resistance in transgenic cabbage and tobacco expressing a glucose oxidase gene from *Aspergillus niger*. Plant Cell Rep 20:857–863

Lennon AM, Neuenschwander UH, Ribas-Carbo M, Giles L, Ryals JA, Siedow JN (1997) The effects of salicylic acid and tobacco mosaic virus infection on the alternative oxidase of tobacco. Plant Physiol 115:783–791

Levine A, Tenhaken R, Dixon R, Lamb C (1994) H_2O_2 from the oxidative burst orchestrates the plant hypersensitive disease resistance response. Cell 79:583–593

L'Haridon F, Besson-Bard A, Binda M, Serrano M, Abou-Mansour E, Balet F, Schoonbeek HJ, Hess S, Mir R, Léon J, Lamotte O, Métraux JP (2011) A permeable cuticle is associated with the release of reactive oxygen species and induction of innate immunity. PLoS Pathog 7: e1002148

Li J, Brader G, Palva ET (2004) The WRKY70 transcription factor: a node of convergence for jasmonate-mediated and salicylate-mediated signals in plant defense. Plant Cell 16:319–331

Liao Y, Tian M, Zhang H, Li X, Wang Y, Xia X, Zhou J, Zhou Y, Yu J, Shi K, Klessig DF (2015) Salicylic acid binding of mitochondrial alpha-ketoglutarate dehydrogenase E2 affects mitochondrial oxidative phosphorylation and electron transport chain components and plays a role in basal defense against *tobacco mosaic virus* in tomato. New Phytol 205:1296–1307

Liao YWK, Shi K, Fu LJ, Zhang S, Li X, Dong DK, Jiang YP, Zhou YH, Xia XJ, Liang WS, Yu JQ (2012) The reduction of reactive oxygen species formation by mitochondrial alternative respiration in tomato basal defense against TMV infection. Planta 235:225–238

Liu PP, Bhattacharjee S, Klessig DF, Moffett P (2010a) Systemic acquired resistance is induced by R gene-mediated responses independent of cell death. Mol Plant Pathol 11:155–160

Liu YK, Liu YB, Zhang MY, Li DQ (2010b) Stomatal development and movement: the roles of MAPK signaling. Plant Signal Behav 5:1176–1180

Lohar DP, Haridas S, Gantt JS, VandenBosch KA (2007) A transient decrease in reactive oxygen species in roots leads to root hair deformation in the legume-rhizobia symbiosis. New Phytol 173:39–49

Macho AP, Boutrot F, Rathjen JP, Zipfel C (2012) ASPARTATE OXIDASE plays an important role in Arabidopsis stomatal immunity. Plant Physiol 159:1845–1856

Marino D, Dunand C, Puppo A, Pauly N (2012) A burst of plant NADPH oxidases. Trends Plant Sci 17:9–15

Mathioudakis MM, Veiga RS, Canto T, Medina V, Mossialos D, Makris AM, Livieratos I (2013) *Pepino mosaic virus* triple gene block protein 1 (TGBp1) interacts with and increases tomato catalase 1 activity to enhance virus accumulation. Mol Plant Pathol 14:589–601

Maxwell DP, Wang Y, McIntosh L (1999) The alternative oxidase lowers mitochondrial reactive oxygen production in plant cells. Proc Natl Acad Sci USA 96:8271–8276

Melotto M, Underwood W, He SY (2008) Role of stomata in plant innate immunity and foliar bacterial diseases. Annu Rev Phytopathol 46:101–122

Melotto M, Underwood W, Koczan J, Nomura K, He SY (2006) Plant stomata function in innate immunity against bacterial invasion. Cell 126:969–980

Mellersh DG, Foulds IV, Higgins VJ, Heath MC (2002) H_2O_2 plays different roles in determining penetration failure in three diverse plant-fungal interactions. Plant J 29:257–268

Mersmann S, Bourdais G, Rietz S, Robatzek S (2010) Ethylene signaling regulates accumulation of the FLS2 receptor and is required for the oxidative burst contributing to plant immunity. Plant Physiol 154:391–400

Miller G, Mittler R (2006) Could heat shock transcription factors function as hydrogen peroxide sensors in plants? Ann Bot 98:279–288

Miller G, Schlauch K, Tam R, Cortes D, Torres MA, Shulaev V, Dangl J, Mittler R (2009) The plant NADPH oxidase RBOHD mediates rapid systemic signaling in response to diverse stimuli. Sci Signal 2:ra45

Mittal D, Madhyastha DA, Grover A (2012) Genome-wide transcriptional profiles during temperature and oxidative stress reveal coordinated expression patterns and overlapping regulons in rice. PLoS One 7:e40899

Mittler R (2002) Oxidative stress, antioxidants and stress tolerance. Trends Plant Sci 7:405–410

Mittler R, Feng X, Cohen M (1998) Post-transcriptional suppression of cytosolic ascorbate peroxidase expression during pathogen-induced programmed cell death in tobacco. Plant Cell 10:461–473

Mittler R, Shulaev V, Seskar M, Lam E (1996) Inhibition of programmed cell death in tobacco plants during a pathogen-induced hypersensitive response at low oxygen pressure. Plant Cell 8:1991–2001

Mittler R, Vanderauwera S, Suzuki N, Miller G, Tognetti VB, Vandepoele K, Gollery M, Shulaev V, Van Breusegem F (2011) ROS signaling: the new wave? Trends Plant Sci 16:300–309

Mizuno M, Tada Y, Uchii K, Kawakami S, Mayama S (2005) Catalase and alternative oxidase cooperatively regulate programmed cell death induced by b-glucan elicitor in potato suspension cultures. Planta 220:849–853

Montillet JL, Leonhardt N, Mondy S, Tranchimand S, Rumeau D, Boudsocq M, Garcia AV, Douki T, Bigeard J, Laurière C, Chevalier A, Castresana C, Hirt H (2013) An abscisic acid independent oxylipin pathway controls stomatal closure and immune defense in Arabidopsis. PLoS Biol 11:e1001513

Morel F, Doussiere J, Vignais PV (1991) The superoxide-generating oxidase of phagocytic cells. Physiological, molecular and pathological aspects. Eur J Biochem 201:523–546

Mou Z, Fan W, Dong X (2003) Inducers of plant systemic acquired resistance regulate NPR1 function through redox changes. Cell 113:935–944

Mur LAJ, Carver TLW, Prats E (2006) NO way to live; the various roles of nitric oxide in plant–pathogen interactions. J Exp Bot 57:489–505

Mustilli AC, Merlot S, Vavasseur A, Fenzi F, Giraudat J (2002) *Arabidopsis* OST1 protein kinase mediates the regulation of stomatal aperture by abscisic acid and acts upstream of reactive oxygen species production. Plant Cell 14:3089–3099

Norman C, Howell KA, Millar AH, Whelan JM, Day DA (2004) Salicylic acid is an uncoupler and inhibitor of mitochondrial electron transport. Plant Physiol 134:492–501
O'Brien JA, Daudi A, Butt VS, Bolwell GP (2012) Reactive oxygen species and their role in plant defence and cell wall metabolism. Planta 236:765–779
Ouf MF, Gazar AA, Shehata ZA, Abdou E-S, Király Z, Barna B (1993) The effect of superoxide anion on germination and infectivity of wheat stem rust (*Puccinia graminis* Pers. f. sp. *tritici* Eriks. and Henn.) uredospores. Cereal Res Commun 21:31–37
Pandey SP, Somssich IE (2009) The role of WRKY transcription factors in plant immunity. Plant Physiol 150:1648–1655
Pastor V, Luna E, Mauch-Mani B, Ton J, Flors V (2013) Primed plants do not forget. Environ Exp Bot 94:46–56
Park H, Miura Y, Kawakita K, Yoshioka H, Doke N (1998) Physiological mechanisms of a sub-systemic oxidative burst triggered by elicitor-induced local oxidative burst in potato tuber slices. Plant Physiol 126:1281–1290
Peng M, Kuć J (1992) Peroxidase-generated hydrogen peroxide as a source of antifungal activity in vitro and on tobacco leaf disks. Phytopathology 82:696–699
Peleg-Grossman S, Melamed-Book N, Levine A (2012) ROS production during symbiotic infection suppresses pathogenesis-related gene expression. Plant Signal Behav 7:409–415
Perez-Bueno ML, Rahoutei J, Sajnani C, Garcia-Luque I, Baron M (2004) Proteomic analysis of the oxygen-evolving complex of photosystem II under biotec stress: studies on *Nicotiana benthamiana* infected with tobamoviruses. Proteomics 4:418–425
Pieterse CMJ, van Wees SCM, Hoffland E, van Pelt JA, van Loon LC (1996) Systemic resistance in *Arabidopsis* induced by biocontrol bacteria is independent of salicylic acid accumulation and pathogenesis-related gene expression. Plant Cell 8:1225–1237
Pogány M, von Rad U, Grün S, Dongó A, Pintye A, Simoneau P, Bahnweg G, Kiss L, Barna B, Durner J (2009) Dual roles of reactive oxygen species and NADPH oxidase RBOHD in an Arabidopsis-*Alternaria* pathosystem. Plant Physiol 151:1459–1475
Proels RK, Oberhollenzer K, Pathuri IP, Hensel G, Kumlehn J, Hückelhoven R (2010) RBOHF2 of barley is required for normal development of penetration resistance to the parasitic fungus *Blumeria graminis* f. sp. *hordei*. Mol Plant Microbe Interact 23:1143–1150
Rahoutei J, García-Luque I, Barón M (2000) Inhibition of photosynthesis by viral infection: effect on PSII structure and function. Physiol Plant 110:286–292
Rhoads DM, Umbach AL, Subbaiah CC, Siedow JN (2006) Mitochondrial reactive oxygen species. Contribution to oxidative stress and interorganellar signaling. Plant Physiol 141:357–366
Rizhsky L, Hallak-Herr E, Van Breusegem F, Rachmilevitch S, Barr JE, Rodermel S, Inzé D, Mittler R (2002) Double antisense plants lacking ascorbate peroxidase and catalase are less sensitive to oxidative stress than single antisense plants lacking ascorbate peroxidase or catalase. Plant J 32:329–342
Robson CA, Vanlerberghe GC (2002) Transgenic plant cells lacking mitochondrial alternative oxidase have increased susceptibility to mitochndria-dependent and -independent pathways of programmed cell death. Plant Physiol 129:1908–1920
Rodriguez-Herva JJ, Gonzalez-Melendi P, Cuartas-Lanza R, Antunez-Lamas M, Rio-Alvarez I, Li Z, Lopez-Torrejon G, Diaz I, Del Pozo JC, Chakravarthy S, Collmer A, Rodriguez-Palenzuela P, Lopez-Solanilla E (2012) A bacterial cysteine protease effector protein interferes with photosynthesis to suppress plant innate immune responses. Cell Microbiol 14:669–681
Ross AF (1961a) Localized acquired resistance to plant virus infection in hypersensitive hosts. Virology 14:329–339
Ross AF (1961b) Systemic acquired resistance induced by localized virus infections in plants. Virology 14:340–358
Sawinski K, Mersmann S, Robatzek S, Böhmer M (2013) Guarding the green: pathways to stomatal immunity. Mol Plant Microbe Interact 26:626–632

Scarpeci T, Zanor M, Bernd Mueller R, Valle E (2013) Overexpression of AtWRKY30 enhances abiotic stress tolerance during early growth stages in *Arabidopsis thaliana*. Plant Mol Biol 83:265–277
Schäfer P, Hückelhoven R, Kogel KH (2004) The white barley mutant *Albostrians* shows a supersusceptible but symptomless interaction phenotype with the hemibiotrophic fungus *Bipolaris sorokiniana*. Mol Plant Microbe Interact 17:366–373
Schiffer R, Görg R, Jarosch B, Beckhove U, Bahrenberg G, Kogel KH, Schulze-Lefert P (1997) Tissue dependence and differential cordycepin sensitivity of race-specific resistance responses in the barley-powdery mildew interaction. Mol Plant Microbe Interact 10:830–839
Schulze-Lefert P (2004) Knocking on the heaven's wall: pathogenesis of and resistance to biotrophic fungi at the cell wall. Curr Opin Plant Biol 7:377–383
Schweizer P, Pokorny J, Abderhalden O, Dudler R (1999) A transient assay system for the functional assessment of defense-related genes in wheat. Mol Plant Microbe Interact 12:647–654
Segal AW (2008) The function of the NADPH oxidase of phagocytes and its relationship to other NOXs in plants, invertebrates, and mammals. Int J Biochem Cell Biol 40:604–618
Sels J, Mathys J, De Coninck BMA, Cammue BPA, De Bolle MFC (2008) Plant pathogenesis-related (PR) proteins: a focus on PR peptides. Plant Physiol Biochem 46:941–950
Serrano M, Coluccia F, Torres M, L'Haridon F, Métraux JP (2014) The cuticle and plant defense to pathogens. Front Plant Sci 5:274
Shang J, Xi DH, Yuan S, Xu F, Xu MY, Qi HL, Wang SD, Huang QR, Wen L, Lin HH (2010) Difference of physiological characters in dark green islands and yellow leaf tissue of *Cucumber mosaic* virus (CMV)-infected *Nicotiana tabacum* leaves. Z Naturforsch 65c:73–78
Sharma P, Jha AB, Dubey RS, Pessarakli M (2012) Reactive oxygen species, oxidative damage, and antioxidative defense mechanism in plants under stressful conditions. J Bot 2012:1–26
Shaw SL, Long SR (2003) Nod factor inhibition of reactive oxygen efflux in a host legume. Plant Physiol 132:2196–2204
Shetty NP, Kristensen BK, Newman MA, Møller K, Gregersen PL, Jørgensen HJL (2003) Association of hydrogen peroxide with restriction of *Septoria tritici* in resistant wheat. Physiol Mol Plant Pathol 62:333–346
Shetty NP, Mehrabi R, Lütken H, Haldrup A, Kema GHJ, Collinge DB, Jørgensen HJL (2007) Role of hydrogen peroxide during the interaction between the hemibiotrophic fungal pathogen *Septoria tritici* and wheat. New Phytol 174:637–647
Shetty NP, Jørgensen HJL, Jensen JD, Collinge DB, Shetty HS (2008) Roles of reactive oxygen species in interactions between plants and pathogens. Eur J Plant Pathol 121:267–280
Shi H, Wang X, Ye T, Cheng F, Deng J, Yang P et al (2014) The Cysteine2/Histidine2-type transcription factor *ZINC FINGER OFARABIDOPSIS THALIANA 6* modulates biotic and abiotic stress responses by activating salicylic acid-related genes and *C-REPEAT-BINDING FACTOR* genes in *Arabidopsis*. Plant Physiol 165:1367–1379
Simons BH, Millenaar FF, Mulder L, Van Loon LC, Lambers H (1999) Enhanced expression and activation of the alternative oxidase during infection of Arabidopsis with *Pseudomonas syringae* pv. tomato. Plant Physiol 120:529–538
Spoel SH, Dong X (2012) How do plants achieve immunity? Defence without specialized immune cells. Nat Rev Immunol 12:89–100
Sticher L, Mauch-Mani B, Métraux JP (1997) Systemic acquired resistance. Annu Rev Phytopathol 35:235–270
Suzuki N, Miller G, Morales J, Shulaev V, Torres MA, Mittler R (2011) Respiratory burst oxidases: the engines of ROS signaling. Curr Opin Plant Biol 14:691–699
Tada Y, Spoel SH, Pajerowska-Mukhtar K, Mou Z, Song J, Wang C, Zuo J, Dong X (2008) Plant immunity requires conformational changes of NPR1 via S-nitrosylation and thioredoxins. Science 321:952–956

Takahashi H, Chen Z, Du H, Liu Y, Klessig D (1997) Development of necrosis and activation of disease resistance in transgenic tobacco plants with severely reduced catalase levels. Plant J 11:995–1005

Talarczyk A, Krzymowska M, Borucki W, Hennig J (2002) Effect of yeast *CTA1* gene expression on response of tobacco plants to tobacco mosaic virus infection. Plant Physiol 129:1032–1044

Tenhaken R, Levine A, Brisson LF, Dixon RA, Lamb C (1995) Function of the oxidative burst in hypersensitive disease resistance. Proc Natl Acad Sci USA 92:4158–4163

Torres MA (2010) ROS in biotic interactions. Physiol Planta 138:414–429

Torres MA, Dangl JL (2005) Functions of the respiratory burst oxidase in biotic interactions, abiotic stress and development. Curr Opin Plant Biol 8:397–403

Torres MA, Jones JD, Dangl JL (2005) Pathogen-induced, NADPH oxidase-derived reactive oxygen intermediates suppress cell death in *Arabidopsis thaliana*. Nat Genet 37:1130–1134

Torres MA, Jones JDG, Dangl JL (2006) Reactive oxygen species signaling in response to pathogens. Plant Physiol 141:373–378

Torres MA, Onouchi H, Hamada S, Machida C, Hammond-Kosack KE, Jones JDG (1998) Six *Arabidopsis thaliana* homologues of the human respiratory burst oxidase ($gp91^{phox}$). Plant J 14:365–370

Thoma I, Loeffler C, Sinha AK, Gupta M, Krischke M, Steffan B, Roitsch T, Mueller MJ (2003) Cyclopentenone isoprostanes induced by reactive oxygen species trigger defense gene activation and phytoalexin accumulation in plants. Plant J 34:363–375

Thordal-Christensen H, Zhang Z, Wei Y, Collinge DB (1997) Subcellular localization of H_2O_2 in plants: H_2O_2 accumulation in papillae and hypersensitive response during the barley-powdery mildew interaction. Plant J 11:1187–1194

Tsugama D, Liu S, Takano T (2012) Drought-induced activation and rehydration-induced inactivation of MPK6 in Arabidopsis. Biochem Biophys Res Commun 426:626–629

Tzeng DD, Lee MH, Chung KR, DeVay JE (1990) Products in light-mediated reactions of free methionine–riboflavin mixtures that are biocidal to microorganisms. Can J Microbiol 36:500–506

Vandenabeele S, Van Der Kelen K, Dat J, Gadjev I, Boonefaes T, Morsa S, Rottiers P, Slooten L, Van Montagu M, Zabeau M, Inzé D, Van Breusegem F (2003) A comprehensive analysis of hydrogen peroxide-induced gene expression in tobacco. Proc Natl Acad Sci USA 100:16113–16118

Van Loon LC, Rep M, Pieterse CM (2006) Significance of inducible defense-related proteins in infected plants. Annu Rev Phytopathol 44:135–162

Van Loon LC, Van Strien E (1999) The families of pathogenesis-related proteins, their activities, and comparative analysis of PR-1 type proteins. Physiol Mol Plant Pathol 55:85–97

Walz A, Zingen-Sell I, Loeffler M, Sauer M (2008) Expression of an oxalate oxidase gene in tomato and severity of disease caused by *Botrytis cinerea* and *Sclerotinia sclerotiorum*. Plant Pathol 57:453–458

Wei Y, Zhang Z, Andersen CH, Schmelzer E, Gregersen PL, Collinge DB, Smedegaard-Pedersen-V, Thordal-Christensen H (1998) An epidermis/papilla-specific oxalate oxidase-like protein in the defence response of barley attacked by the powdery mildew fungus. Plant Mol Biol 36:101–112

Willekens H, Inzé D, Van Montagu M, Van Camp W (1995) Catalases in plants. Mol Breeding 1:207–228

Williams B, Kabbage M, Kim HJ, Britt R, Dickman MB (2011) Tipping the balance: *Sclerotinia sclerotiorum* secreted oxalic acid suppresses host defenses by manipulating the host redox environment. PLoS Pathog 7:e1002107

Wu G, Shortt BJ, Lawrence EB, Levine EB, Fitzsimmons KC, Shah DM (1995) Disease resistance conferred by expression of a gene encoding H_2O_2-generating glucose oxidase in transgenic potato plants. Plant Cell 7:1357–1368

Wu G, Shortt BJ, Lawrence EB, León J, Fitzsimmons KC, Levine EB, Raskin I, Shah DM (1997) Activation of host defense mechanisms by elevated production of H_2O_2 in transgenic plants. Plant Physiol 115:427–435

Yi SY, Yu SH, Choi D (1999) Molecular cloning of a catalase cDNA from *Nicotiana glutinosa L.* and its repression by tobacco mosaic virus infection. Mol Cells 9:320–325

Yi SY, Yu SH, Choi D (2003) Involvement of hydrogen peroxide in repression of catalase in TMV-infected resistant tobacco. Mol Cells 15:364–369

Yoshioka H, Numata N, Nakajima K, Katou S, Kawakita K, Rowland O, Jones JD, Doke N (2003) *Nicotiana benthamiana* gp91^{phox} homologs *NbrbohA* and *NbrbohB* participate in H_2O_2 accumulation and resistance to *Phytophthora infestans*. Plant Cell 15:706–718

Yu IC, Parker J, Bent AF (1998) Gene-for-gene disease resistance without the hypersensitive response in *Arabidopsis dnd1* mutant. Proc Natl Acad Sci USA 95:7819–7824

Zeng W, He SY (2010) A prominent role of the flagellin receptor FLAGELLIN-SENSING2 in mediating stomatal response to *Pseudomonas syringae* pv. *tomato* DC3000 in Arabidopsis. Plant Physiol 153:1188–1198

Zhang Z, Henderson C, Gurr SJ (2004) *Blumeria graminis* secretes an extracellular catalase during infection of barley: potential role in suppression of host defence. Mol Plant Pathol 5:537–547

Zhang L, Xing D (2008) Methyl jasmonate induces production of reactive oxygen species and alterations in mitochondrial dynamics that precede photosynthetic dysfunction and subsequent cell death. Plant Cell Physiol 49:1092–1111

Zhang Z, Collinge DB, Thordal-Christensen H (1995) Germin-like oxalate oxidase, a H_2O_2-producing enzyme, accumulates in barley attacked by the powdery mildew fungus. Plant J 8:139–145

Zipfel C, Robatzek S, Navarro L, Oakeley EJ, Jones JDG, Felix G, Boller T (2004) Bacterial disease resistance in *Arabidopsis* through flagellin perception. Nature 428:764–767

Modulation of the Ascorbate–Glutathione Cycle Antioxidant Capacity by Posttranslational Modifications Mediated by Nitric Oxide in Abiotic Stress Situations

J.C. Begara-Morales, B. Sánchez-Calvo, M. Chaki, R. Valderrama, C. Mata-Pérez, M.N. Padilla, F.J. Corpas, and J.B. Barroso

Contents

Abstract Environmental stresses cause a rapid burst of second messengers belonging to reactive oxygen (ROS) and nitrogen (RNS) species, mainly hydrogen peroxide (H_2O_2) and nitric oxide (NO), respectively. H_2O_2 can act as a signal molecule or become toxic at high levels. Plants have developed different antioxidant tools, such as the ascorbate–glutathione (Asa–GSH) cycle, a key antioxidant system involved in the finely tuned regulation of H_2O_2 in cells, in order to control H_2O_2 overproduction. In recent years, a growing body of evidence points to the existence of a link between NO and physiological and stress responses in plants.

J.C. Begara-Morales • B. Sánchez-Calvo • M. Chaki • R. Valderrama • C. Mata-Pérez • M.N. Padilla • J.B. Barroso (✉)
Biochemistry and Cell Signaling in Nitric Oxide Group, Department of Experimental Biology, Center for Advanced Studies in Olive Grove and Olive Oils, Faculty of Experimental Sciences, University of Jaén, Campus Universitario "Las Lagunillas" s/n, 23071 Jaén, Spain
e-mail: jbarroso@ujaen.es

F.J. Corpas
Antioxidants, Free Radicals and Nitric Oxide in Biotechnology, Food and Agriculture Group, Department of Biochemistry, Cell and Molecular Biology of Plants, Estación Experimental del Zaidín, CSIC, Apartado 419, 18008 Granada, Spain

D.K. Gupta et al. (eds.), *Reactive Oxygen Species and Oxidative Damage in Plants Under Stress*, DOI 10.1007/978-3-319-20421-5_12

NO activity is mainly conveyed through posttranslational modifications (PTMs) such as S-nitrosylation and/or tyrosine nitration. Over the last 10 years, the number of S-nitrosylated and nitrated proteins subjected to physiological and a stress condition has been observed to increase significantly in higher plants, suggesting that NO-PTMs are involved in plant physiology. Emerging evidence shows that ROS and NO interact during plant responses to (a)biotic stress, and proteins linked to ROS metabolism have been reported to be regulated by NO-related PTMs. Furthermore, using proteomic analytical techniques, enzymes involved in the Asa–GSH cycle have been identified as NO targets. However, little information exists on the specific impact of NO-PTMs on the structure and activity of these antioxidant enzymes. In this chapter, we will discuss recent findings concerning the regulation of the Asa–GSH cycle antioxidant capacity by NO-PTMs, particularly in relation to the role played by the NO target residues identified under stress conditions.

Keywords Ascorbate–glutathione cycle • Nitric oxide • Tyrosine nitration • S-nitrosylation • Abiotic stress

1 Introduction

In plants, (a)biotic stress situations trigger signaling pathways that produce secondary messengers, the most important being hydrogen peroxide (H_2O_2) and nitric oxide (NO). A growing body of evidence suggests a link between both pathways through a regulation of the antioxidant systems by NO, including the ascorbate–glutathione cycle (Asa–GSH) (Corpas et al. 2011; Groβ et al. 2013; Yu et al. 2014).

The Asa–GSH cycle is a powerful antioxidant system, essential for the detoxification and regulation of hydrogen peroxide in plant cells (Asada 1992; Noctor and Foyer 1998). This cycle is composed of monodehydroascorbate reductase (MDAR), glutathione reductase (GR), ascorbate peroxidase (APX), and dehydroascorbate reductase (DHAR) in addition to the antioxidant metabolites ascorbate and glutathione, and NADPH as a key electron donor for the oxidoreductase catalyzed reactions taking place in this cycle. In this system, H_2O_2 is reduced to water by APX, using ascorbate as the electron donor and generating monodehydroascorbate (MDA). Having a relatively short lifetime, if MDA is not rapidly reduced, it disproportionates to ascorbate and dehydroascorbate (DHA) (Noctor and Foyer 1998). MDA can also be reduced to ascorbate by MDAR using NADPH as the electron donor. Furthermore, DHA is reduced by DHAR and regenerates ascorbate using GSH as a reducing substrate. This reaction generates oxidized glutathione (GSSG) which is finally reduced by GR with the aid of NADPH as electron donor. Thus, GR is involved in maintaining intracellular levels of GSH. At the subcellular level, these enzymes have been demonstrated to be located in cellular compartments such as the cytosol, chloroplasts, peroxisomes, and mitochondria (Asada 2006; Foyer and Halliwell 1976; Groden and Beck 1979; Jimenez et al. 1998; Romero-Puertas et al. 2006; Reumann and Corpas 2010). The enzymes of the Asa–GSH cycle enhance plant tolerance under various stress situations (Diaz-Vivancos et al. 2013; Liu et al. 2014; Zhang et al. 2013, 2015).

Nitric oxide (NO) has emerged as an essential signaling molecule associated with the modulation of physiological processes (Shapiro 2005; del Río et al. 2004; Lamattina et al. 2003) and plant responses to (a)biotic stress conditions (see Corpas et al. 2011; Siddiqui et al. 2011). NO is a lipophilic-free radical-type molecule with an unpaired electron in its outer orbital, which facilitates its diffusion across cell membranes, and also reacts with macromolecules such as proteins, lipids, and nucleic acids. NO belongs to a family of related molecules known as reactive nitrogen species (RNS) capable of directly or indirectly mediating posttranslational modifications (PTM) such as protein S-nitrosylation and tyrosine nitration. These NO-PTMs may be involved in cell signaling under physiological and stress conditions (Corpas et al. 2011).

As mentioned above, the close relationship between NO and the Asa–GSH cycle in plants has been studied in recent years. Endogenous and exogenous NO has been reported to modulate the function of antioxidant plant systems, such as Asc–GSH cycle enzymes, in stress situations (see Groβ et al. 2013). In addition, proteomic studies have identified certain enzymatic components of the ascorbate–glutathione cycle as potential PTM targets mediated by NO-derived molecules (Reumann et al. 2007). However, little is known about the specific impact of NO-related PTMs on the activity and structure of proteins involved in antioxidative systems (Holtgrefe et al. 2008; Chaki et al. 2011a; Astier and Lindermayr 2012; Begara-Morales et al. 2013a, b).

This chapter presents an overview of the state of the art concerning the modulation by NO of the Asa–GSH cycle's antioxidant capacity, particularly in relation to the physiological significance of NO-PTMs and their role in plant responses to stress situations.

2 S-Nitrosylation and Tyrosine Nitration Under Stress Conditions

As mentioned above, NO exerts its biological function mainly through posttranslational modifications such as tyrosine nitration and S-nitrosylation. Tyrosine nitration adds a nitro group ($-NO_2$) to one of two equivalent ortho-carbons of the tyrosine residue aromatic ring that generates 3-nitrotyrosine (NO_2-Tyr) (Gow et al. 2004; Radi 2004). This converts tyrosine into a negatively charged hydrophilic nitrotyrosine moiety and causes a marked shift in the hydroxyl group's local pK_a from 10.1 in tyrosine to 7.5 in nitrotyrosine (Turko and Murad 2002; Abello et al. 2009). In recent years, a growing body of evidence has shown that tyrosine nitration occurs under physiological conditions and/or in response to abiotic and biotic stress situations regulating key processes in plants (see Corpas et al. 2015). This PTM is an irreversible process which can lead to a loss, gain, or no change in the protein's function (Souza et al. 2008; Radi 2013), although up to now, most studies have shown that nitration usually causes function loss (Álvarez et al. 2011; Chaki et al. 2011a, 2013; Begara-Morales et al. 2013a; Corpas et al. 2013a, b).

On the other hand, S-nitrosylation is a reversible process which adds a NO group to a specific cysteine thiol in the target protein, giving rise to *S*-nitrosothiols (SNOs), which can alter the function of a wide range of proteins (Astier et al. 2011). In recent years, the *S*-nitrosothiol metabolism has become an important issue, with considerable evidence suggesting that protein S-nitrosylation may play a key role in plant biology (Belenghi et al. 2007; Romero-Puertas et al. 2007, 2008; Lindermayr and Durner 2009; Astier et al. 2011; Hu et al. 2015). Moreover, the importance of *S*-nitrosothiols in plant responses to both pathogens (Feechan et al. 2005; Rusterucci et al. 2007; Chaki et al. 2009a) and abiotic stresses (Valderrama et al. 2007; Corpas et al. 2008; Airaki et al. 2011a; Chaki et al. 2011a, b) has recently been demonstrated in several plant species. Among SNOs, *S*-nitrosoglutathione (GSNO), formed by the S-nitrosylation of the antioxidant tripeptide glutathione (GSH), is a key low-molecular-weight *S*-nitrosothiol regarded as an endogenous NO reservoir in cells (Gaston et al. 1993; Durner et al. 1999; Leitner et al. 2009; Airaki et al. 2011b). Furthermore, being phloem mobile, GSNO is regarded as a long-distance NO vehicle that is important for redox signaling mechanisms (Malik et al. 2011). GSNO can also generate transnitrosylation reactions in which a NO group is transferred from an *S*-nitrosothiol to a target protein's cysteine thiol group (Astier et al. 2011). GSNO is at present being used to study the S-nitrosylation of proteins in vitro and is associated with plant responses to biotic stress (Feechan et al. 2005; Chaki et al. 2009a). Recently, the number of S-nitrosylated proteins has been demonstrated to increase considerably (Hu et al. 2015). Interestingly, a link between S-nitrosylation and the ROS metabolism has been observed in reports on NADPH oxidase (Yun et al. 2011), catalase (Ortega-Galisteo et al. 2012), and peroxiredoxin IIE (Romero-Puertas et al. 2007) and IIF (Camejo et al. 2015), among others. A connection between NO and the ROS pathway under different physiological and stress conditions has also been observed (Corpas et al. 2011; Groβ et al. 2013; Procházková et al. 2014).

NO can enhance plant tolerance to stress situations by directly regulating the antioxidant capacity of plant tissues involving Asa–GSH cycle enzymes under aluminum (Sun et al. 2014), cadmium (Wang et al. 2015), and drought (Shan et al. 2015) stress conditions, among others (see Corpas et al. 2011; Procházková et al. 2014). However, although many nitrated and S-nitrosylated proteins have been identified in stress situations in plants, little is known about the role of the NO target residues identified or the specific impact of NO-PTMs on the structure and activity of proteins involved in antioxidant systems such as the ascorbate–glutathione cycle whose APX structure and activity have been most fully characterized (see below).

3 Glutathione Reductase (GR) is Unaffected by NO in Pea Plants

Glutathione reductase (GR) is a key enzyme in the cellular redox metabolism due to its ability to reduce oxidized glutathione (GSSG) to reduced glutathione (GSH) using NADPH as a cofactor. Thus, GR enables GSH/GSSG to be maintained at high levels which is very important as GSH is considered to be the most abundant soluble antioxidant in plants. GR, with isoenzymes located in different compartments (Edwards et al. 1990; Romero-Puertas et al. 2006; Wu et al. 2013), has been reported to be involved in maintaining and regenerating reduced glutathione (GSH) in response to biotic and abiotic stress in plants (Creissen et al. 1992; Foyer and Noctor 2011; Leterrier et al. 2012; Gill et al. 2013; Zhang et al. 2015). NO modulation of GR activity in abiotic stress situations is the subject of a previous study (see Groβ et al. 2013) which highlighted the capacity of this enzyme to respond to (a)biotic stress conditions. In addition, a nitroproteomic study of sunflower hypocotyls has identified GR as a target for tyrosine nitration (Chaki et al. 2009b) and for S-nitrosylation in rice (Lin et al. 2012). Human and bovine GR activity has also been reported to be inhibited by tyrosine nitration. This suggests that Tyr106 and Tyr114, located close to the GSSG binding zone, are responsible for the decrease observed in GR activity after treatment with peroxynitrite (Francescutti et al. 1996; Savvides et al. 2002). Furthermore, GSNO inactivates human GR activity through the S-nitrosylation of two key cysteines Cys63 and/or Cys58 (Becker et al. 1995) involved in catalytic GR mechanisms (Francescutti et al. 1996). However, to our knowledge, little information exists on the effect of NO-PTMs on the GR structure in higher plants. In contrast to animals, chloroplastic and cytosolic GR in pea plants is S-nitrosylated and nitrated by GSNO and peroxynitrite, respectively, although these NO-PTMs do not significantly affect recombinant protein activity. This is unusual in higher plants as most studies show that nitration causes function loss in all proteins identified up to now (Astier and Lindermayr 2012; Corpas et al. 2013a). This could be a mechanism for maintaining GSH regeneration in order to sustain the ascorbate–glutathione cycle's resistance to nitro-oxidative cell conditions.

4 Monodehydroascorbate Reductase (MDAR) is Inactivated by NO-Related PTMs

Monodehydroascorbate reductase (MDAR) is a key enzyme in the ascorbate–glutathione cycle involved in the regeneration of the reduced ascorbate. In plants, MDAR has been reported to be localized in subcellular compartments, such as chloroplasts (Hossain et al. 1984), cytosols, mitochondria (Dalton et al. 1993; Jimenez et al. 1997; Mittova et al. 2003), and peroxisomes (Bowdicht and

Donaldson 1990; Leterrier et al. 2005; Lisenbee et al. 2005). MDAR plays a crucial role in plant responses to situations that trigger nitro-oxidative stress. Pea MDAR activity has been reported to increase under high light intensity and cadmium conditions but is reduced by herbicide 2,4-D (Leterrier et al. 2005). In tomato, MDAR activity is increased by salinity (Mittova et al. 2003) and high light intensity (Gechev et al. 2003), by low temperature in rice (Oidaira et al. 2000), and by UV-B radiation in *Arabidopsis* (Kubo et al. 1999). However, in *Arabidopsis*, stresses such as high temperature (30 °C), enhanced light intensity (200 μE m^{-2} s^{-1}), water deficiency (water deprivation for 2d), and low temperatures (5 °C) did not affect MDAR activity (Kubo et al. 1999). As mentioned above, several proteomic studies have identified MDAR as a potential target for both S-nitrosylation and nitration (Lin et al. 2012; Tanou et al. 2012). However, the specific effects of these NO-PTMs on MDAR functions are unclear. Begara-Morales et al. 2015 have shown that S-nitrosylation and tyrosine nitration of MDAR by GSNO and peroxynitrite, respectively, cause a reduction in peroxisomal recombinant MDAR protein activity. With the aid of mass spectrometry, Tyr213, Tyr292, and Tyr345 were identified as nitration targets following peroxynitrite ($ONOO^-$) treatment. In addition, site-directed mutagenesis confirmed that Tyr345 is the primary nitration site responsible for the inactivation of MDAR activity by $ONOO^-$. Localization analysis of these residues in the pea MDAR structure reveals that Tyr345 is found at 3.3 Å of His-313, which is involved in the NADP-binding site.

5 Effect of NO-Related PTMs on Dehydroascorbate Reductase (DHAR)

Dehydroascorbate reductase (DHAR) is the other enzyme in the ascorbate–glutathione cycle involved in the regeneration of reduced ascorbate for which GSH is used as electron donor. Researchers have devoted less attention to this enzyme which, however, can also be modulated in stress situations and in response to NO donors (Groβ et al. 2013). Using proteomic analyses, DHAR has recently been identified as a S-nitrosylation target in potato, demonstrating that S-nitrosylation at Cys20 and Cys147 inactivates DHAR activity (Kato et al. 2013). DHAR has also been reported to be S-nitrosylated in *Arabidopsis* under physiological conditions at Cys 20 (Fares et al. 2011; Puyaubert et al. 2014), although this protein has not been observed to be S-nitrosylated during cold stress (Puyaubert et al. 2014), a condition which increases DHAR activity (Eltelib et al. 2011). Taking these results together, it is possible to speculate that cold stress could cause the denitrosylation of DHAR and thus increase its activity in response to these environmental conditions.

6 Dual Regulation of Ascorbate Peroxidase (APX): Inactivated by Nitration and Enhanced by S-Nitrosylation

APX is one of the key antioxidant enzymes involved in regulating H_2O_2 levels during plant development and adverse stress conditions (Jiménez et al. 1998; Gomez et al. 2004; Palma et al. 2006). Different APX isoenzymes are distributed in subcellular compartments including the cytosol, chloroplasts, peroxisomes, and mitochondria (see Shigeoka et al. 2002). In higher plants, APX is an essential element in the finely tuned regulation mechanism of H_2O_2 during plant development and under environmental stress conditions. NO is able to modulate APX activity through a process of either inactivation (Clark et al. 2000) or activation (Keyster et al. 2011; Lin et al. 2011). With the aid of proteomic analysis, APX has been identified as a potential NO-PTM target through tyrosine nitration in *Arabidopsis* (Lozano-Juste et al. 2011) and *Citrus aurantium* (Tanou et al. 2012) and through S-nitrosylation in *Arabidopsis* (Fares et al. 2011), *Antiaris toxicaria* (Bai et al. 2011), and *Solanum tuberosum* (Kato et al. 2013). In the next section, we will review the effect of tyrosine nitration and S-nitrosylation on the APX structure in different plant species and will suggest that APX is regulated by both NO-PTMs.

6.1 *APX Is Inactivated by Nitration of Tyr235*

APX has been identified as a tyrosine nitration target (Lozano-Juste et al. 2011; Tanou et al. 2012). However, little information exists on the specific impact of tyrosine nitration on the activity and structure of APX or on the proteins involved in antioxidative systems (Radi 2013). With regard to the effect of tyrosine nitration on the APX structure, a recent study has shown that pea cytosolic APX can be nitrated and inactivated by the peroxynitrite donor SIN-1 (Begara-Morales et al. 2013b). Mass spectrometry shows that tyrosine residues Tyr5 and Tyr235 are nitration targets, with Tyr235 appearing to be the most reliable candidate for causing APX inactivation as this residue is located at the bottom of the catalytic pocket and only 3.6 Å from the heme group (Patterson and Poulos 1995; Jespersen et al. 1997; Mandelman et al. 1998; Begara-Morales et al. 2013b). The addition of the nitro group may therefore disrupt heme group properties and result in a loss of activity. With respect to Tyr5, it has been reported that nitration of this tyrosine, which is conserved in plants, could have physiological consequences, although, given its location, it is difficult to demonstrate that tyrosine nitration affects enzymatic activity or protein structure (Begara-Morales et al. 2013b). Thus, determination of the role played by the nitration of Tyr235 in vivo under physiological conditions and in plant responses to different stress situations could be a useful starting point for future analysis.

6.2 APX Is Enhanced by S-Nitrosylation of Cys32

With the identification of the cysteine(s) and their potential role in different stress situations, information concerning the effect of S-nitrosylation on the APX structure has recently been provided. Correa-Aragunde et al. (2013) have reported that APX1 is S-nitrosylated in vivo and that auxins cause its denitrosylation and partial inhibition in *Arabidopsis* roots. To corroborate these findings, the APX1 recombinant protein was treated with increasing concentrations of a NO donor (CysNO) which showed that S-nitrosylation of APX1 produced a gain-of-function mutation (Correa-Aragunde et al. 2013). This increase in APX activity due to NO has previously been described by Keyster et al. (2011) and Lin et al. (2011). *Arabidopsis* APX1contains five Cys residues, with Cys32 and Cys168 being the most conserved cysteine in plants. Using in silico analysis, it has been suggested that, among the cysteine residues present in the *Arabidopsis* APX1, the Cys168 residue located near the heme group could be the S-nitrosylation target. However, there is no clear evidence to suggest that Cys168 rather than Cys32, located near one of the two ascorbate-binding sites (Correa-Aragunde et al. 2013), is responsible for S-nitrosylation and increased APX activity. Proteomic analysis has also identified Cys32 as an S-nitrosylation target in *Arabidopsis* (Fares et al. 2011). Finally, using *apx1* mutants, the authors suggest that the counterbalance of APX1 S-nitrosylation/denitrosylation mediated by auxins could help to control root development and to determine root architecture (Correa-Aragunde et al. 2013).

In contrast to *Arabidopsis* roots, APX S-nitrosylation during programmed cell death (PCD) induced by H_2O_2 or heat shock (HS) inactivates enzyme activity which is corroborated using partial purified APX and GSNO as NO donor (de Pinto et al. 2013). These findings are in line with the inhibition of APX observed after treatment of cell culture tobacco leaves with GSNO (Clark et al. 2000). It is worth noting that S-nitrosylation has been found to induce cytosolic APX ubiquitination and subsequent proteasomal degradation, suggesting that S-nitrosylation may act as part of a signaling pathway that leads to PCD (de Pinto et al. 2013). Tobacco APX also has five cysteine residues including Cys32 and Cys168 (Correa-Aragunde et al. 2013). Based on the changes observed in the kinetic properties of APX enzymes following S-nitrosylation, it has been suggested that S-nitrosylation of Cys32 could be responsible for the decrease in APX activity as the high-affinity site for ascorbate binding is in adjacent to Cys32 (de Pinto et al. 2013). However, Clark et al. (2000) report that, after treatment with GSNO, tobacco APX enzymes are inhibited by the formation of an iron–nitrosyl complex between NO and the heme group's iron atom. Thus, as Cys168 is located close to the heme group (Correa-Aragunde et al. 2013), it cannot be ruled out that S-nitrosylation of Cys168, and not Cys32, inactivates APX activity.

To determine the way in which APX is affected by NO-derived molecules, another study, using cytosolic APX from pea plants, demonstrated that this antioxidant enzyme is subjected to dual regulation by NO-related PTMs, i.e., inactivated by Tyr235 nitration (see above) and enhanced by S-nitrosylation mediated by

GSNO (Begara-Morales et al. 2013b). The treatment of recombinant APX with increasing concentrations of GSNO causes S-nitrosylation and increased protein activity, while treatment with GSH did not have any effect. In order to evaluate the potential physiological role played by S-nitrosylation under stress conditions, APX activity under salinity stress has been analyzed. The data show that APX activity increases after treatment with 150 mM NaCl, with a concomitant increase in the levels of H_2O_2, malondialdehyde (MDA), NO, and SNOs (Begara-Morales et al. 2013b). In addition, APX was found to be S-nitrosylated in vivo, process that was increased under salinity conditions as a consequence of an increase in both NO and SNOs. Under these circumstances, it has been suggested that a rise in APX activity caused by salinity stress could be partly due to S-nitrosylation where SNOs may alleviate oxidative damage induced by salinity stress. It is worth noting that the sequence of pea APX only contains Cys32, making this cysteine an S-nitrosylation target, which is responsible for the effects, observed under in vivo and in vitro conditions. As mentioned above, Cys32 is close to the heme group's propionate side chain and has been reported to form thiyl radicals through the interaction of APX with H_2O_2 (Kitajima et al. 2008) which would indicate a direct reaction with NO (Martínez-Ruiz and Lamas 2007). Interestingly, Cys32 oxidation causes enzyme deactivation, suggesting that glutathionylation protects the enzyme from irreversible oxidation (Kitajima et al. 2008). Furthermore, it has been hypothesized that Cys32 S-nitrosylation may be protected from deactivation by H_2O_2 oxidation (Begara-Morales et al. 2013b), as a relationship between cysteine oxidation and S-nitrosylation in plants has been observed (Lounifi et al. 2013). In this respect, S-nitrosylation of mouse galectin-2 has recently been reported to prevent oxidative inactivation by hydrogen peroxide (Tamura et al. 2015).

This finding has been corroborated in a very recent study by Yang et al. (2015) who show that S-nitrosylation induces APX activity in *Arabidopsis* seedlings. Cys32 and Cys49 have been identified as S-nitrosylation targets, which, with the mutation of these cysteines, confirm that Cys32 is responsible for the increased activity observed following S-nitrosylation. They also demonstrate that the S-nitrosylation of Cys32 plays a role in plant immunity and plant resistance to oxidative stress. All this data taken together appears to show that S-nitrosylation of Cys32 is involved in plant responses to stress situations and may prevent the oxidation of this cysteine and therefore protect the enzyme from inactivation under adverse conditions.

7 Conclusions

Our knowledge of NO involvement in physiological and stress situations in plants has increased in recent years. It has been established that a link exists between ROS and NO pathways through the control of proteins involved in the ROS metabolism caused by posttranslational modifications associated with NO (NO-PTMs) such as S-nitrosylation and tyrosine nitration. Enzymes in the ascorbate–glutathione cycle,

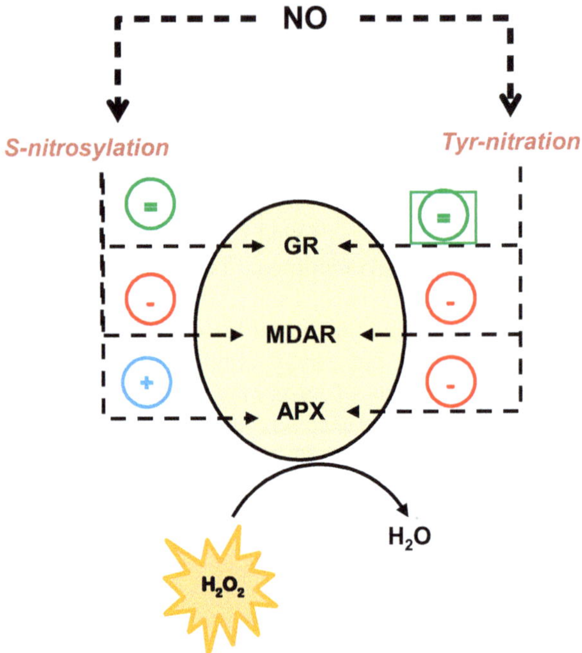

Fig. 1 Regulation of the ascorbate–glutathione cycle by nitric oxide (NO). NO can modulate the antioxidant capacity of the ascorbate–glutathione cycle by posttranslational modifications such as S-nitrosylation and tyrosine nitration. In this sense, during a nitrosative stress situation, tyrosine nitration could inhibit MDAR and APX activity, while S-nitrosylation enhances APX function by S-nitrosylation of Cys32. DHAR is also inactivated by S-nitrosylation. In addition, GR is not affected by these NO-PTMs in an attempt to maintain the cellular redox state under the adverse conditions. *MDAR* monodehydroascorbate reductase, *APX* ascorbate peroxidase, *DHAR* dehydroascorbate reductase, *GR* glutathione reductase

one of the most important H_2O_2-removing antioxidant systems in cells, are regulated by NO. However, little information exists on the impact of NO-PTMs on the structure of these proteins and the role of NO targets in abiotic stress situations. Figure 1 summarizes existing data regarding the effect of NO-PTMs on the ascorbate–glutathione cycle and shows that APX is inactivated by tyrosine nitration and enhanced by S-nitrosylation of Cys32 (dual regulation). MDAR is also inactivated by S-nitrosylation and tyrosine nitration, which may compromise the cycle's antioxidant capacity. However, GR is not affected by any of these modifications in order to maintain GSH levels, and thus, the cellular redox state which could be crucial in nitro-oxidative stress situations. Future research on the identification of endogenous NO targets in the ascorbate–glutathione cycle under (a) biotic stress conditions will be necessary in order to understand how NO modulates this antioxidant system.

Acknowledgments This study was supported by an ERDF grant co-financed by the Ministry of Economy and Competitiveness (projects BIO2012-33904 and RECUPERA2020) and the Junta de Andalucía (groups BIO286 and BIO192) in Spain. J. Begara-Morales would like to thank the Ministry of Science and Innovation for funding the Ph.D. fellowship (F.P.U.). LC/MS/MS analyses were carried out at the Laboratorio de Proteómica LP-CSIC/UAB, a member of ProteoRed network.

References

Abello N, Kerstjens HAM, Postma DS, Bischoff R (2009) Protein tyrosine nitration: selectivity, physicochemical and biological consequences, denitration, and proteomics methods for the identification of tyrosine-nitrated proteins. J Proteome Res 8:3222–3238

Airaki M, Leterrier M, Mateos RM, Valderrama R, Chaki M, Barroso JB, del Río LA, Palma JM, Corpas FJ (2011a) Metabolism of reactive oxygen species and reactive nitrogen species in pepper (*Capsicum annuum* L.) plants under low temperature stress. Plant Cell Environ 35:281–295

Airaki M, Sánchez-Moreno L, Leterrier M, Barroso JB, Palma JM, Corpas FJ (2011b) Detection and quantification of S-nitrosoglutathione (GSNO) in pepper (*Capsicum annuum* L.) plant organs by LC-ES/MS. Plant Cell Physiol 52:2006–2015

Álvarez C, Lozano-Juste J, Romero LC, García I, Gotor C, León J (2011) Inhibition of Arabidopsis O-acetylserine (thiol) lyase A1 by tyrosine nitration. J Biol Chem 286:578–586

Asada K (1992) Ascorbate peroxidase – a hydrogen peroxide-scavenging enzyme in plants. Physiol Planta 85:235–241

Asada K (2006) Production and scavenging of reactive oxygen species in chloroplasts and their functions. Plant Physiol 141:391–396

Astier J, Rasul S, Koen E, Manzoor H, Besson-Bard A, Lamotte O, Jeandroz S, Durner J, Lindermayr C, Wendehenne D (2011) S-nitrosylation: an emerging post-translational protein modification in plants. Plant Sci 181:527–533

Astier J, Lindermayr C (2012) Nitric oxide-dependent posttranslational modification in plants: an update. Int J Mol Sci 13:15193–15208

Bai X, Yang L, Tian M, Chen J, Shi J, Yang Y, Hu X (2011) Nitric oxide enhances desiccation tolerance of recalcitrant *Antiaris toxicaria* seeds via protein S-nitrosylation and carbonylation. PLoS One 6, e20714

Becker K, Gui M, Schirmer RH (1995) Inhibition of human glutathione reductase by S-nitrosoglutathione. Eur J Biochem 234:472–478

Begara-Morales JC, Chaki M, Sanchez-Calvo B, Mata-Parez C, Leterrier M, Palma JM, Barroso JB, Corpas FJ (2013a) Protein tyrosine nitration in pea roots during development and senescence. J Exp Bot 64:1121–1134

Begara-Morales JC, Sánchez-Calvo B, Chaki M, Valderrama R, Mata-Pérez C, López-Jaramillo J, Padilla MN, Carreras A, Corpas FJ, Barroso JB (2013b) Dual regulation of cytosolic ascorbate peroxidase (APX) by tyrosine nitration and S-nitrosylation. J Exp Bot 65:527–538

Begara-Morales JC, Sanchez-Calvo B, Chaki M, Mata-Perez C, Valderrama R, Padilla MN, Lopez-Jaramillo J, Luque F, Corpas FJ, Barroso JB (2015) Differential molecular response of monodehydroascorbate reductase and glutathione reductase by nitration and S-nitrosylation. J Exp Bot. doi:10.1093/jxb/erv306

Belenghi B, Romero-Puertas MC, Vercammen D, Brackenier A, Inza D, Delledonne M, Van Breusegem F (2007) Metacaspase activity of *Arabidopsis thaliana* is regulated by S-nitrosylation of a critical cysteine residue. J Biol Chem 282:1352–1358

Bowdicht ML, Donaldson RP (1990) Ascorbate free-radical reduction by glyoxysomal membranes. Plant Physiol 94:531–537

Camejo D, Ortiz-Espín A, Lázaro JJ, Romero-Puertas MC, Lázaro-Payo A, Sevilla F, Jiménez A (2015) Functional and structural changes in plant mitochondrial PrxII F caused by NO. J Proteomics 119:112–125

Clark D, Durner J, Navarre DA, Klessig DF (2000) Nitric oxide inhibition of tobacco catalase and ascorbate peroxidase. Mol Plant Microbe Interact 13:1380–1384

Corpas FJ, Chaki M, Fernández-Ocaña A, Valderrama R, Palma JM, Carreras A, Begara-Morales JC, Airaki M, del Río LA, Barroso JB (2008) Metabolism of reactive nitrogen species in pea plants under abiotic stress conditions. Plant Cell Physiol 49:1711–1722

Corpas FJ, Leterrier M, Valderrama R, Airaki M, Chaki M, Palma JM, Barroso JB (2011) Nitric oxide imbalance provokes a nitrosative response in plants under abiotic stress. Plant Sci 181:604–611

Corpas FJ, Leterrier M, Begara-Morales JC, Valderrama R, Chaki M, López-Jaramillo J, Luque F, Palma JM, Padilla MN, Sánchez-Calvo B (2013a) Inhibition of peroxisomal hydroxypyruvate reductase (HPR1) by tyrosine nitration. Biochim Biophys Acta 1830:4981–4989

Corpas FJ, Palma JM, Luis A, Barroso JB (2013b) Protein tyrosine nitration in higher plants grown under natural and stress conditions. Front Plant Sci 4:29

Corpas FJ, Begara-Morales JC, Sánchez-Calvo B, Chaki M, Barroso JB (2015) Nitration and S-nitrosylation: two post-translational modifications (PTMs) mediated by reactive nitrogen species (RNS) and their role in signalling processes of plant cells. In: Gupta KJ, Igamberdiev AU (eds) Reactive oxygen and nitrogen species signaling and communication in plants. Springer, Germany

Correa-Aragunde N, Foresi N, Delledonne M, Lamattina L (2013) Auxin induces redox regulation of ascorbate peroxidase 1 activity by S-nitrosylation/denitrosylation balance resulting in changes of root growth pattern in Arabidopsis. J Exp Bot 64:3339–3349

Creissen G, Edwards EA, Enard C, Wellburn A, Mullineaux P (1992) Molecular characterization of glutathione reductase cDNAs from pea (*Pisum sativum* L.). Plant J 2:129–131

Chaki M, Fernández-Ocaña AM, Valderrama R, Carreras A, Esteban FJ, Luque F, Gómez-Rodíguez MV, Begara-Morales JC, Corpas FJ, Barroso JB (2009a) Involvement of reactive nitrogen and oxygen species (RNS and ROS) in sunflower-mildew interaction. Plant Cell Physiol 50:265–279

Chaki M, Valderrama R, Fernández-Ocaña AM, Carreras A, López-Jaramillo J, Luque F, Palma JM, Pedrajas JR, Begara-Morales JC, Sánchez-Calvo B (2009b) Protein targets of tyrosine nitration in sunflower (*Helianthus annuus* L.) hypocotyls. J Exp Bot 60:4221–4234

Chaki M, Valderrama R, Fernández-Ocaña AM, Carreras A, Gómez-Rodríguez MV, López-Jaramillo J, Begara-Morales JC, Sánchez-Calvo B, Luque F, Leterrier M (2011a) High temperature triggers the metabolism of *S*-nitrosothiols in sunflower mediating a process of nitrosative stress which provokes the inhibition of ferredoxin-NADP reductase by tyrosine nitration. Plant Cell Environ 34:1803–1818

Chaki M, Valderrama R, Fernández-Ocaña AM, Carreras A, Gómez-Rodíguez MV, Pedrajas JR, Begara-Morales JC, Sánchez-Calvo B, Luque F, Leterrier M (2011b) Mechanical wounding induces a nitrosative stress by down-regulation of GSNO reductase and an increase in S-nitrosothiols in sunflower (*Helianthus annuus*) seedlings. J Exp Bot 62:1803–1813

Chaki M, Carreras A, López-Jaramillo J, Begara-Morales JC, Sánchez-Calvo B, Valderrama R, Corpas FJ, Barroso JB (2013) Tyrosine nitration provokes inhibition of sunflower carbonic anhydrase (β-CA) activity under high temperature stress. Nitric Oxide 29:30–33

Dalton DA, Baird LM, Langeberg L, Taugher CY, Anyan WR, Vance CP, Sarath G (1993) Subcellular localization of oxygen defense enzymes in soybean (*Glycine max* [L.] Merr.) root nodules. Plant Physiol 102:481–489

de Pinto MC, Locato V, Sgobba A, Romero-Puertas MC, Gadaleta C, Delledonne M, De Gara L (2013) S-nitrosylation of ascorbate peroxidase is part of programmed cell death signaling in tobacco Bright Yellow-2 cells. Plant Physiol 163:1766–1775

del Río LA, Javier Corpas F, Barroso JB (2004) Nitric oxide and nitric oxide synthase activity in plants. Phytochemistry 65:783–792

Diaz-Vivancos P, Faize M, Barba-Espin G, Faize L, Petri C, Hernández JA, Burgos L (2013) Ectopic expression of cytosolic superoxide dismutase and ascorbate peroxidase leads to salt stress tolerance in transgenic plums. Plant Biotechnol J 11:976–985

Durner J, Gow AJ, Stamler JS, Glazebrook J (1999) Ancient origins of nitric oxide signaling in biological systems. Proc Natl Acad Sci U S A 96:14206–14207

Edwards EA, Rawsthorne S, Mullineaux PM (1990) Subcellular distribution of multiple forms of glutathione reductase in leaves of pea (*Pisum sativum* L.). Planta 180:278–284

Eltelib HA, Badejo AA, Fujikawa Y, Esaka M (2011) Gene expression of monodehydroascorbate reductase and dehydroascorbate reductase during fruit ripening and in response to environmental stresses in acerola (*Malpighia glabra*). J Plant Physiol 168:619–627

Fares A, Rossignol M, Peltier JB (2011) Proteomics investigation of endogenous S-nitrosylation in Arabidopsis. Biochem Biophys Res Commun 416:331–336

Feechan A, Kwon E, Yun BW, Wang Y, Pallas JA, Loake GJ (2005) A central role for S-nitrosothiols in plant disease resistance. Proc Natl Acad Sci U S A 102:8054–8059

Foyer CH, Halliwell B (1976) The presence of glutathione and glutathione reductase in chloroplasts: a proposed role in ascorbic acid metabolism. Planta 133:21–25

Foyer CH, Noctor G (2011) Ascorbate and glutathione: the heart of the redox hub. Plant Physiol 155:2–18

Francescutti D, Baldwin J, Lee L, Mutus B (1996) Peroxynitrite modification of glutathione reductase: modeling studies and kinetic evidence suggest the modification of tyrosines at the glutathione disulfide binding site. Protein Eng 9:189–194

Gaston B, Reilly J, Drazen JM, Fackler J, Ramdev P, Arnelle D, Mullins ME, Sugarbaker DJ, Chee C, Singel DJ (1993) Endogenous nitrogen oxides and bronchodilator S-nitrosothiols in human airways. Proc Natl Acad Sci USA 90:10957–10961

Gechev T, Willekens H, Van Montagu M, Inzo D, Van Camp W, Toneva V, Minkov I (2003) Different responses of tobacco antioxidant enzymes to light and chilling stress. J Plant Physiol 160:509–515

Gill SS, Anjum NA, Hasanuzzaman M, Gill R, Trivedi DK, Ahmad I, Pereira E, Tuteja N (2013) Glutathione and glutathione reductase: a boon in disguise for plant abiotic stress defense operations. Plant Physiol Biochem 70:204–212

Gomez JM, Jimenez A, Olmos E, Sevilla F (2004) Location and effects of long-term NaCl stress on superoxide dismutase and ascorbate peroxidase isoenzymes of pea (*Pisum sativum* cv. Puget) chloroplasts. J Exp Bot 55:119–130

Gow AJ, Farkouh CR, Munson DA, Posencheg MA, Ischiropoulos H (2004) Biological significance of nitric oxide-mediated protein modifications. Am J Physiol Lung Cell Mol Physiol 287:L262–L268

Groβ F, Durner J, Gaupels F (2013) Nitric oxide, antioxidants and prooxidants in plant defence responses. Front Plant Sci 4:419

Groden D, Beck E (1979) H_2O_2 destruction by ascorbate-dependent systems from chloroplasts. Biochim Biophys Acta 546:426–435

Holtgrefe S, Gohlke J, Starmann J, Druce S, Klocke S, Altmann B, Wojtera J, Lindermayr C, Scheibe R (2008) Regulation of plant cytosolic glyceraldehyde 3-phosphate dehydrogenase isoforms by thiol modifications. Physiol Planta 133:211–228

Hossain MA, Nakano Y, Asada K (1984) Monodehydroascorbate reductase in spinach chloroplasts and its participation in regeneration of ascorbate for scavenging hydrogen peroxide. Plant Cell Physiol 25:385–395

Hu J, Huang X, Chen L, Sun X, Lu C, Zhang L, Wang Y, Zuo J (2015) Site-specific nitrosoproteomic identification of endogenously S-nitrosylated proteins in Arabidopsis. Plant Physiol 167:1731–1746

Jespersen H, Kjrd I, Stergaard L, Welinder K (1997) From sequence analysis of three novel ascorbate peroxidases from *Arabidopsis thaliana* to structure, function and evolution of seven types of ascorbate peroxidase. Biochem J 326:305–310

Jimenez A, Hernandez JA, del Rio LA, Sevilla F (1997) Evidence for the presence of the ascorbate-glutathione cycle in mitochondria and peroxisomes of pea leaves. Plant Physiol 114:275–284

Jimenez A, Hernandez JA, Pastori G, del Rio LA, Sevilla F (1998) Role of the ascorbate-glutathione cycle of mitochondria and peroxisomes in the senescence of pea leaves. Plant Physiol 118:1327–1335

Kato H, Takemoto D, Kawakita K (2013) Proteomic analysis of S-nitrosylated proteins in potato plant. Physiol Planta 148:371–386

Keyster M, Klein A, Egbich I, Jacobs A, Ludidi N (2011) Nitric oxide increases the enzymatic activity of three ascorbate peroxidase isoforms in soybean root nodules. Plant Signal Behav 6:956–961

Kitajima S, Kurioka M, Yoshimoto T, Shindo M, Kanaori K, Tajima K, Oda K (2008) A cysteine residue near the propionate side chain of heme is the radical site in ascorbate peroxidase. FEBS J 275:470–480

Kubo A, Aono M, Nakajima N, Saji H, Tanaka K, Kondo N (1999) Differential responses in activity of antioxidant enzymes to different environmental stresses in *Arabidopsis thaliana.* J Plant Res 112:279–290

Lamattina L, García-Mata C, Graziano M, Pagnussat G (2003) Nitric oxide: the versatility of an extensive signal molecule. Annu Rev Plant Biol 54:109–136

Leitner M, Vandelle E, Gaupels F, Bellin D, Delledonne M (2009) NO signals in the haze: nitric oxide signalling in plant defence. Curr Opin Plant Biol 12:451–458

Leterrier M, Corpas FJ, Barroso JB, Sandalio LM, Luis A (2005) Peroxisomal monodehydroascorbate reductase. Genomic clone characterization and functional analysis under environmental stress conditions. Plant Physiol 138:2111–2123

Leterrier M, Barroso JB, Valderrama R, Palma JM, Corpas FJ (2012) NADP-dependent isocitrate dehydrogenase from Arabidopsis roots contributes in the mechanism of defence against the nitro-oxidative stress induced by salinity. Sci World J. ID: 694740.

Lin A, Wang Y, Tang J, Xue P, Li C, Liu L, Hu B, Yang F, Loake GJ, Chu C (2012) Nitric oxide and protein S-nitrosylation are integral to hydrogen peroxide-induced leaf cell death in rice. Plant Physiol 158:451–464

Lin CC, Jih PJ, Lin HH, Lin JS, Chang LL, Shen YH, Jeng ST (2011) Nitric oxide activates superoxide dismutase and ascorbate peroxidase to repress the cell death induced by wounding. Plant Mol Biol 77:235–249

Lindermayr C, Durner J (2009) S-nitrosylation in plants: pattern and function. J Proteomics 73:1–9

Lisenbee CS, Lingard MJ, Trelease RN (2005) Arabidopsis peroxisomes possess functionally redundant membrane and matrix isoforms of monodehydroascorbate reductase. Plant J 43:900–914

Liu Z, Bao H, Cai J, Han J, Zhou L (2014) A novel thylakoid ascorbate peroxidase from *Jatrophacurcas* enhances salt tolerance in transgenic tobacco. Int J Mol Sci 15:171–185

Lounifi I, Arc E, Molassiotis A, Job D, Rajjou L, Tanou G (2013) Interplay between protein carbonylation and nitrosylation in plants. Proteomics 13:568–578

Lozano Juste J, Colom Moreno R, León J (2011) In vivo protein tyrosine nitration in *Arabidopsis thaliana*. J Exp Bot 62:3501–3517

Malik SI, Hussain A, Yun BW, Spoel SH, Loake GJ (2011) GSNOR-mediated de-nitrosylation in the plant defence response. Plant Sci 181:540–544

Mandelman D, Jamal J, Poulos TL (1998) Identification of two electron-transfer sites in ascorbate peroxidase using chemical modification, enzyme kinetics, and crystallography. Biochemistry 37:17610–17617

Martínez-Ruiz A, Lamas S (2007) Signalling by NO-induced protein S-nitrosylation and S-glutathionylation: convergences and divergences. Cardiovasc Res 75:220–228

Mittova V, Tal M, Volokita M, Guy M (2003) Up-regulation of the leaf mitochondrial and peroxisomal antioxidative systems in response to salt-induced oxidative stress in the wild salt-tolerant tomato species *Lycopersicon pennellii*. Plant Cell Environ 26:845–856

Noctor G, Foyer CH (1998) Ascorbate and glutathione: keeping active oxygen under control. Annu Rev Plant Biol 49:249–279

Oidaira H, Sano S, Koshiba T, Ushimaru T (2000) Enhancement of antioxidative enzyme activities in chilled rice seedlings. J Plant Physiol 156:811–813

Ortega-Galisteo AP, Rodríguez-Serrano M, Pazmino DM, Gupta DK, Sandalio LM, Romero-Puertas MC (2012) S-nitrosylated proteins in pea (*Pisum sativum* L.) leaf peroxisomes: changes under abiotic stress. J Exp Bot 63:2089–2103

Palma JM, Jiménez A, Sandalio LM, Corpas FJ, Lundqvist M, Gómez M, Sevilla F, del Río LA (2006) Antioxidative enzymes from chloroplasts, mitochondria, and peroxisomes during leaf senescence of nodulated pea plants. J Exp Bot 57:1747–1758

Patterson WR, Poulos TL (1995) Crystal structure of recombinant pea cytosolic ascorbate peroxidase. Biochemistry 34:4331–4341

Procházková D, Sumaira J, Wilhelmová Na, Pavlíková D, Száková J (2014) Reactive nitrogen species and the role of NO in abiotic stress. In: Ahmad P, Rasool S (ed) Emerging Technologies and Managment of Crops Stress Tolerance. Elsevier.

Puyaubert J, Fares A, Rézé N, Peltier JB, Baudouin E (2014) Identification of endogenously S-nitrosylated proteins in Arabidopsis plantlets: effect of cold stress on cysteine nitrosylation level. Plant Sci 215:150–156

Radi R (2004) Nitric oxide, oxidants, and protein tyrosine nitration. Proc Natl Acad Sci USA 101:4003–4008

Radi R (2013) Protein tyrosine nitration: biochemical mechanisms and structural basis of functional effects. Acc Chem Res 46:550–559

Reumann S, Babujee L, Ma C, Wienkoop S, Siemsen T, Antonicelli GE, Rasche N, Luder F, Weckwerth W, Jahn O (2007) Proteome analysis of Arabidopsis leaf peroxisomes reveals novel targeting peptides, metabolic pathways, and defense mechanisms. Plant Cell 19:3170–3193

Reumann S, Corpas FJ (2010) The peroxisomal ascorbate-glutathione pathway: molecular identification and insights into its essential role under environmental stress conditions. In: Anjum NA, Chan MT, Umar S (eds) Ascorbate-glutathione pathway and stress tolerance in plants. Springer, Germany

Romero-Puertas MC, Corpas FJ, Sandalio LM, Leterrier M, Rodríguez-Serrano M, Del Río LA, Palma JM (2006) Glutathione reductase from pea leaves: response to abiotic stress and characterization of the peroxisomal isozyme. New Phytol 170:43–52

Romero-Puertas MC, Laxa M, Matte A, Zaninotto F, Finkemeier I, Jones AME, Perazzolli M, Vandelle E, Dietz KJ, Delledonne M (2007) S-nitrosylation of peroxiredoxin II E promotes peroxynitrite-mediated tyrosine nitration. Plant Cell 19:4120–4130

Romero-Puertas MC, Campostrini N, Matte A, Righetti PG, Perazzolli M, Zolla L, Roepstorff P, Delledonne M (2008) Proteomic analysis of S-nitrosylated proteins in *Arabidopsis thaliana* undergoing hypersensitive response. Proteomics 8:1459–1469

Rusterucci C, Espunya MC, Díaz M, Chabannes M, Martínez MC (2007) S-nitrosoglutathione reductase affords protection against pathogens in Arabidopsis, both locally and systemically. Plant Physiol 143:1282–1292

Savvides SN, Scheiwein M, Bohme CC, Arteel GE, Karplus PA, Becker K, Schirmer RH (2002) Crystal structure of the antioxidant enzyme glutathione reductase inactivated by peroxynitrite. J Biol Chem 277:2779–2784

Shan C, Zhou Y, Liu M (2015) Nitric oxide participates in the regulation of the ascorbate-glutathione cycle by exogenous jasmonic acid in the leaves of wheat seedlings under drought stress. Protoplasma. doi:10.1007/s00709-015-0756-y

Shapiro AD (2005) Nitric oxide signaling in plants. Vitam Horm 72:339–398

Shigeoka S, Ishikawa T, Tamoi M, Miyagawa Y, Takeda T, Yabuta Y, Yoshimura K (2002) Regulation and function of ascorbate peroxidase isoenzymes. J Exp Bot 53:1305–1319

Siddiqui MH, Al-Whaibi MH, Basalah MO (2011) Role of nitric oxide in tolerance of plants to abiotic stress. Protoplasma 248:447–455

Souza JM, Peluffo G, Radi R (2008) Protein tyrosine nitration-functional alteration or just a biomarker? Free Radic Biol Med 45:357–366

Sun C, Liu L, Yu Y, Liu W, Lu L, Jin C, Lin X (2014) Nitric oxide alleviates aluminum-induced oxidative damage through regulating the ascorbate-glutathione cycle in roots of wheat. J Int Plant Biol 57:550–561

Tamura M, Saito M, Yamamoto K, Takeuchi T, Ohtake K, Tateno H, Hirabayashi J, Kobayashi J, Arata Y (2015) S-nitrosylation of mouse galectin-2 prevents oxidative inactivation by hydrogen peroxide. Biochem Biophys Res Commun 457:712–717

Tanou G, Filippou P, Belghazi M, Job D, Diamantidis G, Fotopoulos V, Molassiotis A (2012) Oxidative and nitrosative-based signaling and associated post-translational modifications orchestrate the acclimation of citrus plants to salinity stress. Plant J 72:585–599

Turko IV, Murad F (2002) Protein nitration in cardiovascular diseases. Pharmacol Rev 54:619–634

Valderrama R, Corpas FJ, Carreras A, Fernández-Ocaña A, Chaki M, Luque F, Gómez-Rodríguez MV, Colmenero-Varea P, Luis A, Barroso JB (2007) Nitrosative stress in plants. FEBS Lett 581:453–461

Wang D, Liu Y, Tan X, Liu H, Zeng G, Hu X, Jian H, Gu Y (2015) Effect of exogenous nitric oxide on antioxidative system and S-nitrosylation in leaves of *Boehmeria nivea* (L.) Gaud under cadmium stress. Environ Sci Pollut Res Int 22:3489–3497

Wu TM, Lin WR, Kao YT, Hsu YT, Yeh CH, Hong CY, Kao CH (2013) Identification and characterization of a novel chloroplast/mitochondria co-localized glutathione reductase 3 involved in salt stress response in rice. Plant Mol Biol 83:379–390

Yang H, Mu J, Chen L, Feng J, Hu J, Li L, Zhou JM, Zuo J (2015) S-Nitrosylation positively regulates ascorbate peroxidase activity during plant stress responses. Plant Physiol 167:1604–1615

Yu M, Lamattina L, Spoel SH, Loake GJ (2014) Nitric oxide function in plant biology: a redox cue in deconvolution. New Phytol 202:1142–1156

Yun BW, Feechan A, Yin M, Saidi NBB, Le Bihan T, Yu M, Moore JW, Kang JG, Kwon E, Spoel SH, Pallas JA, Loake GJ (2011) S-nitrosylation of NADPH oxidase regulates cell death in plant immunity. Nature 478:264–268

Zhang X, Quan G, Wang J, Han H, Chen S, Guo S, Yin H (2015) Functional validation of *Phragmites communis* glutathione reductase (PhaGR) as an essential enzyme in salt tolerance. Appl Biochem Biotechnol 175:3418–3430

Zhang Z, Zhang Q, Wu J, Zheng X, Zheng S, Sun X, Qiu Q, Lu T (2013) Gene knockout study reveals that cytosolic ascorbate peroxidase 2 (OsAPX2) plays a critical role in growth and reproduction in rice under drought, salt and cold stresses. PLoS One 8:e57472

ROS–RNS–Phytohormones Network in Root Response Strategy

Urszula Krasuska and Agnieszka Gniazdowska

Contents

Abstract Roots are considered as a "brain" of a plant, with apical meristems pointed as command centers of an organism. Besides nutrient and water uptake, roots act as organs decisive and responsive for environmental factors. Dynamic reorganization of root growth and architecture in reaction to various stressors is commonly observed. Self-recognition and plants' ability to identify neighbors are manifested by root movement. Root physiology is strictly controlled by diverse agents including phytohormones, reactive oxygen species (ROS), and reactive nitrogen species (RNS). ROS and RNS as molecules of bimodal function are known as cellular messengers crucial for regulation of fundamental physiological processes. Most of them depend on auxins; thus, auxin–ROS–RNS cross talk seems to be a typical pattern in root response to different stimuli. This chapter presents information on ROS and RNS contribution in regulation of root movement, growth, and development, described on the basis of auxin and abscisic acid action.

Keywords Auxin • Growth • Nitric oxide • Root apex • Root movement

---

U. Krasuska (✉) • A. Gniazdowska
Department of Plant Physiology, Warsaw University of Life Sciences-SGGW, Nowoursynowska 159, 02-776 Warsaw, Poland
e-mail: urszula_krasuska@sggw.pl

D.K. Gupta et al. (eds.), *Reactive Oxygen Species and Oxidative Damage in Plants Under Stress*, DOI 10.1007/978-3-319-20421-5_13

1 Roots as the Administrative Center of Plant Response to Environmental Signals

Some parts of plant studies refer to the controversial idea called "plant intelligence." This aspect of plant physiology, although questionable, is connected with recognition of the root as the most important organ of the plant body responsible for signal perception. The root apical meristem covered by a root cap is considered as a sensory structure, called also as command/decision center (Baluška et al. 2010) (Fig. 1). During evolution, plants were forced by terrestrial environment to develop well-organized root system. The anatomy of the root apical meristem, constantly producing new cells, depends on plant species (evolutionary or environmental changes in structure) and alteration in its organization, e.g., haustorial roots of parasitic plants are frequently observed. Haustorium penetrates the host tissues to absorb nutrients and water (De Tullio et al. 2010). In plant ontogeny, during the process of seed germination, elongation growth of embryonic root is of great importance, linked to breakage of seed coats. This process also terminates seed *sensu stricto* germination and initiates the next phase—development of seedling (Gniazdowska et al. 2010a, b). Root as a soil-localized fast-growing organ of the entire plant body is responsible for water and nutrient uptake (related to development and growth of lateral roots and root hairs) as well as recognition and response to neighbors (other plants or microorganisms) or the environment (abiotic factors of positive or negative character) (De Tullio et al. 2010; Roy and Bassham 2014). A single root is not sufficient to ensure the survival of the entire plant organism; therefore, one plant produces many roots, able to recognize themselves (self-recognition and communication between roots) and able to grow in different directions to accomplish life strategy. As roots of individual plant and/or roots of many plants share the same soil territory, they developed a course of action to produce, secrete, and respond to chemical compounds (signaling molecules) widespread in the environment (plant–plant communication and communication between plants and other organisms). The response and acclimation/adaptation of roots to environmental factors (concentration of nutrients) determine architecture of this organ. Availability of nutrients concentration of nitrogen and/or phosphorous, strongly affects root growth and development, related particularly to auxin regulation (Correa-Aragunde et al. 2004; Sun et al. 2014). High concentration of nitrate ions affects the nitrate sensor—NITRATE TRANSPORTER1·1 (NTR1·1) protein-dependent auxin transport. As a result, higher auxin accumulation in lateral roots initiates and stimulates their growth (Kazan 2013). Moreover, nutrients, nitrogen or phosphorous, are involved in strigolactone production and exudation from the roots, acting as a part of the nutrient acquisition strategy also linked to the modulation of auxin transport (Ruyter-Spira et al. 2011; Sun et al. 2014). In this context, diverse environmental factors bring into alterations in root architecture length of the primary root, density of lateral roots, and root branching.

Motion sensing of roots is regulated by phototropism (negative) and gravitropism (positive). In primary roots, commonly orthogravitropism is observed.

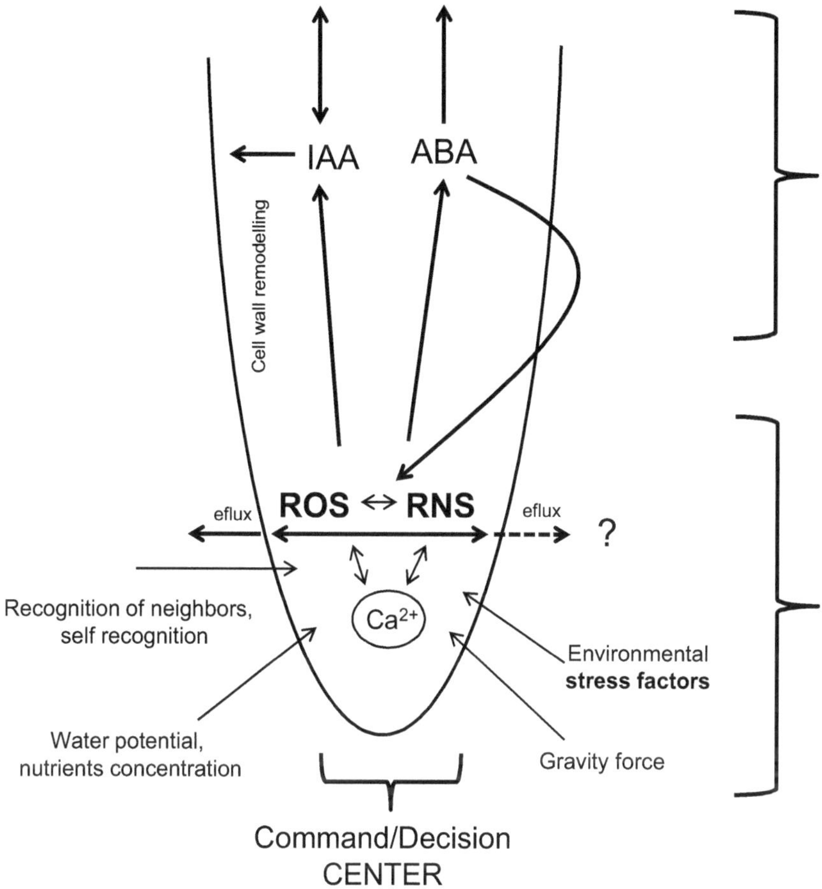

Fig. 1 Schematic view of the root apex. Complex action of ROS and RNS with auxin and Ca^{2+} indicates dependence of root growth on balance between ROS/RNS formation, scavenging, and efflux. Arrows indicate importance of direct and indirect action of ROS/RNS in processes of cell differentiation and elongation in reaction to environmental stimuli

Lateral roots, during maturation process, exhibit temporal variations in the direction of growth. Reorientation of the course of organ growth toward the gravity vector is described as plagiogravitropism (Kuya and Sato 2011). Gravitropic movements of roots strongly impact root architecture, as is associated with asymmetric growth of cells located on opposite sides of the organ. This response starts from seed germination which provides proper orientation of the young plant in the environment. Gravity signal initiates characteristic curvature at the distal elongation zone of the primary root as was shown for germinating seeds of grass pea (*Lathyrus sativus*) (Jiang et al. 2012). Gravitropic response is linked to a root cap, especially to the columella cells, where sedimentation of amyloplasts (gravity susceptors) depends on the gravity vector (Kuya and Sato 2011). Gravitropism is connected with the auxin polar (cell-to-cell) transport. This process is mediated by different auxin carriers: influx carriers AUXIN RESISTANT1 (AUX1) and LIKE AUX1 (LAX1) and efflux carriers PIN-FORMED (PIN) and ATP-binding cassette type B (ABCB) (Kazan 2013). Among them, especially PINs (particularly PIN2) determine root gravitropic movements. Arabidopsis (*Arabidopsis thaliana*) mutants lacking PIN2 have roots that poorly recognize gravity force and are more sensitive to different environmental stress conditions (Baluška et al. 2010). Moreover, modification of PIN2 content allows roots to avoid stress factors, e.g., salinity. In cells of salt-stressed roots, PIN2 selective degradation correlating to root growth alteration was observed (Li and Zhang 2008). PIN phosphorylation is an important modification that regulates membrane binding. Phosphorylation state is modified by accumulation of reactive oxygen species (ROS) (Bartoli et al. 2013). Root apices are characterized by the presence of two pathways of the polar auxin streams, symmetry of which governs the root growth direction. Any terrestrial factor (gravity, light, humidity) influences this symmetry and forces root apices to initiate the response (tropism) (Baluška et al. 2010). In natural conditions, roots are not exposed to light, as they grow in darkness or shadow/twilight maintained by soil layers. Light exposure alters auxin transport and stimulates reaction of light avoidance called as light-escape tropism (Baluška et al. 2010). Roots are also able to sense (and respond) mechanical stimuli (thigmotropism) (Roy and Bassham 2014). Root movement (reaction) is under auxin control (with indole-3-acetic acid (IAA) accepted as the most important metabolic active form). Concentration of this hormone and its undisturbed polar transport regulate various tropisms as described in detail for Arabidopsis plants (Rashotte et al. 2000; Baluška et al. 2010). Auxins are shoot synthesized, then transported to the root tips (root caps), and later relocated laterally. This IAA stream plays a potent role in root development and branching (Ruyter-Spira et al. 2011). Local auxin's gradient in the root elongation zone is mediated by acropetal transport, but accompanied also by basipetal one. Authors of the idea of "plant neurobiology" proposed that auxins could be considered as plant neurotransmitters (Baluška et al. 2010; Ruyter-Spira et al. 2011).

Root growth is under the control of various signaling molecules (ROS, reactive nitrogen species (RNS), Ca^{2+}), mode of action of which will be explained in more details in the next part of this chapter. Besides auxins, other phytohormones and growth regulators (such as ethylene, abscisic acid (ABA), brassinosteroids, and polyamines) affect growth and physiology of roots (Baluška et al. 2010; Yu et al. 2014). These organs are also liable for perception of variations in local water potential (hydrotropism) although hydrotropic movement is independent of gravitropism (Roy and Bassham 2014). Moreover, temporal water deficit in soil impacts root-to-shoot signaling. Information on water stress is mediated by ABA, phytohormone influencing also root growth and root hydrotropic movement (Roy and Bassham 2014; Tracy et al. 2015). ABA is commonly regarded as inhibitory and a stress-related regulator. The mode of action of ABA, including cross talk with ROS and RNS, is described in detail in guard cells and the regulation of stomata movement in stress response. ABA stimulates activity of NADPH oxidase (known also as respiratory burst oxidase homologues—Rbohs), resulting in activation of

Ca^{2+}-permeable channels, and leads to modification in Ca^{2+} concentration (Bartoli et al. 2013). Nevertheless, during stress conditions (water deficiency), ABA acts as stimulator of root growth (Zhang et al. 2014). This stimulatory effect provides a stress avoidance strategy of root growth. Root and nodule development in *Medicago truncatula* is regulated by a nitrate transporter encoded by the *LATERAL ROOT ORGAN DEFECTIVE/NUMEROUS INFECTIONS AND POLYPHENOLICS* (*LATD/NIP*) gene. Mutation of this gene (*latd*) led to disruption in elongation of primary and lateral roots and formation of nodules (Liang et al. 2007). ABA application preserved viability of *latd* apical root meristem (Liang et al. 2007). ABA (10 μM) treatment of mutants lacking *MtLATD/NIP* resulted in declined *RbohC* expression (Zhang et al. 2014) and modulated expression of *Cu/Zn SOD*. Thus, authors proposed that in the regulation of root elongation, ROS may act downstream of ABA signaling. Experiments carried out on wild-type and ABA-deficient mutant of tomato (*Solanum lycopersicum*) confirmed importance of ABA (at physiological level) in root growth and movements. ABA influenced the number of lateral roots providing soil exploration, especially in compacted soil (Tracy et al. 2015). Involvement of ABA in the regulation of redox potential is connected with ABSCISIC ACID INSENSITIVE4 (ABI4) transcriptional factor required for ascorbate-dependent regulation of plant growth (Kerchev et al. 2011). Modification of the redox potential of Arabidopsis ascorbate *vtc1* and *vtc2* mutants led to alteration of gene expression comparable to *abi4* mutants (Kerchev et al. 2011).

Among plant growth regulators, polyamines (PAs) are known as moderators of diverse physiological processes. PAs are polycationic, low molecular weight compounds with positive charge. Due to their structure, they can bind to the biomacromolecules such as proteins, lipids, or nucleic acids. PAs participate in the regulation of gene expression and stabilization (or destabilization) of proteins and membrane structures and are crucial for acclimation to abiotic stresses (Kusano et al. 2008; Tisi et al. 2011). An indirect impact of spermine (Spm) on root growth occurs *via* products of its catabolism. Polyamine oxidases (PAOs), enzymes catalyzing Spm degradation, produce hydrogen peroxide (H_2O_2), as a coproduct of the reaction. H_2O_2 as other ROS is known to contribute to the regulation of the cell wall integrity (Kusano et al. 2008). However, Spm at mM concentration inhibited root growth of Arabidopsis seedlings. Spm is covalently bound to the hydroxycinnamic acid often conjugated with components of the cell wall (hemicelluloses) and acting in lignin–polysaccharide cross-linking in the wall, leading finally to stiffening of the cell wall. Also, overaccumulation of spermidine (Spd) resulted in the restriction of root growth (Tassoni et al. 2000). In maize seedling treated with Spd, inhibition of elongation growth of cells of primary roots and enhanced accumulation of phenolic compounds was shown (Tisi et al. 2011). Development of a primary root is connected with differentiation of the xylem depending on H_2O_2 concentration. This ROS compound regulates also cell wall maturation observed as generation of cross-linking of polysaccharides and proteins. Enhanced cell wall maturation in primary roots of maize seedlings was accompanied by increase of amine oxidase activity (Tisi et al. 2011).

As newly identified phytohormones, strigolactones (SLs) or their derivatives belong to the family of plant regulators participating in the regulation of root growth (Umehara et al. 2008; Sun et al. 2014). Different environmental and developmental stimuli lead to SLs synthesis in roots, resulting in alterations in root branching and development (Rasmussen et al. 2012). There is also evidence for SL–auxin cross talk. In rice (*Oryza sativa*), SLs produced in Arabidopsis affected root growth by decreasing auxin transport from shoots (Ruyter-Spira et al. 2011; Sun et al. 2014). SLs took part in plant nutrient acquisition strategy, detected in soil deficient in phosphate and nitrate ions (Sun et al. 2014). It has been shown that in roots of Arabidopsis, SLs' mode of action was linked to modification of ascorbate and glutathione concentration, influencing cellular redox state. In contrast, SLs had no effect on the activity of antioxidant enzymes such as catalase (CAT) or superoxide dismutase (SOD) (Bartoli et al. 2013). As was mentioned above, root apical meristem can be evolutionarily modified depending on a plant life strategy. In some parasitic plants, induction of haustorium formation depends on SLs presence (De Tullio et al. 2010).

Ca^{2+} plays not only nutritional but also signaling role as a second messenger in signal transduction pathways (Xiong et al. 2006). It has been indicated that Ca^{2+} wave is part of the systemic communication (from one cell or tissue within the entire plant). Thus, Ca^{2+} wave is spread through the entire plant when local environmental factors are recognized by a root tip (signaling recognition) (Gilroy et al. 2014). Development, growth, and plant responses to various stimuli depend on alteration of Ca^{2+} concentration in cellular compartments (so-called Ca^{2+} signature of the stimulus). Cell wall stability also requires the presence of Ca^{2+}, and gravitropic bending is regulated by redistribution of Ca^{2+} (Xiong et al. 2006; Gilroy et al. 2014). Activity of many proteins depends on Ca^{2+} binding ability, e.g., NADPH oxidase, that is the example of protein–ROS–Ca^{2+} interaction. Moreover, Ca^{2+}-dependent protein kinase 5 (CPK5) plays a beneficial role in the regulation of ROS wave (Causin et al. 2012; Gilroy et al. 2014). On the other hand, ROS impact on Ca^{2+} signaling in root growth is observed. It was reported that radical forms of ROS rather than H_2O_2 activated Ca^{2+}-permeable channels (Foreman et al. 2003).

Root organogenesis is connected with cell division and differentiation and is originated from a stem cell center at the tip. Differentiation of the root cells is initiated by the transition from cellular proliferation to elongation. After elongation, cells differentiate into various cell types in the maturation zone. However, the final volume/dimension of the root meristem depends on ROS availability (Tsukagoshi et al. 2010). The shape of the root differs in mono- and dicotyledon plants. In Arabidopsis, belonging to the dicot plants, root system consists of a primary root and lateral roots. Monocots, e.g., maize (*Zea mays*) or rice roots, have a specific shape, recognized as shoot-borne and lateral roots (Tian et al. 2014). Lateral root primordium is formed from cells of the root pericycle that pass differentiation (Correa-Aragunde et al. 2004). It is suggested that the role of administrative center in roots is linked to the part of the root apex which is defined as the transition zone (TZ). Cells of this zone are characterized by the highest rate of auxin transport. This allows integration of the sensory–motoric pathways necessary for various root tropisms (Baluška et al. 2010).

2 Reactive Oxygen Species in Root Responses

Aerobic conditions and existence of O_2 molecule as a component of metabolic pathways, especially electron transport chains, are associated with activation of oxygen atoms and generation of ROS (Demidchik 2015). Singlet oxygen ($^1O_2$), hydroxyl radical ($^{\bullet}OH$), superoxide radical ($O_2^{\bullet -}$), and H_2O_2 are most frequently studied ROS. Other ROS involved in alterations of the redox potential are peroxyl, alkoxy, and hydroperoxyl radicals. Strong oxidative activity is attributed to peroxynitrite ($ONOO^-$), the molecule belonging to the RNS family. This compound is generated by the reaction of nitric oxide ($^{\bullet}NO$) and $O_2^{\bullet -}$ (Yu et al. 2014; Demidchik 2015; Corpas and Barroso 2014).

It is commonly accepted that ROS are molecules of bimodal function. ROS generated at high levels disturb cell metabolism. Toxicity of ROS relates to their ability to react with a high range of biological molecules: nucleic acids, proteins (including enzymes and transcription factors), lipids, and sugars. Uncontrolled oxidation of cell components leads to oxidative stress. Such pathophysiological conditions accompanied by low activity of antioxidant system, in the long run, cause the death of cells. On the other hand, low concentration of ROS is necessary to maintain metabolism at the accurate level. ROS are known as signaling molecules involved in all physiological processes including immune responses, stomata movements, seed germination, seedling growth, root gravitropic response, and control of root and root hair growth (Foreman et al. 2003; Liszkay et al. 2004; Krasuska and Gniazdowska 2012; Demidchik 2015). ROS induce posttranslational modifications (PTM) of protein structure that could be regarded both as positive and negative. Some of ROS-modified (carbonylated) proteins (or peptides) probably function as organellum-specific signals (inter-organellar communication) (Møller and Sweetlove 2010). These peptides acting as messengers from mitochondria could participate in the control of root growth (Causin et al. 2012). These findings strongly highlight that the entire cell metabolism depends on ROS concentration which must be kept at the optimal hormetic window.

At physiological conditions, ROS homeostasis is maintained by the presence and activity of an antioxidant system: enzymatic and nonenzymatic. Enzymes involved in ROS scavenging (modulation) are various isoforms of superoxide dismutase (SOD); catalase (CAT); peroxidases (POx), especially glutathione peroxidase (GPX) and ascorbate peroxidase (APX); and thioredoxins, glutaredoxins, and peroxiredoxins (Krasuska and Gniazdowska 2012; Demidchik 2015). Nonenzymatic antioxidant system involves reduced form of ascorbic acid (ASA) and glutathione (GSH). PAs, proline, betaine, carotenoides, and α-tocopherol are also included in the group of ROS scavengers (Krasuska and Gniazdowska 2012; Demidchik 2015). GSH is not only an antioxidant but also plays an important role in protein PTM by glutathionylation, resulting in modification of signaling transduction by alteration of MAP kinase activity or influencing Ca^{2+} concentration/waving (Bartoli et al. 2013). Elongation of root cells is under the control of ASA and dehydroascorbic acid (DHA). Gravitropic response of Arabidopsis roots depends on oxidation of ASA that leads to the formation of DHA in the reaction catalyzed by ASA oxidase (Lee et al. 2011). Moreover, the gravistimulation enhanced the expression of gene coding ASA oxidase (*AAO1*). Expression of this gene was stimulated by application of brassinosteroids and correlated to growth expansion (Lee et al. 2011).

ROS are molecules of a short half-life. It is well documented that various ROS have diverse physiological properties, for example, $O_2^{\bullet -}$ by itself is not able to modify macromolecules, but reacting with H^+ forms highly reactive hydroperoxyl radical ($HO_2^{\bullet -}$). Both compounds in reaction of dismutation generate H_2O_2, molecule of relatively long half-life. Also, mobility of H_2O_2, which is able to pass through membranes *via* aquaporins, makes this compound a good candidate for systemic signaling molecule (Mori and Schroeder 2004). H_2O_2 at low level (below 1 μM) participates in the creation of balanced physiological metabolic state of the cells. On the other hand, at high (mM) level H_2O_2 initiates and points out on oxidative stress acting as one of the markers of oxidative stress (Demidchik 2015). Apoplastic space (cell wall surrounding) is one of the cellular sites of H_2O_2 generation. H_2O_2 can be synthesized also by reactions depending on the activity of NADPH oxidases, extracellular heme-containing Class III peroxidases (POx), PAO, and oxalate oxidase (Tsukagoshi et al. 2010; Demidchik, 2015). In barley (*Hordeum vulgare*) roots, POx were indicated as essential components for initiation of root hair formation (Kwasniewski et al. 2013). Moreover, it was proposed that root hair-specific POx and root hair-specific NADPH oxidase work in a team and are necessary to provide production of $^{\bullet}OH$ in the cell wall of trichoblast required for root hair formation. Treatment of barley roots with inhibitors of POx drastically decreased root ability for the formation of root hairs (Kwasniewski et al. 2013). Moreover, negative impact of H_2O_2 on root hair formation and growth was shown for willow (*Salix nigra*) seedlings; 72 h of H_2O_2 (1 mM) treatment led to complete blockage of root hair formation (Causin et al. 2012). Treatment of Arabidopsis seedlings with H_2O_2 (0.1 mM) for 24 h resulted in decline of root growth and shortened root meristem (Tsukagoshi et al. 2010). Addition of H_2O_2 (30 or 40 mM) into the germination medium of grass pea seeds resulted in further disturbance in gravitropic response of the primary root (Jiang et al. 2012). Prolonged exposure of these roots to H_2O_2 led to root curling, maintenance of horizontal curvature, and finally blocked growth. Removal of ROS donor from root medium resulted after 24 h in the recovery of growth consistent with the action of the gravity vector. Thus, it can be assumed that exogenous H_2O_2 negatively but temporarily and reversibly controls primary root growth (Jiang et al. 2012). Similarly, inhibition of embryonic root growth was observed in apple (*Malus domestica*) seedlings developed from embryos shortly (3 h) pretreated with H_2O_2 (1 mM) just after seed coat removal (Gniazdowska et al. 2010b). Jiang et al. (2012) suggested that restriction in growth and root horizontal curvature is rather a response to exogenous H_2O_2 (at high level) application than the result of excessive accumulation of this compound. Gravitropic bending of the roots is dependent on different endogenous H_2O_2 concentrations in the lower and upper cortex of roots (Joo et al. 2001). But it should be mentioned that the horizontal curvature of primary root growth depends on the plant species and was observed in faba bean (*Vicia faba*), pea (*Pisum sativum*), and dwarf chickling (*Lathyrus cicera*) but not in *Arabidopsis*, soybean (*Glycine max*), rice, maize, tobacco (*Nicotiana tabacum*) (Jiang et al. 2012). Inhibition of root growth of maize and tomato caused by cyanamide—allelopathic compound produced by hairy vetch (*Vicia villosa* Roth.)—was due to overaccumulation of H_2O_2 (observed particularly in the apical region) inducing earlier differentiation of cells and restriction in cells' division and elongation (Soltys et al. 2012, 2014).

Radical forms of ROS react with polysaccharides (including pectin and hemicelluloses) of the cell wall, resulting in loosening of their structure (Carol and Dolan 2006; Müller et al. 2009). Thus, root growth is associated with elongation of cells, as the consequence of initially loosening and finally stiffening of the cell wall. H_2O_2 acts as an important stiffening agent which leads to cross-linking of cell wall polymers (Müller et al. 2009). Cell wall loosening enables radicle protrusion during seed germination and was deeply examined in cress (*Lepidium sativum*). Incubation of cress seeds with H_2O_2 inhibited endosperm rupture (Müller et al. 2009). As was mentioned above, one of the important apoplastic sources of H_2O_2 is catabolism of PAs mediated by the activity of amine oxidases. Among them, copper-containing amine oxidases (DAO, CuAO), oxidizing diamines, flavin-containing PAO, and oxidizing Spm and Spd should be distinguished. High level of H_2O_2 produced by enhancement of DAO or PAO activity correlates with cell wall stiffening and lignification (Delis et al. 2006; Kusano et al. 2008; Tisi et al. 2011). Putrescine (diamine) catabolism mediated by DAO seems to be involved in cell expansion as was proposed for soybean seedlings. These findings are supported by organ-specific gene (coding DAO) localization, mainly in regions of cell elongation in primary and secondary roots. The highest level of transcripts of *GmCuAO1* was identified in the root tips, the root hair region, and the lateral root emergence region (Delis et al. 2006).

It was reported in maize, cucumber (*Cucumis sativus*), or Arabidopsis that the root apex is rich in radical forms of ROS (Liszkay et al. 2004; Renew et al. 2005; Dunand et al. 2007). The growing zone (narrow zone behind the apical meristem) of primary maize roots was identified as the region of $O_2^{\bullet-}$ production. Generation of this molecule is localized in the epidermis and vascular tissues of the growing zone. These findings come from experiment conducted on Arabidopsis, tomato, and soybean and indicated that the apoplastic ROS production in the growing zone of the roots is a common feature of seed plants (Liszkay et al. 2004). In addition, restriction of root elongation correlating with decline in $O_2^{\bullet-}$ concentration was observed as a result of application of IAA (Liszkay et al. 2004).

NADPH oxidases lead to increased production of oxygen free radicals. The catalytic domain of plant Rbohs shows homology to the mammalian phagocyte respiratory burst oxidase subunit gp91^{phox} (Müller et al. 2012). In Arabidopsis, an amount of ten *Rbohs* (*AtrbohA–AtrbohJ*) was identified. Physiological importance of this enzyme was shown in *Rboh* knockout mutants characterized by reduction of their size (Jones et al. 2007). Rbohs seem to participate not only in seed germination (Müller et al. 2009) but also in regulation of growth of cells of root tips (Foreman et al. 2003), mechanosensing (Monshausen et al. 2009), Casparian strip formation (Lee et al. 2013), and root response to phytotoxins/allelochemicals (Oracz et al. 2012). ROS action in allelopathic and phytotoxic interaction (defined as root–root communication) was recently described in detail (Gniazdowska et al. 2015). The importance of Rboh in the regulation of root growth in cress seedling was demonstrated using transgenic plants with reduced expression of *Rboh* genes. Roots in mutants were shorter compared to the wild-type plants (Müller et al. 2012). Initially, roots had thickened root tips, and then abnormal (irregular) formation of lateral root was observed. Moreover, these malformations were also characteristic of auxin mutants (e.g., auxin overproducer's phenotypes). Based on these results, authors suggested that there is a link between Rboh activity and auxin signaling. Morphologically intact architecture of roots and their proper function are associated with the ability of plant for precise control of ROS generation and maintenance of redox state in specific roots' zones. This is in agreement with findings in Arabidopsis that *RbohB* is specifically expressed in the root tip below the elongation zone (Vernoux et al. 2000; Müller et al. 2012). Activity of NADPH oxidases is also regulated by ROP (Rho of plant) GTPases (Jones et al. 2007). This particular ROS modulation is linked to the stimulation of the Ca^{2+} influx via plasma membrane hyperpolarization-activated Ca^{2+} channels (HACC). A high intracellular Ca^{2+} gradient is necessary for growth of root hair tips. Elongation of root hair is also connected to ROS formation *via* activity of ROP GTPase (Jones et al. 2007). The differences in Ca^{2+} gradients play a role not only in the control of growth of root hairs but also in the growth of the whole root system. Importance of Ca^{2+} in root growth was proved by application of the Ca^{2+} channel blocker—lithium ($LaCl_3$) (Demidchik et al. 2003; Foreman et al. 2003). Moreover, it was demonstrated that NADPH oxidases also undergo *S*-nitrosylation (reversible PTM) in the presence of nitric oxide (NO) (Yun et al. 2011). This finding points out the strong correlation of ROS and RNS in the growth process. ROS participate in the regulation of Ca^{2+} concentration (by indirect creation of Ca^{2+} wave) and also stimulate mitogen-activated protein kinase (MAPK) cascades (Pitzschke and Hirt 2006).

Fenton reaction depending on the simultaneous presence of the reductant, transition metal ions, and H_2O_2 is the source of $^{\bullet}OH$ in the apoplastic space. $^{\bullet}OH$ is also generated in peroxidase-mediated Haber–Weiss reactions (Liszkay et al. 2003; Müller et al. 2009; Demidchik, 2015). Significance of $^{\bullet}OH$ in the weakening of endosperm and finally growth of embryo axis in germinating cress seed was proven using a method corresponding to puncture force measurement. It is not the random production and high reactivity of $^{\bullet}OH$ with cell wall polysaccharides that suggest its specific action in apoplastic space. Reactivity of this compound is temporal (of narrow time window) and depends on tissue, developmental progress, and ABA concentration, which inhibits $^{\bullet}OH$ production (Müller et al. 2009). On the other hand, growth of roots of willow seedlings was associated with $O_2^{\bullet -}$ rather than $^{\bullet}OH$. These findings were confirmed by experiment with ROS scavengers. Inhibition of root growth of willow was observed after application of $O_2^{\bullet -}$ scavenger, while scavenging of $^{\bullet}OH$ had no effect (Causin et al. 2012). The authors postulated that generation of $O_2^{\bullet -}$ in the meristematic root zone is probably related to some other physiological processes including growth.

Transcription factor UPBEAT1 (UPB1), repressing a set of POx and modulating ROS concentration and also H_2O_2/O_2^- ratio, affects the size of the root meristem (Tsukagoshi et al. 2010). Thus, differentiation and proliferation of root cells depends on accumulation of specific ROS. Increase in H_2O_2 level is required for cell differentiation, whereas high $O_2^{\bullet -}$ concentration is necessary for cell proliferation. In consequence, cell elongation in the root elongation zone occurs after $O_2^{\bullet -}/H_2O_2$ ratio reaches a certain level. In the elongation zone, UPB1 repressed POx expression. Moreover, *UPB1* expression is controlled by H_2O_2 *via* feedback loop. It was shown that H_2O_2 application caused upregulation of *UBP1* expression (Tsukagoshi et al. 2010). It was also demonstrated that one of the direct targets of UPB1 was the gene coding thioredoxin reductase (*NTRA*). The involvement of UPB1 in lignin synthesis, thus in cell wall modification, is also suggested.

Cellular redox potential impacts development of the root apical meristem. It depends on homeostasis of ASC/DHA and GSH/GSSG (oxidized form of GSH) (De Tullio et al. 2010). The balance between reduced/oxidized states is mediated by ROS modulating system. It is commonly accepted that both ASA and GSH regulatory roles in meristem development refer to the impact of these cellular antioxidants on mitotic activity (Sanchez-Fernandez et al. 1997). As was mentioned above, ASA content regulates cell cycle progression. Modification in ASA/DHA ratio is central in root gravitropic response that correlates to ROS level (Lee et al. 2011). Cellular redox homeostasis is controlled by thioredoxins. In root meristem, NTR impacts POx activity and thus the transition from cell proliferation to cell differentiation (Tsukagoshi et al. 2010).

There are suggestions of ROS involvement in auxin metabolism. ROS accumulation influences auxin response probably due to its conjugation or catabolism to inactive products (Peer et al. 2013).

A link of H_2O_2 to hormone signaling was also reported for ethylene (ET). This phytohormone is known as root growth regulator acting by protein kinases (MAPK3 and MAPK6) which are also elements of ROS transduction cascades (Bartoli et al. 2013). It is well known that this gaseous hormone modifies root architecture. Characteristic growth responses of Arabidopsis roots to ET are the enhancement of the thickness of the organ and reduction of cell elongation (Huang et al. 2013). Mung bean (*Vigna radiata*) seedlings treated with ET donor showed shortened length of primary roots and lowered number of lateral roots. These results were correlated with increased lignin content and increased POx activity (enhanced H_2O_2 level) (Huang et al. 2013). ET precursor 1-aminocyclopropane-1-carboxylic acid (ACC) acts synergistically to auxin in inhibition of cell expansion in the elongation zone of roots (Muday et al. 2012). Any modification of ET concentration in root tips affects auxin transport and signaling. However, in lateral root initiation, antagonistic action of auxin and ET is observed. Also, this hormone negatively impacts gravitropic response (Muday et al. 2012). There is evidence that increased ET nonenzymatic formation by oxidation of ACC can be achieved in conditions of enhanced ROS (high H_2O_2 concentration) (Gniazdowska et al. 2010a). Inhibition of elongation of roots of apple seedling after embryo treatment with H_2O_2 could be due to elevated ET production. Also, increase of ET emission from seedlings was observed after NO treatment, suggesting contribution of RNS in biosynthetic pathway of this hormone.

3 Reactive Nitrogen Species Contribution in Root Responses

RNS, in addition to ROS, are considered as important morphogens. Nitric oxide (NO) at different redox states modulates various physiological and morphological processes including also root growth and development or molding of root architecture. NO is generated via enzymatic or nonenzymatic pathways and at low concentration acts as a signaling molecule (Yu et al. 2014). This gaseous compound takes part in signal transduction cascades as activator of cyclic guanosine 3′,5″-monophosphate (cGMP) synthesis, modulator of Ca^{2+} signaling, and activator of protein kinase cascades and participates in posttranslational modification of proteins (e.g., *S*-nitrosylation and/or nitration) (Pagnussat et al. 2004; Kasprowicz et al. 2009; Yu et al. 2014). As was recently presented for roots of pepper (*Capsicum annuum*) seedlings, their growth and development corresponded to the decrease of the level of nitrated proteins (Airaki et al. 2015). Important modulators of developmental processes are also fatty acid amides, e.g., alkamides (Méndez-Bravo et al. 2010). It was shown that NO treatment of *Arabidopsis* seedlings mimics variation of root system architecture typical to alkamides. In addition, NO acts as a second messenger in signaling pathway initiated by alkamides (Méndez-Bravo et al. 2010). NO affects actin cytoskeleton and thus can influence root architecture (Kasprowicz et al. 2009). The experiments carried out on maize root apices indicated a strong, but reversible, effect of different NO donors on the actin cytoskeleton assembly and organization (Kasprowicz et al. 2009). NO treatment influenced endocytosis, formation of endocytic vesicle, and trafficking in cortex cells of the transition zone of the maize root apex. Moreover, NO affected membrane trafficking and recycling of cell wall polysaccharides. NO synthesized in the pericycle cells (during lateral root formation), similarly to auxin-induced transcription of genes coding cyclins *CYCD3;1* and *CYCA2;1* responsible for G1-to-S phase transition (Correa-Aragunde et al. 2006). It is also known that NO can act as antioxidant by prevention of $^{\bullet}OH$ generation or by affecting activity of antioxidant enzymes (Tewari et al. 2008; Krasuska and Gniazdowska 2012). As observed in mountain ginseng (*Panax ginseng*), NO application during adventitious root formation enhanced activity of SOD, CAT, POx, APX, and glutathione reductase (GR). In the contrary, application of NO stimulated activity of NADPH oxidase and increased $O_2^{\bullet-}$ level (Tewari et al. 2008). However, NG-Nitro-L-arginine methyl ester (L-NAME, an inhibitor of mammalian NO synthase) did not alter $O_2^{\bullet-}$ level and did not affect the number of proliferated ginseng rootlets (Tewari et al. 2008) that points out NO origin in this process. On the other hand, in the literature, it can be found that *S*-nitrosylation of NADPH oxidase due to high concentration of *S*-nitrosothiols is coupled rather to inhibition of enzyme activity. This inhibition was observed in plant immunology (Yu et al. 2014). These findings indicate the regulatory role of NO, which modulates concentration of specific ROS in particular root regions. Impact of NO on ROS metabolism was analyzed also in roots of young apple seedlings, wherein stimulation of CAT, POx, and GPX activity after NO treatment was observed (Krasuska and Gniazdowska 2012). APX utilizing ASA as electron donor is involved in ROS scavenging and maintenance of redox homeostasis. Cytosolic isoform of APX (cAPX) was shown to be modified by ROS and NO; it underwent carbonylation, nitration, or *S*-nitrosylation (Begara-Morales et al. 2014). Although carbonylation and nitration which are irreversible inhibited cAPX activity, it was reported that *S*-nitrosylation prevented carbonylation of cAPX and even stimulated its activity (Correa-Aragunde et al. 2015).

With increasing amount of data focusing on NO action in plant morphogenesis, much more attention was focused on its interaction with auxins. It was shown that during adventitious root, lateral root, and root hair formation, auxin treatment led to NO accumulation (Pagnussat et al. 2002; Correa-Aragunde et al. 2004; Lombardo et al. 2006; Terrile et al. 2012). Moreover, conversion of indole-3-butyric acid (IBA), localized in peroxisomes, into IAA was followed by NO formation, an important process of IBA-induced lateral root formation (Schlicht et al. 2013). In soybean root tips, gravitropic bending mediated by NO was due to its asymmetric localization similarly as noted for auxins, accompanied by enlargement of cGMP (Hu et al. 2005). NO as the "molecule of the 1992 year" was shown to participate in the initiation of trichoblast differentiation of developing root hairs of lettuce (*Lactuca sativa*) and Arabidopsis seedlings. NO application increased number of root hairs and promoted their elongation (Lombardo et al. 2006). Similarly as observed for other plant growth regulators or signaling compounds, NO concentration has to fit in a range of hormetic window, characteristic for the maintenance of the specific physiological process (Monzón et al. 2014). Application of NO scavengers, including 2-(4-carboxyphenyl)-4,4,5,5-tetramethylimidazoline-1-oxyl-3 oxide (cPTIO), unblocked inhibition of primary root growth and inhibited lateral root development (Correa-Aragunde et al. 2004; Monzón et al. 2014). Reduction of the number of lateral roots in sunflower (*Helianthus annuus*) seedlings after cPTIO application was accompanied by upregulation of POx, cytochrome P450, or GR gene expression (Monzón et al. 2014). NO takes part in lateral root development but rather negatively impacts their length. Application of cPTIO resulted in enhancement of lateral root length by 95 % compared to the control (Monzón et al. 2014). This data point out NO–ROS cross talk in the design of root architecture.

Involvement of NO in mediation of the randomization of cortical microtubules that impacted cell polarity during the morphogenic events is strongly suggested. In Arabidopsis roots, NO altered auxin interaction with the TRANSPORT INHIBITOR RESPONSE1/AUXIN SIGNALING F-BOX (TIR1/AFB) proteins (auxin receptors) and AUX/IAA proteins (transcriptional repressors) that regulated expression of auxin-induced genes. This NO-related modulation referred to *S*-nitrosylation of TIR1 protein at two specific Cys residues, probably promoting its interaction to auxin/IAA protein (Terrile et al. 2012). PIN1-dependent acropetal auxin transport was also considered as the process controlled by NO (Fernández-Marcos et al. 2011). Elevated NO concentration resulted in reduction of PIN1 protein level. Moreover, it is believed that PIN1 loss was connected with PTM (Fernández-Marcos et al. 2011).

As was mentioned before, Class III POx activity is linked to root shaping. NO application altered expression of genes coding several extracellular POx involved in lignin polymerization, as well as some other genes involved in lignin biosynthesis (Monzón et al. 2014). Moreover, NO donors can inhibit POx activity (Krasuska and Gniazdowska 2012; Monzón et al. 2014). In sunflower roots, it was proposed that NO is a transcriptional regulator of the lignin biosynthesis (Monzón et al. 2014). Modification of lignification (lignin content and activity of related enzymes) after NO treatment was also demonstrated for soybean seedlings (Böhm et al. 2010).

Recently, NO action in root development was confirmed using NO-deficient Arabidopsis mutants. Reduced length of the primary root and shortened meristem with abnormal pattern of cell division were characteristic features of NO-deficient mutants (*nitric oxide associated1* [*noa1*], *nitrate reductase1* [*nia1*], and *nia2*) (Sanz et al. 2014). In Arabidopsis, accumulation of ROS and flavonol, mostly quercetin, enhanced by NO-deficient mutants in root region of developmental transition was reported (Sanz et al. 2014). These findings strongly indicate RNS–ROS cross talk in the mode of action of root response to different (positive or negative) environmental stimuli.

4 Conclusions

Plants, as other organisms, are able to modify their metabolism and development according to the conditions of the environment, including also stimuli of negative or positive nature. Believable examples of adaptation of plants modifying their movement, size, and growth rate in the presence of various stress factors by activation of complicated signaling network led to the rise of the concept of plant intelligence (Trewavas 2005). Perception of gravity, light, nutrient, and soil moisture allows to point on roots (particularly root tips) as space-extended guide center—"a brain" of the plant (Darwin's view of the plant body) (Baluška and Mancuso 2009). The ability of plants to integrate in real time the environmental data that are simultaneously provided and are of a variety of parameters and strength depends on root function and results in modification of root growth and architecture or movement toward or against the stressor. Such behavior of plant organism needs complex network of phytoneurotransmitters with auxins accepted as key messengers (Fig. 1). A detailed explanation of plant susceptibility implies the existence of short-life and membrane-permeable signaling molecules acting in orchestra with classical phytohormones. Thus, ROS and RNS occur as natural candidates; all the more, their mode of action requires oxidative window and "nitrosative door," respectively (Krasuska et al. 2015), or similarly to Ca^{2+} signature obliges formation of waves, special zones, or puffs, but this is another story.

Acknowledgments The manuscript was prepared during realization of grant no. 2014/13/B/NZ9/02074 financed by National Science Centre, Poland.

References

Airaki M, Leterrier M, Valderrama R, Chaki M, Begara-Morales JC, Barroso JB, del Rio LA, Palma JM, Corpas FJ (2015) Spatial and temporal regulation of the metabolism of reactive oxygen and nitrogen species during the early development of pepper (*Capsicum annuum*) seedlings. Ann Bot. doi:10.1093/aob/mcv023

Baluška F, Mancuso S, Volkmann D, Barlow PW (2010) Root apex transition zone: a signalling-response nexus in the root. Trends Plant Sci 15:402–408

Baluška F, Mancuso S (2009) Plant-environment interactions: from sensory plant biology to active plant. In: Baluska F, Vivanco J (eds) Signaling and communication in plants. Springer, Berlin

Bartoli CG, Casalongué CA, Simontacchi M, Marquez-Garcia B, Foyer CH (2013) Interactions between hormone and redox signaling pathways in the control of growth and cross tolerance to stress. Environ Exp Bot 94:73–88

Begara-Morales JC, Sanchez-Calvo B, Chaki M, Valderrama R, Mata-Pérez C, López-Jaramillo J, Padilla MN, Carreras A, Corpas FJ, Barroso JB (2014) Dual regulation of cytosolic ascorbate peroxidase (APX) by tyrosine nitration and S-nitrosylation. J Exp Bot 65:527–538

Böhm FMLZ, Ferrarese MLL, Zanardo DIL, Magalhaes JR, Ferrarese-Filho O (2010) Nitric oxide affecting root growth, lignification and related enzymes in soybean seedlings. Acta Physiol Planta 32:1039–1046

Carol RJ, Dolan L (2006) The role of reactive oxygen species in cell growth: lessons from root hairs. J Exp Bot 57:1829–1834

Causin HF, Roqueiro G, Petrillo E, Lainez V, Pena LB, Marchetti CF, Gallego SM, Maldonado SL (2012) The control of root growth by reactive oxygen species in *Salix nigra* Marsh. seedlings. Plant Sci 183:197–205

Corpas FJ, Barroso JB (2014) Peroxynitrite ($ONOO^-$) is endogenously produced in *Arabidopsis* peroxisomes and is overproduced under cadmium stress. Ann Bot 113:87–96

Correa-Aragunde N, Graziano M, Lamattina L (2004) Nitric oxide plays a central role in determining lateral root development in tomato. Planta 218:900–905

Correa-Aragunde N, Graziano M, Chevalier C, Lamattina L (2006) Nitric oxide modulates the expression of cell cycle regulatory genes during lateral root formation in tomato. J Exp Bot 57:581–588

Correa-Aragunde N, Foresi N, Lamattina L (2015) Nitric oxide is an ubiquitous signal for maintaining redox balance in plant cells: regulation of ascorbate peroxidase as a case study. J Exp Bot. doi:10.1093/jxb/erv073

Delis C, Dimou M, Flemetakis E, Aivalakis G, Katinakis P (2006) A root- and hypocotyl-specific gene coding for copper-containing amine oxidase is related to cell expansion in soybean seedlings. J Exp Bot 57:101–111

Demidchik V (2015) Mechanisms of oxidative stress in plants: from classical chemistry to cell biology. Environ Exp Bot 109:212–228

De Tullio M, Jiang K, Feldman L (2010) Redox regulation of root apical meristem organization: connecting root development to its environment. Plant Physiol Biochem 48:328–336

Dunand C, Crèvecoeur M, Penel C (2007) Distribution of superoxide and hydrogen peroxide in *Arabidopsis* root and their influence on root development: possible interaction with peroxidases. New Phytol 174:332–341

Fernández-Marcos M, Sanz L, Lewis DR, Muday GK, Lorenzo O (2011) Nitric oxide causes root apical meristem defects and growth inhibition while reducing PIN-FORMED 1 (PIN1)-dependent acropetal auxin transport. Proc Natl Acad Sci USA 108:18506–18511

Foreman J, Demidchik V, Bothwell JHF, Mylona P, Miedema H, Torres MA, Linstead P, Costa S, Brownlee C, Jones JD, Davies JM, Dolan L (2003) Reactive oxygen species produced by NADPH oxidase regulate plant cell growth. Nature 422:442–446

Gilroy S, Suzuki N, Miller G, Choi WG, Toyota M, Devireddy AR, Mittler R (2014) A tidal wave of signals: calcium and ROS at the forefront of rapid systemic signaling. Trends Plant Sci 19:623–630

Gniazdowska A, Krasuska U, Bogatek R (2010a) Dormancy removal in apple embryos by nitric oxide or cyanide involves modifications in ethylene biosynthetic pathway. Planta 232:1397–1407

Gniazdowska A, Krasuska U, Czajkowska K, Bogatek R (2010b) Nitric oxide, hydrogen cyanide and ethylene are required in the control of germination and undisturbed development of young apple seedlings. Plant Growth Regul 61:75–84

Gniazdowska A, Krasuska U, Andrzejczak O, Soltys D (2015) Allelopathic compounds as oxidative stress agents: YES or NO. In: Gupta KJ, Igamberdiev AU (eds) Reactive Oxygen and Nitrogen Species Signaling and Communication in Plants. Springer, Germany

Hu X, Neill SJ, Tang Z, Cai W (2005) Nitric oxide mediates gravitropic bending in soybean roots. Plant Physiol 137:663–670

Huang WN, Liu HK, Zhang HH, Chen Z, Guo YD, Kang YF (2013) Ethylene-induced changes in lignification and cell wall-degrading enzymes in the roots of mungbean (*Vigna radiata*) sprouts. Plant Physiol Biochem 73:412–419

Jiang J, Su M, Wang L, Jiao C, Sun Z, Cheng W, Li F, Wang C (2012) Exogenous hydrogen peroxide reversibly inhibits root gravitropism and induces horizontal curvature of primary root during grass pea germination. Plant Physiol Biochem 53:84–93

Jones MA, Raymond MJ, Yang Z, Smirnoff N (2007) NADPH oxidase-dependent reactive oxygen species formation required for root hair growth depends on ROP GTPase. J Exp Bot 58:1261–1270

Joo JH, Bae YS, Lee JS (2001) Role of auxin-induced reactive oxygen species in root gravitropism. Plant Physiol 126:1055–1060

Kasprowicz A, Szuba A, Volkmann D, Baluska F, Wojtaszek P (2009) Nitric oxide modulates dynamic actin cytoskeleton and vesicle trafficking in cell type-specific manner in root apices. J Exp Bot 60:1605–1617

Kazan K (2013) Auxin and the integration of environmental signals into plant root development. Ann Bot 112:1655–1665

Kerchev PI, Pellny TK, Vivancos PD, Kiddle G, Hedden P, Driscoll S, Vanacker H, Verrier P, Hancock RD, Foyer CH (2011) The transcription factor ABI-4 is required for the ascorbic acid-dependent regulation of growth and regulation of jasmonate-dependent defense signaling pathways in *Arabidopsis*. Plant Cell 23:3319–3334

Krasuska U, Gniazdowska A (2012) Nitric oxide and hydrogen cyanide as regulating factors of enzymatic antioxidant system in germinating apple embryos. Acta Physiol Planta 34:683–692

Krasuska U, Ciacka K, Andryka P, Bogatek R, Gniazdowska A (2015) "Nitrosative door" in seed dormancy alleviation and germination. In: Gupta KJ, Igamberdiev AU (eds) Reactive oxygen and nitrogen species signaling and communication in plants, Seria: signaling and communication in plants. Springer, Berlin

Kusano T, Berberich T, Tateda C, Takahashi Y (2008) Polyamines: essential factors for growth and survival. Planta 228:367–381

Kuya N, Sato S (2011) The relationship between profiles of plagiogravitropism and morphometry of columella cells during the development of lateral roots of *Vigna angularis*. Adv Space Res 47:553–562

Kwasniewski M, Chwialkowska K, Kwasniewska J, Kusak J, Szarejko I (2013) Accumulation of peroxidase-related reactive oxygen species in trichoblasts correlates with root hair initiation in barley. J Plant Physiol 170:185–195

Lee Y, Park CH, Kim AR, Chang SC, Kim SH, Lee WS, Kim SK (2011) The effect of ascorbic acid and dehydroascorbic acid on the root gravitropic response in *Arabidopsis thaliana*. Plant Physiol Biochem 49:909–916

Lee Y, Rubio MC, Alassimone J, Geldner N (2013) A mechanism for localized lignin deposition in the endodermis. Cell 153:402–412

Li X, Zhang WS (2008) Salt-avoidance tropism in *Arabidopsis thaliana*. Plant Signal Behav 3:351–353

Liang Y, Mitchell DM, Harris JM (2007) Abscisic acid rescues the root meristem defects on the *Medicago truncatula* latd mutant. Dev Biol 304:297–307

Liszkay A, Kenk B, Schopfer P (2003) Evidence for the involvement of cell wall peroxidase in the generation of hydroxyl radicals mediating extension growth. Planta 217:658–667

Liszkay A, Van der Zalm E, Schopfer P (2004) Production of reactive oxygen intermediates ($O_2^{\bullet -}$ H_2O_2, and $^{\bullet}OH$) by maize roots and their role in wall loosening and elongation growth. Plant Physiol 136:3114–3123

Lombardo MC, Graziano M, Polacco J, Lamattina L (2006) Nitric oxide functions as a positive regulator of root hair development. Plant Signal Behav 1:28–33

Méndez-Bravo A, Raya-González J, Herrera-Estrella L, Lopez-Bicio J (2010) Nitric oxide is involved in alkamide-induced lateral root development in Arabidopsis. Plant Cell Physiol 51:1612–1626

Monshausen GB, Bibikova TN, Weisenseel MH, Gilroy S (2009) Ca^{2+} regulates reactive oxygen species production and pH during mechanosensing in Arabidopsis roots. Plant Cell 21:2341–2356

Monzón GC, Pinedo M, Di Rienzo J, Novo-Uzal E, Pomar F, Lamattina L, de la Canal L (2014) Nitric oxide is required for determining root architecture and lignin composition in sunflower. Supporting evidence from microarray analyses. Nitric Oxide 39:20–28

Mori IC, Schroeder JI (2004) Reactive oxygen species activation of plant Ca^{2+} channels. A signaling mechanism in polar growth, hormone transduction, stress signaling, and hypothetically mechanotransduction. Plant Physiol 135:702–708

Møller IM, Sweetlove LJ (2010) ROS signalling – specificity is required. Trends Plant Sci 15:370–374

Muday GK, Rahman A, Binder BM (2012) Auxin and ethylene: collaborators or competitors? Trends Plant Sci 14:181–195

Müller K, Linkies A, Vreeburg RAM, Fry SC, Liszkay AK, Leubner-Metzger G (2009) In vivo cell wall loosening by hydroxyl radicals during cress seed germination and elongation growth. Plant Physiol 150:1855–1865

Müller K, Linkies A, Leubner-Metzger G, Kermode AR (2012) Role of a respiratory burst oxidase of *Lepidium sativum* (cress) seedlings in root development and auxin signaling. J Exp Bot 63:6325–6334

Oracz K, Voegele A, Tarkowská D, Jacquemoud D, Tureckova T, Urbanova T, Strnad M, Sliwinska E, Leubner-Metzger G (2012) Myrigalone A inhibits *Lepidium sativum* seed

germination by interference with gibberellin metabolism and apoplastic superoxide production required for embryo extension growth and endosperm rupture. Plant Cell Physiol 53:81–95
Pagnussat GC, Simontacchi M, Puntarulo S, Lamattina L (2002) Nitric oxide is required for root organogenesis. Plant Physiol 129:954–956
Pagnussat GC, Lanteri ML, Lombardo MC, Lamattina L (2004) Nitric oxide mediates the indole acetic acid induction activation of a mitogen-activated protein kinase cascade involved in adventitious root development. Plant Physiol 135:279–286
Peer WA, Cheng Y, Murphy AS (2013) Evidence of oxidative attenuation of auxin signaling. J Exp Bot 64:2629–2639
Pitzschke A, Hirt H (2006) Mitogen-activated protein kinases and reactive oxygen species signaling in plants. Plant Physiol 141:351–356
Rashotte AM, Brady SR, Reed RC, Ante AJ, Muday GK (2000) Basipetal auxin transport is required for gravitropism in root of *Arabidopsis*. Plant Physiol 122:481–490
Rasmussen A, Mason MG, De Cuyper C et al (2012) Strigolactones suppress adventitious rooting in Arabidopsis and pea. Plant Physiol 158:1976–1987
Renew S, Heyno E, Schopfer P, Liszkay A (2005) Sensitive detection and localization of hydroxyl radical production in cucumber roots and Arabidopsis seedlings by spin trapping electron paramagnetic resonance spectroscopy. Plant J 44:342–347
Roy R, Bassham DC (2014) Root growth movements: waving and skewing. Plant Sci 221–222:42–47
Ruyter-Spira C, Kohlen W, Charnikhova T (2011) Physiological effects of the synthetic strigolactone analog GR24 on root system architecture in Arabidopsis: another belowground role for strigolactones? Plant Physiol 155:721–734
Sanchez-Fernandez R, Fricker M, Corben LB, White NS, Sheard N, Leaver CJ, Montagu MV, Inze D, May MJ (1997) Cell proliferation and hair tip growth in the *Arabidopsis* root are under mechanistically different forms of redox control. Proc Natl Acad Sci USA 94:2745–2750
Sanz L, Fernández-Marcos M, Modrego A (2014) Nitric oxide plays a role in stem cell niche homeostasis through its interaction with auxin. Plant Physiol 166:1972–1984
Schlicht M, Ludwig-Müller J, Burbach C, Volkmann D, Baluska F (2013) Indole-3-butyric acid induces lateral root formation via peroxisome-derived indole-3-acetic acid and nitric oxide. New Phytol 200:473–482
Soltys D, Rudzińska-Langwald A, Kurek W, Szajko K, Sliwinska E, Bogatek R, Gniazdowska A (2014) Phytotoxic cyanamide affects maize (*Zea mays*) root growth and root tip function: from structure to gene expression. J Plant Physiol 171:565–575
Soltys D, Rudzinska-Lagwald A, Gniazdowska A, Wiśniewska A, Bogatek R (2012) Inhibition of tomato (*Solanum lycopersicum* L.) root growth by cyanamide is due to altered cell division, phytohormone balance and expansin gene expression. Planta 236:1629–1638
Sun H, Tao J, Liu S, Huang S, Chen S, Xie X, Yoneyama K, Zhang Y, Xu G (2014) Strigolactones are involved in phosphate- and nitrate-deficiency-induced root development and auxin transport in rice. J Exp Bot 65:6735–6746
Tassoni A, van Buuren M, Franceschetti M, Fornale S, Bagni N (2000) Polyamine content and metabolism in *Arabidopsis thaliana* and effect of spermidine on plant development. Plant Physiol Biochem 38:383–393
Terrile MC, París R, Calderón-Villalobos LIA, Iglesias MJ, Lamattina L, Estelle M, Casalonque CA (2012) Nitric oxide influences auxin signaling through *S*-nitrosylation of the Arabidopsis TRANSPORT INHIBITOR RESPONSE 1 auxin receptor. Plant J 70:492–500
Tewari RK, Hahn EJ, Paek KY (2008) Function of nitric oxide and superoxide anion in the adventitious root development and antioxidant defence in *Panax ginseng*. Plant Cell Rep 27:563–573
Tian H, De Smet I, Ding Z (2014) Shaping a root system: regulating lateral versus primary root growth. Trends Plant Sci 19:426–431

Tisi A, Federico R, Moreno S, Lucretti S, Moschou PN, Roubelakis-Angelakis KA, Angelini R, Cona A (2011) Perturbation of polyamine catabolism can strongly affect root development and xylem differentiation. Plant Physiol 157:200–215

Tracy SR, Black CR, Roberts JA, Dodd IC, Mooney SJ (2015) Using X-ray computed tomography to explore the role of abscisic acid in moderating the impact of soil compaction on root system architecture. Environ Exp Bot 110:11–18

Trewavas A (2005) Green plants as intelligent organisms. Trends Plant Sci 10:413–419

Tsukagoshi H, Busch W, Benfey PN (2010) Transcriptional regulation of ROS controls transition from proliferation to differentiation in the root. Cell 143:606–616

Umehara M, Hanada A, Yoshida S (2008) Inhibition of shoot branching by new terpenoid plant hormones. Nature 455:195–200

Vernoux T, Wilson RC, Seeley KA, Reichheld JP, Muroy S, Brown S, Maughan SC, Cobbett CS, van Montagu M, Inze D, May MJ, Sung ZR (2000) The *ROOT MERISTEMLESS1/CADMIUM SENSITIVE2* gene defines a glutathione-dependent pathway involved in initiation and maintenance of cell division during postembryonic root development. Plant Cell 12:97–110

Xiong TC, Bourque S, Lecourieux D, Amelot N, Grat S, Briere C, Mazars C, Pugin A, Ranjeva R (2006) Calcium signaling in plant cell organelles delimited by a double membrane. Biochim Biophys Acta 1763:1209–1215

Yu M, Lamattina L, Spoel SH, Loake GJ (2014) Nitric oxide function in plant biology: a redox cue in deconvolution. New Phytol 202:1142–1156

Yun BW, Feechan A, Yin M, Saidi NBB, Bihan TL, Yu M, Moore JW, Kang JG, Kwon E, Spoel SH, Pallas JA, Loake GJ (2011) *S*-nitrosylation of NADPH oxidase regulates cell death in plant immunity. Nature 478:264–268

Zhang C, Bousquet A, Harris JM (2014) Abscisic acid and LATERAL ROOT ORGAN DEFECTIVE/NUMEROUS INFECTIONS AND POLYPHENOLICS modulate root elongation via reactive oxygen species in *Medicago trunculata*. Plant Physiol 166:644–658

Relationship Between Changes in Contents of Nitric Oxide and Amino Acids Particularly Proline in Plants Under Abiotic Stress

David W.M. Leung

Contents

Abstract Studies on the physiological response of plants to abiotic stress have identified an array of changes including nitric oxide (NO) generation, accumulation of free proline, reactive oxygen species, antioxidants and oxidative damages. Little is known about the relationships between two of the concurrent changes, NO and proline metabolism. Here, the insights obtained so far and the important research gaps about this were explored.

Keywords Crop productivity • Morphological alterations • L-proline • Sodium nitroprusside • Stress tolerance

D.W.M. Leung (✉)
School of Biological Sciences, University of Canterbury, Private Bag 4800, Christchurch 8140, New Zealand
e-mail: david.leung@canterbury.ac.nz

D.K. Gupta et al. (eds.), *Reactive Oxygen Species and Oxidative Damage in Plants Under Stress*, DOI 10.1007/978-3-319-20421-5_14

1 Introduction

The agricultural productivity of crop plants could be severely limited by adverse environmental factors or abiotic stressors. Drought and elevated soil salinity are the two main abiotic stressors that can threaten crop production worldwide (Farooq et al. 2012; Munns et al. 2006). Research into the physiological impacts of these and other adverse plant growth conditions (Deinlein et al. 2014; He et al. 2014; Iqbal et al. 2014; Li et al. 2014; Shin et al. 2015; Talbi et al. 2015) has yielded information about a diverse array of changes in the plant body under abiotic stress (Fig. 1). It is frequently thought that abiotic stress-induced oxidative damages including unfavourable alterations to protein structure and activity and destabilisation of cell membranes (Gill and Tuteja 2010) are closely associated with the stress-induced visual symptoms or morphological alterations to the plant body (Potters et al. 2009; Leung 2015).

The goal of developing crop plants with enhanced adaptation to abiotic stress (via improved avoidance or tolerance mechanisms) remains an important grand research challenge in contemporary plant science. A way toward this goal is to gain a comprehensive understanding of the interrelationships among all the changes already shown to take place in plants reacting to different abiotic stressors (Fig. 1). Here, the few insights already gained and the important research gaps about the relationship between nitric oxide (NO) and accumulation of amino acids particularly proline in plants under abiotic stress were explored.

Abiotic Stress

⇩

Hormonal/Signal transducing/Biochemical/Molecular Changes

Hormones (notably abscisic acid level)

Signals (notably nitric oxide generation and polyamines)

Water content

Ion (mineral) contents

'Biocompatible solutes' including soluble sugars and amino acids (notably proline)

Reactive oxygen species

Antioxidants

Gene expression

Oxidative Damages

⇩

Morphological Alterations

Fig. 1 Scheme of major changes in plants under abiotic stress

2 Nitric Oxide Generation and Proline Accumulation are Concurrent Biochemical Changes

Proline accumulation is frequently observed as a response of plants to different stresses (Table 1). For example, in response to a low temperature treatment, a cold-resistant sugarcane cultivar exhibited a higher level of free proline than a cold-sensitive cultivar (Huang et al. 2015). This was one of the many biochemical differences found between the two contrasting sugarcane cultivars.

In comparison, there are relatively fewer studies that determined NO generation in plants under abiotic stress (Table 1). For example, increased NO emission by poplar roots, leaves and stems was found under oxygen-deficit conditions (Liu et al. 2015a). Changes in NO levels were also found in response to cold (Cantrel et al. 2011), high temperature stress (Hasanuzzamann et al. 2014), salinity stress (Gould et al. 2003), water-deficit stress (Signorelli et al. 2013b; Planchet et al. 2014) and copper (heavy metal) stress (Hu et al. 2015). Therefore, it would seem that similar to free proline level, NO level in plants is also likely modulated in response to different stresses. In addition, there are many more studies implicating NO as a signal molecule involved in the response of plants to a variety of abiotic stresses (Table 1; Hu et al. 2015; Leung 2015). These were based on experiments involving (a) inhibition of nitric oxide synthase (NOS, an enzyme that is associated with NO synthesis in plants) with a NOS inhibitor such as *N*-nitro-L-arginine methyl ester (L-NAME) or manipulating the nitrate reductase pathway for NO production (Boldizsar et al. 2013), (b) exogenous application of a nitric oxide donor such as sodium nitroprusside (SNP) (He et al. 2014) and (c) a NO-specific scavenger 2-(4-carboxyphenyl)-4,4,5,5-tetramethylimidazoline-1-oxyl-3 oxide (cPTIO) (Planchet et al. 2014).

It was clearly confirmed in a recent study that endogenous NO generation and free proline accumulation were concurrent, correlated changes in roots of *Lotus japonicus* (Signorelli et al. 2013b) and the embryo axes during seedling establishment in *Medicago truncatula* under water-deficit stress (Planchet et al. 2014). However, in a literature survey, it was found that proline accumulation and nitric oxide generation in plants under abiotic stress have been mainly studied separately (see the few sample studies shown in Table 1). Little is known about the relationships or precise mode of interaction, if any, between nitric oxide and proline in contributing to stress-induced oxidative damages and morphological alterations. This is hardly an ideal situation on the road map to genetic improvement of crop plants with enhanced abiotic stress tolerance.

Table 1 Examples of investigations into nitric oxide generation and free proline accumulation in plants under abiotic stress

Abiotic stressor	Plant material	NO generation	Proline accumulation	Reference
Cold	Sugarcane seedlings	n.d.	Stimulated	Huang et al. (2015)
Cold	*Arabidopsis thaliana* L. Heynh. wild type (WT) in the Columbia (Col-0) background 4-week-old plants	Measurement based on fluorometric and chemiluminescence methods. A significant increase in leaves after exposure of plants at 4 °C for 1–4 h	n.d.	Cantrel et al. (2011)
Cold	Different plants	Measurement based on a variety of methods: fluorometric NO determination using the fluorophore 4,5-diaminofluorescein diacetate (DAF-2DA) haemoglobin assay, chemiluminescence sensor-coupled NO metres, most studies showing an increase in NO level following treatment of plant materials at low temperatures	n.d. But recognised as a critical component in stress response	Puyaubert and Baudouin (2014)
High temperatures	Leaf peels and mesophyll cell suspensions of *Nicotiana tabacum* cv. Xanthi	Increased Determined using DAF-2DA, a NO-specific fluorophore	n.d.	Gould et al. (2003)
High temperature stress	10-day-old rapeseed (*Brassica napus* L. cv. BINA Sarisha 3) seedlings treated at 38 °C for 24–48 h	n.d.	Increased	Hasanuzzamann et al. (2014)
Hypoxia	Roots of intact 4-month-old poplar (*Populus* × *canescens*) cuttings	NO gas emission measure using NO analyzers (chemiluminescence) NO emission from stems of young tress if the roots experienced a deficit in oxygen supply	n.d.	Liu et al. (2015a)

Drought	2-month-old *Oudneya africana* (a Saharian plant), grown in pots under greenhouse conditions	Determined using a NO-specific fluorophore (diaminofluorescein) Increased with duration and severity of drought conditions; returned to lower levels upon 5 days of re-watering	n.d.	Talbi et al. (2015)
Drought	Two adult field-grown tea [*Camellia sinensis* (L.) O. Kuntze] plant genotypes including drought-susceptible 'Zhuyeqi' (T1) and drought-tolerant 'Ningzhou 2' (T2)	n.d.	Increased significantly as water deficit progressed and then rapidly decreased following re-watering in both drought-susceptible and drought-tolerant genotypes	Liu et al. (2015b)
Water stress	Roots of cucumber plants (*Cucumis sativus* cv. Dar) Plants treated at the third fully expanded leaf stage	NO-specific fluorescence was enhanced more than eightfold after 17 h from start of water withdrawal in root segments above the elongation zone using confocal laser scanning microscopy	Progressive water stress caused more than fourfold increase at 17 h 100 μM SNP caused a significant decrease in free proline content by about 30 % 200 μM cPTIO + SNP did not reduce free proline content Implication is that SNP pretreatment protected the roots, but this was accompanied by a reduction in free proline content, suggesting that the osmolyte might not be essential in water stress tolerance	Arasimowicz-Jelonek et al. (2009)
Water stress	*Lotus japonicus* plants	NO content was increased in roots under water stress compared to well-watered plants	23-fold increase in proline content of roots under water stress compared to well-watered plants	Signorelli et al. (2013b)
Water-deficit stress (PEG induced)	During seedling establishment in *Medicago truncatula*	Based on fluorometric NO determination NO fluorescence increased in the embryo axes of 3-day-old seedlings	Increased with duration of PEG-induced stress	Planchet et al. (2014)

(continued)

Table 1 (continued)

Abiotic stressor	Plant material	NO generation	Proline accumulation	Reference
		with the increased duration (2–24 h) of PEG-induced stress		
Salinity (150 mM NaCl)	Maize seedlings (*Zea mays* L. cv. Silverking) at the 3-leaf stage were grown hydroponically Measured in the youngest fully developed leaves	n.d. But the study involved the use of inhibitors of NO synthesis (5 μM nitro-L-arginine (L-NNA)) and exogenous application of NO donors (5 μM DETA/NO [*N*-(2-ammonioethyl)amino]diazen-1-ium-1,2-diolate])	Proline accumulation increased during NaCl treatment	Boldizsar et al. (2013)
Salinity (50 mM NaCl)	Cucumber seedlings (cv. Jinchun 2) at the 2-leaf stage or 21 days old	n.d. But NO donor (100 μM sodium nitroprusside or SNP) was used	Increased in leaves and roots with increasing time under salt stress (0–8 days) A greater increase in response to SNP application	Fan et al. (2012)
Salinity (200 mM NaCl)	Spinach (*Spinacia oleracea* L.) seedlings with three true leaves grown hydroponically	n.d. But application of 200 nl/l NO gas partially counteracted salt-induced growth inhibition	Increased in the leaves after plants were under salt stress for 21 days compared to control (absence of NaCl); there was no difference in proline level in leaves of the plants grown in the absence of NaCl with or without exposure to NO In the NaCl + NO treatment, the proline content was more than twice to that in the NaCl treatment alone	Du et al. (2015)
Cadmium (Cd)	2-week-old wheat seedling (*Triticum aestivum* L.) variety 'Inqalab-91' grown in liquid nutrient culture or	Determined using a chemiluminescence NO analyser NO production in cut root sections	n.d.	Mahmood et al. (2009)

	exposed to 1 μM Cd for 4 weeks Fairly Cd resistant with ninefold increase in Cd accumulation (2.53 mM) without exhibiting any visual stress-related growth reduction	more than double; respiration about halved after 4 weeks of Cd treatment compared to the control (in the absence of Cd)		
Cd stress	*Brassica juncea* L. (Czern. & Coss.) (25-day-old Indian mustard plants) Treated with 200 mg/l $CdSO_4 \cdot 8H_2O$	n.d.	Increased significantly and almost doubled the level of the control (no Cd stress) Even enhanced more by a calcium treatment which reversed the Cd toxicity implication Proline was involved in salt stress protection	Ahmad et al. (2015)
Cu ($CuSO_4$) stress	3-day-old seedlings of Hulles barley (*Hordeum vulgare* L. var. nude)	NO generation based on haemoglobin assay Peaked at 24 h after start of Cu stress and then began to decline	n.d.	Hu et al. (2015)
Pb [0–200 μM Pb $(NO_3)_2$]	*Potamogeton crispus* L. (Potamogetonaceae), a submerged macrophyte	n.d.	Increased in response to 50 μM Pb but decreased at higher Pb concentrations	Qiao et al. (2015)

n.d. = not determined

3 Involvement of NO and Proline Interactions in Abiotic Stress Tolerance Mechanisms

Accumulation of free proline is determined by activation of proline biosynthesis and catabolic degradation in plants under stress, and the details of the metabolic pathways and regulation in relation to reactive oxygen species (ROS) have been discussed in a recent review (Rejeb et al. 2014). The immediate precursor for proline biosynthesis is glutamate. A key enzyme involved in the formation of proline is pyrroline-5-carboxylate synthetase (P5CS), while proline dehydrogenase (ProDH) is a key proline catabolic enzyme. The expression of these genes has been shown to be altered in plants under stress (Planchet et al. 2014).

It has been frequently speculated that accumulation of free proline in plants under abiotic stress compared to absence of stress has physiological significance. Specifically proline is thought to be a critical component of abiotic stress tolerance mechanistic network (Deinlein et al. 2014). Osmotic adjustment seems to be an important adaptive reaction to stress. Elevated accumulation of 'compatible solutes' such as free proline and soluble sugars has been frequently observed, and they are thought to be important osmolytes (Deinlein et al. 2014). In addition, in in vitro assays, proline seems to be capable of free radical scavenging (Kaul et al. 2008). However, the physiological significance of increased accumulation of free proline in antioxidant defence of plants under abiotic stress has been considered to be unlikely (Signorelli et al. 2013a, 2014).

There are studies with findings putting the notion of proline involvement in abiotic stress tolerance in doubt, however. For example, the impact of osmotic stress (induced by treatment with 5 % PEG) on winter oilseed rape (*Brassica napus* L.) seedlings (26 d from sowing seed) was different between the drought-sensitive and drought-resistant genotypes (Hatzig et al. 2014). In response to osmotic stress, the sensitive genotypes exhibited significant shoot and root growth inhibition which was accompanied by typical accumulation of free proline. However, these effects of osmotic stress were not observed in the resistant genotypes, suggesting that accumulation of proline for being an osmolyte involved in osmotic adjustment is not the critically important mechanism underpinning osmotic stress tolerance.

The controversy surrounding the role of proline accumulation in abiotic stress tolerance can be examined from another line of investigations by probing the relationship between nitric oxide level and proline metabolism in plants under abiotic stress. For example, in response to 50 mM NaCl for 8 days, the proline content in the cucumber leaves was 356 % that of the leaves of cucumber plants grown in the absence of NaCl (Fan et al. 2012). In treatment of cucumber plants with 50 mM NaCl and 100 μM SNP (a NO donor), leaf proline content was about 12 % higher than that in NaCl alone (Table 1). It was suggested that the higher proline content was associated with protection of the SNP treatment against salinity stress. However, it should be noted that there were no data showing any growth improvement with the SNP treatment. Moreover, SNP treatment or exogenous application of other NO donor solutions which would presumably supply more bioactive NO molecules to the experimental material under question may not necessarily change or lead to an increase in proline accumulation (Planchet et al. 2014). Exogenous SNP application in the absence of abiotic stress was found to result in a reduction in proline content in the calli of a halophyte *Nitraria tangutorum* Bobr. instead (Yang et al. 2014).

In spinach, there was more than a fourfold increase in proline content under NaCl stress (Du et al. 2015; Table 1), and the plant biomass was concomitantly reduced. However, application of NO gas promoted about a tenfold increase in proline content and counteracted partially reduction in plant biomass under NaCl stress. Based on manipulations of the response of maize seedlings under NaCl stress using NO donors and a NOS inhibitor (Table 1), it was concluded that the much higher levels of free proline content were associated with NO-induced salt tolerance (Boldizsar et al. 2013).

Water-deficit stress induced by treatment of *Medicago truncatula* seedlings with polyethylene glycol (PEG 8000 at −0.75 MPa) had a substantial inhibitory effect on embryo axis (hypocotyl and root) elongation. A correlative temporal and localisation relationship was found between inhibition of axis elongation and NO production which was increased about 2.2-fold in the axis after 24 h under PEG-induced water-deficit stress (Table 1). Both nitrate reductase-dependent pathway and NOS activity were involved in NO generation under water-deficit stress. However, a scavenger of NO production (cPTIO) used at a concentration that was very effective in eliminating endogenous NO level in the seedling axis could not counteract the inhibition of seedling growth induced by the PEG treatment (Planchet et al. 2014). Therefore, NO generation in the embryo axis of *M. truncatula* was just coincidental with inhibition of axis growth under water-deficit stress. Interestingly, unlike many other studies, exogenous application of a NO donor also did not protect the axis under PEG-induced water stress.

During seedling establishment of *M. truncatula* in the absence of drought stress, total amino acids or proline content in the embryo axis decreased, but the opposite pattern was exhibited under PEG-induced water stress (Planchet et al. 2014). Elevation in proline content induced by PEG was contributed by an up-regulation of *P5CS2* gene expression and a concomitant downregulation in *ProDH* gene expression. Interestingly, the content of glutamate (the precursor for proline biosynthesis) in the embryo axis remained unchanged during 24 h of water stress. Altered NO and proline metabolism were correlative biochemical changes, but application of cPTIO was not able to alter the response of the *M. truncatula* seedlings to PEG treatment as far as total amino acid or proline content elevation in the embryo axis under water stress is concerned. Therefore, elevation in proline content and NO production can occur in the embryo axis at the same time induced by a 24-h PEG treatment, but they are independent biochemical responses to water stress. This is a variation to the observations leading to the following proposed sequence of changes under NaCl or cold stress (Boldizsar et al. 2013; Puyaubert and Baudouin 2014): abiotic stress → modification in NO level → changes in amino acid contents, notably free proline.

4 Other Amino Acids

Levels of total amino acids are often elevated under abiotic stress (Boldizsar et al. 2013; Planchet et al. 2014), although most studies have only put emphasis on free proline levels. In maize seedlings under NaCl stress, for example, besides proline, amino acids of different properties or chemical categories including the nonprotein amino acid γ-aminobutyric acid (GABA) were generally increased to levels of several-fold higher than in the absence of stress (Boldizsar et al. 2013). The effects of a NO donor and a NOS inhibitor on several amino acids besides free proline during stress and upon recovery from stress clearly warrant a closer and further examination in maize and other plants. For example, exogenous application of a NO donor to maize seedlings under NaCl stress for 11 days increased GABA content more than those under NaCl stress alone. Interestingly, exogenous application of a NOS inhibitor to the seedlings under NaCl stress for 11 days brought about an even higher level of GABA than in the treatment of NO donor + NaCl. Based on these findings, the precise interaction between NO and the metabolism of this amino acid remains unclear until endogenous NO generation is determined as well.

Soil drench with 0.3 mM β-aminobutyric acid was found to improve drought tolerance of potato plants (Sos-Hegedus et al. 2014). This was accompanied by an increase in generation of NO and ROS in the roots but not in the leaves. The protective effect, however, appeared to last for a few weeks only.

5 Concluding Perspectives

Generally, changes in NO generation and signal transduction and changes in amino acid metabolism are concurrent events in plants under stress. The paucity of studies on their precise interactions is clear. In particular, the implications and connections of the following are in need of urgent further research: (a) application of NO donors is frequently linked to the effects of antioxidants and (b) the effects of ROS signals inducing proline accumulation and the influence of proline metabolism on ROS generation (Rejeb et al. 2014).

Our understanding of the changes in NO level and accumulation of individual amino acids apart from free proline in plants under stress is in a highly unsatisfactory state. Given that free proline accumulation may or may not be associated with abiotic stress tolerance, the idea that modulation of NO levels could change the accumulation of other amino acids which might be associated with abiotic stress tolerance is worthy of further investigations.

References

Ahmad P, Sarwat M, Bhat NA, Wani MR, Kazi AG, Tran LSP (2015) Alleviation of cadmium toxicity in *Brassica juncea* L. (Czern. & Coss.) by calcium application involves various physiological and biochemical strategies. PLoS One 10:e0114571

Arasimowicz-Jelonek M, Floryszak J, Kubis J (2009) Involvement of nitric oxide in water stress-induced responses in cucumber roots. Plant Sci 177:682–690

Boldizsar A, Simon-Sarkadi L, Szirtes K, Soltesz A, Szalai G, Keyster M et al (2013) Nitric oxide affects salt-induced changes in free amino acid levels in maize. J Plant Physiol 170:1020–1027

Cantrel C, Vazquez T, Puyauberr J, Reze N, Lesch M, Kaiser WM, Dutilleul C, Guillas I, Zachowski A, Baudouin E (2011) Nitric oxide participates in cold-responsive phosphosphingolipid formation and gene expression in *Arabidopsis thaliana*. New Phytol 189:415–427

Deinlein U, Stephan AB, Horie T, Luo W, Xu G, Schroeder JL (2014) Plant salt-tolerance mechanisms. Trends Plant Sci 19:371–379

Du ST, Liu Y, Zhang P, Liu HJ, Zhang XQ, Zhang RR (2015) Atmospheric application of trace amounts of nitric oxide enhances tolerance to salt stress and improves nutritional quality in Spinach (*Spinacia oleracea* L.). Food Chem 173:905–911

Fan HF, Du CX, Guo SR (2012) Effect of nitric oxide on proline metabolism in cucumber seedlings under salinity stress. J Am Soc Hort Sci 137:127–133

Farooq M, Hussain M, Wahid A, Siddique KHM (2012) Drought stress in plants: an overview. In: Aroca R (ed) Plant responses, From morphological to molecular features. Springer, New York, NY

Gill SS, Tuteja N (2010) Reactive oxygen species and antioxidant machinery in abiotic stress tolerance in crop plants. Plant Physiol Biochem 48:909–930

Gould KS, Lamotte O, Klinguer A, Pugin A, Wendehenne D (2003) Nitric oxide production in tobacco leaf cells: a generalized stress response? Plant Cell Environ 26:1851–1862

Hatzig S, Zaharia I, Abrams S, Hohmann M, Legoahec L, Bouchereau A, Nesi N, Snowdon RJ (2014) Early osmotic adjustment responses in drought-resistant and drought-sensitive oilseed rape. J Integr Plant Biol 56:797–809

Hasanuzzamann M, Nahar K, Alam MM, Fujita M (2014) Modulation of antioxidant machinery and the methylglyoxal detoxification system in selenium-supplemented *Brassica napus* seedlings confers to high temperature stress. Biol Trace Elem Res 161(Special Issue):297–307

He JY, Ren YF, Chen XL, Chen H (2014) Protective roles of nitric oxide on seed germination and seedling growth of rice (*Oryza sativa* L.) under cadmium stress. Ecotoxicol Environ Saf 108: 114–119

Hu Y, You J, Liang X (2015) Nitrate reductase-mediated nitric oxide production is involved in copper tolerance in shoots of hulless barley. Plant Cell Rep 34:367–379

Huang X, Chen MH, Yang LT, Li YR, Wu JM (2015) Effects of exogenous abscisic acid on cell membrane and endogenous hormone contents in leaves of sugarcane seedlings under cold stress. Sugar Tech 17:59–64

Iqbal N, Umar S, Khan NA, Khan MIR (2014) A new perspective of phytohormones in salinity tolerance: regulation of proline metabolism. Environ Exp Bot 100:34–42

Kaul S, Sharma SS, Mehta IK (2008) Free radical scavenging potential of L-proline: evidence from in vitro assays. Amino Acids 34:315–320

Leung DWM (2015) Regulatory role of nitric oxide (NO) in alterations of morphological features of plants under abiotic stress. In: Khan MN, Mobin M, Mohammad F, Corpas FJ (eds) Nitric oxide action in abiotic stress responses in plants. Springer, New York

Li X, Gong B, Xu K (2014) Interaction of nitric oxide and polyamines involves antioxidants and physiological strategies against chilling-induced oxidative damage in *Zingiber officinale* Roscoe. Scientia Hort 170:237–248

Liu B, Rennenberg H, Kreuzwieser J (2015a) Hypoxia induces stem and leaf nitric oxide (NO) emission from poplar seedlings. Planta 241:579–589

Liu SC, Yao MZ, Ma CL, Jin JQ, Ma JQ, Li CF, Chen L (2015b) Physiological changes and differential gene expression of tea plant under dehydration and rehydration conditions. Scientia Hort 184:129–141

Mahmood T, Gupta KJ, Kaiser WM (2009) Cadmium stress stimulates nitric oxide production. Pak J Bot 41:1285–1290

Munns R, James RA, Lauchli A (2006) Approaches to increasing the salt tolerance in wheat and other cereals. J Exp Bot 57:1025–1043

Planchet E, Verdu I, Delahaie J, Cukier C, Girard C, Morere-Le Paven MC, Limami AM (2014) Abscisic acid-induced nitric oxide and proline accumulation in independent pathways under water-deficit stress during seedling establishment in *Medicago truncatula*. J Exp Bot 65: 2161–2170

Potters G, Pasternak TP, Guisez Y, Jansen MAK (2009) Different stresses, similar morphogenic responses: integrating a plethora of pathways. Plant Cell Environ 32:158–169

Puyaubert J, Baudouin E (2014) New clues for a cold case: nitric oxide response to low temperature. Plant Cell Environ 37:2623–2630

Qiao X, Zheng Z, Zhang L, Wang J, Shi G, Xu X (2015) Lead tolerance mechanism in sterilized seedlings of *Potamogeton crispus* L.: subcellular distribution, polyamines and proline. Chemosphere 120:179–187

Rejeb KB, Abdelly C, Savoure A (2014) How reactive oxygen species and proline face stress together. Plant Physiol Biochem 80:278–284

Shin JH, Vaughn JN, Abdel-Haleem H, Chavarro C, Abernathy B, Do Kim K, Jackson SA, Li Z (2015) Transcriptome changes due to water deficit define a general soybean response and accession-specific pathways for drought avoidance. BMC Plant Biol 15:26

Sos-Hegedus A, Juhasz Z, Poor P, Kondrak M, Antal F, Tari I (2014) Soil drench treatment with beta-aminobutyric acid increases drought tolerance of potato. PLoS One 9:e114297

Signorelli S, Arellano JB, Melo TB, Borsani O, Monza J (2013a) Proline does not quench singlet oxygen: evidence to reconsider its protective role in plants. Plant Physiol Biochem 64:80–83

Signorelli S, Corpas FJ, Borsani O, Barroso JB, Monza J (2013b) Water stress induces a differential and spatially distributed nitro-oxidative stress response in roots and leaves of *Lotus japonicus*. Plant Sci 201–202:137–146

Signorelli S, Coitino EL, Borsani O, Monza J (2014) Molecular mechanisms for the reaction between ${}^{\bullet}OH$ radicals and proline: insights on the role as reactive oxygen species scavenger in plant stress. J Phys Chem 118:37–47

Talbi S, Romero-Pueras MC, Hernandez A, Terron L, Ferchichi A, Sandalio LM (2015) Drought tolerance in a Saharian plant *Oudneya africana*: role of antioxidant defences. Environ Exp Bot 111:114–126

Yang F, Ding F, Duan XH, Zhang J, Li XN, Yang YL (2014) ROS generation and proline metabolism in calli of halophyte *Nitraria tangutorum* Bobr. to sodium nitroprusside treatment. Protoplasma 251:71–80

Transgenic Plants and Antioxidative Defense: Present and Future?

Sarma Rajeevkumar, Hema Jagadeesan, and Sathishkumar Ramalingam

Contents

Abstract Environmental stresses considerably limit plant growth, distribution, and productivity. Biological systems generate a range of different reactive oxygen species (ROS) like superoxide ($O_2^{\bullet-}$), hydroxyl radical ($^{\bullet}OH$), and hydrogen peroxide (H_2O_2), during the course of normal metabolic reactions. If it is not effectively and rapidly removed, ROS damages a wide range of macromolecules, ultimately leading to cell death. ROS are also generated in response to various biotic and abiotic stresses. Cells have evolved both enzymatic and nonenzymatic defense mechanisms to protect cells from lethal effects of free radicals. ROS-scavenging enzymes, such as superoxide dismutase (SOD), catalase (CAT), and peroxidase (POD) that play a crucial role in regulating ROS accumulation in cells. Transgenic plants expressing specific gene candidates have been proven to

S. Rajeevkumar
Plant Biotechnology Division, Central Institute of Medicinal and Aromatic Plants Research Centre, Allalasandra, GKVK post, Bangalore 560065, Karnataka, India

H. Jagadeesan
Department of Biotechnology, PSG College of Technology, Coimbatore, India

S. Ramalingam (✉)
Plant Genetic Engineering Laboratory, Department of Biotechnology, Bharathiar University, Coimbatore 641046, India
e-mail: rsathish@buc.edu.in

D.K. Gupta et al. (eds.), *Reactive Oxygen Species and Oxidative Damage in Plants Under Stress*, DOI 10.1007/978-3-319-20421-5_15

increase the tolerance to environmental stresses significantly. In recent years, several efforts have been made to improve the oxidative stress tolerance in plants by over-expressing plant or bacterial genes coding either for ROS-scavenging enzymes or for enzymes modulating the cellular antioxidant capacity. In this chapter, we have discussed some of the significant reports on transgenic plants with altered antioxidant capacity mainly focusing on the new insight into the antioxidant defense mechanisms. Finally, future focus of transgenic research to combat oxidative stress has been briefed.

Keywords Reactive oxygen species • Antioxidants • Superoxide dismutase • Ascorbate peroxidase • Glutathione peroxidase • Glutathione reductase • Catalase • Alternative oxidase • Uncoupling protein

1 Introduction

The rise in global population along with reduction in agricultural practices and diminishing availability of resources poses serious challenges to present community for continuous supply of agricultural products. To feed 9 billion people that is expected by 2050, a significant increase in grain yield of approximately 44 million metric tons per year is required (Tester and Langridge 2010). Global warming and other stress factors negatively influence the plant growth and yield, that is making it even more challenging (Reguera et al. 2012). Thus, the environmental stress is one of the major limiting factors that determine the crop yield. Stress injuries on plants cause oxidative damages both at the cellular and molecular levels.The critical role of reactive oxygen intermediates (ROI) in cellular damage has been well studied (Rennenberg and Polle 1994).

Over the last few decades, deep understanding of the role played by of ROS lead to discovery of signaling molecules regulating various physiological processes in plants (Mylona and Polidoros 2010) (Fig. 1). Recently, several reports proved that low concentration of ROS is acting as redox signal molecules playing a crucial role in various signal transduction pathways in plant cells (Foyer and Noctor 2005a, b). Environmental stresses like salinity, temperature, or drought result in a marked increase in ROS level leading to oxidative damage in plant cells. During the course of evolution, plant system also developed potential antioxidant defense system to protect itself from oxidative damage (Asada 1999; Mittler et al. 2004). Till date, different ROS-scavenging enzymes like superoxide dismutase (SOD), ascorbate peroxidase (APX), glutathione peroxidase (GPX), glutathione reductase (GR), and catalase (CAT) have been discovered, which are localized in different cellular compartments (Table 1). In general, there are more than one type of enzyme in a cell for scavenging different kind of ROS (Mittler et al. 2004).

ROS are continuously generated as by-products of various cellular processes like photosynthesis, photorespiration, fatty acid oxidation, electron transport chain, etc.

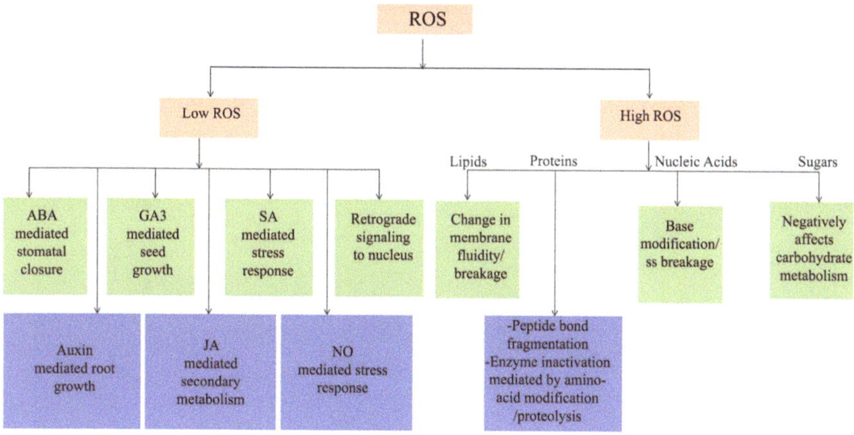

Fig. 1 Reactive oxygen species (ROS) at low levels act as secondary messengers in several physiological processes, whereas at high levels they negatively act on biomolecules

Table 1 Major ROS-scavenging enzymes, reaction mechanism, and cellular localization

Sl. No	ROS-scavenging enzymes	Mechanism	Localization	References
1	SOD	$O_2^{\bullet-} + O_2^{\bullet-} + 2H^+ \rightarrow 2H_2O_2 + O_2$	Cytosol, golgi, plastids, and mitochondria	Mittler et al. (2004)
2	APX	$AA + H_2O_2 \rightarrow DHA + 2H_2O$	Cytosol, plastids, peroxisomes, glyoxysomes, and mitochondria	Foyer and Noctor (2005a, b)
3	GPX	$H_2O_2 + 2GSH \rightarrow H_2O + GSSG$	Cytosol, plastids, nucleus, peroxisomes, and mitochondria	Ursini et al. (1995), Rodriguez et al. (2003)
4	GR	$GSSG + NAD(P)H \rightarrow 2GSH + NAD(P)^+$	Cytosol, plastids, peroxisomes, and mitochondria	Edwards et al. (1990), Creissen et al. (1995)
5	CAT	$2H_2O_2 \rightarrow O_2 + 2H_2O$	Peroxisomes, glyoxysomes	Mhamdi et al. (2010)

Up to 1 % of the molecular oxygen consumed by plant is metabolized to form ROS in different cellular compartments (Bartoli et al. 2000). ROS molecules include free radicals like superoxide ($O_2^{\bullet-}$), hydroxyl radical ($^{\bullet}OH$), etc., and non-radical molecules like hydrogen peroxide (H_2O_2), singlet oxygen ($^1O_2$), etc. (Shafi et al. 2015). Under normal conditions, ROS generated is efficiently metabolized by the multicomponent antioxidant system and protects them from potentially deleterious effects of ROS (Fig. 2). However, during stress (both biotic and abiotic),

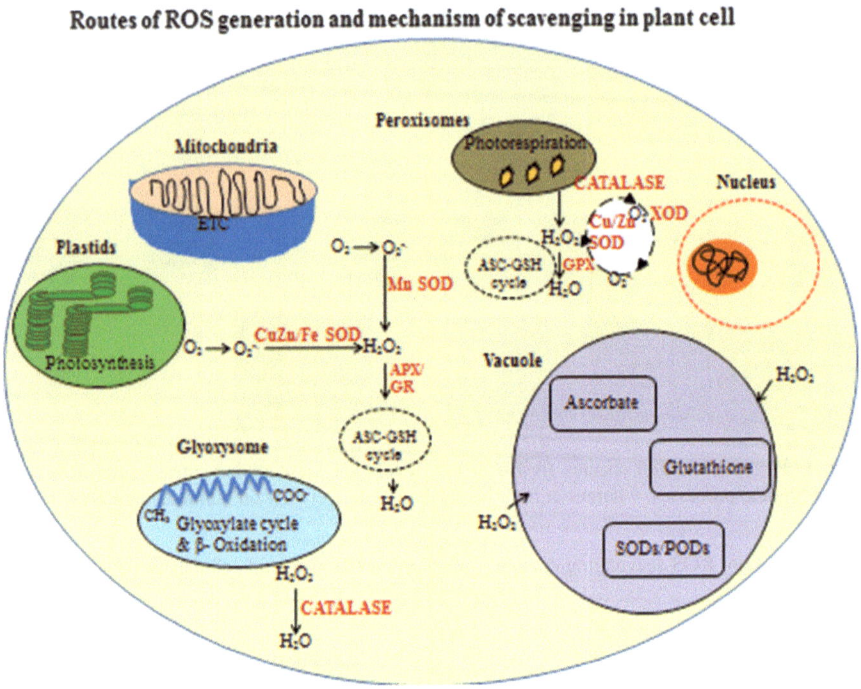

Fig. 2 Overview of cellular process generates ROS in different compartments and their scavenging in cells [Modified from Kuzniak (2002)]

ROS levels are elevated and thus the cellular antioxidant capacity can be overwhelmed leading to oxidative stress (Robinson and Bunce 2000). It is well known that much of plant injury as a result of environmental stresses are related to ROS-initiated oxidative damage. Hence, enhancing tolerance to several environmental stresses could be achieved through modulation of gene expression to different ROS-scavenging enzymes ultimately reducing ROS population in cells (Fig. 3).

Transgenic approaches are widely used for crop improvement programs in recent years. Gene discovery and functional genomics have revealed infinite mechanisms and led to the identification of many potential gene families, which could confer adaptation and improved productivity during adverse environmental conditions (Kumar et al. 2012). To improve oxidative stress tolerance in plants, overexpression of genes encoding either ROS-scavenging enzymes or enzymes modulating the cellular antioxidant capacity from various sources has been proved effective (Liu et al. 2013, 2014; Zhai et al. 2013; Zhang et al. 2013). The rationale behind these approaches is to minimize massive loss in productivity of crop plants by various environmental constraints. The study of transgenic lines expressing antioxidant genes revealed a complete understanding of the roles of individual enzymes in scavenging free radicals and imparting tolerance to various stress factors.

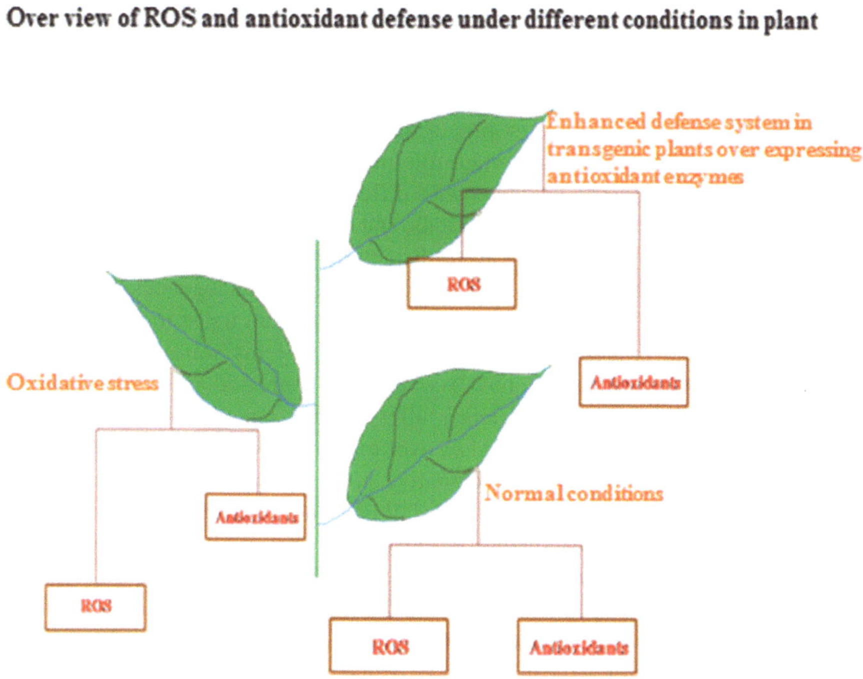

Fig. 3 Modulating ROS scavenging in transgenic plants by overexpression of antioxidant enzymes

Besides, transgenic lines with altered antioxidant capacity also provided new insights into the interrelations between various enzymatic and nonenzymatic antioxidant components and their signaling network. It has been already reported that the transgenic lines generated through gene pyramiding or co-expression of several antioxidant defense genes imparted more stress tolerance than the plants overexpressing either of genes alone (Lee et al. 2007; Wei et al. 2011; Diaz-Vivancos et al. 2013; Xu et al. 2014). In this chapter, we focus mainly on recent studies on transgenic plants with altered ROS-scavenging capacity by overexpression of antioxidant genes; readers are requested to refer to Gill and Tuteja (2010) and reference therein for transgenic plants for ROS scavenging by non-enzymatic antioxidants. Future focus is on the new insight in antioxidant defense mechanism by expressing genes like alternative oxidase, uncoupling protein has also been discussed briefly. In the end, future of transgenic research under field conditions has been emphasized.

2 Superoxide Dismutase (SOD)-Expressing Transgenic Plants

SOD and APX constitute the first line of defense response against ROS (Alscher et al. 2002), thus playing an important role in protecting cells against superoxide radicals. SODs are metallo-enzymes with different isoforms localized in various cellular compartments like cytosol, plastids, and mitochondria. Irrespective of their compartmentalization, they all catalyze dismutation of superoxide ions ($O_2^{\bullet -}$) to hydrogen peroxide (H_2O_2) (Gill and Tuteja 2010). The endogenous plant genes encoding different SOD isoforms have been used for plant transformation studies. Initial attempts to express SOD in heterologous systems were not successful. *Petunia hybrida* plastidial Cu/ZnSOD when over-expressed in tobacco did not increase the photosynthetic efficiency after methyl viologen (MV) treatment (Tepperman and Dunsmuir 1990). Also, over-expression of *Petunia* SOD in transgenic tobacco lines did not alter the oxidative stress tolerance (Pitcher et al. 1991). Later, several reports revealed reasons for failure of initial attempts using plastidial Cu/Zn/Fe SOD are due to inactivation of SOD by peroxides, whereas MnSOD is not susceptible to peroxide inactivation (Allen et al. 1997). A chimeric MnSOD (native mitochondrial transit peptide was replaced by plastidal peptide) from *Nicotiana plumbaginifolia* when over-expressed in tobacco lead to reduced levels of membrane damage following exposure to MV, ozone, and photooxidative stress (Van Camp et al. 1994). Transgenic alfalfa lines expressing same chimeric MnSOD enhanced freezing tolerance and water deficit (McKersie et al. 1993, 1996).

Sen Gupta et al. (1993a, b) reported some interesting results with respect to expression of different isoforms of SOD in tobacco. Transgenic tobacco lines over-expressing plastidial Cu/ZnSOD or a chimeric chloroplastic MnSOD showed different tolerance level to MV. MnSOD-expressing lines showed superior protection after MV exposure, whereas Cu/ZnSOD-expressing plants revealed low level of tolerance. Since expression levels of SOD in both lines were similar, it was reported that peroxide sensitivity of Cu/ZnSOD could account for low level of tolerance relative to the chloroplastic MnSOD-expressing plants. It was also shown that tobacco lines expressing plastidial Cu/ZnSOD were least susceptible to photo-oxidative damage induced by excess light and also chilling temperatures; however, tobacco lines expressing chloroplastic MnSOD were susceptible. Adding to this, chloroplastic MnSOD-expressing lines did not protect MV-induced damage to photosystem II (PSII). Since plastidial Cu/ZnSOD is located on the thylakoid membrane, close association with photosystem I (PSI) enzyme rapidly scavenges $O_2^{\bullet -}$ before significant damage occurs. Since MnSOD is not native to the chloroplast, it is possible that it may not form a productive association with PSI. Hence, expression of peroxide insensitive MnSOD in plastids alleviate the damage to cellular membranes possibly by reducing the leakage of ROI from plastids to cytosol, but MnSOD does not appear to provide considerable protection to photosynthetic components.

Over-expression of Cu/ZnSOD from *Kandelia candel* in tobacco resulted in reduced ROS generation in plastids and enhanced tolerance to salinity stress (Jing et al. 2015). Transgenic *Arabidopsis* co-expressing *Potentilla atrosanguinea SOD* and *Rheum austral APX* enhanced lignin deposition along with higher biomass production and yield under salinity stress (Shafi et al. 2015). The involvement of SOD and peroxides in cell wall lignification has been also previously reported (Karpinska et al. 2001; Kim and Barbara 2008). It was also reported that H_2O_2 signaling enhanced genes involved in lignin biosynthesis indirectly through other secondary messengers like MAP kinases and transcription factors like NAC, MYB, and WRKY (Fujita et al. 2006). The same combination of genes (*PaSOD* and *RaAPX*) also conferred tolerance to chilling stress in *Arabidopsis* and had higher levels of total antioxidant enzymes and chlorophyll content and lower levels of ROS (Shafi et al. 2014). Transgenic rice either over-expressing or knocked down with *OsMnSOD* was tolerant and susceptible to heat stress, respectively. Plants over-expressing MnSOD upregulated ROS-scavenging enzymes, chaperone, and quality control systems in rice grain under heat stress (Shiraya et al. 2014). It was also shown that over-expressing lines had better quality rice grains compared to knock down lines revealing that constitutive expression of golgi/plastid targeted MnSOD is effective in regulating formation of perfect grains under unfavorable conditions like heat stress. Over-expression of both yeast MnSOD and pea mitochondrial MnSOD in plastids of rice conferred tolerance to salt and oxidative stress (Tanaka et al. 1999; Wang et al. 2005). Tolerance to oxidative and chilling stresses in transgenic Cassava plants co-expressing Cu/ZnSOD and APX was mediated through improved ROS scavenging and resulted in reduced peroxide accumulation and improved chilling stress resistance (Xu et al. 2014).

Transgenic potato lines over-expressing tomato plastidal Cu/ZnSOD had reduced levels of damage after MV treatment (Perl et al. 1993). Transgenic rice transformed with mangrove cytosolic Cu/ZnSOD showed superior tolerance to drought stress (Prashanth et al. 2008). *E. coli* MnSOD gene with a soybean-derived leader sequence over-expressed in tobacco chloroplasts showed superior tolerance to photooxidative stress (Van Assche et al. 1989). A FeSOD over-expressed in transgenic tobacco chloroplasts conferred protection from MV damage as well as PSII inactivation (Van Camp et al. 1996). However, expression of FeSOD did not enhance tolerance to chilling-induced photoinhibition in Poplar (Arisi et al. 1998) and salinity stress in tobacco (Van Camp et al. 1996). This could be due to that response of different stress factors is mediated through different routes, and superoxide anion scavenging capacity of the chloroplasts may not be rate limiting for each of them. Further physiological, biochemical, and molecular analyses of transgenic lines along with transgenic lines with silenced SOD activities will provide deeper insight into the functions of SOD in plant cells.

3 Ascorbate Peroxidase (APX) Expressing Transgenic Plants

Ascorbate peroxidase (APX) is the primary H_2O_2-scavenging enzyme in both cytosol and plastids (Asada 1992). APXs have also been reported from glyoxysome, where it is presumed to act with catalase to remove peroxides (Yamaguchi et al. 1995; Bunkelmann and Trelease 1996). Transgenic tobacco lines expressing either pea cytosolic APX or pea cytosolic APX fused with pea chloroplastic transit peptide of Cu/ZnSOD to target plastids showed 3 and 16-fold increase in SOD activity, respectively, with enhanced membrane stability. Also, both transgenic lines were tolerant to photooxidative stress, which is attributed to increase in scavenging of peroxides (Pitcher et al. 1994; Webb and Allen 1995, 1996). *Arabidopsis* over-expressing rice *OsAPXa* or *OsAPXb* enhanced salinity tolerance to different levels. Over-expression of *OsAPXb* enhanced and maintained APX expression to a much higher extent than *OsAPXa* (Lu et al. 2007). Transgenic tobacco lines expressing *C. annuum APX*-like 1 gene enhanced tolerance to oxidative stress and resistance to fungal pathogen *Phytophthora nicotianae*. Interestingly, transgenic lines showed different levels of susceptibility to other pathogens like *Pseudomonas syringae* and *Ralstonia solanacearum*. Tolerance to oxidative stress and *Phytophthora* is associated with overproduction of APX, which could have further increased the POD activity further strengthening an ROS-scavenging system (Sarowar et al. 2005).

Transgenic *A. thaliana* over-expressing peroxisomal *APX* from *Puccinellia tenuiflora* enhanced salt tolerance by reducing H_2O_2 accumulation (Guan et al. 2015). Similarly, peroxisomal *APX* from *Salicornia brachiata* imparted salinity tolerance when over-expressed in *Arachis hypogaea* (Singh et al. 2014a, b). Peroxisomal-specific *Sb APX* not only conferred salinity and drought tolerance but also enhanced vegetative growth and germination rate in transgenic tobacco lines (Singh et al. 2014a). A cytosolic APX from *Lycium chinense* enhanced salinity tolerance and photosynthetic rate and reduced peroxide accumulation in transgenic tobacco lines during salt stress (Wu et al. 2014). Transgenic tobacco expressing thylakoid targeted APX from *Jatropha curcas* enhanced ROS-scavenging system by reducing malondialdehyde (MDA) and H_2O_2 accumulation during salt stress (Liu et al. 2014). *Anthurium andraeanum APX* conferred chilling tolerance in transgenic tobacco lines, which corroborated with enhanced APX activity, reduced MDA content, and enhanced membrane stability (Liu et al. 2013). Rice *OsAPX 2* mutant negatively affected growth and development of rice seedlings and was sensitive to abiotic stress like cold, salinity, and drought. ROS metabolites like H_2O_2 and MDA levels were high in *OsAPX2* mutants; however, they were low in transgenic lines over-expressing *OsAPX2* after stress treatments (Zhang et al. 2013). Similarly, *Arabidopsis* cytosolic *APX* knockout mutants were highly sensitivity to wounding and MeJA treatment. In control plants, peroxide accumulated only in the vicinity of wound, but in leaves of *APX* knockouts, it accumulated both in damaged and undamaged regions. Transgenic *Medicago sativa* over-expressing rice *APX2* had longer roots; ~3-fold more APX activity, less MDA content as compared to control plants (Guan et al. 2012). All the above studies clearly indicate that over-expression of different isoforms of *APX* improved detoxification of ROS and imparted enhanced tolerance in transgenic host system.

4 Glutathione Peroxidase (GPX) Expressing Transgenic Plants

GPX in mammals reduces H_2O_2 and other organic hydroperoxides to water and corresponding alcohols using glutathione (GSH), thus protecting cells from oxidative damage. Plant GPX family shares the highest sequence homology to animal phospholipid hydroperoxide GPXs (PHGPXs) and possess a cysteine (Cys) residue in catalytic site rather than Se-Cys seen in PHGPXs (Zhai et al. 2013). In plants, GPXs are localized in subcellular organelles like cytosol, nucleus, chloroplast, mitochondria, peroxisome, etc., and use thioredoxin (Trx) as a reducing agent (Ursini et al. 1995; Rodriguez et al. 2003). Plant GPXs are expressed in response to different kinds of abiotic factors like salinity, mechanical injury, high-light stress or treatment with Paraquat, etc. (Holland et al. 1993; Avsian-Kretchmer et al. 2004; Herbette et al. 2011; Gaber et al. 2012). *Arabidopsis* knockout mutants of *AtGPX8* were highly sensitive to oxidative damage by Paraquat treatment in roots, and transgenic lines over-expressing *AtGPX8* were less sensitive to oxidative damage than the wild-type controls (Gaber et al. 2012). *Mus musculus* GPX5 over-expressing transgenic tomato lines were more tolerant to oxidative damage after chilling exposure (Herbette et al. 2005). *MmGPX5* transgenic tomato lines were resistant to mechanical injury; however, transgenic plants were sensitive to biotic agents like *Oidium neolycopersici* and *Botrytis cinerea* (Herbette et al. 2011). Two wheat *GPXs* when expressed in plastids of *Arabidopsis* enhanced growth and survival rate during salt stress. Transgenic lines also had increased peroxide-scavenging capacity and enhanced tolerance to H_2O_2 (Zhai et al. 2013). Over-expression of a tobacco glutathione S-transferase having both GST and GPX activity lead to enhanced activities of both enzymes (GST/GPX) in transgenic tobacco seedlings. Also transgenic plants had elevated levels of monodehydro-ascorbate reductase activity and higher levels of glutathione and ascorbate compared to control plants (Roxas et al. 2000). Transient over-expression of tomato phospholipid like GPX (PHGPx) in tobacco leaves suppressed apoptotic-like features during severe salt and high-temperature stresses, and stable tobacco lines expressing *LePHGPx* also conferred protection against the fungus *B. cinerea* (Chen et al. 2004).

5 Glutathione Reductase (GR) Expressing Transgenic Plants

Maintenance of ascorbate and glutathione levels in a reduced state is a crucial factor in ROS-scavenging system in which glutathione reductase (GR) plays a critical role. A bacterial GR over-expressed in tobacco enhanced GR activity to ~4-fold, further reduced visible damage after MV exposure (Aono et al. 1991). Targeting of the *E. coli* GR to plastids in transgenic tobacco also showed similar level of resistance to MV and sulfur dioxide (which was not observed in above study), but not to ozone in both cases (Aono et al. 1993). Transgenic poplar plants over-expressing plastid directed *E. coli* GR resulted in ~500-fold transgene activity with close to two fold higher levels of both glutathione and ascorbate pools compared to control and cytosolic GR expressing lines. Transgenic lines were only resistant to photoinhibition by high-light intensity and chilling temperature (which is attributed to increased level of both GSH and ascorbate) and did not increase MV tolerance (Foyer et al. 1991). A pea cDNA when over-expressed in tobacco elevated GR activity in different sites like cytosol, plastid, and mitochondria, which was attributed to leaky targeting of GR due to presence of N-terminal transit peptide (Creissen et al. 1995). Transgenic lines with pea GR showed different levels of tolerance to ozone; however, none of them significantly reduced MV-induced damage (Broadbent et al. 1995). Surprisingly, few of transgenic lines with pea GR either with the native transit peptide or without any signal sequence showed tolerance to MV; again there was no correlation with the level of protection with the levels of GR expression. Payton et al. (1996) reported that tobacco over-expressing *Arabidopsis* GR had higher levels of GSH and also enhanced MV tolerance; however, some lines showed up to 20% reduction in GR activity, which could be due to sense suppression. Antisense glutathione reductase tomato lines accumulated more peroxides, leaked more electrolyte, and showed lower net photosynthetic rate as compared to WT plants (Feng Shu et al. 2011). Aono et al. (1995) generated antisense tobacco with reduced GR activity, which increased their susceptibility to MV-induced damage. Co-expression of both GR and SOD in cytosol of tobacco plants enhanced protection from MV treatment than transgenic lines with either gene expressed individually (Aono et al. 1995). Hence, it is clear that expressing combinations of different antioxidant enzymes can scavenge different ROS with synergistic effects further enhancing stress tolerance in transgenic lines.

6 Catalase Expressing Transgenic Plants

Catalases (CAT) are tetrameric heme containing enzymes, which directly dismutate H_2O_2 into water and molecular O_2, and they are indispensable for ROS detoxification during stressed conditions (Garg and Manchanda 2009). CAT has high turnover rates with one molecule of CAT converting approximately 6 million molecules of peroxide per minute (Gill and Tuteja 2010). CAT plays an important role in scavenging peroxides generated in peroxisomes by oxidases involved in β-oxidation of fatty acids, photorespiration, and purine catabolism. Transgenic rice over-expressing endogenous type 1 metallothionein *OsMT1a* showed enhanced CAT activity and tolerance to drought stress (Yang et al. 2009). *B. juncea CAT 3* (*BjCAT3*) enhanced cadmium tolerance with two fold higher activity when compared to control plants (Guan et al. 2009). Over-expression of maize CAT and SOD in plastids of *B. campestris* not only enhanced tolerance to sulfur dioxide and salinity stress but also enhanced endogenous K^+, Mg^{2+}, and Ca^{2+}, which could maintain ion homeostasis and secondary signaling molecule (Tseng et al. 2007). Co-expression of *E. coli CAT* and Mn*SOD* in *B. campestris* also enhanced tolerance to sulfur dioxide and showed increased activity of APX and GR, which could reduce ROS accumulation. Peroxide accumulation in rice as a result of dismutation of $O_2^{\bullet-}$ is accelerated during chilling stress as SOD activity is normally enhanced during low temperature (Saruyama and Tanida 1995). A rapid increase in generation and accumulation of peroxides during chilling exposure was reported in wheat (Okuda et al. 1991). Although peroxides themselves are not lethal, they become highly toxic when they react with $O_2^{\bullet-}$ to form the highly reactive $OH^\bullet$ radical, which negatively act on various cellular components like proteins, DNA, and membrane lipids (Bowler et al. 1992). Expression of wheat CAT in rice reduced peroxide accumulation during chilling stress, and its superior tolerance is attributed to effective detoxification of peroxides in transgenic lines (Matsumura et al. 2002). Transgenic Cassava expressing endogenous catalase and Cu/ZnSOD had higher catalase, SOD activities, and lower MDA content than control plants after cold stress (Xu et al. 2014). Also, after harvest, the storage roots of over-expressing cassava lines showed a delay in postharvest physiological deterioration (PPD) response for approximately 10 days, which was accompanied by less mitochondrial oxidation and H_2O_2 accumulation. This was attributed to the combined expression of catalase and Cu/ZnSOD leading to an improved synergistic ROS-scavenging capacity in roots of transgenic plants (Xu et al. 2013). The physiological responses of tobacco with catalase-deficient plants revealed higher SOD activity (less catalase activity) in response to cadmium treatment (Iannone et al. 2015). On the other hand, a type 3 catalase from *B. juncea* enhanced tolerance to cadmium stress in tobacco. At 100 μM Cd, control plants became chlorotic, while *BjCAT*3 transgenic tobacco lines remained green and phenotypically normal. Transgenic lines had lower level of Cd-induced peroxide accumulation (Guan et al. 2009). In catalase-deficient *A. thaliana*, reduced peroxisomal catalase expression leads to increased sensitivity to ozone and photorespiratory peroxide-induced cell death (Vandenabeele et al. 2004). Transgenic tobacco expressing maize catalase 2 gene exhibited ~1.3-fold higher catalase activity and was not negatively affected by MV damage at moderate and high MV concentrations (Polidoros et al. 2001).

7 Current Limitations and Future Directions for the Usage of Transgenic Plants to Combat Oxidative Stress

It is clear from the above studies that modulating the expression of genes coding for ROS-scavenging enzymes should provide a better way to study their specific role, in stress tolerance, and interactions with other enzymes or nonenzymatic antioxidants or osmolytes. Although, modification of ROS-scavenging systems can cause considerable changes in oxidative stress tolerance, there are also several reports on the variance between levels of enzymes and plant tolerance to oxidative stress, which indicate that other cellular factors might be also involved in achieving such increased oxidative stress tolerance in transgenic plants. Also, reports of different transgenic plants expressing single enzymes has achieved only a slight increase in stress tolerance and failed to tolerate different kinds of stress factors. One of the major reason is the need for a balanced interaction of different ROS enzymes at similar levels. Besides, some of initial attempts did not increase stress tolerance (Tepperman and Dunsmuir 1990; Pitcher et al. 1991) due to enzyme sensitivity to products or lack of proper knowledge in targeting. Hence, from the enormous set of data on expression of ROS-scavenging enzymes known till date, it is obvious that an appropriate physiological balance of all the components of the antioxidative defense is crucial to achieve stress tolerance without affecting plants' normal physiological function. Moreover, studies with SOD and glutathione system have clearly indicated that in the future, enzyme should be engineered to overcome feedback inhibition or inhibition by-product.

Plants with the ability to scavenge and/or control the level of cellular ROS may be useful in the future to withstand harsh environmental conditions. Hence, alternate approaches like engineering of plants to reduce ROS levels using proteins like mitochondrial alternative oxidase (AOX), uncoupling protein (UCP), plastid targeted DnaJ protein, etc., are few of the promising genes, which have been proven already to reduce ROS in plants. AOX plays a role in lowering ROS formation in plant mitochondria. Besides function of thermogenesis in plants, the possible explanation for reduced ROS by AOX is explained below. AOX lowers ROS levels by acting on a second oxidase downstream of the ubiquinone pool maintains upstream components of electron-transport chain in a more oxidized state and hence lowering ROS generation by over reduced electron carriers (Maxwell et al. 1999; Umbach et al. 2005; Sugie et al. 2006; Zidenga et al. 2012; Vanlerberghe 2013).

Mitochondrial UCP is a specialized protein that uncouples electron transport from ATP synthesis (Nicholls and Rial 1999). Although thermogenesis was initially attributed to UCPs also, their ubiquitous presence suggests that this protein may have other functions including reducing ROS generation or maintaining its level. It has been already reported that UCP expression in transgenics positively regulated important physiological process like reduced ROS, stomatal conductance, transpiration rates, net photosynthesis, enhance seed viability, improved water balance, higher biomass, etc. (Begcy et al. 2011).

DnaJ proteins act as molecular chaperones (along with Hsp 70) and play a crucial role in maintaining protein conformation and cellular protein homeostasis under harsh environments like heat, high light, MV, chilling stresses, etc. (Piippo et al. 2006; Scarpeci et al. 2008; Rajan and D'Silva 2009). *Arabidopsis* knockout lines of DnaJ proteins induced global response to oxidative stress (Chen et al. 2010). Transgenic tomato lines over-expressing DnaJ accumulated lower ROS level during chilling stress (Kong et al. 2014).

8 Conclusion

New approaches using systems biology are opening doors to generate an all-inclusive transgenic line that are able to maintain its crop productivity even under stressed and changing environmental conditions. As already described above, numerous genes associated with plant response(s) to oxidative stress have been identified and used for generation of transgenic plants to overcome the stress. However, most of these studies were conducted only under controlled laboratory conditions applying artificial stress conditions using model plants like tobacco, *Arabidopsis*, *Medicago*, etc., with a main focus only on recovery from a particular type of stress. In fact plants in real time situations are exposed to different stresses under field conditions. Therefore, in the future, more emphasis should be on the development of transgenic crops (rather than model plants), and the testing should mimic the field condition (i.e., combination of environmental stresses) and focus should be on the crop yield.

References

Allen RD, Webb RP, Schake SA (1997) Use of transgenic plants to study antioxidant defenses. Free Radic Biol Med 23:473–479

Alscher RG, Erturk N, Heath LS (2002) Role of superoxide dismutases (SODs) in controlling oxidative stress in plants. J Exp Bot 53:1331–1341

Aono M, Kubo A, Saji H, Tanaka K, Kondo N (1993) Enhanced tolerance to photooxidative stress of transgenic *Nicotiana tabacum* with high chloroplastic glutathione reductase activity. Plant Cell Physiol 34:129–136

Aono M, Kubo A, Saji H, Natori T, Tanaka K, Kondo N (1991) Resistance to active oxygen toxicity of transgenic *Nicotiana tabacum* that expresses the gene for glutathione reductase from *Escherichia coli*. Plant Cell Physiol 32:691–698

Aono M, Saji H, Fujiyama K, Sugita M (1995) Decrease in activity of glutathione reductase enhances paraquat sensitivity in transgenic *Nicotiana tabacum*. Plant Physiol 107:645–648

Arisi ACM, Cornic G, Jouanin L, Foyer CH (1998) Overexpression of iron superoxide dismutase in transformed poplar modifies the regulation of photosynthesis at low CO_2 partial pressures or following exposure to the prooxidant herbicide methyl viologen. Plant Physiol 117:565–574

Asada K (1992) Ascorbate peroxidase – a hydrogen peroxide-scavenging enzyme in plants. Physiol Planta 85:235–241

Asada K (1999) The water-water cycle in chloroplasts: scavenging of active oxygens and dissipation of excess photons. Annu Rev Plant Physiol Plant Mol Biol 50:601–639

Avsian-Kretchmer O, Gueta-Dahan Y, Lev-Yadun S, Gollop R, Ben-Hayyim G (2004) The salt-stress signal transduction pathway that activates the gpx1 promoter is mediated by intracellular H_2O_2, different from the pathway induced by extracellular H_2O_2. Plant Physiol 135:1685–1696

Bartoli CG, Pastori GM, Foyer CH (2000) Ascorbate biosynthesis in mitochondria is linked to the electron transport chain between complexes III and IV. Plant Physiol 123:335–344

Begcy K, Mariano ED, Mattiello L, Nunes AV, Mazzafera P, Maia IG, Menossi M (2011) An Arabidopsis mitochondrial uncoupling protein confers tolerance to drought and salt stress in transgenic tobacco plants. PLoS One 6:e23776

Bowler C, Montagu MV, Inzé D (1992) Superoxide dismutase and stress tolerance. Annu Rev Plant Physiol Plant Mol Biol 43:83–116

Broadbent P, Creissen GP, Kular B, Wellburn AR, Mullineaux P (1995) Oxidative stress responses in transgenic tobacco containing altered levels of glutathione reductase activity. Plant J 8: 247–255

Bunkelmann JR, Trelease RN (1996) Ascorbate peroxidase – a prominent membrane protein in oilseed glyoxysomes. Plant Physiol 110:589–598

Chen KM, Holmström M, Raksajit W, Suorsa M, Piippo M, Aro EM (2010) Small chloroplast-targeted DnaJ proteins are involved in optimization of photosynthetic reactions in *Arabidopsis thaliana*. BMC Plant Biol 43:1471–2229

Chen S, Vaghchhipawala Z, Li W, Asard H, Dickman MB (2004) Tomato phospholipid hydroperoxide glutathione peroxidase inhibits cell death induced by Bax and oxidative stresses in yeast and plants. Plant Physiol 135:1630–1641

Creissen G, Reynolds H, Xue Y, Mullineaux P (1995) Simultaneous targeting of pea glutathione reductase and of a bacterial fusion protein to chloroplasts and mitochondria in transgenic tobacco. Plant J 8:167–175

Diaz-Vivancos P, Faize M, Barba-Espin G, Faize L, Petri C, Hernández JA, Burgos L (2013) Ectopic expression of cytosolic superoxide dismutase and ascorbate peroxidase leads to salt stress tolerance in transgenic plums. Plant Biotechnol J 11:976–985

Edwards EA, Rawsthorne S, Mullineaux PM (1990) Subcellular distribution of multiple forms of glutathione reductase in leaves of pea (*Pisum sativum* L.). Planta 180:278–284

Foyer CH, Lelandais M, Galap C, Kunert KJ (1991) Effects of elevated cytosolic glutathione reductase activity on the cellular glutathione pool and photosynthesis in leaves under normal and stress conditions. Plant Physiol 97:863–872

Foyer CH, Noctor G (2005a) Redox homeostasis and antioxidant signalling: a metabolic interface between stress perception and physiological responses. Plant Cell 17:1866–1875

Foyer CH, Noctor G (2005b) Oxidant and antioxidant signalling in plants: a re-evaluation of the concept of oxidative stress in a physiological context. Plant Cell Environ 28:1056–1071

Fujita M, Fujita Y, Noutoshi Y, Takahashi F, Narusaka Y, Yamaguchi-Shinozaki K, Shinozaki K (2006) Crosstalk between abiotic and biotic stress responses: a current view from the points of convergence in the stress signalling networks. Curr Opin Plant Biol 9:436–442

Gaber A, Ogata T, Maruta T, Yoshimura K, Tamoi M, Shigeoka S (2012) The involvement of Arabidopsis glutathione peroxidase 8 in the suppression of oxidative damage in the nucleus and cytosol. Plant Cell Physiol 53:1596–1606

Garg N, Manchanda G (2009) ROS generation in plants: boon or bane? Plant Biosyst 143:8–96

Gill SS, Tuteja N (2010) Reactive oxygen species and antioxidant machinery in abiotic stress tolerance in crop plants. Plant Physiol Biochem 48:909–930

Guan Q, Takano T, Liu S (2012) Genetic transformation and analysis of rice *OsAPx2* gene in *Medicago sativa*. PLoS One 7:e41233

Guan Q, Wang Z, Wang X, Takano T, Liu S (2015) A peroxisomal APX from *Puccinellia tenuiflora* improves the abiotic stress tolerance of transgenic *Arabidopsis thaliana* through decreasing of H_2O_2 accumulation. J Plant Physiol 175:183–191

Guan Z, Chai T, Zhang Y, Xu J, Wei W (2009) Enhancement of Cd tolerance in transgenic tobacco plants overexpressing a Cd-induced catalase cDNA. Chemosphere 76:623–630

Herbette S, de Labrouhe DT, Drevet JR, Roeckel-Drevet P (2011) Transgenic tomatoes showing higher glutathione peroxydase antioxidant activity are more resistant to an abiotic stress but more susceptible to biotic stresses. Plant Sci 180:548–553

Herbette S, Menn AL, Rousselle P, Ameglio T, Faltin Z, Branlard G, Eshdat Y, Julien JL, Drevet JR, Roekel-Drevet P (2005) Modification of photosynthetic regulation in tomato overexpressing glutathione peroxidase. Biochim Biophys Acta 1724:108–118

Holland D, Ben-Hayyim G, Faltin Z, Camoin L, Strosberg AD, Eshdat Y (1993) Molecular characterization of salt-stress-associated protein in citrus: protein and cDNA sequence homology to mammalian glutathione peroxidases. Plant Mol Biol 21:923–927

Iannone MF, Groppa MD, Benavides MP (2015) Cadmium induces different biochemical responses in wild type and catalase-deficient tobacco plants. Environ Exp Bot 109:201–211

Jing X, Hou P, Lu Y, Deng S, Li N, Zhao R, Sun J, Wang Y, Han Y, Lang T, Ding M, Shen X, Chen S (2015) Overexpression of copper/zinc superoxide dismutase from mangrove *Kandelia candel* in tobacco enhances salinity tolerance by the reduction of reactive oxygen species in chloroplast. Front Plant Sci 6:23

Karpinska B, Karlsson M, Schinkel H, Streller S, Suss KH, Melzer M, Wingsle G (2001) A novel superoxide dismutase with a high isoelectric point in higher plants: expression, regulation and protein localisation. Plant Physiol 126:1668–1677

Kim HJ, Barbara T (2008) Involvement of extracellular Cu/Zn superoxide dismutase in cotton fiber primary and secondary cell wall biosynthesis. Plant Signal Behav 3:1119–1121

Kong F, Deng Y, Zhou B, Wang G, Wang Y, Meng Q (2014) A chloroplast-targeted DnaJ protein contributes to maintenance of photosystem II underchilling stress. J Exp Bot 65:143–158

Kumar SR, Sivalingam A, Selvaraj D, Ahmed Z, Ramalingam S (2012) Isolation and characterization of cold inducible genes in carrot by suppression subtractive hybridization. Biol Plant 57:97–104

Kuzniak E (2002) Transgenic plants: an insight into oxidative stress tolerance mechanisms. Acta Physiol Planta 24:97–113

Lee SH, Ahsan N, Lee KW, Kim DH, Lee DG, Kwak SS, Kwon SY, Kim TH, Lee BH (2007) Simultaneous overexpression of both CuZn superoxide dismutase and ascorbate peroxidase in transgenic tall fescue plants confers increased tolerance to a wide range of abiotic stresses. J Plant Physiol 164:1626–1638

Liu HC, Tian DQ, Liu JX, Ma GY, Zou QC, Zhu ZJ (2013) Cloning and functional analysis of a novel ascorbate peroxidase (APX) gene from *Anthurium andraeanum*. J Zhejiang Univ Sci B 14:1110–1120

Liu Z, Bao H, Cai J, Han J, Zhou L (2014) A novel thylakoid ascorbate peroxidase from *Jatropha curcas* enhances salt tolerance in transgenic tobacco. Int J Mol Sci 15:171–185

Lu ZQ, Liu D, Liu SK (2007) Two rice cytosolic ascorbate peroxidases differentially improve salt tolerance in transgenic *Arabidopsis*. Plant Cell Rep 26:1909–1917

Maxwell DP, Wang Y, McIntosh L (1999) The alternative oxidase lowers mitochondrial reactive oxygen production in plant cells. Proc Natl Acad Sci USA 6:8271–8276

McKersie BD, Bowley SR, Harjanto E, Leprince O (1996) Water deficit tolerance and field performance of transgenic alfalfa overexpressing superoxide dismutase. Plant Physiol 111: 1177–1181

McKersie BD, Chen Y, De Beus M, Bowley SR, Bowler C (1993) Superoxide dismutase enhances tolerance of freezing stress in transgenic alfalfa (*Medicago sativa* L.). Plant Physiol 103: 1155–1163

Mhamdi A, Queval G, Chaouch S, Vanderauwera S, Van Breusegem F, Noctor G (2010) Catalase function in plants: a focus on Arabidopsis mutants as stress-mimic models. J Exp Bot 61:4197–4220

Mittler R, Vanderauwera S, Gollery M, Van Breusegem F (2004) The reactive oxygen gene network of plants. Trends Plant Sci 9:490–498

Mylona PV, Polidoros AN (2010) ROS regulation of antioxidant genes. In: Gupta SD (ed) Reactive oxygen species and antioxidants in higher plants. CRC Press, New York

Nicholls DG, Rial E (1999) A history of the first uncoupling protein, UCP1. J Bioenerg Biomembr 31:399–406

Okuda T, Matsuda Y, Yamanaka A, Sagisaka S (1991) Abrupt increase in the level of hydrogen peroxide in leaves of winter wheat is caused by cold treatment. Plant Physiol 97:1265–1267

Payton P, Allen RD, Holaday AS (1996) Expression of Arabidopsis glutathione reductase in tobacco. Plant Physiol (Suppl 111):120 abstract

Perl A, Perl-Treves R, Galili S, Aviv D, Shalgi E, Malkin S, Galun E (1993) Enhanced oxidative stress defense in transgenic potato expressing tomato Cu, Zn superoxide dismutases. Theor Appl Genet 85:568–576

Piippo M, Allahverdiyeva Y, Paakkarinen V, Suoranta UM, Battchikova N, Aro EM (2006) Chloroplast-mediated regulation of nuclear genes in *Arabidopsis thaliana* in the absence of light stress. Physiol Genom 25:142–152

Pitcher LH, Brennan E, Hurley A, Dunsmuir P, Tepperman JM, Zilinskas BA (1991) Overproduction of petunia copper/zinc superoxide dismutase does not confer ozone tolerance in transgenic tobacco. Plant Physiol 97:452–455

Pitcher LH, Repetti P, Zilinskas BA (1994) Overproduction of ascorbate peroxidase protects transgenic tobacco plants against oxidative stress. Plant Physiol Suppl 105:116

Polidoros AN, Mylona PV, Scandalios JG (2001) Transgenic tobacco plants expressing the maize Cat2 gene have altered catalase levels that affect plant pathogen interactions and resistance to oxidative stress. Transgen Res 10:555–569

Prashanth SR, Sadhasivam V, Parida A (2008) Over expression of cytosolic copper/zinc superoxide dismutase from a mangrove plant *Avicennia marina* in indica rice var Pusa Basmati-1 confers abiotic stress tolerance. Transgen Res 17:281–291

Rajan VB, D'Silva P (2009) *Arabidopsis thaliana* J-class heat shock proteins: cellular stress sensors. Funct Integr Genom 9:433–446

Reguera M, Peleg Z, Blumwald E (2012) Targeting metabolic pathways for genetic engineering abiotic stress-tolerance in crops. Biochim Biophys Acta 1819:186–194

Rennenberg H, Polle A (1994) Protection from oxidative stress in transgenic plants. Biochem Soc Trans 22:936–940

Robinson M, Bunce JA (2000) Influence of drought-induced water stress on soybean and spinach leaf ascorbate-dehydroascorbate level and redox status. Int J Plant Sci 161:271–279

Rodriguez MM, Maurer A, Rodriguez HA, Gustafson JP (2003) Glutathione peroxidase genes in Arabidopsis are ubiquitous and regulated by abiotic stresses through diverse signaling pathways. Plant J 36:602–615

Roxas VP, Lodhi SA, Garrett DK, Mahan JR, Allen RD (2000) Stress tolerance in transgenic tobacco seedlings that overexpress glutathione S-transferase/glutathione peroxidase. Plant Cell Physiol 41:1229–1234

Sarowar S, Kim EN, Kim YJ, Ok SH, Kim KD, Hwang BK, Shin JS (2005) Overexpression of a pepper ascorbate peroxidase-like 1 gene in tobacco plants enhances tolerance to oxidative stress and pathogens. Plant Sci 1:55 63

Saruyama H, Tanida M (1995) Effect of chilling on activated oxygen-scavenging enzymes in low temperature-sensitive and tolerant cultivars of rice (*Oryza sativa* L.). Plant Sci 109:105–113

Scarpeci TE, Zanor MI, Carrillo N, Mueller-Roeber B, Valle EM (2008) Generation of superoxide anion in chloroplasts of *Arabidopsis thaliana* during active photosynthesis: a focus on rapidly induced genes. Plant Mol Biol 66:361–378

Sen Gupta A, Heinen J, Holaday AS, Burke JJ, Allen RD (1993a) Increased resistance to oxidative stress in transgenic plants that over-express chloroplastic Cu/Zn superoxide dismutase. Proc Natl Acad Sci USA 90:1629–1633

Sen Gupta A, Webb RP, Holaday AS, Allen RD (1993b) Overexpression of superoxide dismutase protects plants from oxidative stress. Plant Physiol 103:1067–1073

Shafi A, Chauhan R, Gill T, Swarnkar MK, Sreenivasulu Y, Kumar S, Kumar N, Shankar R, Ahuja PS, Singh AK (2015) Expression of SOD and APX genes positively regulates secondary cell wall biosynthesis and promotes plant growth and yield in Arabidopsis under salt stress. Plant Mol Biol 87:615–631

Shafi A, Dogra V, Gill T, Ahuja PS, Sreenivasulu Y (2014) Simultaneous over-expression of PaSOD and RaAPX in transgenic *Arabidopsis thaliana* confers cold stress tolerance through increase in vascular lignifications. PLoS One 9:e110302

Shiraya T, Mori T, Maruyama T, Sasaki M, Takamatsu T, Oikawa K, Itoh K, Kaneko K, Ichikawa H, Mitsui T (2014) Golgi/plastid-type manganese superoxide dismutase involved in heat stress tolerance during grain filling in rice. Plant Biotechnol J 1–13

Shu DF, Wang LY, Duan M, Deng YS, Meng QW (2011) Antisense-mediated depletion of tomato chloroplast glutathione reductase enhances susceptibility to chilling stress. Plant Physiol Biochem 49:1228–1237

Singh N, Mishra A, Jha B (2014a) Over-expression of the peroxisomal ascorbate peroxidase (SbpAPX) gene cloned from halophyte *Salicornia brachiata* confers salt and drought stress tolerance in transgenic tobacco. Mar Biotechnol 16:321–332

Singh N, Mishra A, Jha B (2014b) Ectopic over-expression of peroxisomal ascorbate peroxidase (SbpAPX) gene confers salt stress tolerance in transgenic peanut (*Arachis hypogaea*). Gene 547:119–125

Sugie A, Naydenov N, Mizuno N, Nakamura C, Takumi S (2006) Overexpression of wheat alternative oxidase gene Waox1a alters respiration capacity and response to reactive oxygen species under low temperature in transgenic Arabidopsis. Genes Genet Syst 81:349–354

Matsumura T, Tabayashi N, Kamagata Y, Souma C, Saruyama H (2002) Wheat catalase expressed in transgenic rice can improve tolerance against low temperature stress. Physiol Planta 116: 317–327

Tanaka Y, Hibino T, Hayashi Y, Tanaka A, Kishitani S, Takabe T, Yokota S, Takabe T (1999) Salt tolerance of transgenic rice overexpressing yeast mitochondrial Mn-SOD in chloroplasts. Plant Sci 148:131–138

Tepperman JM, Dunsmuir P (1990) Transformed plants with elevated levels of chloroplastic SOD are not more resistant to superoxide toxicity. Plant Mol Biol 14:501–511

Tester M, Langridge P (2010) Breeding technologies to increase crop production in a changing world. Science 327:818–822

Tseng MJ, Liu CW, Yiu JC (2007) Enhanced tolerance to sulfur dioxide and salt stress of transgenic Chinese cabbage plants expressing both superoxide dismutase and catalase in chloroplasts. Plant Physiol Biochem 45:822–833

Umbach AL, Fiorani F, Siedow JN (2005) Characterization of transformed Arabidopsis with altered alternative oxidase levels and analysis of effects on reactive oxygen species in tissue. Plant Physiol 139:1806–1820

Ursini F, Maiorino M, Brigelius-Flohe R, Aumann KD, Roveri A, Schomburg D, Flohé L (1995) Diversity of glutathione peroxidases. Methods Enzymol 252:38–53

Van Assche CJ, Davies HM, O'Neal JK (1989) Superoxide dismutase expression in plants. European Patent Application. Publication number EP 0356061A2

Van Camp W, Capiau K, Van Montagu M, Inzé D, Slooten L (1996) Enhancement of oxidative stress tolerance in transgenic tobacco plants overproducing Fe-superoxide dismutase in chloroplasts. Plant Physiol 112:1703–1714

Van Camp W, Willekens H, Bowler C, Van Montague M, Inze D (1994) Elevated levels of superoxide dismutase protect transgenic plants against ozone damage. BioTechnology 12: 165–168

Vandenabeele S, Vanderauwera S, Vuylsteke M, Rombauts S, Langebartels C, Seidlitz HK, Zabeau M, Van Montagu M, Inzé D, Van Breusegem F (2004) Catalase deficiency drastically affects gene expression induced by high light in *Arabidopsis thaliana*. Plant J 39:45–58

Vanlerberghe GC (2013) Alternative oxidase: a mitochondrial respiratory pathway to maintain metabolic and signaling homeostasis during abiotic and biotic stress in plants. Int J Mol Sci 14:6805–6847

Wang FZ, Wang QB, Kwon SY, Kwak SS, Su WA (2005) Enhanced drought tolerance of transgenic rice plants expressing a pea manganese superoxide dismutase. J Plant Physiol 162:465–472

Webb RP, Allen RD (1995) Overexpression of pea cytosolic ascorbate peroxidase in *Nicotiana tabacum* confers protection against the effects of paraquat. Plant Physiol 108(Suppl):64

Webb RP, Allen RD (1996) Overexpression of pea cytosolic ascorbate peroxidase confers protection against oxidative stress in transgenic *Nicotiana tabacum*. Plant Physiol 111(Suppl):48

Wei A, He C, Li B, Li N, Zhang J (2011) The pyramid of transgenes TsVP and BetA effectively enhances the drought tolerance of maize plants. Plant Biotechnol J 9:216–229

Wu G, Wang G, Ji J, Gao H, Guan W, Wu J, Guan C, Wang Y (2014) Cloning of a cytosolic ascorbate peroxidase gene from *Lycium chinense* Mill. and enhanced salt tolerance by overexpressing in tobacco. Gene 543:85–92

Xu J, Duan X, Yang J, Beeching JR, Zhang P (2013) Enhanced reactive oxygen species scavenging by overproduction of superoxide dismutase and catalase delays postharvest physiological deterioration of cassava storage roots. Plant Physiol 161:1517–1528

Xu J, Yang J, Duan X, Jiang Y, Zhang P (2014) Increased expression of native cytosolic Cu/Zn superoxide dismutase and ascorbate peroxidase improves tolerance to oxidative and chilling stresses in cassava (*Manihot esculenta* Crantz). BMC Plant Biol 14:208

Yamaguchi K, Mori H, Nishimura M (1995) A novel isozyme of ascorbate peroxidase localized on glyoxysomal and leaf peroxisomal membranes in pumpkin. Plant Cell Physiol 36:1157–1162

Yang Z, Wu Y, Li Y, Ling HQ, Chu C (2009) OsMT1a, a type 1 metallothionein, plays the pivotal role in zinc homeostasis and drought tolerance in rice. Plant Mol Biol 70:219–229

Zhai CZ, Zhao L, Yin LJ, Chen M, Wang QY, Li LC, Xu ZS, Ma YZ (2013) Two wheat glutathione peroxidase genes whose products are located in chloroplasts improve salt and H_2O_2 tolerances in Arabidopsis. PLoS One 8:e73989

Zhang Z, Zhang Q, Wu J, Zheng X, Zheng S, Sun X, Qiu Q, Lu T (2013) Gene knockout study reveals that cytosolic ascorbate peroxidase 2(OsAPX2) plays a critical role in growth and reproduction in rice under drought, salt and cold stresses. PLoS One 8:e57472

Zidenga T, Leyva-Guerrero E, Moon H, Siritunga D, Sayre R (2012) Extending cassava root shelf life via reduction of reactive oxygen species production. Plant Physiol 159:1396–1407

MIX
Papier aus verantwortungsvollen Quellen
Paper from responsible sources
FSC® C105338

If you have any concerns about our products,
you can contact us on
ProductSafety@springernature.com

In case Publisher is established outside the EU,
the EU authorized representative is:
Springer Nature Customer Service Center GmbH
Europaplatz 3, 69115 Heidelberg, Germany

Printed by Libri Plureos GmbH
in Hamburg, Germany